D1700625

Springer Tracts in Advanced Robotics 120

Editors

Prof. Bruno Siciliano
Dipartimento di Ingegneria Elettrica
e Tecnologie dell'Informazione
Università degli Studi di Napoli
Federico II
Via Claudio 21, 80125 Napoli
Italy
E-mail: siciliano@unina.it

Prof. Oussama Khatib
Artificial Intelligence Laboratory
Department of Computer Science
Stanford University
Stanford, CA 94305-9010
USA
E-mail: khatib@cs.stanford.edu

Editorial Advisory Board

Nancy Amato, Texas A & M, USA
Oliver Brock, TU Berlin, Germany
Herman Bruyninckx, KU Leuven, Belgium
Wolfram Burgard, Univ. Freiburg, Germany
Raja Chatila, ISIR - UPMC & CNRS, France
Francois Chaumette, INRIA Rennes - Bretagne Atlantique, France
Wan Kyun Chung, POSTECH, Korea
Peter Corke, Queensland Univ. Technology, Australia
Paolo Dario, Scuola S. Anna Pisa, Italy
Alessandro De Luca, Sapienza Univ. Rome, Italy
Rüdiger Dillmann, Univ. Karlsruhe, Germany
Ken Goldberg, UC Berkeley, USA
John Hollerbach, Univ. Utah, USA
Lydia Kavraki, Rice Univ., USA
Vijay Kumar, Univ. Pennsylvania, USA
Bradley Nelson, ETH Zürich, Switzerland
Frank Park, Seoul National Univ., Korea
Tim Salcudean, Univ. British Columbia, Canada
Roland Siegwart, ETH Zurich, Switzerland
Gaurav Sukhatme, Univ. Southern California, USA

More information about this series at http://www.springer.com/series/5208

Andreas Pott

Cable-Driven Parallel Robots

Theory and Application

Deinventarisiert

 Springer

Andreas Pott
Stuttgart
Germany

ISSN 1610-7438 ISSN 1610-742X (electronic)
Springer Tracts in Advanced Robotics
ISBN 978-3-319-76137-4 ISBN 978-3-319-76138-1 (eBook)
https://doi.org/10.1007/978-3-319-76138-1

Library of Congress Control Number: 2018932992

© Springer International Publishing AG, part of Springer Nature 2018
This work is subject to copyright. All rights are reserved by the Publisher, whether the whole or part
of the material is concerned, specifically the rights of translation, reprinting, reuse of illustrations,
recitation, broadcasting, reproduction on microfilms or in any other physical way, and transmission
or information storage and retrieval, electronic adaptation, computer software, or by similar or dissimilar
methodology now known or hereafter developed.
The use of general descriptive names, registered names, trademarks, service marks, etc. in this
publication does not imply, even in the absence of a specific statement, that such names are exempt from
the relevant protective laws and regulations and therefore free for general use.
The publisher, the authors and the editors are safe to assume that the advice and information in this
book are believed to be true and accurate at the date of publication. Neither the publisher nor the
authors or the editors give a warranty, express or implied, with respect to the material contained herein or
for any errors or omissions that may have been made. The publisher remains neutral with regard to
jurisdictional claims in published maps and institutional affiliations.

Printed on acid-free paper

This Springer imprint is published by the registered company Springer International Publishing AG
part of Springer Nature
The registered company address is: Gewerbestrasse 11, 6330 Cham, Switzerland

"Theoria cum praxi"
(Unity of theory and application)
Maxim of Gottfried W. Leibniz

Dedicated to
Eva and Jana.

Acknowledgements

The work was developed within my post-doc research in Stuttgart at the Fraunhofer Institute for Manufacturing Engineering and Automation IPA as well as at the Institute for Control Engineering of Machine Tools and Manufacturing Units ISW, University of Stuttgart.

First of all, I am most grateful for the support and inspiration of Prof. Alexander Verl who encouraged me to start the work group on cable-driven parallel robots in Stuttgart and who acted as main reviewer of my habilitation thesis.

I thank all members of the examination board for my habilitation thesis, where very special thanks are dedicated to Prof. Remco Leine for reviewing this lengthy documents. I was very happy that Dr. Jean-Pierre Merlet served as external reviewer and I want to deeply acknowledge his contribution to my work since some 15 years. His seminal work as well as his reference book were a role model for me.

The research projects on cable robots would not have been possible without an excellent and tireless team: Thank you to Dr. Werner Kraus, Dr. Valentin Schmidt, Philipp Miermeister, Philipp Tempel, Christian Schenk, and Christoph Martin. My scientific and technical developments were intensively inspired by fruitful discussion with my current and former colleagues from Fraunhofer IPA as well as from ISW. I want to express my very special thanks to Thomas Dietz, Martin Hägele, Dr. Christian Meyer, Dr. Christian Connette, Hendrik Mütherich, Dr. Matthias Palzkill, and Dr. Jens Kubacki whose feedback and support guided me along this journey.

The research on cable robots started on the previous works encouraged by my doctoral advisor Prof. Manfred Hiller as well as Dr. Tobias Bruckmann, Dr. Lars Mikelsons, Dr. Richard Verhoeven, Dr. Shiqing Fang, and Prof. Andrés Kecskeméthy.

Many of my work on cable robotics benefited from the enlightening discussions with other great researcher in the field: Dr. Marc Gouttefarde, Prof. Clément Gosselin, Prof. Marco Carricato, and Prof. Sunil Agrawal. All of them made pathbreaking contribution to the field and they are hereby also acknowledged for their unconditional support when starting the CableCon conference series. Furthermore, I found support and great collaborators at our partner institutes in

Berlin, Germany and Gwangju, Korea with Dr. Dragoljub Surdilovic, Prof. Jong-Oh Park, Prof. Seong Young Ko, and Dr. Jinwoo Jung.

Industrial control systems are a key enabler bringing robot to life. I am grateful for the insights I received from the discussions with Dr. Armin Lechler, Hannes Richter, and Dr. Dieter Scheifele.

I owe my technical knowledge on cable technology the kind and fruitful discussions with Prof. Karl-Heinz Wehking, Dr. Konstantin Kühner, Dr. Martin Wehr, and Prof. Markus Michael.

Dedication to application of cable robots is the very heart of all my theoretical work on cable robots and application was a considerable driver for my research directions. I have received many hours of constructive discussion with partners and customers of the cable robots that have been set up. I like to thank Prof. Heinrich Bülthoff, Christoph Bauer, Dr. Mathias Dobner, Prof. Peter Eberhard, Fabian Schnelle, Ingo Kaske, and Ulrich Kunkel.

I found encouraging and open-minded colleagues within the SimTech cluster as well as at the University of Stuttgart. Many thanks go to Prof. Bernard Haasdonk, Andreas Weiß, Prof. Jörg Fehr, Prof. Oliver Röhrle, and Dr. Simon Eugster.

The applied research on cable robots relies heavily on experimental investigations and software implementations which have been carried out by an army of students working together with me and my colleagues in Duisburg and Stuttgart. In the hope of acknowledging all valuable contributions, I list them in alphabetic order: Corina Abraham, Abdulrahman Alshawakri, Moriz Arns, Jens Birkenbach, Thomas Bock, Ricardo Esteves Borges, Lars Brandt, Simon Breunig, Erwin Brosch, Tim Brosch, Max Daibert-Huppert, Omar Diab, Florian Dóleschal, Daniel Frerich, Dirk Förstner, Mark Gauglitz, Matthias Göcks, Manuel Gruber, Tran Hieu, Clemens Honold, Philipp Janssen, Matthias Kapica, Thomas Kaufmann, Manuel Koch, Julian Koller, Daniel Küthe, Ragnar Lodwig, Navid Maghadamnejad, Alexander Mangold, Carissa Michalkowski, Eric Mittler, Bertram Müller, Thomas Münchow, Lukas Neudorfer, Johannes Neumann, Frank Pirkl, Abhinav Prakash, Puneeth Rajendra, Kai Salzmann, Dominik Scheible, Lukas Schelbert, Verena Schmidt, Marco Schröter, Florian Sieber, Christian Sigle, Tobias Simon, Patric Skalecki, Patrick Stolz, Tim Willkens, Jing Yu, Hui Yuan, Da Zhu, Su Zhu, and Silke Zipfel.

Thank you to Luzia Schuhmacher for endless language editing and revising most of my papers, slides, and especially this book. Even more, I am pretty sure that she found more errors in my equations than any other single person while tirelessly working through some one thousand pages.

Beside all personal and social efforts, I thank all organizations that provided the financial support for my research: The European Commission, Fraunhofer-Gesellschaft zur Förderung der angewandten Forschung, German Research Foundation (DFG), Stuttgart Research Center for Simulation Technology (SimTech), Hans L. Merkle-Stiftung, Baden-Württemberg Stiftung, Baden-Württemberg Ministry of Science, Research and the Arts (MWK), as well as the University of Stuttgart.

I am most grateful to the endless support and patience of my wife Ariane and my daughters Eva and Jana who accepted night shifts and weekends full of work and writing.

Sindelfingen, Germany   Andreas Pott
September 2017

Contents

Symbols

$\mathbf{A}^{\mathrm{T}}$	Pose-dependent structure (wrench) matrix $\in \mathbb{R}^{m \times n}$
$\widehat{\mathbf{A}}^{\mathrm{T}}$	Non-normalized structure matrix $\in \mathbb{R}^{m \times n}$
$\mathbf{A}^{+\mathrm{T}}$	Moore–Penrose pseudo-inverse of $\mathbf{A}^{\mathrm{T}}$
A_{C}	Cross section of the cable
$\mathbf{a}_i$	Position vector $\in \mathbb{R}^3$ of i-th proximal anchor point of the machine frame
$\mathbf{b}_i$	Position vector $\in \mathbb{R}^3$ of i-th distal anchor point of the mobile platform
β_{R}	Wrapping angle of the cable on the pulley
$\mathbf{c}$	Vector of calculation variables in the constraint satisfaction problem
$\mathbf{C}$	Compliance matrix $\in \mathbb{R}^{6 \times 6}$ of the cable robot in operational space
$\mathcal{C}$	Set $\subset \mathbb{R}^m$ of the m-dimensional hypercube of feasible forces in the cables
γ_{R}	Rotation of the panning pulley about the z-axis
d_{D}	Diameter of the drum
E_{C}	Young's modulus of the cable
$\mathbf{f}$	vector $\in \mathbb{R}^m$ collecting all cable forces as generalized forces
$\mathcal{F}$	Set $\subset \mathbb{R}^m$ with feasible solutions for the force distribution problem
$f_{\min}$	Required pretension in the cable
$f_{\max}$	Feasible maximum tension in the cable
f_{H}	Horizontal part of the cable force $\mathbf{f}_i$ when considering cable sagging
$\mathbf{f}_{\mathrm{P}}$	Force applied to the platform
$\boldsymbol{\varphi}^{\mathrm{IK}}$	$\mathbb{R}^n \to \mathbb{R}^m$ the mapping defining the inverse kinematics transformation
$\boldsymbol{\varphi}^{\mathrm{DK}}$	$\mathbb{R}^m \to \mathbb{R}^n$ the mapping defining the forward kinematics transformation
$\boldsymbol{\Phi}^{\mathrm{C}}$	$\mathbb{R}^2 \to \mathbb{R}^2$ the mapping of the sagging cable model
$\boldsymbol{\Phi}^{\mathrm{G}}$	$\mathbb{R}^{n_{\mathrm{D}}} \to \mathbb{R}^{6m}$ mapping for parameterization for robot geometry
$\boldsymbol{\Phi}$	Mapping of inequality constraints in the constraint satisfaction problem
g	Gravity acceleration
g_{C}	Specific gravity force of the cable per length
$\mathbf{g}^{\mathrm{C}}$	Vector $\in \mathbb{R}^n$ of generalized centripetal and Coriolis forces
h_{D}	Pitch of the groove of the drum
$\mathbf{H}$	Matrix $\in \mathbb{R}^{m \times r}$ with a spanning base of the kernel of $\mathbf{A}^{\mathrm{T}}$
$\mathbf{h}_i$	Spanning base vector $\in \mathbb{R}^m$ of the kernel $\mathbf{H}$ of $\mathbf{A}^{\mathrm{T}}$

$\mathbf{I}$	Square identify matrix
I_D	Moment of inertia of the drum
I_M	Moment of inertia of the servo motor
$\mathbf{I}_P$	Inertia tensor $\in \mathbb{R}^{3\times3}$ of the mobile platform
I_{PG}	Moment of inertia of the planetary gearbox
I_R	Moment of inertia of the pulley
I_S	Moment of inertia of the spooling unit
I_W	Effective overall moment of inertia of the winch
$\mathbf{i}_{dq}$	Current of the servo motor
$\mathbf{J}$	Kinematic Jacobian $\in \mathbb{R}^{n\times m}$ of the robot
$\mathbf{J}_A$	Jacobian of the closure constraints v w.r.t. changes in the pose
$\mathbf{J}_B$	Jacobian of the closure constraints v w.r.t. changes in the cable length
$\mathbf{J}_v$	Jacobian of forward kinematics constraints v w.r.t. changes in the pose
$\mathbf{K}_C$	Stiffness matrix $\in \mathbb{R}^{m\times m}$ of the robot in configuration space
$\mathbf{K}_G$	Geometric stiffness matrix $\in \mathbb{R}^{6\times6}$ of the robot in operational space
$\mathbf{K}_O$	Linear elastic stiffness matrix $\in \mathbb{R}^{6\times6}$ of the robot in operational space
$\mathbf{K}_{OS}$	Stiffness matrix $\in \mathbb{R}^{6\times6}$ of the robot in operational space
$\mathcal{K}_0$	Fixed world coordinate system
$\mathcal{K}_i$	Coordinate system associated with point i
$\mathcal{K}_P$	Reference coordinate system attached to the mobile platform
$\mathcal{K}_{A,i}$	Local coordinate system of the i-th proximal anchor point A_i
$\mathcal{K}_{B,i}$	Local coordinate system of the i-th distal anchor point B_i
k_i	Stiffness coefficient of the i-th cable
k'_C	Specific stiffness of the cable
k_A	Stiffness of the winch drivetrain
$\mathbf{l}_i$	Vector $\in \mathbb{R}^3$ of the i-th cable
l_i	Length of the i-th cable
$\mathbf{l}$	Vector $\in \mathbb{R}^m$ collecting all cable lengths as generalized coordinates
$\mathbf{L}$	Diagonal matrix $\in \mathbb{R}^{m\times m}$ with the cable lengths as elements
L	Lagrangian function
$\mathbf{L}_{12}$	Winding induction of the servo motor
m	Number of cables of the robot
$\mathbf{m}$	Vector $\in \mathbb{R}^3$ of the center of the workspace used for hull projection
m_G	Mass of linear traveling carriage (spooling unit) of the winch
m_P	Mass of the mobile platform
n	Degrees-of-freedom of the mobile platform
n_R	Number of pulleys in the drivetrain
n_W	Number of windings on the drum
n_D	Number of design parameters
v	$\mathbb{R}^m \to \mathbb{R}^n$ mapping defining the robot's kinematic closure constraints
v_C	Effective density of a cable cross section
v_{PG}	Gear ratio of the winch's planetary gearbox
v_W	Overall transmission ratio of the winch
$\mathbf{P}$	Transformation matrix $\in \mathbb{R}^{3\times4}$ for quaternion in dynamics

$\mathcal{P}$	Set of measurement poses for calibration
ψ_{dq}	Flux linkage of the servo motor
$\mathcal{Q}$	Set $\subset \mathbb{R}^n$ of wrenches to be generated by the robot
$\mathbf{Q}$	Quaternion for parameterization of SO_3
$\mathbf{q}$	Vector of the generalized coordinates in dynamics
r	Degree-of-redundancy of a robot
$\mathbf{r}$	Position of the platform where $\mathbf{r} \in \mathbb{R}^3$
$\mathbf{r}_M$	Mobile platform's center of gravity $\in \mathbb{R}^3$
r_C	Radius of the cable
r_D	Radius of the drum of the winch
r_R	Radius of the pulley
$\mathbf{R}$	Orientation of the platform where $\mathbf{R} \in SO_3$
$\mathcal{R}_0$	Set $\subset SO_3$ of orientation matrices
R_{12}	Ohmic resistance of the servo motor windings
ϱ_C	Density of the cable
ϱ'_C	Linear density of the cable per length
s	Sagging of the cable
$\mathcal{S}$	Set $\subset \mathbb{R}^m$ of the solution space of feasible force distributions
SO_3	Set of the special orthogonal group with all spatial rotations
SE_3	Euclidian displacement group with all spatial rigid body transformations
T	Kinematic energy of the cable robot
T_P	Kinematic energy of the mobile platform
T_W	Torque in the winch
T_M	Servo motor torque
T_F	Friction torque in the winch
$\mathbf{T}$	Transformation matrix $\in \mathbb{R}^{6 \times 6}$
$\Theta_{eff,i}$	Shaft/drum angle of the i-th motor/winch
Θ_D	Rotation angle of the winch drum
$\mathbf{\Theta}_D$	Vector $\in \mathbb{R}^m$ collecting the rotation angles of all winch drums
Θ_M	Rotation angle of the servo motor
τ_P	Torque vector $\in \mathbb{R}^3$ applied to the mobile platform
U	Potential energy of the robot
U_P	Potential energy of the mobile platform
U_i	Potential energy of the i-th cable
$\mathbf{u}_i$	Unit vector $\in \mathbb{R}^3$ of the i-th cable
$\widehat{\mathbf{u}}_i$	Unit vector $\in \mathbb{R}^3$ of the effective direction of i-th cable under sagging
$\mathbf{u}_{dq}$	Input voltage of the servo motor
$\mathbf{v}$	Vector of verification variables in the constraint satisfaction problem
$\mathbf{w}_P$	Vector $\in \mathbb{R}^n$ of the applied wrench of the platform
$\mathcal{W}$	Set of poses in the respective motion group representing the workspace
$\mathcal{X}_c$	Solution set of the constraint satisfaction problem
$\mathcal{X}_s$	Search space is a set used with constraint satisfaction problem

$\mathcal{X}_v$	Verification domain is a set for the constraint satisfaction problem
$\mathbf{y}$	Pose as generalized coordinate vector $\in \mathbf{R}^n$ of the platform
Z_P	Pole pair number of the servo motor

Subscripts

A	Proximal anchor point on the machine frame
B	Distal anchor point on the mobile platform
C	Cable
D	Drum
M	Motor
P	Platform
R	Pulley (roller), not to be confused with platform
W	Winch
DK	Direct kinematics/forward kinematics
IK	Inverse kinematics
FD	Forward dynamics
ID	Inverse dynamics
OS	Operational space
CO	Constant orientation (workspace)
TO	Total orientation (workspace)

Chapter 1
Introduction

Generating a defined motion is one of the fundamental tasks for a machine. By definition, a robot is a universal machine dedicated to creating motion that can be freely programmed. A large variety of mechanisms are known that are capable of creating different kinds of motion which can be characterized by the degree-of-freedom of the motion, its dynamic characteristics such as velocity and acceleration, and its accuracy. Whenever objects are moved or manipulated, the payload or strength of the machine becomes an additional topic of interest. Once a technology to achieve the desired effect is known, the technological development aims at optimizing the task in some sense, for example by increasing the payload, velocity, or accuracy. At the same time, economical factors drive the development into decreasing costs for the machine while maintaining the performance level. Therefore, there is a persistent trend in robotics to develop technical solutions that are superior in some of these aspects.

In the field of robotics and motion generators, the idea of a robot that is purely suspended and driven by cables was introduced in the 1980s. Since then, some researchers all over the world have taken up the idea to create a new generation of robotic systems that exploit the outstanding potentials of using cables to actuate a robot. The use of cables to constrain a mobile platform in space offers a number of promising advantages that mostly follows from the ultra light-weight design of the robot.

1.1 From Serial Robots to Cable Robots

Serial robots are inspired by the structure and function of the human arm which consists of a series of joints and bones connecting the torso with the hand. Using the muscles as actuators, the hand can be freely moved in space to manipulate objects in the environment. For a robot as technical system, muscles and bones are replaced by

© Springer International Publishing AG, part of Springer Nature 2018
A. Pott, *Cable-Driven Parallel Robots*, Springer Tracts in Advanced
Robotics 120, https://doi.org/10.1007/978-3-319-76138-1_1

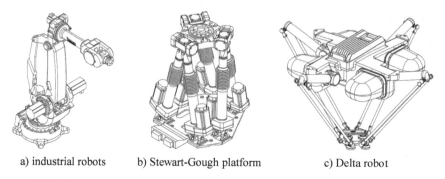

a) industrial robots   b) Stewart-Gough platform   c) Delta robot

Fig. 1.1 Conventional robot architectures: Industrial robots, Stewart–Gough platform, and delta robot

links and actuated joints in order to mimic the human's motion capacities (Fig. 1.1a). The overall kinematic chain is called *serial*, meaning that there is one unique sequence of links and actuated joints that create the desired motion. Serial robots are often referred to as *robotic arms* or articulated robots due to their kinematic similarity with the human arm.

The difference of a *parallel robot* with respect to a serial robot is to add more kinematic chains to connect the end-effector with the base (see Fig. 1.1b, c). In order to allow for independent mobility in such a structure, most of the joints are left passive without motors. Although there are many possible choices to distribute the motors on the parallel robot, a beneficial solution is to put one motor on each kinematic chain to distribute the load amongst all kinematic chains. Again, this concept of parallel actuation can be found in the human body. Although the overall kinematic structure of human arms and legs is serial, the actuation of the joints has a parallel topology. This becomes evident especially for multi degree-of-freedom joints such as the hip or the shoulder, where a number of muscles drive the joint in parallel. The actuation of the human arms is done through muscles and sinews, where both elements are only able to generate and transmit tensile forces and thus cannot push. Therefore, all joints are driven by an antagonistic principle requiring two actuators, one for each direction of motion. This unilateral actuation scheme was adopted in robots for hands, serial, and parallel robots. However, a cable robot is not a bionic approach since both structure and actuation are achieved through the cables.

When discussing the differences between serial and parallel robots, it is important to note that *parallelism* relates to the *topological structure* of the robot and not to its geometry (Fig. 1.2). The difference can be understood when comparing the robot structure with electrical networks. A parallel circuit as an electrical network has nothing to do with geometrical parallelism of the elements on the board but with the electric current divided amongst the resistors, whereas the current flows through all resistors in a series connection (Fig. 1.3). If we exchange the resistor with joints and the current with force, respectively, the terms serial robot and parallel robot become clear; parallel and serial relate rather to the topology of the system than to its geometry.

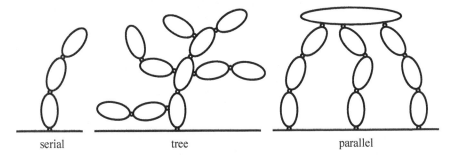

serial   tree   parallel

Fig. 1.2 Comparison between the topological structure of serial, tree, and parallel robots

Fig. 1.3 Analogy to electrical networks concerning serial and parallel topology with current i and resistance R_1, R_2

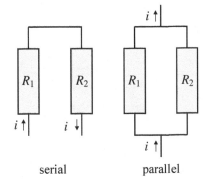

serial   parallel

1.1.1 Cable Robots as Intelligent Cranes

The idea of using cables to manipulate loads is very old. Even in the ancient world, cables were used to lift loads in construction and similar applications. Until today, cranes are widely used in construction and industrial production since they are versatile and cost efficient machines. When looking closely at typical handling and large-scale assembly scenarios, the seed for using cable robots can be found: When the load needs to be placed accurately or even assembled, human workers fix some handy ropes on the load to drag the load into position, to suppress swaying of the load, or to counteract perturbation caused by wind (Fig. 1.4). Every cable added to the platform constrains one degree-of-freedom and in practice, it is common that some workers cooperate by dragging a load into position. Having noticed that a coordinated motion of the additional cables grants control over the load's motion, it is only a small step to use motor winches to do the dragging. In the second step, a computer is applied to synchronize the motion of the winches and to perform predefined motions with the winches such that the load is moved in the desired way. A computer controlled crane is nothing else but a cable-driven parallel robot.

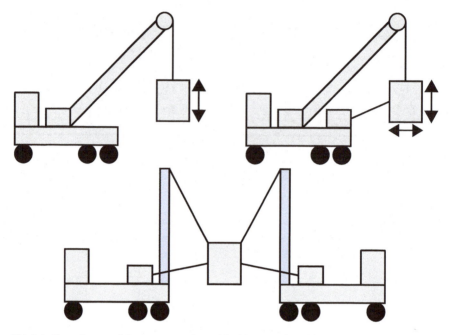

Fig. 1.4 Extending a mobile crane towards a cable-driven parallel robot. Adding one winch allows to control one direction of the load's motion. Using some winches the motion can be fully-constrained

1.1.2 Cable Robots as Ultra Light-Weight Designs

The weight of a robot mostly results from the weight of its actuators as well as from the weight of the robot structure. For most serial robots, the weight of electric cables, sensors, and tools comes second to the machine structure and the drive-trains. A major advantage of parallel robots is that each of the legs is driven by a single actuator and therefore no actuator has to support the weight of any other actuator. Moreover, there are different designs where the actuators can be fixed on the machine frame and thus, only the passive machine structure must be lifted, balanced, and accelerated. The advantage of this design is twofold: One can use smaller actuators and lighter links.

The weight of the robot's structure depends on the load case that has to be used for dimensioning. Serial robots are mostly driven by the torque in their joints and therefore the arms of serial robots are subject to bending (left in Fig. 1.5). To withstand bending, large cross sections have to be used leading to heavy machine parts. This limits, in turn, the dynamic performance as well as the maximum payload. The structure of parallel robots like Stewart–Gough-platforms or Delta robots can ease the load case used to dimension their parts. The critical load case changes from bending to buckling, which allows to reduce the weight of the robot's legs compared to serial robots. Delta robots impressively demonstrate the effectiveness of this approach and achieve very high accelerations [103]. The basic idea of cable-driven parallel robots

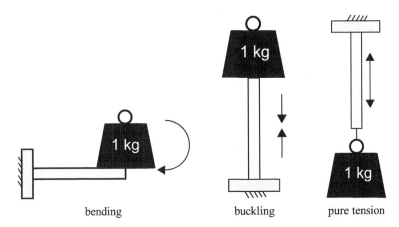

bending   buckling   pure tension

Fig. 1.5 Comparison of the critical load case in the design of a robot. Serial robots have to be designed for bending, Stewart–Gough-platforms are designed to withstand buckling of the legs, and cables are subject to pure tension

is to use a structure where the load case on the structure elements is optimal, i.e. pure tension is applied to the elements which allows the minimization of the transmission element's weight. This can be achieved by thin bars, belts, chains, and cables. Cables offer the advantage that they can be coiled for both actuation and transport. From the point of view of the structural design, cable robots are probably the most light-weight structure for a manipulator.

1.2   State of the Art

So far, although cable robots are hardly used in practical applications, there are a number of prototypes. In the following, an overview of the development of cable robots is presented along with a list of many laboratory prototype systems. As far as possible, the targeted application and related literature are presented. Clearly, the overview has to be concise and details of the robot system are omitted due to limited space here.

1.2.1   History and Prototypes

As early as 1984, Landsberger and Sheridan presented some first ideas on cable-controlled parallel linked robotic devices in the master thesis [280], as well as in [281, 282]. Their robots were completely actuated by cables but rely on an additional strut. Around the same time, the patent of the famous cable-based camera system

Skycam was filed by Brown [67, 100] which is at the same time one of the few commercial applications of cable robots. A first kinetostatic study of Skycam-type systems was presented by Tanaka [456] and Higuchi [209] proposed to use such multi-cable cranes as robots for construction.

Later, the RoboCrane system at NIST [106] was presented in 1989, and it seems to be the first larger prototype. A first patent on the RoboCrane was filed already in 1988 by Albus [5]. The RoboCrane was inspired by the structure of the well-known Stewart–Gough-platform where the arrangement is somewhat inverted: To keep the cables under tension, the mobile platform is located under the fixed anchor points. Therefore, only pulling forces have to be exerted by the connecting legs and cables can be used instead of rigid struts. RoboCrane was evaluated for large-scale handling and one targeted application was ship building [6, 7, 106]. At the same time, a cable robot was proposed as haptic interface in space robots [290]. Kurtz [267] studied the problem of force distribution already in 1991 and proved that $n+1$ cables are required to fully constrain a rigid body with n degrees-of-freedom. Kawamura [238] outlined first ideas of a seven cable parallel robot for teleoperation. At this time, Ming and Higuchi [349] introduced a classification for cable robots to characterize different designs depending on the number of cables and degrees-of-freedom. The prototype Falcon was published some years later by Kawamura [237]. Already in 1995, it was experimentally shown that accelerations > 400 m/s^2 can be generated by cable robots. Kawamura designed the Falcon system for fast pick-and-place operations to take advantages of outstanding dynamic capabilities of cable robots. Some years later, Tadokoro developed a mobile cable robot system for rescue after earthquakes [305, 452] and underlined the possibility to build a relatively light and portable robot.

The research on cable robots broadened with the turn of the millennium. Beside the works in Japan and the USA, groups in Europe, China, and Iran also started their research. Lafourcade applied the ultra-light structure of a cable robot for motion generation in wind tunnels [270, 271]. A light-weight prototype named Segesta was developed at the University of Duisburg-Essen, Germany [68, 139, 210, 473], as a research system for kinematics, control, and design studies. For a small payload of 150 g an acceleration of up to 200 m/s^2 was experimentally realized. At the same time, the under-constrained robot Cablev was developed as scaled prototype for handling and automated container cranes at the University of Rostock, Germany [206, 306]. When parallel robots were intensively analyzed in a priority program in Germany for machine tools, some efforts were undertaken to build a cable robot to be used as machine tool [258, 467]. A bit later, the robot String-Man was used at Fraunhofer IPK (Berlin, Germany) for gait rehabilitation with focus on force control and safety considerations [448, 449]. At INRIA in France, the robot family Marionet included a small size prototype for high-speed applications, a portable crane for rescue, and components for person assistance [324]. There are a couple of recent works from Canada where Otis developed a locomotion system [372, 374]. A motion simulator for sport devices was developed at the ETH Zurich, Switzerland [411, 531, 532]. In China, researchers are building the world's largest cable robot for positioning the reflector of the telescope FAST (Five hundred meter Aperture Spherical Telescope) [27]. In Iran, the KNTU cable robots were studied [15]. Beside these, many other prototypes have been developed that we cannot discuss here in detail. An overview of systems that have been identified is given in Table 1.1 and a detailed overview of applications is presented in Sect. 2.4.

1.2.2 Overview

Most works on cable robots were published as contributed papers or as theses. Currently, some 700 contributions in journals and on conferences have been identified. Taking into account the difficulty to find all contributions, it is estimated that around 1000 papers might exist. The available literature mostly deals with specific aspects such as kinematics, statics, dynamics, and control of cable robots rather than giving a thorough introduction to the topic. The early works mentioned above deal with most of the basic effects. However, their approach and terminology are frequently difficult to understand and mostly not in line with recent publications. Especially in the first papers, cable robots are often considered as special forms of collaborative cranes [106] or parallel robots [267, 281]. Merlet's reference book on parallel robots [322] contains many aspects that hold true for cable robots but only slightly touches the topic of which theorems on conventional parallel robots need extension to cable robots and which aspects are completely different. Longer introductions related to a whole field of problems can be found in a number of PhD theses, by e.g. Verhoeven [473] (statics, force distribution, and workspace), Maier [306] (control of under-constrained robots), Bosscher [45] (kinematics and workspace), Fang [139] (control of fully-constrained robots and motion planning), Heyden [206] (control of under-constrained robots), Arsenault [16] (tensegrity), Pusey [405] (design, workspace, and control of suspended robots), Diao [115] (HIL simulation), Tavolieri [461] (rehabilitation), Bouchard [59] (wrench set, interference, and design), Otis [371] (haptic interface), Borgstrom [41] (force distribution), Bruckmann [68] (statics and application), and Gouttefarde [178] (workspace), Liu [292] (design, statics, and dynamics), Riehl [415] (design), Azizian [19] (parameter design), Lamaury [274] (force control), Nguyen [361] (kinematics and cable modeling), Berti [34] (kinematics of under-constrained robots), Yuan [513] (stiffness and sagging), Kraus [259] (force control), Schmidt [431] (kinematisc and accuracy), and Wehr [488] (fatigue of synthetic fiber cables). In contrast, little was published on practical issues such as the mechatronic design of the hardware, procedure for planning a cable robot system, or on economical feasibility studies.

Only a small number of contributions presented the framework of cable robots in a somewhat introductive and overview style [72, 73]. A recent review on open problems was given by Gosselin [170] and Merlet [329] as well as by Kino and Kawamura [247]. Therefore, a major goal of this work is to provide a solid foundation of the theoretical framework with the latest results in some domains.

Table 1.1 Overview of cable-driven parallel robot prototypes and demonstrators

Prototype	Researcher, Affiliation, Country	Classification	Application
Cable master robot	Kawamura/Ito, Ritsumeikan University, Japan	Redundantly constrained	Master-slave teleoperation
Cablecam	N/A, Cablecam Inc., USA	Incompletely constrained	Aerial cameras at sport or entertainment events
Cablev	Woernle, Rostock University, Germany	Incompletely constrained	Precise handling at construction sites and shipyards
CALOWI	Ottaviano, University of Cassino, Italy	Incompletely constrained	Hospital applications
CAREX	Mao/Agrawal, University of Delaware, USA	Unknown	Neural rehabilitation
CaTraSys	Ceccarelli et al., University of Cassino, Italy	Completely constrained	Tracking devices
CaTraSys II	Tavolieri, University of Cassino, Italy	Incompletely constrained	Tracking devices
CHARLOT-TE	N/A, McDonnell Douglas Corporation, USA	Unknown	Space and terrestrial applications
CoGiRo	Gouttefarde et al., LIRMM, Montpellier, France	Redundantly constrained	Handling
Copacabana	Pott et al., ISW, Stuttgart, Germany	Redundantly constrained	Manufacturing
CSHI	Williams et al., Ohio University, USA	Various	Haptic interface
DeltaBot	Khajepour, AEMK Systems, Waterloo, Canada	Fully-constrained	Handling, Pick-and-place
Expo robots	Pott et al., University of Stuttgart, Germany	Redundantly constrained	Entertainment
Falcon-7	Kawamura, Ritsumeikan University, Japan	Redundantly constrained	Ultra fast assembly
FAST	Duan et al., Xidian University, China	Completely constrained	Next generation large radio telescope
FeRiBa3	Gallina/Rosati, University of Padova, Italy	Redundantly constrained	Haptic display
Flight simulator	Bülthoff et al., MPI for Cybernetics, Tubingen, Germany	Completely constrained	Motion simulation

(continued)

Table 1.1 (continued)

Prototype	Researcher, Affiliation, Country	Classification	Application
iFeel6	Hu et al., Beihang University, China	Redundantly constrained	Haptic interface
IPAnema family	Pott et al., Fraunhofer IPA, Germany	Redundantly constrained	Handling, assembly, haptic, inspection
LAR	Bouchard et al., Laval University, Canada	Various	Next generation large radio telescope
LCDR	Alikhani et al., Amirkabir University of Technology, Iran	Redundantly constrained	Automated machining of large workpieces, material handling, construction
MACARM	Mayhew et al., IAI and RIC, USA	Redundantly constrained	Upper limb neuro-rehabilitation
Mantis Duo	N/A, Mimic Technologies, USA	Unknown	Haptic interface
MariBot	Rosati/Rossi, University of Padova, Italy	Redundantly constrained	Rehabilitation
Marionet-Assist	Merlet, INRIA, France	Completely constrained	Lifting crane for assistance robotics at home
Marionet-Crane	Merlet, INRIA, France	Completely constrained	Rescue operations and manipulation of large
Marionet-Rehab	Merlet, INRIA, France	Completely constrained	Rehabilitation tasks and fast pick-and-place operation
Marionet-VR	Merlet, INRIA, France	Completely constrained	Virtual reality (motion provider and haptic device)
NeReBot	Rosati/Rossi, University of Padova, Italy	Fully-constrained	Neural rehabilitation
NIMS3D	Borgstrom, University of California, USA	Completely constrained	3-dimensional actuated sensing device for different environments
NIMS-PL	Borgstrom, University of California, USA	Redundantly constrained	Planar actuated sensing device for aquatic environments
NIMS-RD	Borgstrom, University of California, USA	Unknown	Actuated sensing device
Flying carpet	Bostelman et al., National Institute of Standards and Technology (NIST), USA	Redundantly constrained	Positioning system for large manufacturing

(continued)

Table 1.1 (continued)

Prototype	Researcher, Affiliation, Country	Classification	Application
RoboCrane	Bostelman/Albus, National Institute of Standards and Technology (NIST), USA	Completely constrained	Construction, lunar mission, metrology, large scale manufacturing
Reactive rope robot	von Zitzewitz, ETH Zurich, Switzerland	Various	Haptic interface for sport simulation
Robot calibrator	Bostelman, National Institute of Standards and Technology (NIST), USA	Unknown	Calibration of position of PUMA robots
SACSO-7/-9	Lafourcade, ONERA-CERT, France	Redundantly constrained	Wind tunnel suspension system for aircraft models
Segesta	Hiller et al., University of Duisburg-Essen, Germany	Redundantly constrained	Vibration testing
Shelf robot	Bruckmann et al., University of Duisburg-Essen, Germany	Completely constrained	Warehousing
Skycam	N/A, Skycam, USA	Unknown	Aerial camera at sport or entertainment events
Sophia-3/4	Rosati et al., University of Padova, Italy	Redundantly constrained	Post-stroke upper limb rehabilitation
SPIDAR-G	Kim et al., Tokyo Institute of Technology, Japan	Redundantly constrained	Haptic interface
SpiderBot	Capua/Shapiro, Ben Gurion University, Israel	Incompletely constrained	Imitation of fictional Spiderman
Spydercam	N/A, Spidercam GmbH, Austria	Incompletely constrained	Aerial camera at sport or entertainment events
String-Man	Surdilovic et al., Fraunhofer IPK, Germany	Various	Gait rehabilitation
String-Pot 1/2	Bostelman/Ferguson, National Institute of Standards and Technology (NIST), USA	Various	Metrology system for sculpting assistance

(continued)

Table 1.1 (continued)

Prototype	Researcher, Affiliation, Country	Classification	Application
Texas 9-String	Kawamura et al., Ritsumeikan University, Japan	Redundantly constrained	Haptic interface
VIDET	Melchiorri et al., University of Bologna, Italy	Various	Mobility assisting system for visually impaired users (partly haptic interface)
WARP	Tadokoro, Kobe University, Japan	Redundantly constrained	Creation of virtual acceleration by illusion
WDPSS	Zheng, Xiamen University, China	Redundantly constrained	Wind tunnel suspension system for aircraft model
WireMan	Bonivento et al., University of Bologna, Italy	Completely constrained	Haptic interface
N/A	Ko et al., Robot Research Initiative, Korea	Completely constrained	Handling
N/A	Carricato et al., University of Bologna, Italy	Completely constrained	N/A

1.3 Scope of this Book

The aim of this work is to consolidate the state of the art in cable-driven parallel robots by presenting a consistent theory and well-defined terminology. Based on this foundation, an overview of the latest and ongoing research fields is given. Additionally, in-depth results are presented on selected topics.

The presentation of the theoretical and practical results are intended to allow for building and operating cable robots for real world applications. However, since the field of cable robots is quickly evolving, not every aspect of cable robots is considered at length. Most aspects of the closed-loop control of cable robots are not considered in this book. The remaining chapters of the book are structured as follows:

Chapter 2 introduces the terminology used for cable robots and also defines the technical terms used throughout this book. Then, different categories of classification are presented for cable robots. As reference for the following chapters, some reference designs from the literature are presented which are used for the experimental and simulation examples. Finally, the chapter gives an overview of existing and foreseen fields of application for cable robots.

In Chap. 3, the so-called *standard model* for statics and kinematics is introduced together with the assumptions made. For the standard model, most problems in the field of kinetostatics are solved and a consolidated theory is presented. A special emphasis is placed on methods to compute force distributions for over-constrained cable robots which is a rather specific challenge for cable robots.

Chapter 4 deals with algorithms to compute the kinematics transformation of cable robots. When considering over-constrained cable robots, the inverse kinematics is trivial for the standard model. In contrast, forward kinematics for over-constrained robots is involved to dealing with. The chapter aims at presenting numerical algorithms called *kinematic codes* to efficiently solve the task.

In Chap. 5, the workspace of the cable robot is considered. Firstly, different types of workspace are defined and criteria for determining the workspace are discussed. An overview of existing methods to compute the workspace is introduced and the used data models to describe the computed workspace are discussed. Then, algorithms based on interval analysis are presented to compute the workspace in a continuous way. This method is very reliable but may be slow. Contrary, discretization methods for grids, cross sections, or hulls are straightforward and fast but may be inaccurate for special designs. Detailed workspace studies are presented that unveil the influence of typical criteria such as payload, workspace criterion, orientation, and force limits on the shape as well as on the size of the workspace.

Chapter 6 presents the modeling of the dynamics of a cable robot. Therefore, the dynamics of the platform, the cables, and the winches are considered. Some extensions are made with respect to the standard model which leads to the consideration of dynamic effects caused by pulleys and linear elastic cables. To complete the dynamic model of the robot, the control cascade and the electrodynamics of the actuators must be considered. To facilitate further use, parameters of the dynamic model that have been validated on the prototypes are presented.

Based on the kinematic foundation presented before, Chap. 7 aims at advanced cable models and their respective use in kinematic codes. Some of the assumptions made in the standard model are relaxed. Firstly, robots with pulleys are addressed and kinematic codes to compute forward and inverse kinematics are presented. Secondly, the effect of the mass of the cables is addressed and the resulting sagging cable is modeled. Finally, the effect of elastic cables is considered for inverse and forward kinematics.

In Chap. 8, different aspects in the design of cable robots are considered. An approach is introduced to design an entire robot to fulfill a given task. Then, different structural architectures for cable robots are described and compared. A parameter design approach using interval analysis and optimal design is proposed to perform parameter synthesis. Finally, the mechanical design of cable robots is discussed including the construction of winches as actuation units, cable end-point connectors, and pulley systems for cable guidance. Also, specific design aspects for the integration of force sensors are addressed.

The last Chap. 9 is dedicated to practical aspects of cable robots. Initially, calibration of cable robots is discussed. The focus lies on issues that arise from experimental work with the IPAnema robot family as well as related cable robots. Therefore, the hardware and controller design of some reference robots are discussed and experimental results are compiled.

In the appendix, the notation used for the mathematical expressions is introduced and a short introduction to interval analysis (Appendix B) is given for better reference. Some additional parameters are provided in tables.

Chapter 2
Classification and Architecture

Abstract This chapter deals with terminology and criteria for the classification of cable robots. Different architectures which have been proposed in the literature and presented prototypes are described. Fields of application are presented at the end of this chapter.

2.1 Terminology

Cable-driven parallel robots belong to the large group of multi-body systems. This includes, beside robots, also other mechanisms with coupled motion of their bodies. Within the multi-body system, robots or manipulators are a subgroup designed to generate motion that can be defined by a program. The norm ISO 8373 [125] defining industrial robots states that a robot has at least three degrees-of-freedom, the motion generated by the robot is programmable, and the robot is universal with respect to the application. Disregarding variants of robots which are kinematically redundant, over-, or under-actuated, the degree-of-freedom of the end-effector motion is (roughly) equal to the number of actuators. In other words, in a nonsingular configuration, each actuator contributes a mostly unique part to the generation of the motion at the end-effector.

Based on the topology, robots are subdivided into serial and parallel manipulators. Serial manipulators consist of a sequence of joints and links, where every articulated joint is actuated. Such mechanical structures are called open kinematic chains. If one connects more than one kinematic chain to the end-effector, the resulting mechanism is called a parallel robot.

If the number of chains is equal to the number of actuators, the robot is called fully parallel [322]. Cable-driven parallel robots are a special kind of parallel kinematic machines or parallel robots. There is no unique or standardized technical term for cable robots in the literature but a number of different wordings with synonymical meaning can be identified:

© Springer International Publishing AG, part of Springer Nature 2018
A. Pott, *Cable-Driven Parallel Robots*, Springer Tracts in Advanced
Robotics 120, https://doi.org/10.1007/978-3-319-76138-1_2

Table 2.1 Common terminology for cable-driven parallel robots extracted from the titles of 578 papers. For the common terms, the number of found instances is given in percent, for the seldom used terms the absolute numbers are given

Transmission element	Actuation principle	System
Cable (64%)	Driven	Robot (48%)
Wire (23%)	Based	Manipulator (22%)
Tendon (22)	Suspended	Mechanism (11%)
Rope (4)	Actuated	(Stewart–Gough) platform
String (6)		Crane

- Cable-controlled parallel link manipulator [280]
- Parallel link robot crane [106]
- Positioning mechanism using wires [349, 350]
- Robot using wire drive system [237]
- (Parallel) wire (driven) robot [97]
- Cable-driven parallel manipulator [451]
- Cable suspension robot [307]
- Tendon-driven Stewart(-Gough) platform [474, 477]
- Cable-driven parallel robot [305]
- Cable direct driven robot [492]
- Cable array robot [160, 468].

The frequently used terminology was analyzed based on 578 paper titles.[1] Most authors (64%) used the term *cable* for the transmission element where another 23% papers prefer *wire*. Only a very small number of authors call the cable *tendon* (22), *string* (6), or *rope* (4). However, it seems worthwhile to remark that material scientists working on synthetic fibers often use the technical term *rope* which was hardly taken up by the robotics community. For the system, the term *robot* is used by 48% of the papers where another 22% call it *manipulator* instead. Only 11% of the works refer to *mechanisms*, whereas *platform* and *crane* are each used by less than 4% of the authors. An overview of the terminology can be found in Table 2.1. In newer contributions, there is an increasing tendency towards the term *cable-driven parallel robot* or, as a handy shortcut, *cable robot* which we will use throughout this work. Also, the abbreviation CDPR (cable-driven parallel robot) is wide-spread (Fig. 2.1).

A cable robot can be decomposed into a *mobile platform*,[2] a fixed *machine frame*, *m* cables attached to the mobile platform on their *distal* end and attached to the

---

[1] Although this analysis is based on a relative large number of papers, it is not necessarily statistically significant. There are some reasons that could decrease the quality of the data base in terms of representation. The papers in the used data base reflect the author's research interests and the selection may thereby be biased. Also, the assumptions on the used terminology influence the literature research. Furthermore, rarely cited publications might be overseen in the review of literature because of their differing terminology.

[2] The mobile platform is sometimes called traveling platform or end-effector, although the end-effector is only the part that performs the actual process.

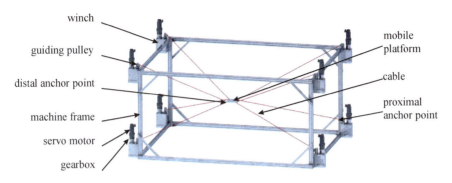

Fig. 2.1 Concept and components of a cable-driven parallel robot

machine frame on their *proximal* end. The lengths of the cables (and sometimes also the positions where the cables are attached to the frame) are changed by an actuation system which is called *winch* for simplicity although there are other mechanisms for actuation. For most robots, the winches are fixed to the machine frame to simplify the electric connection with the power and control system. However, designs where the winches are located on the mobile platform were proposed [6]. Many cable robots use sensors to indirectly measure the effective length of the cables: e.g. through encoders on the drum or with a linear measurement system on a pulley tackle. The direct determination of the cable length is difficult to practically achieve. Alternatively, the position and orientation of the mobile platform are directly measured [108]. For a couple of applications, it is necessary to determine the tension in the cables as well. This is mostly done by force sensors that are connected to one end of the cable or to some pulleys in between. When using a winch with a drum, one can also measure the cable force with a torque sensor in the drive-train. Finally, one can rely on internal sensors such as current sensors of the electric motor to estimate the cable forces. The sensor signals are connected to a *controller* system which generates the set values (typically positions, velocities, or forces) for the actuation system. Typical controller systems can be decomposed into the closed-loop control for current, torque, velocity, and position. The set values for position, velocity and/or force are generated in an open-loop controller structure that deals with trajectory generation and path planning. Higher level controller functions such as task or process control are often also included into the open-loop control system.

The mobile platform may largely vary in its size. On the one hand, the platform may have a weight of some grams and dimensions of a couple of millimeters. Contrary, there are examples of huge platforms, e.g. the collector of the Arecibo telescope, with a size of some dozens of meters and a weight of more than 800 tons. For cable robots, the most important properties of the platform are the relative location of the distal anchor points with respect to the reference point of the platform, the center of gravity, the mass, and the inertia tensor.

The machine frame is the mechanical structure that carries the winches or the proximal anchor points. In many laboratory and industrial setups, the machine frame is a closed framework structure made from steel or aluminum bars. Especially for larger robots, the winches might as well be attached to decentralized structures such as towers or buildings. It was also proposed to use winches or cables on multiple flying or swimming structures such as helicopters [370], balloons, ships, off-shore platforms [135], and submarines [315]. Integrated cable robots use whatever is appropriate from the surrounding machinery or building as supporting structure [223].

The cables (or wires, tendons, seldom ropes, or strings) can be made of different materials. The most widely used materials for the cables are steel and synthetic fibers such as high-modulus polyethylene fiber, aramid (kevlar), or polyester. However, other materials such as hemp can also be employed. Lately, so-called smart cables with integrated electric wires were proposed and used in prototype in order to supply the mobile platform with electric energy or fieldbus signals.

2.2 Classification

There is not one unique classification of cable-driven parallel robots. Moreover, cable robots can be classified by means of the following criteria:

- The number of cables m and degrees-of-freedom n of the mobile platform: This kinematic classification was proposed by Ming and Higuchi [349] to distinguish between under-constrained, fully-constrained, and over-constrained cable robots (see Sect. 2.2.1).
- *Degree-of-parallelism*: A cable robot might by fully or partially constrained through the cables. Fully-constrained cable robots solely employ cables to connect the mobile platform with the base, whereas partially constrained or hybrid cable robots have passive joints to connect some or all degrees-of-freedom of the mobile platform to the base. There is a related family of serial mechanisms which are actuated by cables as well as robotic hands, that are driven by cables. However, serial cable-driven robots and robot hands are out of the scope of this book.
- The *Motion pattern* or mobility of the mobile platform characterize the kind of independent motions of the platform and were proposed by Verhoeven [473] (see Sect. 2.2.2).
- The concept of the *actuation system*, e.g. winches, pulley tackles, linear actuators, can be used to distinguish different types of cable robots. Also, more exotic concepts such as twisting the cables to change the length were described (Sect. 2.2.3).
- The *function* of the cable robots, e.g. as programmable positioning system, as a force generating system, or as measurement system for position or force (Sect. 2.2.4).

In the following sections, we detail the different classifications. The classification is very important for many methods and algorithms described in this work, since many methods are limited to a subset of cable robots.

2.2.1 Kinematic Classification

An obvious criterion for classification is to consider the number of cables denoted by m and the controllable degrees-of-freedom of the mobile platform denoted by n. The first classification of this type was introduced by Ming and Higuchi [349]. Furthermore, the *degree-of-redundancy* $r = m - n$ is introduced. One can distinguish between the following classes:

- $m < n \leq 6$: The robot is under-constrained and in general cannot withstand arbitrary applied wrenches $\mathbf{w}_P$. Taking into account gravity or other applied forces and torques, one or more poses in which the robot is in stable or unstable equilibrium may exist. Still, some degrees-of-freedom cannot be controlled through the cables in general. The number and direction of the controllable degrees-of-freedom vary throughout the workspace. This class of robots is called *incompletely restrained positioning mechanism* (IRPM) [349], see Fig. 2.2a.
- $n = m$: The robot is kinematically fully-constrained but the force equilibrium depends on the applied forces such as gravity. There is a limited range of forces and torques the robot can withstand depending on the magnitude and direction of the applied force. In Ming and Higuchi's classification, these robots also belong to

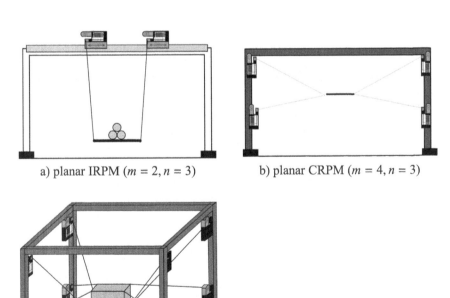

a) planar IRPM ($m = 2, n = 3$)   b) planar CRPM ($m = 4, n = 3$)

c) spatial RRPM ($m = 8, n = 6$)

Fig. 2.2 Examples of the different kinematic classes

the IRPM class since they also rely on external forces. Other authors [139] define an own class for such robots.

- $n + 1 = m$: The robot can be fully-constrained through the cables in certain poses. Different types of motion patterns are possible (see Sect. 2.2.2). The forces that the robot can withstand depend on the minimum and maximum forces in the cables that can be generated by the robot, see Fig. 2.2b. Robots of this class are referred to as *completely restrained positioning mechanisms* (CRPM).
- $n + 1 < m$: The robot is redundantly constrained and forces have to be distributed between the cables. These robots are called *redundantly restrained positioning mechanisms* (RRPM), see Fig. 2.2c. As pointed out by Merlet, these robots are not kinematically redundant since they have only one solution to the inverse kinematics problem. The redundancy relates to the number of kinematic constraints and thus also to their actuation since there are more kinematic constraints than degrees-of-freedom. Therefore, the static forces of the robot are generally undefined.

Beside this, let $\mathbf{u}_i$ be the direction of the ith cable, then a pose of the platform is called *suspended*, if for all cable i

$$\mathbf{u}_i \cdot \mathbf{g} < 0, \quad i = 1, \ldots, m \tag{2.1}$$

holds true, where $\mathbf{g}$ is the direction of gravity. If the robot is predominantly operated in suspended configurations, it is common to also call the cable robot *suspended* or in *crane configuration*, indicating that the robot relies on gravity to be balanced. The attribute *suspended* provides a handy way of describing that the workspace of the robot is mostly below the robot frame and the robot is operated in a *crane-like* way. Therefore, a significant influence of gravity was taken into account in the design of the robot, mostly to allow for a considerable workspace. A cable robot design itself is hardly suspended but the attribute *suspended* depends on how the cable robot is used. Some robot designs can only be operated in a suspended configuration while other robots allow for both a fully-constrained and a suspended operation mode. Note, that all classes of robots (IRPM, CRPM, and RRPM) listed above can be operated in a suspended configuration.

2.2.2   Motion Patterns for Cable Robots

The motion pattern of a robot characterizes a subset of the generalized virtual displacements $\delta \mathbf{y}$ that can be executed with the end-effector and that is consistent with its kinematic constraints. In this general setting, a virtual displacement $\delta \mathbf{y}$ is some superposition of translation and rotation of a rigid body motion in three-dimensional space. In the spatial Euclidean motion group SE_3, there exists six independent virtual displacements and therefore, the maximum number of degrees-of-freedom n of a mobile platform is six. In the following, a motion pattern is represented by an abbreviation of the form $n_R R n_T T$, where R represents n_R rotational degrees-of-freedom and T stands for n_T translational degrees-of-freedom, respectively.

A robot may not be able to generate all six independent virtual displacements. Simple examples of robots with less than six degrees-of-freedom are planar ($n = 2$ or $n = 3$) or spatial translational robots ($n = 3$). As it can be seen from these simple examples, the degree-of-freedom is not a unique characteristic to distinguish between the motion patterns of a platform. As an example, we consider $n = 3$ in the following. We find, amongst others, the planar rigid body motion (1R2T) with two translational and one rotational degrees-of-freedom, the motion of a point in space (3T) with three translational degrees-of-freedom, and also the spherical motion (3R) which is the special orthogonal group SO$_3$.

The motion pattern of the platform is in general a pose-dependent property, i.e. it can change throughout the workspace. A well-known defect in the motion pattern is a singular configuration where the robot loses or gains degrees-of-freedom. Beside changes in the number of degrees-of-freedom of the end-effector motion, the direction of the available virtual displacements may also depend on the configuration. In practice this is not desirable and robot designers usually concentrate on designs with a well-defined motion pattern.

The term *motion pattern* is quite general. Therefore, we restrict in the following the discussion to motion patterns that can be expressed by superposition of the three purely translational displacements and three purely rotational displacements, where we arbitrarily identify these directions with translation along and rotation about the axes of an Euclidian coordinate system.

In the notation, we abbreviate the translational motions with T and the rotational motions with R, leading to shortcuts such as 1R2T for the planar rigid body motion with one rotational degree-of-freedom and two translational degrees-of-freedom.

Verhoeven [473] created an exhaustive list of possible motion patterns for fully parallel cable robots (Table 2.2, Fig. 2.3) and proved that this list is complete. However, the underlying assumptions are strict since it is assumed that each cable is independently actuated. Especially, it was shown that no designs without translational degrees-of-freedom exist (i.e. robots with the motion patterns 1R, 2R, and

Table 2.2 Complete list of all possible motion patterns for fully parallel cable-driven robots. T represents translational degrees-of-freedom, while R represents rotational degrees-of-freedom in the acronyms of the motion patterns

Symbol	Description
1T	Trivial one degree-of-freedom purely translational cable robot
2T	Planar pure translational cable robot
3T	Spatial pure translational cable robot
1R2T	Planar cable robot with rotation
2R3T	Spatial robot with two rotational degrees-of-freedom
3R3T	Spatial robot with three rotational degrees-of-freedom

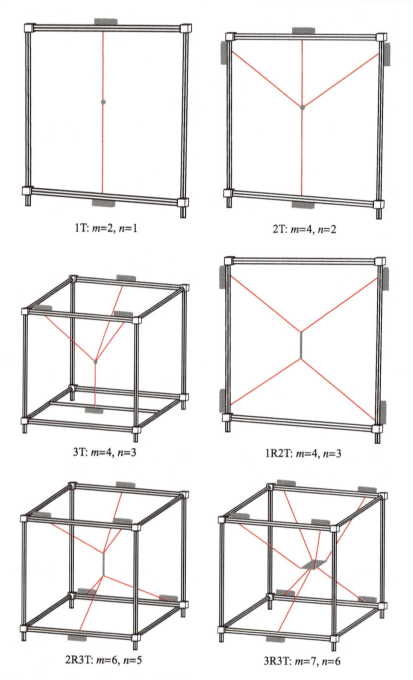

Fig. 2.3 All possible motion patterns for fully parallel cable robots with the number of cables m and the degree-of-freedom n

3R are impossible) and that it is not possible to create Schönflies-motion (1R3T) generators based on fully parallel cable robots.

Beyond the assumption of Verhoeven, it is possible to add cables without actuation to the robot to generate e.g. Schönflies-motion. This can be done either by appropriate control schemes or by connecting two or more cables to one actuator. A simple way of achieving this effect is by coiling two or more cables onto the same drum or connecting them to the same linear drive. The motion can also be constrained with mechanical elements such as prismatic joints, where in this case the robot is no longer a fully cable-driven robot in the sense of the definition used in Verhoeven's proof. Fixing some cables to a constant length allows for a pure rotation without translation (3R). To construct such a cable robot, one can combine the 3T and the 3R3T robot in Fig. 2.3 into one robot with $m = 11$ cables. Now, we consider the four cables of the 3 T robot to be of constant length without actuation. Using the remaining cables of the 3R3T robot, we can still change the orientation of the platform where no translation is possible due to the fixed cable length of the 3 T part.

2.2.3 Classification of Actuation

Most cable robots are actuated by winches which coil the cables on drums [6, 210, 238, 404]. This concept was adopted from cranes and it allows for dealing with very long cables. Winches are well-established construction elements with a very compact housing (Fig. 2.4). Their mechanical design is simple and cost-efficient. In most winches used for cable robots, servo motors are used to control the cable length in cascaded position control. Since the winches are normally fixed to some bigger structure, there is nearly no upper limit on the size of the motors that can be applied, except for cost reasons. Thus, winches are a good choice for high forces and long cables.

Another straightforward concept to control the cable length is to use linear drives coupled to a pulley tackle [324, 336, 448]. Either the end of the cable or one or more pulleys are moved along a line generating the desired change in the length of the cable. Examples of such actuation systems are shown in Fig. 2.5. Some kind of gear can be implemented through the pulley tackles which also enables linear actuated cable robots to realize a large workspace. Linear drives with cable tackles possibly offer the highest velocities and accelerations of the cables where the maximum forces may be limited due to available linear actuation systems.

It was also proposed [439] and patented [440] to change the cable length by twisting the cables. This concept allows for very small increments of the changes in the length of the cables where the maximum change in length is rather limited. The twisting of the cables behaves like a very high gear reduction leading to very high forces with relatively small velocities. Twisting the cables also involves very high wear in the cables. Furthermore, it may be applicable to generate very precise motion. Twisted cables generate aside from the contraction force also a torque on the platform that might have to be taken into account.

Segesta winch with one pulley with integrated force sensing and a panning pulley as cable guidance system. (Courtesy of Chair of Mechatronics, University of Duisburg-Essen, Germany)

Cable winch with pulleys for guiding the cable. The second pulley is equipped with a rotation sensor to estimate the direction of the cable. (Courtesy of Chair of Mechanics, University of Rostock, Germany)

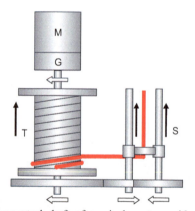

Conceptual draft of a winch system with a servo drive (M), a gearbox (G), a single layer and a unidirectional pulley mechanism drum (T), and a guidance system for the cable (S)

an IPAnema1 winch with cable guidance system

Fig. 2.4 Different concepts for actuation of cable robots with winches

Finally, kinematic concepts similar to Delta robots [98] were proposed. The cables of fixed length are connected to rotating levers imitating the kinematic structure of the Delta robot. Since Delta robots are mostly used for small design size and small payloads, a telescopic strut with a spring is connected to the mobile platform in order to maintain tension in the cables. Maeda [305] presented the well-known demonstrator WARP that was actuated in this manner.

Linear actuation of the IPAnema 2 planar (Bandroboter) using linear direct drives along with a pulley system for transmission and guiding of the cables. T: carriage, X: drag chain, Y: drag chain holder, W: drag chain housing, U: linear motor's primary part, Z: linear motor's secondary part, V: prismatic guideways

Linear actuation of the String-Man (Courtesy of Fraunhofer IPK, Berlin, Germany)

Fig. 2.5 Cable robot with linear actuator. Left: IPAnema planar with linear direct drives. Right: String-Man used by Fraunhofer IPK for gait rehabilitation

There are also other methods for actuation. The Cablev system has winches that are movable on a guideway [306, 310, 496]. The planar IPAnema 2 system [404] uses cable-driven pulleys combined with linear direct actuators to maintain an isotropic configuration of the platform throughout a rectangular workspace [392]. Bruckmann proposed a robotic system very similar to cable robots where the cables were replaced by slim rigid legs [77]. Here, the proximal anchor points of the winches are moved with linear guideways. This architecture is well-known as Linaglide from parallel robots but here the forces in the legs are kept positive or only very low pushing forces are allowed. However, despite an encouraging result from a conceptual study, the built machine employed cables and was used as manipulator in a wind-tunnel. A cable robot consisting of restraining cables and linear springs was lately put forward [131], where the springs were used to shape the workspace through an artificial potential field.

2.2.4 Classification of Function

The name *robot* implies a device to handle or manipulate other objects. One can distinguish between the following functions that can be performed by the cable robot:

- *Motion generation*: Most cable robots are designed to create a well-defined motion of the mobile platform in order to reach programmed positions and orientations in a given sequence. Depending on the control system, the trajectories between these poses are also well-defined in terms of velocity and acceleration. This is the typical control scheme for robots and machine tools. Especially, fully- and redundantly constrained cable robots usually have a well-defined motion behavior since the control systems have to solve their over-constrained equations anyway (Fig. 2.6a).
- *Force-torque generation*: The robot can be controlled to generate defined forces and torques at the mobile platform, e.g. to perform production tasks or act as a force feedback system. The motion of the platform is then depending on the interaction of the robot with its environment, especially the motion depends on the wrench applied to the mobile platform (Fig. 2.6b).
- *Force-torque measurement*: The cable forces can be measured e.g. through the motors, through additional sensors in the cables, or through the cables themselves. These measurements can be used to derive the applied forces and torques to the mobile platform creating a multi-directional force/torque sensor (Fig. 2.6c).
- *Motion measurement*: The platform of the robot can be moved around by a human, by other machinery, or by the environment and the robot uses its sensors to estimate the current pose of the platform. If the motors are replaced for example with contracting springs, one can apply a position measurement system to the winches in order to determine the length of the cables. Just using the forward transformation, one can reconstruct the pose of the mobile platform from the measurements. Since cable robots can have a large workspace and the inertial mass of the mobile platform is small, the robots can be used as large scale spatial measurement systems. Applications in motion tracking and calibration were proposed, see Figs. 2.6d and 2.22.

In practice, some of these functions may be leveraged at the same time. For example, a positioning system used for an assembly task could use its force-torque measurement capabilities for advanced process control.

2.2.5 Size, Payload and Dynamics

Next, we consider some limiting factors for size, payload, and dynamics. Ultimate limits for the length of the cables are given by:

- The *specific strength* l_R of the cable material gives an upper limit on length of a suspended cable under gravity and is determined in meter. If a cable is longer than l_R, its weight is higher than its breaking load and it cannot carry any external load. Zylon and Dyneema (polyethylene) have very high breaking lengths $l_{R,PES}$, around 350 and 400 km, respectively. Carbon fibers are somewhat weaker with 250 km, whereas the breaking length of steel is only around 25 km. Compared to Dyneema, roughly ten times higher values could eventually be reached using graphene and

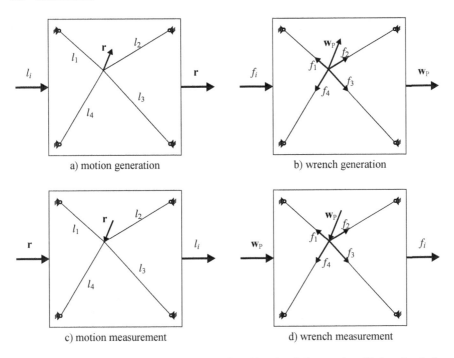

Fig. 2.6 Illustration of the four basic functions of a cable robot. l_i denotes the cable length, $\mathbf{r}$ is the motion of the platform, f_i are the cable forces, and $\mathbf{w}_P$ is the wrench applied to the platform

carbon nano tubes (CNT). However, grapheme and CNT are rather expensive and, to the best of the author's knowledge, cables are currently not manufactured from CNT.

- In contrast, cables made from Dyneema and Zylon are available on the market at acceptable costs and it seems that this puts today's practical limit on size to around 350 km as long as no additional payload needs to be supported and without safety factors. Clearly, such limitation does not apply for space robotics application where gravity is not an issue.

There is a cross relation between size, material, and dynamics when it comes to really large robots. If the length of the cable is large compared to the sonic speed within the cable material, there is a decreasing response time to fast motion at the actuator since the cable is a spring and thus acts like a first order low pass filter.

Some key figures are depicted in Fig. 2.7 where the ultimate limits on the size arising from the strength of the fibers are indicated with the grey zone. The zone is increased by an order of magnitude to reflect the typically required safety factor of 10.

Concerning the maximum payload, today's largest cranes allow for maximum loads of some thousand tons which can be considered as a technological limitation. Such huge loads are realized through pulley tackles with a high number of pulleys.

Fig. 2.7 Comparison of different robots and cable structures with respect to payload and size in logarithmic scale. The region of the right side is related to the specific strength of typical materials for cables. The axis for the payload is bounded by the largest cranes

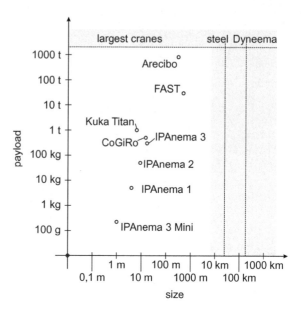

The operation of pulley tackles for the suspending cables of a cable robot makes the construction of such a system involved and, to the best of the author's knowledge, no such design was presented in the literature. Therefore, effective limits for the payload are expected to be lower.

Some limits on the maximum acceleration are given through the available motors. Using standard servo drive acceleration in the range of 20 and $1000\,\mathrm{ms}^{-2}$ can be reached on the surface of the motor's shaft. Theoretically, one can coil very light cables directly onto this surface to achieve the given acceleration for the cables. However, a realistic mechanical design adds additional inertia to realizing the drum. Using larger drums also scales up the inertia of the motors although a larger diameter increases the acceleration. Since the inertia of a solid cylinder rises to the fourth power of the diameter where the velocities only scales linearly with the diameter, smaller drums are to be preferable with respect to maximum acceleration.

2.3 Architectures

The geometry of a cable robot is mostly characterized by the relative position of its anchor points on the fixed machine frame and on the mobile platform. Firstly, we will distinguish between so-called *generic* and *non-generic* geometries. A non-generic design is subject to geometric relations such as all points of the platform lie in a plane or form a line. A generic design usually maintains its properties if its geometrical parameters undergo infinitesimal changes. Contrary, a non-generic design only has certain properties as long as well-defined constraints are fulfilled. Typical examples for such constraints are:

• All platform or base anchor points lie in a common plane
• Equal lengths or distances between the anchor points
• Point or axis symmetric designs
• Two or more coinciding anchor points on the platform or on the frame.

In mechanism analysis, such constraints can be very interesting. There are a lot of special cases known where such constraints lead to simpler kinematic equations or even closed-form solutions of equations that cannot be solved in the general case. From a practical point of view, non-generic designs only exist in theory, since they assume that the robot is manufactured and assembled without errors. Furthermore, all kinds of disturbance are neglected such as elasticity, clearance, wear of components, and thermal effects. In contrast, non-generic designs can be very useful as design templates, i.e. as initial guess that can be further adapted to the requirements.

Non-generic design imposes assumptions and, therefore, for algorithms it is desirable that they work with generic designs. Such algorithms are often more robust against imperfect modeling or imperfect manufacturing and these algorithms can be tuned by calibration in the widest sense. Methods for non-generic designs can exploit more specific assumptions. Thus, as long as such assumptions hold true, the resulting algorithms are easier to implement or faster.

2.3.1 Notation for Coinciding Anchor Points

As mentioned above, a widely spread assumption is that two or more cables share a common anchor pointed either on the base or on the mobile platform. Such a classification reflects important conditions used in kinematic analysis. To describe configurations with common anchor points, the $X - Y$ notation is used, meaning that there exist X different points on the frame and Y different points on the platform where $X \leq m$ and $Y \leq m$, m being the number of cables. For example, an $8 - 4$ robot is a typical design with eight different anchor points on the machine frame and 4 different anchor points on the platform. Note that this notation is ambiguous because it does not indicate which cables share a common anchor point and how many cables share one point. In the example, the $8 - 4$ configuration could be realized, amongst others, by a two-to-one connection (Fig. 9.4) as well as by an uneven distribution where five cables are connected to the first anchor point and the remaining connections are connected in an one-to-one scheme between the base and the mobile platform.

For cable robots, it is often assumed that common anchor points only occur on the mobile platform. Again, a notation with hyphens is common indicating how many cables share a common point on the platform. Without loss of generality, the number can be put in descending order. For example, the Segesta robot (Fig. 9.16) has a 4-2-2 configuration for its platform, meaning that four cables share the first vertex of

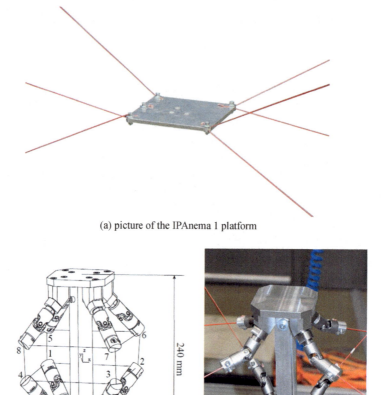

(a) picture of the IPAnema 1 platform

(b) drawing of the IPAnema 1.5 platform   (c) picture of the IPAnema 1.5 platform

Fig. 2.8 Mobile platforms used with the first generation of the IPAnema robot. **a** The cables are directly clamped to the IPAnema 1 platform so that two cables share a common anchor point. **b/c** The cables are connected in a *crossed cable* configuration to the universal joints of the IPAnema 1.5 platform

the triangular platform whereas the two other vertices are connected to two winches each. Contrary, the IPAnema 1 system (Fig. 2.8) has a 2-2-2-2 configuration where each of the four corners of the platform is connected to two different winches.

2.3.2 Fixed Machine Frame

The anchor points of the winches are restricted to geometric primitives in most robot prototypes. Prismatic forms have the advantage that they can be easily built as frame-works and the bars made from aluminum or steel are well suitable to fix the winches. As pointed out in a couple of publications, cable robots can be easily reconfigured. Therefore, some test-beds allow to move the winches along the bars of the frame-work. Frequently, the cable robot prototypes have a box-shaped frame (e.g. Segesta, IPAnema, and CoGiRo), where all base anchor points are aligned on the surface of the box or at least close to that surface to avoid the need for complicated fixtures. Many of these designs restrict the position of the anchor points to the edges or even to the corners of the box. Beside prismatic designs, polar arrangements are preferred where the position of the anchor points is characterized by a radius and angles to distribute the anchor points around a center. For example, a trisymmetric structure called ReelAx 8 was presented by Izard [224], where all winches are mounted on three or four vertical poles. Planar robots are mostly operated in a closed rectangular frame.

2.3.3 Mobile Platform

For the mobile platform, it is common to use planar platforms, where some cables share identical anchor points. As pointed out by Verhoeven [473], this signifi-cantly reduces the influence of cable-cable interferences, and thus allows for larger workspace. Furthermore, common anchor points affect the location of singularities and simplify the computation of the forward kinematics [389, 465]. Spatial platforms are often prismatic or cylindric. Star shaped platforms are also common.

2.4 Fields of Application

Cable robots have been proposed to be used in a very wide range of applications. As many other robotic systems, the development of new ideas for applications is mostly driven by replacing a mostly manual or mechanized process with a robotic solution that allows for fully automatic operation. As proved in many industrial applications, robot systems are well suited to reduce labor costs in production, increase the quality of process execution, or shorten the cycle time. Thus, cable robots may open new fields of application where industrial robots cannot be applied due to restrictions with respect to the size of the workspace, the payload, or the required cycle time.

However, cable robots present some other advantages like minimal installa-tion footprint, simple transportation and deployment, or improved quality. Thus, cable robots can be employed as assistant devices where other robotic systems are inappropriate.

2.4.1 Production Engineering

Most robots installed worldwide are applied in the field of production and manufacturing, where robots succeeded in efficiently automating industrial processes, especially for handling, welding, painting, and assembly. Therefore, numerous research projects on cable robots were dedicated to investigate production tasks.

At the University of Berlin, a machine tool was built from a cable robot [258]. The platform with the spindle was suspended to eight cables and moved around the workpiece. The research aims at performing high-speed motion with the cable robot but the system was found to lack the required stiffness for machining.

Bosscher proposed to use cable robots for counter crafting [49, 51] where a mobile frame was designed to move between construction sites. In the filed patent [51], Bosscher describes a cable robot with up to twelve cables where a group of eight lower cables are mounted on vertical guideways. By continuously changing the configuration, interferences between the lower cables and the currently built structure are avoided.

A handling and assembly system for large-scale products, like collectors for concentrated solar power (CSP) plants, was studied by Pott [401] and presented during the trade fair Automatica 2010 in Munich, Germany (Fig. 2.9).

A number of production tasks require the positioning of specialized equipment around a large workpiece or product such as ships, airplanes, blades of windmills, as well as steel structures such as motors, generators, and gearboxes. Typical tasks to perform are painting, welding, grinding, or blast cleaning. Especially noncontact processes seem very adequate for cable robots since taking measurements for inspection, maintenance, or quality control is easy to realize. Amongst others, the research project CableBOT, founded by the European Commission[3], proposed to use cable robots for painting aircrafts (Fig. 2.11).

A conceptual study for additive manufacturing was undertaken by Pott and Grzesiak and is depicted in Fig. 2.10. Such concepts were recently taken up due to the increasing interest in 3-D printing.

2.4.2 Logistics

Handling and logistics are promising fields of application for cable robots. Two of the main advantages of cable robots can be fully exploited: cost-efficient robot designs for a very large workspace and very high dynamics allow for high throughput in

---

[3]The web site of the CableBOT project can be found under http://www.cablebot.org.

Fig. 2.9 Cable robot
IPAnema 2 for large-scale
handling of collector
modules shown at
Automatica 2010 trade fair,
Munich, Germany

Fig. 2.10 Concept for
additive manufacturing of
mockups using a large-scale
cable robot equipped with an
extruder. Source: Pott and
Grzesiak

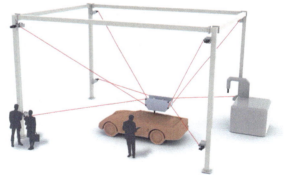

handling, sorting, and (de-)palletizing. Already in the 1990s, the idea of building
ultra-high speed pick-and-place manipulators by means of a cable-driven robot was
addressed with the FALCON robot (Fig. 2.13) by Kawamura [237, 239], as well
as with the Warp system proposed by Tadokoro [305]. For very high payloads, the
connection to automated cranes showed up. The cable robot Cablev [206, 208] at the
University of Rostock, Germany, aimed at automatic performance of container cranes
using an under-constrained cable robot with additional linear axes for translating the
winches. The CABLAR system was developed by Bruckmann (Fig. 2.12) as storage

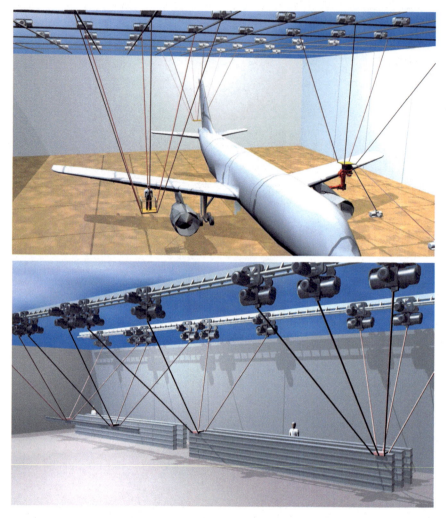

Fig. 2.11 Application concept for painting, cleaning, and maintenance of aircrafts with cable robots (top) and using cable robots as modular handling devices for steel beams (bottom)

retrieval machine [70, 71, 273]. Later, Merlet proposed a portable crane for heavy load handling and rescue where some kind of aerostat was employed to fix pulleys in mid-air [338]. A civil engineering problem of transporting persons across a river was addressed by Castelli [92–94].

Due to their light-weight structure, cable robots were proposed to be used as sensor platforms in different scenarios. The patent proposed by Bauer [30] exploits the huge workspace to move optical and radio sensors through shelf storage systems in order to inspect and locate the stored goods.

Fig. 2.12 The large-scale cable robots as storage retrieval machine CABLAR by University Duisburg-Essen (Courtesy of Chair of Mechatronics, University of Duisburg-Essen, Germany)

Fig. 2.13 The cable robots Falcon (left) and Dolphin (right) from Kawamura (Courtesy of Sadao Kawamura, Ritsumeikan University, Japan)

2.4.3 Construction

The requirements for construction are somewhat related to the demands of handling applications. Again, cable robots benefit from the cost-efficient large scale systems. However, for large scale systems, the potential of fast motion is secondary. Instead, cable robots can profit from their build-in flexibility and can adapt to actual construction tasks by geometrical configuration. One of the first application proposed for the RoboCrane was ship building and bridge building [6, 7, 52–57].

Cable robots were proposed to be used as robotic cranes for the construction of large scale solar power plants (Fig. 2.14) [401]. More recently, Izard studied the installation of a cable robot on the facade of a building [223, 224]. Here, different tasks were considered including cleaning and advertisement. A slightly different approach was discussed by Voss [481, 482], where a cleaning or inspection unit shall be moved over large glass surfaces. Emmens [136] proposed to use a surface

Fig. 2.14 Vision of assembly of parabolic reflector panels with a mobile large-scale cable-driven parallel robot

constraint cable robot for cleaning facades and buildings. In order to avoid relative motion between the cables and the surface, the motors are proposed to be installed on the mobile platform. First results were received from a small scale mock up.

2.4.4 Motion Simulation

A cluster of applications aims at simulating or measuring motion. One field of application is medical or rehabilitation applications, where a cable robot as ultra lightweight system can be applied to guide and measure the motion of limbs. Also, the stabilization of the upper body for gait rehabilitation is a promising application. Ishii [219] introduced a simple robot with four cables as 3D haptic interface. Surdilovic proposed the *String-Man* system [448] where the proband is held by a harness which in turn is suspended by cables (Fig. 2.15). Advanced force control virtually reduces the effective mass and allows gait training with smaller stress on the body and legs. Furthermore, the cable robot hinders the patient to fall down. A similar idea was later studied by Castelli [92, 93] who proposed a device for lifting elderly people for standing up from a wheel chair. Agrawal has studied the application of cable robots as rehabilitation devices in a number of applications for more than one decade [64, 312, 313], where both cable robots and cable-driven exoskeletons were under investigation. Merlet performed a number of practical tests with the Marionet robot for lifting elderly and disabled humans in an ambient assisted living environment (Fig. 2.16). Using cable robots for motion generation and tracking of sportsmen while performing sports such as rowing was studied at the ETH Zurich by Zitzewitz and Rauter [411, 531, 532]. A recently founded company offers a rail-based cable robot as an assistant device for rehabilitation called *the float* (Free Levitation for Overground Active Training) [302, 469].

Some authors proposed to build motion simulator platforms from cable robots. The possibly large workspace along with high dynamic capacities have been identified as

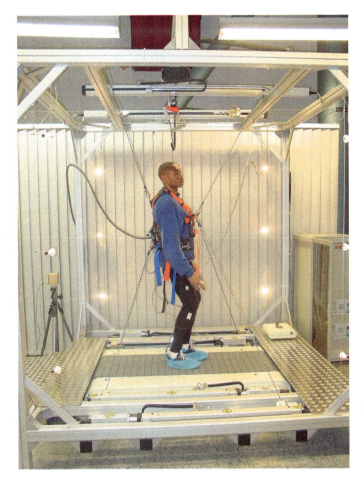

Fig. 2.15 The cable robot String-Man is designed for use in gait rehabilitation. The proband is held by a harness and becomes part of the mobile platform of the robot (Courtesy of Fraunhofer IPK, Berlin)

key benefits of cable robots over other structures. Motion generation and positioning in aerodynamic and hydrodynamic test facilities also received attention. Lafourcade proposed using a cable robot in a wind tunnel for airplanes [270] making use of the very small disturbance in the airflow caused by the cables. The same approach was later followed by Bruckmann, who presented a cable robot (Fig. 2.17) for positioning hulls of ships [77]. Most recently, a cable robot is under design as motion base of experimental research of human vestibular senses as well as for usage as flight simulator (Fig. 2.18).

Using the cable robot both for motion measurement and force generation leads to research on haptic displays and devices for virtual reality. Otis [373] proposed a haptic interface for walking. Kraus [261] presents a control framework to allow for different haptic interaction states between a cable robot and a human operator by

Fig. 2.16 The cable robots
of the Marionet family for
rehabilitation from Merlet
(Courtesy of INRIA
Sophia-Antipolis, France)

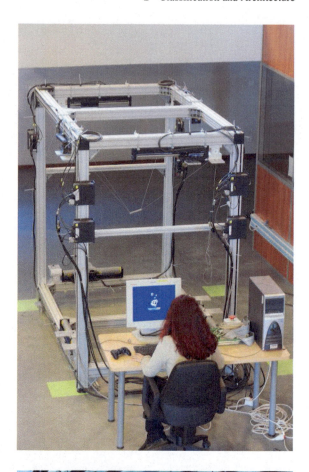

Fig. 2.17 The large-scale
cable robots as manipulator
for wind tunnels (right) by
University Duisburg-Essen
(Courtesy of Institute for
Fluid Dynamics and Ship
Theory, Hamburg University
of Technology, Germany)

Fig. 2.18 A large-scale motion simulator with a 12 by 12 m footprint for tests with proband was initially operated in 2015 at Max Planck Institute for Biological Cybernetics, Tubingen, Germany. A passenger cabin is constrained by eight steel cables to perform flight maneuvers, e.g. for helicopter training or for experimental measurements of human vestibular system. In contrast to most other cable robots, a safety design was used for the basic control to allow for motion with persons on the platform

means of admittance control on the IPAnema 3 Mini robot (Fig. 2.19). Use in virtual reality was also proposed by Merlet as one of the applications of the Marionet robot family.

2.4.5 Entertainment

The movement of a camera suspended by cables was already proposed in the 1920s by l'Argent, who puts the camera along with its operator on a traveling platform suspended by cables. Clearly, the cables were operated manually at this time. However, the basic idea along with the benefits have been understood for almost one hundred years. Much later, but still at an early state of the cable robots' history, the patent for the SkyCam [100] was filed by Brown [67]. Brown, also inventor of the famous steady cam for stabilizing the motion of a camera, proposed a computer controlled system for the original SkyCam. The patent describes a suspended camera system with three cables and includes a sophisticated mechanism on the platform to orient the camera to the desired direction. Therefore, the SkyCam is a 3T cable robot system. Later, a family of patents was filed on the CableCam by the company CableCam Limited (see e.g. [419, US5224426]).

Thrill rides based on cable robots were filed as patents from different parties such as Disney [104, US8147344B2]. A draft of a thrill ride with a suspended cable robot is depicted in Fig. 2.20 and a number of safety mechanisms for such systems were

Fig. 2.19 IPAnema 3 Mini
for haptic interaction during
the Automatica 2014 trade
fair, Munich, Germany

Fig. 2.20 Concept for a
thrill ride with winches
moving along a track

described in a patent [479, EP2572766A1]. The passengers are sitting on the mobile
platform of the cable robot and the winches are mounted on bogeys moving along
roller coaster rails.

Recently, two large-scale cable robots were installed at the German Pavilion at the
EXPO 2015 in Milan (Fig. 2.21), which were flying above the heads of the visitors
[463]. The special achievement of these two cable robots is an advanced safety system
fulfilling recent norms which allows to manipulate significant payloads over persons.

Fig. 2.21 Two large-scale cable robots are core elements of the show at the German pavilion on the Expo 2015 fair in Milan, 2015. © German Pavilion Expo Milan 2015 / B. Handke

2.4.6 Measurement Devices

Already in the late 1990s, a position tracking system was proposed by Jeong [226, 227]. In order to improve the accuracy of this device, Jeong proposed to compensate for the length error induced by sagging of the cables. Later, Ottaviano and Thomas [465] studied the kinematics of a similar tracking system. A similar approach for use in calibration of robots was followed by Pott [404] by proposing the IPAnema measurement device built from industrial cable length sensors (see Fig. 2.22).

2.4.7 Other Applications

Due to their simple and light-weight structure, cable robots were proposed for use in space robotics [82, 466]. Some authors studied the usage of cable robots for radio astronomy. Lambert [278] proposed to use a somewhat inverted cable robot for positioning an aerostat. Instead of relying on gravity, a helium-filled balloon applies an ascending force on the platform and the ground located winches are used to control the position of the aerostat. A project of building the world's largest radio telescope FAST (Five hundred meter Aperture Spherical Telescope) was proposed in China in 1994 and became visible in the literature on cable robots around 2008 [27, 127, 130, 508]. The FAST has its dish suspended over a natural hollow. The collector is suspended from six towers and can be moved by changing the length of the suspending cables in order to target the telescope in the desired direction. The

Fig. 2.22 Cable
sensor-based 6-D-pose
measurement system applied
to measure the pose of an
industrial robot: (1) base
with winch arrangement (2)
cable length sensors (3)
mobile platform (4)
embedded IPC with
real-time pose estimation

FAST is anticipated to be also the world's largest cable robot in terms of size and payload. The concept studies around the FAST telescope discuss a number of issues related to large-scale systems including cable mass, additional loads on the cables, and elasticity, as well as thermal effects and outdoor disturbances like wind.

In the 1990s, Tadokoro [452] studied using cable robots for rescue after natural disasters such as earthquakes. Such considerations were also taken up by Merlet [338] as shown in Fig. 2.23.

2.4.8 Summary

Cable robots have seen a wide area of possible usage. Three main performance advantages allow radically new applications: Huge workspace, very high payloads, and outstanding dynamic capacities. Depending on the field of application, individ-

Fig. 2.23 The cable robots of the Marionet family for rehabilitation purpose and for rescue from Merlet (Courtesy of INRIA Sophia-Antipolis, France)

ually balanced combinations of these three features can be realized with a relatively simple and thus cost-effective mechanical design. Therefore, from customer's point of view, costs and simplicity might be seen as the key advantages for using cable robots as a replacement for conventional solutions.

However, only a few products have made it into the market. Beyond special purpose machinery such as the aforementioned telescopes, only the stadium camera systems and a rehabilitation device for gait training showed some commercial success. The demonstrator systems presented since around 2010 become more and more mature in terms of mechanical designed and industrialization. Therefore, it is foreseen that other fields of application will follow soon.

Chapter 3
Geometric and Static Foundations

Abstract This chapter deals with the geometric and static foundations for the standard model of the cable robot. Firstly, the geometric equations are derived and the static equilibrium is considered. Afterwards, methods for computing force distributions are extensively studied. Finally, the stiffness properties are introduced and analyzed.

3.1 Introduction

Kinematics is closely related to the geometry of a robotic system and studies how the motion of different parts of a robot is coupled and how forces in different parts of the robot are related. However, in the kinematic analysis, it is disregarded how the motion is generated as well as how forces are the cause of motion. These connections are subject to the dynamics addressed in Chap. 6.

For cable robots, the connection between kinematics and statics has to be considered closely compared to conventional robots like industrial robots or parallel kinematic machines. Only if the cables are under tension, one can model them as bilateral constraints because small changes in the cable force maintain the state of inner tension in the robot. Thus, a feasible tension in the cables is a prerequisite also for the kinematic modeling. If the cable tension is too low, the kinematic relations change significantly and we may not expect that the rigid body model of the robot provides a meaningful description at all. It must be underlined that this is an essential condition: Aside from a well-defined tension state, the behavior of the robot is totally different as soon as cables become slack and all methods discussed in this chapter become invalid.

The geometric foundation of cable robots is straightforward. Although the solution of the equations may be involved, the mathematical models can be easily set up in terms of vector loops and algebraic constraints. Firstly, we discuss the *standard kinematic model* which is used as basis of many recent works. The standard model follows from the conventional assumption of a multi-body system with rigid bodies.

© Springer International Publishing AG, part of Springer Nature 2018
A. Pott, *Cable-Driven Parallel Robots*, Springer Tracts in Advanced
Robotics 120, https://doi.org/10.1007/978-3-319-76138-1_3

In the standard model, the cables are idealized as exact linear distance between two points in space, i.e. the proximal anchor point on the fixed frame and the distal anchor point on the mobile platform. Thus, it is assumed that the cable length matches exactly this distance. In the terms of multi-body systems, we model the cables to be prismatic joints. Both ends of the cables are modeled by spherical joints. In practice, these assumptions may be inaccurate due to the following effects that are neglected by the standard model:

- *Unilateral constraint*: The cables can only pull, not push the platform. The same applies on the position level since cables only limit the maximal distance between the frame and the platform. If we take into account the unilateral character of the cables, we have to deal with inequalities instead of equalities in kinematics and statics. Still, the assumptions of unilateral constraints are based on infinite stiffness of the cable.
- *Elastic cables*: The length of the cable depends on its tension. There might be also elastic effects in the winch, that disturb the measurement of the exact length of the cable.
- *Temperature effects*: The length of the cable varies with the temperature of the cable, where changes in the temperature are caused both by the environment as well as by friction in the winch and in the pulleys. Also, the bending of the cable on pulleys and drums creates nonnegligible heat in some cases.
- *Undefined cable guidance*: Imperfect coiling of the cable onto the groove of the drum, multi-layer winding on the drum, or misplacement of the cable on pulleys in the winch cause perturbation in the effective length of the cable.
- *Nontrivial winch and platform kinematics*: The kinematic model of the winch or platform may imply more complex kinematic relations than the ideal assumption of a spherical joint that implies the simple distance between two points in space. In this setting, the effective length of the cable may depend on the actual direction of the cable with respect to the winch, on the orientation of the platform, and on the current length of the cable. This occurs for example when pulleys are used to guide the proximal end of the cable.
- *Hefty cables*: Due to the mass of the cables, the cables are subject to *sagging* and get a curved form which can significantly differ from the straight line of the standard model. This must be taken into account when modeling large-scale cable robots with very long cables, when the cables are relatively heavy, when cables are kept under low tension, or when high accuracy of the robot shall be achieved.
- *Creeping and hysteresis effects* in the material of the cables: Depending on the material and type of the cables, changes in their lengths can occur that are not caused by the current tension in the cable. This effect can also be induced by the manufacturing method of the cables, such as binding. Many cables have been pre-stretched after manufacturing to reduce this effect. However, creeping can show up because of aging, wear, or humidity. Especially synthetic fiber ropes suffer from creeping effects.
- *Vibration of the cable*: Both longitudinal and transversal vibrations cause a deviation of the cable length from the effective geometric distance between the anchor points. Especially for large-scale robots, vibration has a significant effect.

- *Coiling the cable* with varying tension onto a drum results in a hysteresis effect where the length of the cable depends on the history of the length-tension distribution when coiling the cable. The result of this effect is similar to hysteresis effects although the physical origin is different. For relatively soft materials used for the cables, one faces flattening and ovalization of the cable's cross section on drums and pulleys changing the effective radii of coiling and redirection.
- *Elastic reactions* of the mobile platform or the fixed frame: With high performance winches, one can easily apply very high forces to light-weight platforms to achieve extreme accelerations. This leads to elastic deformations of the platform and sometimes also of the machine frame. Static displacements as well as vibrations can be observed in practice in the presence of high forces and large jerks. Contrary, large-scale automated cranes may consist of steel frames that will deform when lifting loads in the range of several tons.

In this chapter, we study an idealized model of the cable robot and we refer to it as the *standard model*. The standard model neglects all these effects listed above. For the standard model, most of the kinematic, static, and dynamic problems have been formulated and solved. Thus, it serves well to introduce the basic concepts and as a starting point for taking into account the additional effects as listed above. The standard model is relatively simple to study since the assumptions mostly separate the kinematic, static, and dynamic issues. Thus, the equations to be solved are simpler sometimes allowing to reveal their structure and properties. For the more involved effects as given in the list, some of the problems like forward kinematics and workspace are still open issues. Many theoretical problems such as the number of possible solutions are unknown and only a little number of algorithms is known to efficiently compute solutions to numerical problems. Some advanced models to cope with these effects are subject of Chap. 7.

3.1.1 Literature Overview

Firstly, the state of the art in statics and kinematics for cable-driven parallel robots is reviewed. In the late 1990s, Roberts [417, 418] presented necessary and sufficient conditions for suspended cable robots to be in a static equilibrium. This is one of the first works that presents clear conditions for static equilibrium based on nullspace vectors of the structure matrix.

Verhoeven studied the problem of force distributions [475] and provides both the theoretical basis as well as algorithms for the computation of force distributions [473] of CRPM and RRPM typed robots. His main result was a mathematical proof for a theorem stating that optimizing p-norms of the force vector yields continuous force distributions along a trajectory. Then, he focuses on numerical schemes for higher p-values since the usage of higher p-norms allows for a larger workspace.

The possibly most comprehensive study on available wrench sets was presented by Bouchard, giving a detailed discussion on the structure of the wrench set [62]. This contribution reveals the mathematical models of the available wrench set, i.e. shows that the available wrench set is a convex set described by a zonotope or Minkowski sum. Both forms can be found by linear projection of a hypercube into a lower-dimensional space. Furthermore, applicable algorithms for checking are presented if a desired wrench set given by a hyper-ellipsoid is included in the available wrench set.

Merlet [326] studied force distributions in redundantly actuated cable robots and pointing out that it is not possible to control forces if the cables are perfectly rigid. Furthermore, alternative approaches are presented how redundantly actuated cables can be operated. The formal arguments in this paper are stringent if the cables are perfectly stiff. However, the assumption that the stiffness of the cables is sufficiently high to allow for force distribution without considering the geometric change may justify the analysis presented in a number of other papers.

A good recent overview on the problem of force distributions using p-norms is presented by Gosselin [168]. Later, a formula for the case of the $p = 4$ norm was formulated as an optimization problem [174]. A remarkable series of contributions related to force distribution was presented by Hassan [199–201], who used the Dykstra method for computation of force distributions. Although the iterative scheme can hardly be used in real-time, it serves well as reference for other methods.

Mikelsons [348] proposes a new approach to solve the cable force distribution problem of RRPM. Instead of considering the problem as an optimization problem, the full feasible nullspace is computed using triangulation. Then, a geometric procedure was used to determine a unique force distribution by considering the barycenter of the triangulated solution space. This approach was taken up by Lamaury [275, 277] with more efficient algorithm. The presented results are restricted to the case of eight cables and six degrees-of-freedom.

An early work on using linear programming to compute force distributions can be found in [438]. Two numerical schemes for force computation are introduced by Borgstrom [43] using linear programming and also quadratic programming. Discontinuities are reported for linear programming and an example is discussed. Experimental results are shown both for the four cable planar NIMS-PL robot (2T, $m = 4$) as well as for the WiRo-6.3 robot (3R3T, $m = 9$).

A closed-form formula to compute the minimum Euclidian norm force distribution was proposed by Pott [396] for RRPM with arbitrarily many cables allowing for easy use in a real-time controller system and also straightforward implementation. An improved version of that algorithm is presented in [392], where some shortcomings in the feasible region are overcome.

Yi [504] presents some force considerations for planar robots with four cables. Li [286] proposed a quadratic programming approach for cable force distribution for under-constrained cable robots with six cables. In [386], a recursive dimension reduction algorithm is introduced to check for wrench-closure and wrench-feasibility. In [380], a linear complementarity formulation is proposed to deal with the unilateral constraints of a cable robot.

An algorithm for the numerical evaluation of forward kinematics is presented by Fang [139] that is based on an integral formulation. Merlet [324] deals with the kinematics of cable robots driven by linear actuators and elastic cables. Furthermore, an approach using interval analysis to solve the forward kinematics was adopted from conventional parallel robots [321] to cable robots [324]. Aref [13] proposes to use a numerical scheme to solve the forward kinematics of the eight cable KNTU robot with Gauss-Newton and Levenberg-Marquardt algorithms. However, no details are given on the application of the numerical scheme. Pott [390] presents a real-time capable kinematic code for over-constrained cable robots and its implementation into the controller.

For special non-generic geometries, one can exploit the specific structure. Guilin [191] presents an interesting idea for the kinematic analysis of a planar three degrees-of-freedom robot with four cables. By exploiting the over-constrained nature of the mechanism, a fourth order polynomial is derived that can be basically solved in closed-form. Jaeung [225] presents a six cable-based pose measurement device with its respective forward kinematics for 6-3 configuration with planar machine frame where both sagging of the cables and elastic elongation are addressed. Ferraresi [146] presents a closed-form solution of a suspended cable robot in 6-3 configuration, where the geometry of the robot corresponds to the simplified symmetric manipulator (SSM) design that is well-known from conventional parallel robots.

It seems that relatively little work was done on experimental bases for revealing the real conditions in cable robots. Kraus [262] performs an experimental study of the influence of elastic cables on the force distribution. More experimental evaluations can be found in [259] where especially the connections between measurable performance of the robot and the control algorithms are investigated.

A more exotic approach was presented by Ghasemi [164] who employed neural networks for forward kinematics. Later, neural networks [433] were used to estimate the pose before iterating with conventional methods.

3.1.2 Effects beyond the Standard Model

The interaction between force distribution and disturbance in the structure was mostly neglected yet. Pott [391] analyzed the influence of pulleys on the force distribution as well as on the workspace. Neglecting the effect of pulleys can significantly change the tension for a pose although the size and shape of the workspace is hardly influenced. Based on this pulley model, a real-time capable forward kinematics code was presented [434].

Jeong considered the effect of sagging cables for his measurement device [226]. Su [446] presents a kinematic analysis of the FAST telescope with six cables taking into account sagging of the cables where the shape of the sagging cables is assumed to have a hyperbolic form. Kozak [256] addresses sagging of cables induced by cable mass and analyzes the influence on the robot's stiffness. The model used by Kozak is mainly based on Irvine's reference book *Cable structures* [218] and assumes that the cables form a catenary. However, it seems that the catenary model known from civil engineering was firstly applied to cable robots. Yao used a parabola as estimation for the cable form to model huge robots [507]. An extension to the catenary model including nonuniform load on the cables was added later [506]. Here, also experimental results from tests on a large robot were added. A stiffness model for robots with sagging cables was proposed by Arsenault and applied to a suspended 2T robot [18]. A kinetostatic model for inductile cables with mass using a catenary line can be found in [179]. A more elaborated model on sagging of hefty cables also takes into account elastic effects in the cables [360]. Lately, Gouttefarde combined the modeling approach for elastic and hefty cables with the kinematic effects of pulleys [189].

Capua [83, 84] presents the kinematic analysis and motion planning of a suspended climbing cable robot where the actuation system is located on the mobile platform and the proximal anchor points can be changed during operation in order to perform climbing motions. Suspended robots present a number of challenging problems and a series of contributions on this field was published. Merlet [337] presents the kinetostatic modeling of spatial under-constrained cable robots with four cables. For such robots, the inverse kinematics is more involved to solve than the forward kinematics since the inverse kinematics in general does not have a stable configuration. Carricato and Merlet [88] study both forward and inverse kinematics of suspended cable robots with less than six cables using screws and a so-called geometric-static model. Later, these authors analyzed the forward kinematics of under-constrained robots with three cables showing that the problem has 156 complex solutions [89]. Analysis of the equilibrium positions of an under-constrained cable robot with three cables was studied by Abbasnejad [1]. This paper aims at finding the upper bound for real solutions of this robot type. From theory, it is known that the upper bound for complex solutions is 156 and it is conjectured that the upper bound for real solutions is 54. According to this paper, a formal proof was not yet found. A combined kinematic and static analysis with two to four cables in suspended under-constrained configuration is discussed in [91]. The work was extended in [2] to under-constrained cable robots with five cables and a univariate polynomial of degree 140 is derived. Merlet [328] analyzed the kinematics of suspended cable robots with three and four cables in 4-2 and 3-2 configurations, both where two or three cables share a common anchor point on the platform.

Barrette [28] analyzed singular configurations for a planar cable robot. Bouchard [60] presents a kinematic sensitivity analysis of the inverted cable robot LAR where the platform is supported by an aerostat and the cables are suspended to the ground to manipulate the platform. A kinematic analysis of the LAR robot is presented by Taghirad and Nahon [454].

3.2 Standard Geometric Model

The fundamental geometric relations between the fixed base, the mobile platform, and the cables are the foundation of the kinematic and static modeling. These non-linear equations formulate velocity and acceleration transmission of the robot which are employed to characterize the behavior of the robot. The transmission of forces and torques is the basis to study the static equilibrium of the robot. By means of kinetostatic duality, force transmission is closely coupled to the velocity transmission.

Fig. 3.1 Geometry and
kinematics of a general cable
robot

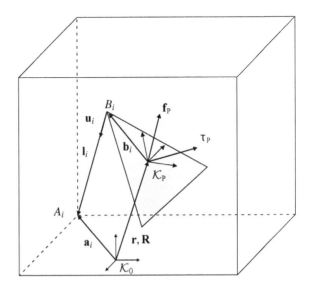

The geometric foundations summarized below have been developed and discussed
by many authors, see e.g. [72, 238, 418, 473]. Equations to solve the inverse kinematics were already given in the year 1984 by Landsberger in his pioneering work
[280]. In the following, we review the geometric modeling and introduce the relevant
symbols. Figure 3.1 shows the structure of a spatial cable robot of general geometry, where the position vectors $\mathbf{a}_i$ denote the proximal attachment points A_i on the
frame in world coordinates $\mathcal{K}_0$, the vectors $\mathbf{b}_i$ are the relative positions of the distal
attachment points B_i on the mobile platform given in local coordinates of the frame
$\mathcal{K}_P$, and $\mathbf{l}_i$ denotes the vector of the cables in $\mathcal{K}_0$. In the standard model, it is assumed
that the cables are straight lines and under tension. Furthermore, it is assumed that
both $\mathbf{a}_i$ and $\mathbf{b}_i$ do not depend on the pose of the platform, i.e. the effect of possible
guiding pulleys or fixture elements on the platform is neglected. Applying a vector
loop as shown in Fig. 3.1, the closure constraint ν_i reads

$$\nu_i : \mathbf{a}_i - \mathbf{r} - \mathbf{R}\mathbf{b}_i - \mathbf{l}_i = 0 \quad \text{for} \quad i = 1, \ldots, m \quad , \tag{3.1}$$

where the vector $\mathbf{r} \in \mathbb{R}^3$ is the Cartesian position of the mobile platform and the
rotation matrix $\mathbf{R} \in SO_3$ represents the orientation of the mobile platform with
respect to the world coordinate frame. Thus, the transformation between $\mathcal{K}_0$ and $\mathcal{K}_P$
is described by the pair $(\mathbf{r}, \mathbf{R}) \in SE_3$ and we denote that pair the *pose* of the platform.
The closure constraint can be easily rewritten to

$$\mathbf{l}_i = \mathbf{a}_i - \mathbf{r} - \mathbf{R}\mathbf{b}_i \quad \text{for} \quad i = 1, \ldots, m \tag{3.2}$$

and the unit vector $\mathbf{u}_i$ along the cable becomes

$$\mathbf{u}_i = \frac{\mathbf{l}_i}{||\mathbf{l}_i||_2} \quad . \tag{3.3}$$

The unit vector $\mathbf{u}_i$ is pointing from the platform towards the base by convention. This convention is introduced by intuition and is used in most of the literature on cable robots since pulling forces in the cables lead to positive values of the cable force. Anyway, the convention for the unit vector $\mathbf{u}_i$ implies that positive forces point into the direction that shortens the cables and thereby reduces the value of the associated generalized coordinate l_i. This leads to an uncommon situation compared to other fields of robotics and care must be taken when applying formulas from other domains to cable robots.

3.3   Statics

In contrast to conventional serial or parallel robots, it is not clear if a cable robot is statically stable for a given pose. In order to study the stability of different poses where the platform can be statically balanced by the cables, we have to consider the mechanical equilibrium of the mobile platform. Firstly, we study the most general case of a spatial robot with 3R3T motion pattern. For force and torque equilibrium (Fig. 3.1) of the mobile platform, one has to consider all forces and torques acting on the platform. Thus, it holds true [349, 418, 473]

$$\left.\begin{array}{c} \displaystyle\sum_{i=1}^{m} \mathbf{f}_i + \mathbf{f}_{\mathrm{P}} = \mathbf{0} \\[2mm] \displaystyle\sum_{i=1}^{m} \mathbf{b}_i \times \mathbf{f}_i + \boldsymbol{\tau}_{\mathrm{P}} = \mathbf{0} \end{array}\right\} \tag{3.4}$$

where $\mathbf{f}_{\mathrm{P}}, \boldsymbol{\tau}_{\mathrm{P}}$ are the applied forces and torques, respectively, and where the cable forces are $\mathbf{f}_i = f_i \mathbf{u}_i$. Recall that we consider the cable's normal vector $\mathbf{u}_i$ pointing from the platform towards the robot frame. From this intuitive definition for $\mathbf{u}_i$, it follows that positive tension in the cables leads to positive values for f_i. However, the definition implies that the positive values for the cable force cause a motion that reduces the cable length l_i. We address this unusual definition later when we consider the connection between differential kinematics and statics. Rewriting Eq. (3.4) to matrix form yield the linear system

$$\underbrace{\begin{bmatrix} \mathbf{u}_1 & \cdots & \mathbf{u}_m \\ \mathbf{b}_1 \times \mathbf{u}_1 & \cdots & \mathbf{b}_m \times \mathbf{u}_m \end{bmatrix}}_{\mathbf{A}^{\mathsf{T}}(\mathbf{r},\mathbf{R})} \underbrace{\begin{bmatrix} f_1 \\ \vdots \\ f_m \end{bmatrix}}_{\mathbf{f}} + \underbrace{\begin{bmatrix} \mathbf{f}_{\mathrm{P}} \\ \boldsymbol{\tau}_{\mathrm{P}} \end{bmatrix}}_{\mathbf{w}_{\mathrm{P}}} = \mathbf{0} \quad , \tag{3.5}$$

where $\mathbf{f} = [f_1, \ldots, f_m]^{\mathsf{T}}$ is the vector of the cable forces and the wrench $\mathbf{w}_{\mathrm{P}}$ composed from the applied force $\mathbf{f}_{\mathrm{P}}$ and the applied torque $\boldsymbol{\tau}_{\mathrm{P}}$. This can be abbreviated in a compact matrix-vector form as

$$\mathbf{A}^{\mathrm{T}}(\mathbf{r}, \mathbf{R})\mathbf{f} + \mathbf{w}_{\mathrm{P}} = \mathbf{0} \quad . \tag{3.6}$$

The matrix $\mathbf{A}^{\mathrm{T}}$ is the transpose of the Jacobian matrix and referred to as *structure matrix* in this work. In other publications, the matrix $\mathbf{A}^{\mathrm{T}}$ is also called *wrench matrix* or simply *Jacobian*. It transforms cable forces from the actuator's joint space into the end-effector wrench in the operational space. As we will see in the following sections and chapters, the structure matrix $\mathbf{A}^{\mathrm{T}}$ is of major importance for kinematic analysis (Chap. 4), workspace analysis (Chap. 5), and control as well as for dynamics (Chap. 6).

At first glance, dealing with Eq. (3.5) seems to be rather simple: Depending on the degree-of-redundancy e.g. $r = m - n$, one has an under-constrained ($r > 0$), a fully-constrained ($r = 0$), or an over-constrained ($r < 0$) system of linear equations and there is a rich library of mathematical tools and algorithms in linear algebra to analyze the properties. If $\mathbf{A}^{\mathrm{T}}$ has full rank, in the over-constrained case, one expects no solutions in general but only for special poses. In the fully-constrained case, the system is quadratic and there is exactly one solution. In the under-constrained case, there exist infinitely many solutions. Taking a closer look reveals a key challenge, since only a subset of the well-known techniques can be applied because one must also take into account the unilateral nature of the force transmission in the cables: Obviously, we have to assume in general that the cable forces are positive $\mathbf{f} > \mathbf{0}$. Therefore, the structure equation is in general a kind of constrained linear system, where the system is over-constrained for cable robots of IRPM type, fully-constrained for cable robots with $m = n$, and under-constrained for both CRPM and RRPM cable robots.

For some calculations, it is useful to separate the normalizing length $\|\mathbf{l}_i\|_2$ used in Eq. (3.3) of the cables from the geometry dependent part of structure matrix (see [473, p. 39]). Thus, we can decompose the structure equation as follows

$$\widehat{\mathbf{A}}^{\mathrm{T}} \mathbf{L}^{-1} \, \mathbf{f} + \mathbf{w}_{\mathrm{P}} = \mathbf{0} \quad , \tag{3.7}$$

where $\mathbf{L}^{-1} = \mathrm{diag}(|\mathbf{l}_1|^{-1}, \dots, |\mathbf{l}_m|^{-1})$ is a diagonal positive-definite matrix containing the reciprocal cable lengths and the non-normalized structure matrix reads

$$\widehat{\mathbf{A}}^{\mathrm{T}} = \begin{bmatrix} \mathbf{l}_1 & \cdots & \mathbf{l}_m \\ \mathbf{b}_1 \times \mathbf{l}_1 & \cdots & \mathbf{b}_m \times \mathbf{l}_m \end{bmatrix} \quad . \tag{3.8}$$

Substituting the vector loop closure condition Eq. (3.1) into $\widehat{\mathbf{A}}^{\mathrm{T}}$ and simplifying the expressions yields

$$\widehat{\mathbf{A}}^{\mathrm{T}} = \begin{bmatrix} \mathbf{a}_1 - \mathbf{r} - \mathbf{R}\mathbf{b}_1 & \cdots & \mathbf{a}_m - \mathbf{r} - \mathbf{R}\mathbf{b}_m \\ \mathbf{b}_1 \times (\mathbf{a}_1 - \mathbf{r}) & \cdots & \mathbf{b}_m \times (\mathbf{a}_m - \mathbf{r}) \end{bmatrix}. \tag{3.9}$$

If we consider the product $\mathbf{L}^{-1} \mathbf{f}$ as a correction factor for the cable forces, some structural information can be derived from the $\widehat{\mathbf{A}}^{\mathrm{T}}$. Since all diagonal elements of $\mathbf{L}^{-1}$

are positive, the product $\mathbf{L}^{-1} \mathbf{f}$ is a scaling without changes in the sign of the cable force. The non-normalized structure matrix is composed of simpler mathematical expressions compared to $\mathbf{A}^T$. Especially, no divisions and square roots of the generalized coordinates $(\mathbf{r}, \mathbf{R})$ of the mobile platform show up the equation. If we choose e.g. a quaternion to parameterize the rotation matrix $\mathbf{R}$, we find a representation of the structure matrix which is even free of transcendental functions such as $\sin(\cdot)$, $\cos(\cdot)$, and $\tan(\cdot)$. Thus, it can be analyzed as a purely algebraic equation in the pose parameters.

In the following sections, we investigate the structure equations for robots with the remaining motion patterns that are restricted in some ways. The following considerations are based on the assumption that the mobile platform can only move according to the motion pattern. Often, it makes sense to consider a robot as if it could only move according to some constraints e.g. perform only planar motion. In the following consideration of the motion pattern, it is ignored why the robot's motion is restricted. If the assumption does not perfectly hold true in practice, additional effects must be expected. For example, planar cable robots tend to vibrate normal to their plane if their motion is not constrained otherwise. For some robots (2T, 3T, and 2R3T), it is assumed that two or more cables exactly share the same anchor point on the platform. Since this cannot be fulfilled exactly in practice, parasitic motion can occur that is neither described nor postulated with the specific models. Anyway, these models can be very useful as approximation or as model system.

All motion patterns other than 3R3T can only be achieved by non-generic geometries, i.e. special constraints must be fulfilled for the geometric parameters. In a certain sense, all these robots are architecturally singular designs that lead to a very specific degeneration of the robot's motion pattern.

3.3.1 Purely Translational Robots (2T and 3T Case)

Firstly, the planar and spatial case for a point mass is addressed where no rotation motion occurs (Fig. 3.2). To set up a planar or spatial robot without rotational motion, the cables have to meet in a common point on the platform. Without loss of generality, we can assume that all vectors $\mathbf{b}_i$ are equal and we can translate the platform's reference point such that all $\mathbf{b}_i = \mathbf{0}$. Thus, we consider a pure force equilibrium and the structure matrix becomes

$$\mathbf{A}^T = \begin{bmatrix} \mathbf{u}_1 \ \dots \ \mathbf{u}_m \end{bmatrix} \quad , \tag{3.10}$$

where the vectors $\mathbf{u}_i = [u_{i,x}, u_{i,y}]^T \in \mathbb{R}^2$ for the 2T case and $\mathbf{u}_i = [u_{i,x}, u_{i,y}, u_{i,z}]^T \in \mathbb{R}^3$ for the 3T case. The applied wrench becomes $\mathbf{w}_P = [f_x, f_y]^T \in \mathbb{R}^2$ and $\mathbf{w}_P = [f_x, f_y, f_z]^T \in \mathbb{R}^3$, respectively. The factorized matrix takes then the form

$$\widehat{\mathbf{A}}^T = \begin{bmatrix} \mathbf{a}_1 - \mathbf{r} \ \dots \ \mathbf{a}_m - \mathbf{r} \end{bmatrix} \quad . \tag{3.11}$$

Fig. 3.2 Statics of a cable
robot of type 2T with four
cables and 3T with purely
translational motion pattern

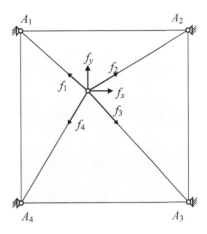

Clearly, 2T and 3T robot can neither withstand nor generate torques. Still, this class
of robots is not a pure mathematical model but can be used in practice. These robots
can be used as pure actuation system on a mechanical structure that is capable to
withstand applied torques. Furthermore, one can connect a platform to the central
point of the cable robot and accept possible oscillations of the platform. Consider,
for example, a crane which hook is guided in all translational directions where the
load on the crane can sway under the hook. Finally, we can combine two or more
translational robots by connecting different points on a rigid platform each to such
a robot structure. Therefore, this class of robots can be an interesting module in a
multi-robot scenario or as a building block in structural synthesis.

3.3.2 Planar Robots (1R2T Case)

The 1R2T motion pattern equals the planar case where two translations and one
rotation of the platform are considered (Fig. 3.3). Without loss of generality, we
consider the xy-plane with rotations about the z-axis as reference. Thus, analyzing
the respective equilibrium conditions leads to

$$\mathbf{A}^{\mathrm{T}} = \begin{bmatrix} \mathbf{u}_1 & \cdots & \mathbf{u}_m \\ h_1 & \cdots & h_m \end{bmatrix} \quad , \tag{3.12}$$

where vectors $\mathbf{u}_i = [u_{i,x}, u_{i,y}]^{\mathrm{T}} \in \mathbb{R}^2$ and $h_i = b_{i,x} u_{i,y} - b_{i,y} u_{i,x}$. The wrench takes
the form $\mathbf{w}_\mathrm{P} = [f_x, f_y, M_z^\mathrm{P}]^{\mathrm{T}} \in \mathbb{R}^3$.

Fig. 3.3 Statics of 1R2T
planar cable robots

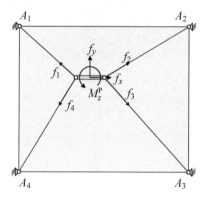

Fig. 3.4 Statics of 2R3T
robots with $m = 8$ cables
and with a needle-shaped
mobile platform in an 8-2
configuration

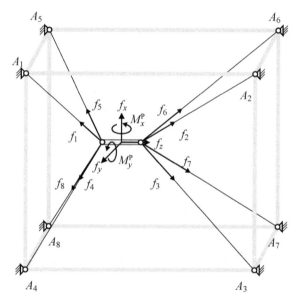

3.3.3 Spatial Robots (2R3T Case)

The 2R3T is a non-generic design, since the geometry of the platform has to follow
strict geometric relations. All distal anchor points $\mathbf{b}_i$ have to lie on a common line
(Fig. 3.4). Thus, one can define the platform coordinate systems such that all points
are located on its z-axis and the vectors take the form $\mathbf{b}_i = [0, 0, b_{i,z}]^{\mathrm{T}}$. Then, the
structure matrix takes the form

$$\mathbf{A}^{\mathrm{T}} = \begin{bmatrix} \mathbf{u}_1 \ldots \mathbf{u}_m \\ \mathbf{h}_1 \ldots \mathbf{h}_m \end{bmatrix} \quad , \tag{3.13}$$

where $\mathbf{u}_i = [u_{i,x}, u_{i,y}, u_{i,z}]^T \in \mathbb{R}^3$ and $\mathbf{h}_i = [-b_{i,z}u_{i,y}, b_{i,z}u_{i,x}]^T \in \mathbb{R}^2$. In contrast to all other cases, we use the moving platform frame $\mathcal{K}_P$ as reference frame for the applied wrench to simply provide a non-redundant definition of the applied wrench. In this moving frame, we maintain the decoupling between the two independent torques. Thus, we have $\mathbf{w}_P = [f_x, f_y, f_z, M_x^P, M_y^P]^T \in \mathbb{R}^5$.

3.4 Force Distributions

Determining the forces in the cables of over-constrained robots of the classes CRPM and RRPM is a problem that is inherent to cable robots when one wishes to fully control the motion. Amongst others, the problem of determining the cable forces occurs in different tasks such as control, workspace determination, and design. In this section, the following issues are addressed: Firstly, the general questions arising for force distribution are discussed. Afterwards, the concepts of wrench-closure and wrench-feasibility are introduced to characterize the static equilibrium of the platform. Then, we derive in detail the convex set of force distributions that characterize all robots with more cables than degrees-of-freedom. Finally, the concept of available wrench sets is discussed which is again a convex set with all wrenches a robot can generate at a given pose.

Most of the robots are statically either under-constrained or over-constrained making the determination of cable forces difficult. Only if the number of cables m matches the degree-of-freedom n, the known solutions for rigid parallel robots can be used. The problem of having positive tension in the cables is then only a matter of determining the workspace.

For cable robots of the CRPM and RRPM type, the problem is related to over-actuated parallel robots [112, 246] where rigid parallel robots also allow for negative forces in their links. Therefore, additional restrictions must be introduced to apply these techniques to cable robots. Ebert-Uphoff and Voglewede [134] pointed out that the determination of cable forces for over-constrained cable robots is equivalent to determining the forces in multi-fingered grasp planning. Both problems share the over-determined mechanical structure (over-actuation) and the unilateral constraints caused by cable and contact conditions, respectively. However, a cable robot has no equivalent to the friction forces acting between robotic fingers and the grasped object.

3.4.1 General Approach

The pioneers in cable robotics already identified that determining force distributions is a key problem for cable robots. The classification from Ming and Higuchi [349] revealed the problem of force distributions and gives criteria on the number of cables and the degree-of-freedom to distinguish between different cases (see Sect. 2.2.1).

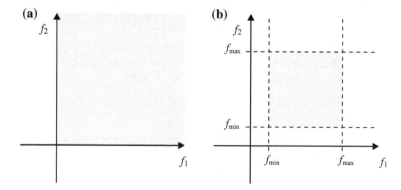

Fig. 3.5 (a) Infinite region of allowed forces for force-closure and (b) the finite force-feasible configurations for wrench-feasibility

An algorithm to solve this problem was already presented by Kawamura [238] to determine feasible cable forces for his robot. Since then, a couple of methods and algorithms have been presented. For over-constrained robots, there are a number of steps to be taken:

- Identify the set $\mathcal{F}$, i.e. the set of all feasible force distributions $\mathbf{f}$ for a pose $(\mathbf{r}, \mathbf{R})$: By analyzing the mathematical structure of the underlying problem, it was found that the solutions for the force distribution problem are defined from intersections of the nullspace defined by the structure matrix $\mathbf{A}^{\mathrm{T}}$ and half-spaces defined by the minimum and possibly also maximum cable force $f_{\min}$ and $f_{\max}$, respectively (see also Sect. 3.4.5). All sets $\mathcal{F}$ defined by these conditions are convex and therefore specialized methods can be applied to analyze their structure [201, 475]. Both sets are visualized in Fig. 3.5.
- Decide if the set $\mathcal{F}$ is nonempty: Proofing nonemptiness of the set $\mathcal{F}$ is equivalent to determining if the pose belongs to the wrench-feasible workspace. This problem is discussed in more detail in Chap. 5.
- Determining any single solution $\mathbf{f} \in \mathcal{F}$: In general, it may be easier to compute randomly any solution in the solution space than computing a solution with specific properties.
- Choose a solution with given properties: Many authors studied methods to choose a solution that is optimal or unique in a certain sense, where authors mostly tried to minimize or maximize different norms on the chosen force distribution [43, 76, 168, 199, 396, 475]. Depending on the used norm, the optimization problem can be solved in closed-form or with an iterative algorithm. Some authors propose to choose the minimal cable tension in order to save energy in the system. Using the maximum cable tension can increase the stiffness of the system, and maintaining an average level of tension allows for robustness with respect to disturbance and errors in the system. Other properties of interest may be minimal variance amongst the components of the vector $\mathbf{f}$.

- Choose and track solutions along a trajectory such that force distributions are continuous: In motion planning and force control, one has to ensure that there is a smooth transition between the force distributions in neighboring poses. It turns out that different methods to choose the force distribution may lead to discontinuous forces along a continuous trajectory [473].
- Find numerically efficient or even real-time capable methods: Most work on efficient algorithms and closed-form solutions are restricted to special cases like degree-of-redundancy of one [475] or two [275, 277, 348]. Also, the degree-of-redundancy of three was analyzed [356]. The closed-form solution presented in [396] can be efficiently applied for any degree-of-redundancy $r > 0$ but may fail to compute a solution although it exists.

In the literature, different approaches were introduced to calculate force distributions, and each one of them delivers force distributions with different characteristics while requiring different computational effort, see [392] for an overview:

- General optimization using a p-norm [473]
- Interval analysis to optimize a p-norm [79]
- Optimization approach for p-norm with $p = 4$ [168]
- Constrained l_1-norm optimization [441]
- Minimizing p-norm with Dykstra method [199]
- Closed-form solution for $p = 2$ [392, 396]
- Linear programming [43, 366, 438]
- Quadratic programming [79, 286][1]
- Barycentric approach [348] and improved implementations [275]
- Kernel translation method [473] for $r = 1$
- Gradient projection onto the kernel [288]
- Weighted sum of solution space vertices [68]
- Available wrench set [62]
- Puncture method [80, 356]

In the following sections, we classify and define the conditions describing the characteristics that are induced by using cables as transmission elements. Different definitions have been introduced to characterize a pose [134] and the most important concepts are reviewed below.

3.4.2 Wrench-Closure Poses

A pose $(\mathbf{r}, \mathbf{R})$ of a cable robot is said to be in *wrench-closure* [134, 183] or *controllable* [474] if for each wrench $\mathbf{w}_\mathrm{P} \in \mathbb{R}^n$ there exists at least one distribution of cable forces $\mathbf{f} \in \mathbb{R}^m$ such that

$$\mathbf{A}^\mathrm{T}(\mathbf{r}, \mathbf{R})\mathbf{f} + \mathbf{w}_\mathrm{P} = \mathbf{0} \quad \text{with} \quad \mathbf{f} > \mathbf{0} \ . \tag{3.14}$$

---

[1]Li [286] only deals with the non redundant case $r = 0$, i.e. six cables and six degrees-of-freedom.

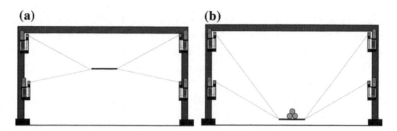

Fig. 3.6 Example of a 1R2T robot: **(a)** Pose with force-closure **(b)** pose without force-closure. Gravity forces are required to balance the platform

In other words, every wrench $\mathbf{w}_P$ can be balanced with positive cable forces (see Fig. 3.6a). In contrast, the structure equation for a pose $(\mathbf{r}, \mathbf{R})$ may be such that one can only find solutions for special wrenches $\mathbf{w}_P \in \mathbb{R}^n$ (Fig. 3.6b). This property is sometimes called *force-closure* when cable robots with purely translational motion (2T and 3T) are considered.[2]

This concept of *wrench-closure* or *controllability* is a rather theoretical concept disregarding any technical upper or lower limits on the cable force beside forces being positive. The notion is a purely geometrical problem without the need for introducing parameters for the desired wrenches on the platform and for the feasible forces in the cables. The definition of wrench-closure yields some principle limits of cable robots even when we accept infinitely large cable forces. The concept of wrench-closure presents a handy formulation to introduce the unilateral nature of force transmission through cables into the mathematical formulation of statics. The problem of checking for the possibility of positive tension in the cables is connected to solving linear systems that are subject to inequalities. Although this formulation seems to be simple and one may expect to find ready to use methods in the literature on linear algebra, it has kept robotic researchers busy for a decade to come up with practical methods to perform the required computations.

3.4.3   Wrench-Feasible Poses

A pose $(\mathbf{r}, \mathbf{R})$ of a cable robot is called *wrench-feasible*[3] [134, 188] or *acceptable* [474] for a given applied wrench $\mathbf{w}_P$ if

$$\mathbf{A}^{\mathrm{T}}(\mathbf{r}, \mathbf{R})\mathbf{f} + \mathbf{w}_P = \mathbf{0} \quad \text{for} \quad 0 < f_{\min} \le f_i \le f_{\max}, \quad i = 1, \dots, m \quad , \tag{3.15}$$

---

[2]There is no need for defining torque-closure since there exists no cable robot with pure rotational degrees-of-freedom, see Sect. 2.2.2.

[3]The term *wrench-feasible* originates from the multi-fingered grasping where a similar unilateral problem of having positive contact forces between the fingers and the manipulated object is studied.

where $f_{\min}$ and $f_{\max}$ are the lower and upper limits for the feasible forces in the cables, respectively. A pose $(\mathbf{r}, \mathbf{R})$ is wrench-feasible for a set of wrenches $\mathcal{Q}$, if Eq. (3.15) holds true for every $\mathbf{w}_P \in \mathcal{Q}$. Clearly, wrench-feasibility depends on the $\mathbf{w}_P$ or $\mathcal{Q}$. Therefore, care must be taken to name the considered wrench $\mathbf{w}_P$ or the considered wrench set $\mathcal{Q}$ when discussing the wrench-feasibility of a cable robot. Every pose of a cable robot is wrench-feasible for some $\mathbf{w}_0$. This can be easily shown by substituting an arbitrary cable force into the structure matrix. Consider, for example, the minimum cable force $\mathbf{f}_{\min}$ and compute $\mathbf{w}_0 = -\mathbf{A}^T \mathbf{f}_{\min}$. Obviously, this wrench exactly balances the cable forces and therefore the robot is wrench-feasible for that special wrench $\mathbf{w}_0$. This even holds true if the robot is in a singular configuration. In any case, if the wrench is not specified, it is often assumed that wrench-feasibility relates to the unloaded robot with $\mathbf{w}_P = \mathbf{0}$ or to the robot balancing its gravitational load. For the latter case, the pose is called *static equilibrium pose* [134]. Wrench-feasibility is a major practical criterion to evaluate the feasibility of a pose for applications.

3.4.4 Stability of Equilibrium

The concepts wrench-closure and wrench-feasibility address the problem whether feasible cable forces exist. Such equilibrium can be unstable for a robot pose, if the platform undergoes instability for infinitesimal changes in the cables forces, the applied forces, and displacements. The effect that causes such instability depends on the classification of the robot. For fully-constrained cable robots of the classes CRPM and RRPM, instability can occur due to deficits in the stiffness matrix [33] and depends on the current cable forces (see Sect. 3.8). In contrast, for under-constrained cable robots, there exists only one cable force distribution and the kinematic and static problems cannot be analyzed separately. As the influence of gravity is essential for under-constrained cable robots, one has to take these forces into account to analyze if the disturbance of the pose is stable or instable. The problem is tackled by Bosscher [46] and more in detail by Carricato and Merlet [88]. The latter derive a form with Lagrangian multiplier as well as a specific Hessian to explore stability of under-constrained cable robots.

The problem of stability of a given pose is targeted in Sect. 6.3.5 where the connection between kinematics, dynamics, and the energy of the cable robot is discussed. In fact, the problem is related to the Hessian of the energy function which can be employed for stability analysis (Sect. 6.3.5). Using appropriate cable models to set up the system energy, one can handle both under-constrained and fully-constrained cable robots within a common framework. However, this issue it not further discussed in this chapter.

3.4.5 Limits on Cable Forces

As pointed out above, the lower and upper limit on the forces of the cables must be taken into account when considering a pose as wrench-feasible. A number of authors [473] carried out that a minimum tension f_{min} must be maintained to hinder the cables from being slack and a maximum tension f_{max} must not be violated to prevent cables and motors from overload. Although there is a rich development for algorithms to compute tension distributions for given limits [168, 199, 396], only little was published on how to derive the actual limits f_{min} and f_{max}. In practice, it turns out that there are much more limiting factors both for the minimum and maximum force which are barely discussed in the literature [393]. Therefore, we dedicate the following section to review and where possible quantify these limits. We investigate what effects must be taken into account when choosing the lower f_{min} and upper f_{max} bound for the cable force.

Analysis, design, and control of cable robots are mostly based on some simplifications. Typical tasks to be solved are forward and inverse kinematics (Chap. 4), workspace computation (Chap. 5), and tension distribution (Sect. 3.6) as well as control and calibration of cable-driven parallel robots. The basic approaches assume the validity of the standard model where the cables are perfect line segments. During the recent years, it becomes clear that such assumptions do not hold under some circumstances such as large-scale systems like IPAnema 3 [259], CoGiRo [276], or FAST [286]. Therefore, it is a matter of design and configuration of the robot to choose the tension f_{min} and f_{max} limits such that the error induced by the simplifications is acceptably small.

3.4.5.1 Upper Bounds on the Tension

In this section, effects are investigated that must be taken into account when choosing the upper bound f_{max} for the cable force. To determine the maximum tension in the cables, the following effects must be considered.

Mechanical Limits and Safety

A limitation is the static breaking load of the cable. Clearly, if the tension exceeds the cable's breaking load, the robot cannot be safely operated. For applications, one has to additionally take into account a safety factor where relatively high factors are typical in the magnitude around 10 in lifting applications such as cranes, elevators, cable cars, and thrill rides.

We have also to take into account that the breaking load of the cable applies to ideally applied forces at the end of the cable. Fixing the cable by clamping, inappropriate knots, too small bending radii, etc. can significantly reduce the cable's breaking load, in some cases such as clamping and bending to less 2% (!) of its nominal load. Note, that in presence of such extreme effects of the damage mechanics, a safety factor of 10 does not prevent the robot from a failure. This effect can be seen from the evaluation of the cable's breaking load in Fig. 3.7.

Fig. 3.7 Cable force f [N] over the absolute elastic elongation Δl [mm] for a Dyneema cable D-PRO with a diameter $d = 2.5$ mm and a probe length of $l_0 = 300$ mm. The diagram reveals the nonlinear characteristics with a progressive stiffness behavior. The cable has a specified breaking load of $f_{max} = 5800$ N where around 3300 N are measured in this experiment. The reduced load is often caused by imperfect force transfer into the cable at its ends

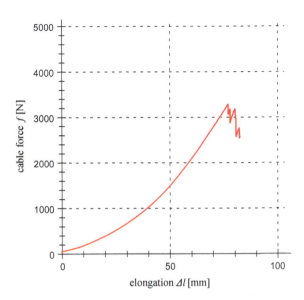

Another mechanical limitation to consider is the maximum load on the winches and pulley mechanisms used to guide the cable. Clearly, the cable may not damage the mechanical structure. In typical applications, the frame, winches, and pulleys can be appropriately dimensioned after the maximum tension of the cables was chosen. Typically, the cable's breaking load is chosen to be at least twice of the winches' load.

The maximum forces exerted by cables on the mobile platform should not deform or damage the platform. This is especially important for robots that can be reconfigured by exchanging the platform. Different platforms may be designed for largely different processes e.g. inspection with light sensors and handling heavy loads. A light-weight sensor platform may get destroyed when applying crane-like forces that are necessary for the handling operation. This effect is mostly a matter of the configuration of the controller and the safety system.

Fatigue

The maximum feasible dynamic tension of the cable and also of the machine elements like the pulleys differs heavily from the static forces when fatigue is taken into account. Even if the breaking load of a certain cable may be very high, one has to take into account the conditions under which the robot is operated. While the robot moves along a trajectory, the cable gets bent every time when it is guided around a pulley or coiled onto the drum. For linear actuators with pulley mechanisms, each motion cycle causes a multiple of bending cycles through the pulley tackle in the cable. Given a certain lifetime for the cable, the maximum feasible tension in the cable must be chosen to avoid failure due to fatigue. Since the fatigue of different cables of the same type may heavily vary, a second safety factor must be chosen to care for the statistical variations of the cable parameters. The first factor depends on mechanical and material properties of the cable, while the second safety factor additionally depends on the level of safety required for the targeted application. Today, figures on fatigue can only be determined by empirical studies [149, 489], while models and simulations such as finite elements method are not reliable enough to compute such data without experimental validation. Clearly, such experimental data are cost intensive to generate. For steel cables, a wide range of data is available in the literature, where for synthetic fibers very little is known.

Actuators

The maximum force of the motor often puts the effective upper limit of the cable force since the actuators and their drive-trains are usually the most costly component in the robot system. Industrial servo drives with integrated brakes are widely spread. If the robot has an emergency braking system, one has to additionally consider the maximum force that can be generated by the brakes for deceleration. From the principles of a safety system, it is clear that the brakes must be stronger than the motor, since the brakes must be able to stop the system even if the motors are still in full operation. In case of an emergency stop, the forces generated by the brakes can get significantly higher than the motors torque and thus present a severe danger to break the cables. A general method to safely predict upper limits on the braking forces is still an open problem, especially since the dynamic characteristics of the robot also influence this behavior. Furthermore, brakes are subject to wear. Therefore, subsequent emergency stops show usually a reduced braking force. Summing up, it must be stated that a significant safety margin between the rated breaking force of the brakes and the static breaking load of the cables must be maintained.

Other Effects

Depending on the environment, other effects might interfere. A change in temperature can contract the cables and thus increase the tension in the cables. Therefore, such thermal effects must be taken into account to avoid overloading of the cables. For synthetic fiber cables, humidity also effects the effective length and stiffness of the cable. This may lead to slackness or overload. Additionally, one has to consider a safety margin caused by control errors. The width of the safety margin depends on the quality of the control systems and the safety requirements of the application.

Summary

Determining the maximum cable force is mostly a matter of mechanical design. Since the costs for the drive-train presents often the highest costs in the robot design (around 50%), this is in most cases the limiting factor due to economic reasons. More precisely, the maximum force generated by the motor should be the limit because the other criteria presented above have to be chosen in an appropriate relation to the motor forces. However, if we reconfigure a cable robot and reuse a winch with a different cable, the effective maximum force f_{max} must be lowered according to the criteria presented above.

Fig. 3.8 Sagging s in the middle of a horizontal cable over distances l in logarithmic scaling. Different cable forces f [N], for a diameter $d = 2.5$ mm of a Dyneema cable

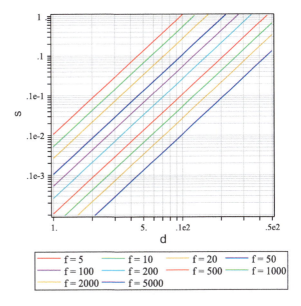

3.4.5.2 Lower Bounds on the Tension

The lower limit on the cable force is considered in the next paragraph. On first sight, one might consider any positive tension feasible. In practice, the following effects require to keep a minimum pretension in the cables.

Slackness

Ensuring pretension and thus preventing the cables from *slackness* is a premise to assume straight cables. However, every cable is subject to sagging under the effect of gravity where the cable's physical parameters length l_i, density ϱ_c, and tension f are the main influence factors. If the tension in the cables is too low, sagging of the cable cannot be neglected. This is often not acceptable because the effective length of the cable and the distance between A_i and B_i heavily differ in this case. Furthermore, the dynamic transition between a tensed and a slack cable induces transversal vibrations in the cables. In the presence of large sagging s, the real direction $\widehat{\mathbf{u}}_i$ of the cable force vector diverges from the ideal direction $\mathbf{u}_i$ (Fig. 3.9). Furthermore, the robot becomes insensitive to control changes in slack cables: Changes in the cable length lead to smaller changes in the cable tension. Therefore, slack cables can hardly be used to control the motion of the mobile platform. This effect can be observed in some measurements taken on the IPAnema 2 robot (see Fig. 3.10). In the left part of the plot, one can see a mostly linear decrease in the cable force. In a transition region between 10 and 15 mm, the curve falls mostly flat. Beginning from that point, one observes a constant value on our sensor independently from the actual cable length.

To find the minimum cable tension, one has to define the maximum error for the cable length or the maximum displacement from the ideal linear form. We consider

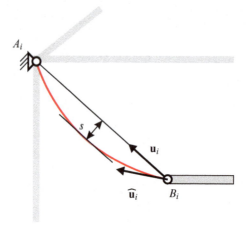

Fig. 3.9 Sagging of a cable with the two major effects: The actual cable length is longer than the straight line and the effective force direction $\widehat{\mathbf{u}}_i$ differs from the idealized direction $\mathbf{u}_i$

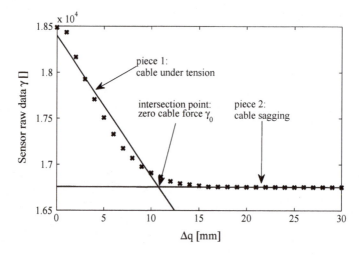

Fig. 3.10 Raw data measured for the cable forces f from the force sensor of the IPAnema 2 system over the relaxed cable length Δq

two criteria. Firstly, the minimum tension required to keep the maximum sagging below a given upper bound. Secondly, the minimum tension required to limit the deviation in the real and effective length by an upper limit.

A thorough consideration of the kinematics under sagging is subject in Chap. 7. In the following sections, we concentrate on some simplifications to derive the limits on the cable force that are required to ignore sagging. If the computed lower force yields too high values, one must use more involved models for cable and also for the control algorithms to cope with the additional effects. Some algorithms are presented in Chap. 7.

The maximum sagging s for a horizontal cable with a distance between the ends of length l is

$$s = \frac{f_{\mathrm{H}}}{g_{\mathrm{C}}} \left(\cosh \left(\frac{g_{\mathrm{C}} l}{2 f_{\mathrm{H}}} \right) - 1 \right) \ , \qquad (3.16)$$

where f_{H} is the horizontal cable force and g_{C} is the gravity force caused by the cable's mass per length. Unfortunately, it is difficult to solve for the sought force f_{H} but a double logarithmic plot reveals simple figures (Fig. 3.8) and one can easily pick the required pretension $f_{\min}$ for a given length l, a given specific cable gravity force g_{C}, and an acceptable sagging of $s_{\max}$ either from diagrams or by numerical solving the implicit equation. Similar computations can be employed if the cable is not horizontal but we restrict this consideration to the basic case.

The geometric deformation of the sagging cable also leads to different effective length l_{s} of the cable since the real curved shape of the cable is obviously longer than the straight line between the end-points A_i and B_i. Considering the length error is straightforward using the equation of the sagging cable. The actual length l_{s} of the cable from the catenary line can be solved in closed-form and we receive the simple equation

$$l_{\mathrm{s}} = \frac{2 f_{\mathrm{H}}}{g_{\mathrm{C}}} \sinh \left(\frac{g_{\mathrm{C}} l}{2 f_{\mathrm{H}}} \right) \qquad (3.17)$$

for the horizontal cable. For selecting the minimal force $f_{\min}$, one has to face the same situation as above. While the equation for length can be written in closed-form, we cannot solve in closed-form for the sought tension f_{H} but numerical determination is simple and robust.

We consider sagging of the cable in more detail in Sect. 7.3.1 and we present a more elaborated derivation of the equations above taking into account the general case without assuming horizontal cables.

Cable Weight

Even if sagging is acceptable, there is a lower bound for the cable force. The minimum tension depends on the weight of the cable and the tension cannot be reduced below a value that is coupled to the density and length of the cable. Consider the following situation: The distal end of the cable is fixed in space and the cable is uncoiled starting from a perfect line. Then, the additional length of the cable increases the overall effective weight of the cable acting on its anchor points. Therefore, the cable force depends for large sagging only on the weight of the cable. For long cables, we find the force to be proportional to the length of the uncoiled cable. In between, a minimal positive cable force cannot be undercut. This limit depends on the distance between A_i and B_i as well as on the density of the cable. Clearly, this limit is important for cable robots with a relatively small platform weight and long cables.

Reliable Operation

Pretension is required for reliable operation of the winch. Without tension, the cable can suddenly slide from the pulley or cross on the drum. There is also a tendency to wrap around inner parts of the winch or to get jammed. If the tension of the cable is very low, the bending stiffness of the cable cannot be neglected. The effect is more important for steel cables but even fiber ropes have a finite bending stiffness that may cause uncontrollable coiling errors in the winches. Also, for steel cables, a very low tension causes the cables to leave the pulleys or drum grooves. Such coiling errors also affect the accuracy of the robot. In the presence of low tension, the cable might also leave the guiding pulleys which in turn causes severe safety problems when cables get jammed at the pulleys or are subject to excessive abrasive wear while sliding on machine elements such as housings.

Practical values are difficult to predict without considering the mechanical design. As reference, in the field of stage equipment for use in theaters 1–2% of the cables breaking load is desirable as minimal tension for coiling.

Elasticity

The elastic elongation of the cables is in general nonlinear (see Fig. 3.7), even if there is a nearly linear region around the operational point. To operate the cables within this desired interval, a certain pretension is required and at the same time a maximum tension may not be exceeded to stay within this preferred state. Since little compensation techniques for nonlinear elongation of the cables can be found in the literature, this presents a relevant limitation on the maximum ratio between $f_{\min}$ and $f_{\max}$.

Considering elastic relations in the cables, it is worth mentioning that only those cables that are in tension contribute the robot's stiffness (see Sect. 3.8). It the cables are not sufficiently tensed, the stiffness is decreased. Thinking about the kinematic constraints, a slack cable does not contribute to the kinematic equations. In this case, a change in the classification and in the topology of the robot is triggered, i.e. a RRPM robot may change to IRPM requiring large different control approaches.

Vibration

Tension in the cables can reduce the vibration of the cables by increasing the cable's first eigenfrequency. This effect is well known from string instruments. It was also shown that pretension can be used to influence the vibration of the platform of planar robots [450]. For an ideal string, the first eigenfrequency f_E can be computed from basic physics

$$f_E = \frac{1}{2l} \sqrt{\frac{f}{\varrho_c A_c}} \quad , \tag{3.18}$$

where ϱ_c is the density of the cable and A_c is the cable cross section. The product $\varrho_c A_c$ called specific weight is often given as parameter for a cable by its manufacturer. We can easily compute the tension $f_{\min}$ for a desired minimum first eigenfrequency

$$f_{\min}(f_E) = 4f_E^2 l^2 \varrho_C A_C \quad . \tag{3.19}$$

To achieve higher eigenfrequencies f_E for large robots, high pretension is required. The contribution of the cable's tension to the stiffness is discussed more in detail in Sect. 3.8.2.

Cable Slip on the Pulleys

Pulleys have a finite inertia I_R although their moment of inertia is mostly neglected. When the cable is accelerated or decelerated, a fraction of the force is consumed to accelerate the pulleys and in order to transmit this force, some pretension must be applied to the cable in order to avoid slip between the cable and the pulley. Slip between cable and pulley is not desired since it leads to additional wear of the cable and possible errors in the measurement if for control purpose the rotation of the pulley is tracked by a sensor. In the following, we compute the required pretension to avoid slipping.

The dynamic cable force equilibrium is depicted in Fig. 3.11. To consider the dynamic equilibrium, one has to take into account that the cable force on the winch side f_W differs from the cable force of the free cable f_S. We can now consider torque for the pulley as follows

$$\tau_R = \frac{I_R a_C}{r_R} \quad , \tag{3.20}$$

where τ_R is the torque accelerating the pulley inertia I_R, a_C is the acceleration of the cable, and r_R is the radius of the pulley. If we consider the pulley to be a full homogeneous cylinder, its moment of inertia is

$$I_R = \frac{1}{2} m_R r_R^2 \quad \text{with} \quad m_R = r_R^2 \pi b_R \varrho_R \quad , \tag{3.21}$$

with the width of the pulley b_R and the average density of the pulley ϱ_R. If we consider more complicated pulley geometries, one can replace the homogeneous cylinder by more accurate values deduced e.g. from CAD data. According to Euler-Eytelwein's

Fig. 3.11 Force equilibrium for the pulley considering pulley inertia and cable-pulley friction

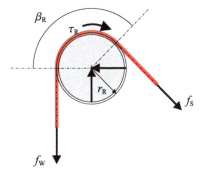

formula (or capstan equation), one computes the maximum difference between f_w and f_s that can be transmitted through friction by

$$f_w = f_s e^{\mu_R \beta_R} \quad , \tag{3.22}$$

where μ_R is the friction coefficient for the cable on the pulley and β_R is the angle around the pulley (Fig. 3.11). We can compute this angle using some advanced kinematic codes as carried out in Sect. 7.2.1. However, for the problem of computing the pretension, one assesses the lowest allowed value for β_R rather than using pose-dependent values. Finally, torque equilibrium related to the center of the pulley yields

$$f_w = \frac{\tau_R}{r_R} + f_s \quad . \tag{3.23}$$

Substituting all equations above in the latter, we find the minimum pretension f_{min} required to avoid the cable from slipping over the pulley to be

$$f_{min} = \frac{1}{2} \frac{r_R^2 \pi b_R \varrho_R a_C}{e^{\mu_R \beta_R} - 1} \quad . \tag{3.24}$$

The minimum cable force required to avoid slip depends on a number of parameters. Firstly, we have some dependency with the geometry and inertia of the pulley expressed by the parameters width b_R, radius r_R, and density ϱ_R. We find a linear dependency for the cable acceleration a_C, making slip an issue for dynamically operated robots. Finally, we have to note that the angle β_R located in the exponential function gives some importance to that parameter. If the cable is hardly deflected by the pulley, one has a poor force transmission from the cable to the pulley leading to high required pretensions.

Some example data are presented in Fig. 3.12. It can be seen that robots operated with acceleration a_C below gravitation acceleration g and with reasonable deflection of at least a right angle $\beta_R > \frac{\pi}{2}$, sufficient pretension in the cable is rather low and pretension can be chosen below 10 N. However, if one want to exploit the excellent dynamic performance of the cable robot, significant pretension must be ensured.

Requirements for Force Sensing

When using force sensors to measure the tension in the cables, it might be necessary to maintain a minimum tension, since some force sensors provide low quality signals close to zero tension. For very small forces, the measurement is also subject to errors caused by mechanical parts for including the sensors. Therefore, small cable forces are hard to distinguish from friction. Also, for the lower cable forces f_{min}, one has to consider a safety margin in order to prevent control errors to trigger one of the effects listed above.

Summary

Taking all these above-mentioned effects into account, it becomes apparent that the ratio between minimum and maximum cable forces may be large but cannot be

Fig. 3.12 Required
pretension f_{min} for different
values of the wrapping angle
β_R and the desired cable
accelerations a_C. The other
values have been chosen to
reflect the values of the
IPAnema 3 winch as follows:
$\mu_R = 0.1$,
$\varrho_R = 2700\,\mathrm{kg\,m^{-3}}$
(aluminum), $b_R = 20\,\mathrm{mm}$,
and $r_R = 35\,\mathrm{mm}$. The lines
in the diagram correspond to
minimum cable force $f_{min} \in$
$\{1, 2, 5, 10, 20, 50, 100, 200, 300\}$
respectively

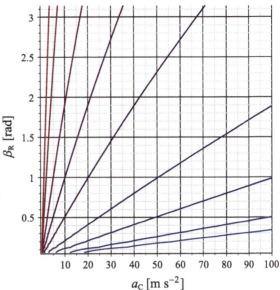

increased to arbitrary values. As it will be apparent in the workspace studies subject
to Chap. 5, the effective ratio from f_{min} and f_{max} has a nonnegligible influence on the
robot. Thus, the concept of wrench-feasibility must be applied rather than wrench-
closure in most practical cases.

3.4.5.3   Examples

The effective limits of the cable forces depend on a large number of physical parameters, design decisions, and application requirements. Since a number of components
such as motors, cable material, and cable diameter are subject to a reflected design
procedure, we consider these parameters to be given. In the following, the considerations on the cable force limits are exemplified based on case-studies performed for
some prototypes, i.e. the IPAnema family prototypes [391]. In particular, we consider
the use-cases described in Table 3.1 which are applied on the following robot setup:

- IPAnema 1 spatial system, medium size robot used for a fast pick-and-place task.
- IPAnema 2 spatial system, medium size robot used for handling and assembly for
  solar collectors.
- IPAnema 3 mini spatial system, small size robot, high-dynamics laboratory system
  for testing of kinematic codes, control algorithms, and calibration.
- IPAnema 3 spatial system, large size robot ($16 \times 6 \times 5\,\mathrm{m}$), $5.0\,\mathrm{kW}$ for handling in
  a logistic scenario.

Table 3.1 Overview of some use-cases of the IPAnema robot family

Robot	Frame length (m)	Width (m)	Height (m)	Winch power (kW)	Cable material	Diameter (mm)	Weight/length (kg/m)	Breaking load (N)
IPAnema 1	4	3	2	1.8	Dyneema	2.5	0.0035	5800
IPAnema 2	8	6	5.5	1.8	Dyneema	2.5	0.0035	5800
IPAnema 3 mini	1	1	1	0.2	Dyneema	1.5	0.0013	2300
IPAnema 3 planar	20	5	0	5.0	Dyneema	6.0	0.023	43000
IPAnema 3 IZS setup	16	6	5	5.0	steel	6.0	0.129	21100
IPAnema 3 winch A	45	0	0	7.5	Dyneema	2.5	0.0035	5800
IPAnema 3 winch B	23	0	0	7.5	Dyneema	6.0	0.023	43000

Beside the application inspired scenario, we consider additionally the nominal design parameter of the IPAnema 3 winches for 2.5 mm cable (IPAnema 3 winch A) and 6.0 mm cable (IPAnema 3 winch B). In order to compute feasible tension, we have to fix some requirements. We checked the following criteria:

- The sagging s shall be smaller than 0.1% of the length of the cable or smaller than the cable's diameter (column 3 and 4 of Table 3.2).
- The length error caused by sagging shall be smaller than 0.1% of the cable length or smaller than 1 mm (column 5 and 6 of Table 3.2).
- The eigenfrequency of the cable should be at least 10 Hz (column 7 of Table 3.2).

The results are given in absolute numbers in Table 3.2. The presented bounds shall assure that the assumptions of the standard model are sufficiently fulfilled. Taking all these issues into account leads to a surprisingly restricted interval for the feasible cable forces in some applications. The use-cases presented above reveal significant restriction. In optimistic use-cases, a ratio was found between minimum and maximum tension of 600.

It can easily be seen from the equations that tension limits are scale-dependent. Therefore, we have to distinguish between small, medium, and large applications, where the latter typically allow for a smaller ratio and thus for less workspace size compared to the theoretical limits. However, since cable robots are evolving towards commercial applications, such limitation needs to be taken into account to allow for safe and reliable operation. Furthermore, some criteria discussed in this section are pose-dependent. To the best of the author's knowledge, it was not yet considered in literature to adjust the cable force limits to the pose e.g. for workspace computation or for control. Relaxing and tightening these bounds may have either positive or negative influence on the workspace since it can present additional potentials at some poses as well as the need for higher pretension in other poses. Pose-dependent limits may also lead to different limits for each cable, since e.g. the actual length depends largely on the current pose. In order to possibly exploit pose-dependent force limits for each cable, force distribution tests need to be developed that work efficiently for individual force limits in each cable.

3.4.6 Force Distributions for CRPM and RRPM

To study the static properties of cable robots of CRPM and RRPM type, one has to analyze the structure equation (3.5) more in detail. In the following, we focus on *wrench-feasible* configurations where we have positive lower $f_{\min}$ and upper $f_{\max}$ force limits for the forces $\mathbf{f}$ in the cables. For redundant cable robots, there might be infinitely many solutions for the force distribution in the cables [141]. It is a well-known result from linear algebra that the solutions of an under-constrained

Table 3.2 Tension limits for the IPAnema use-cases

Robot	Nominal (given) f_{max} [N]	Sagging $s < l/1000$ f_{min} [N]	Sagging $s < d$ f_{min} [N]	Length error $\Delta l < l/1000\ f_{min}$ [N]	Length error $\Delta l < 1\,\text{mm}\ f_{min}$ [N]	Eigenfrequency $f_E > 10\,\text{Hz}\ f_{min}$ [N]
IPAnema 1	180	2.6	5.1	1.2	2.8	40.6
IPAnema 2	720	5.0	22.8	2.5	8.5	182.4
IPAnema 3 mini	60	2.8	3.2	0.1	0.2	1.6
IPAnema 3 planar	3000	59.3	203.7	26.3	136.4	3910.0
IPAnema 3 IZS setup	3000	287.0	851.5	26.0	613.2	16357.2
IPAnema 3 winch A	580	192.9	3472.0	10.0	66.9	2835.0
IPAnema 3 winch B	4300	648.6	2486.4	33.5	160.6	4866.8

linear system such as the structure equation form an r-dimensional linear vector space $\mathcal{S}$ that is embedded in $\mathbb{R}^m$ and that contains all solution vectors. Let $\mathbf{H} = [\mathbf{h}_1, \mathbf{h}_2, \ldots, \mathbf{h}_r] \in \mathbb{R}^{m \times r}$ be a spanning vector basis of the nullspace or kernel of the structure matrix $\mathbf{A}^{\mathrm{T}}$. Then, each base vector $\mathbf{h}_i$ fulfills $\mathbf{A}^{\mathrm{T}}\mathbf{h}_i = \mathbf{0}$ and all vectors $\mathbf{h}_i$ are linearly independent. The vectors $\mathbf{h}_i$ can be chosen such that they are perpendicular to each other, i.e. $\mathbf{h}_i \cdot \mathbf{h}_j = 0$ for any $i \neq j$. Using this definition for $\mathbf{H}$, we can write all solutions of the structure equation in the parameter form

$$\mathcal{S} = \left\{ \mathbf{f} = -\mathbf{A}^{+\mathrm{T}}\mathbf{w}_{\mathrm{P}} + \mathbf{H}\boldsymbol{\lambda} \mid \boldsymbol{\lambda} \in \mathbb{R}^r \right\} \quad , \tag{3.25}$$

where $\boldsymbol{\lambda}$ are the parameters of the solution space and $\mathbf{A}^{+\mathrm{T}}$ is the Moore-Penrose pseudo-inverse $\mathbf{A}^{+\mathrm{T}} = \mathbf{A}(\mathbf{A}^{\mathrm{T}}\mathbf{A})^{-1}$. Then, $-\mathbf{A}^{+\mathrm{T}}\mathbf{w}_{\mathrm{P}}$ is the projection of the applied wrench $\mathbf{w}_{\mathrm{P}}$ onto the solution space $\mathcal{S}$. On the other hand, we can easily define the set

$$\mathcal{C} = \left\{ \mathbf{f} \in \mathbb{R}^m \mid 0 < f_{\min} \leq f_i \leq f_{\max}, \quad i = 1, 2, \ldots, m \right\} \tag{3.26}$$

of all feasible force distributions that form an axis-aligned hypercube in the m-dimensional space. The set $\mathcal{F}$ collecting all solutions of the structure equation is given by the intersection (Fig. 3.13)

$$\mathcal{F} = \mathcal{C} \cap \mathcal{S} \quad . \tag{3.27}$$

A pose is wrench-feasible for a given wrench $\mathbf{w}_{\mathrm{P}}$ if and only if $\mathcal{F}$ is nonempty. If $\mathcal{F}$ is nonempty, it forms a convex polyhedron [473]. From the convexity, one can directly conclude that if $\mathbf{f}_{\mathrm{A}}$ and $\mathbf{f}_{\mathrm{B}}$ are feasible solutions, than every solution

$$\mathbf{f} = \lambda \mathbf{f}_{\mathrm{A}} + (1 - \lambda)\mathbf{f}_{\mathrm{B}} \quad \forall \; \lambda \in [0; 1] \tag{3.28}$$

is also a feasible solution.

3.4.7 Available Wrench Sets

For a given pose of the robot, it is interesting to study which wrenches can be generated by the robot. Any cable robot is able to generate some wrenches. This becomes clear when submitting an arbitrary cable force vector into the structure equation. Take for example the maximum cable forces $f_{\max}$ for each cable force. Then, the robot clearly generates the wrench $\mathbf{w}_{\max} = -\mathbf{A}^{\mathrm{T}}\mathbf{f}_{\max}$. This also holds true both for over-constrained and under-constrained robots. If the robot is at a singular pose, it still generates some wrenches although the dimension of the wrench set may degenerate. Now, we investigate the wrench set that can be generated by a given robot.

Fig. 3.13 Illustration of the
intersection between the
solution space $\mathcal{S}$ of the
structure equation and the
cube $\mathcal{C}$ of feasible force
distributions

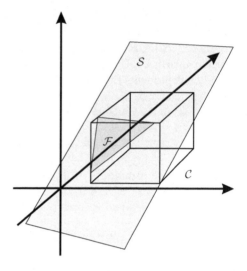

Given a structure matrix $\mathbf{A}^{\mathrm{T}}$ and bounds on the cable forces $f_{\min}$, $f_{\max}$, one can ask
for the set of wrenches $\mathcal{Q}$ that can be generated by the cable robot at this pose. This
problem was addressed in a couple of papers [47, 48] where the contribution from
Bouchard [62] is the most complete one. A detailed analysis of the structure equations
for feasible cable forces shows that the *available wrench set*, i.e. all wrenches $\mathbf{w}_{\mathrm{P}}$
that can be exerted by a cable robot for a given pose are a convex set $\mathcal{Q}$ that can
be received by projecting the m-dimensional hypercube $\mathcal{C}$ into the n-dimensional
wrench space. Summarizing the findings of the paper [62], it is clear, that the linear
projection of a convex set transforms an m-dimensional space into an n-dimensional
space and the linearity of the transformation preserves the convexity of $\mathcal{C}$ after the
projection. Moreover, it turns out that the wrench set is a zonotope, i.e. it is bounded
by pairs of parallel $(n-1)$-dimensional hyperplanes.

Therefore, the following properties hold true [62]:

- It follows from the convexity of the set $\mathcal{Q}$ that if a cable robot can generate two
  wrenches $\mathbf{w}_1$ and $\mathbf{w}_2$ it can also generate every wrench

$$\mathbf{w}_{\mathrm{P}} = \lambda \mathbf{w}_1 + (1 - \lambda)\mathbf{w}_2 \quad \text{for} \quad \lambda \in [0; 1] \tag{3.29}$$

  on the connecting line segment. More generally speaking, if a discrete set $\mathcal{Q} = \{\mathbf{w}_1, \ldots, \mathbf{w}_n\}$ of wrenches can be generated, every wrench $\mathbf{w}_{\mathrm{P}}$ in the convex hull
  of the wrenches $\mathbf{w}_i$ can also be generated.
- Thus, to check if a hypercube of wrenches can be generated, it is sufficient to check
  all of its corners. Still, in n-dimensional space, the hypercube has 2^n corners but
  checking all corners is straightforward. Therefore, for the most general 3R3T
  robot, one has to consider 64 corners.

Fig. 3.14 For the zonotope of the available wrench set $\mathcal{Q}$, we can check if the ellipsoid $\mathcal{E}$ with the desired wrenches is fully contained

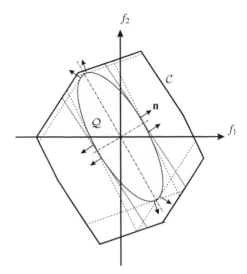

- In a similar way, one can check if every wrench $\mathbf{w}_P$ within an ellipsoid $\mathcal{E}$ of wrenches can be generated by the robot. To do so, one has to check only $2m$ characteristic points on the surface of the ellipsoid (Fig. 3.14). The characteristic points are determined on the surface of the ellipsoid $\mathcal{E}$ through the normals of the hyperplanes which generate the zonotope [62].
- The zonotope defining the wrench set $\mathcal{Q}$ can be constructed from the Minkowski sum of m generating lines (Fig. 3.15).
- Once the zonotope forming the wrench set has been calculated, it is simple to check for wrench-feasibility for any given wrench by testing if the wrench $\mathbf{w}_P$ or a set of wrenches $\mathcal{Q}$ is fully enclosed by the zonotope.
- If the zonotope degenerates, i.e. does not span n dimensions, the robot is singular and it cannot generate or withstand wrenches in the direction in which the degeneration occurs.

We have seen that the available wrench set $\mathcal{Q}$ is convex where for a given wrench $\mathbf{w}_P$ the possible force distributions are a convex set $\mathcal{F}$. Although there are some similarities between force distributions and available wrench sets, these are two different properties of the robot. The set of possible force distributions $\mathcal{F}$ is a matter of the configuration space where the wrench set $\mathcal{Q}$ is a concept in the operational space. Based on this discussion about the structure of the statics, we devote the next sections to the computation of actual cable force distributions from the set $\mathcal{F}$.

Fig. 3.15 Graphical scheme
to compute the Minkowski
sum of four line segments

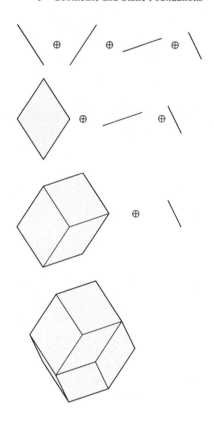

3.5   Force Computation for CRPM

The determination of force distributions for completely constrained cable robots
is comparably simple. If the set $\mathcal{F}$ is nonempty, it forms a line segment that has
well-defined end-points. In this section, we review a method proposed by Verhoeven
[267, 268, 473] to compute force distributions for cable robots of the CRPM type,
i.e. for robots having exactly $m = n + 1$ cables. It follows from the general structure
discussed in the previous section that all possible force distributions form a one-
dimensional subspace in the m-dimensional space of cable forces or, in other words,
all solutions are on a common line. The direction of the line is defined by the kernel
of the structure matrix $\mathbf{A}^T$ and one finds a special point on that line using the pseudo-
inverse matrix $\mathbf{A}^{+T}$. The solution set given by Eq. (3.25) simplifies in this case to

$$\mathcal{S}_1 : \mathbf{f} = -\mathbf{A}^{+T}\mathbf{w}_P + \mathbf{h}\lambda = \mathbf{f}_0 + \mathbf{h}\lambda \quad , \tag{3.30}$$

where $\mathbf{h}$ is any nonzero element of the kernel of $\mathbf{A}^T$ and $\mathbf{f}_0 = -\mathbf{A}^{+T}\mathbf{w}_P$ is a particular
solution on the line. Note that λ is a scalar parameter in this case.

Fig. 3.16 Illustration of the feasible interval $\mathcal{F}$ for the cable forces bounded by $\lambda_{\min}$ and $\lambda_{\max}$

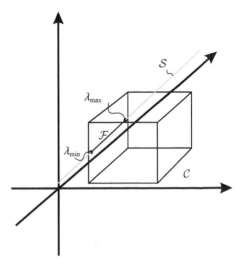

To check wrench-closure (controllability), it is sufficient to consider the vector **h**: A positive force distribution **f** exists, if and only if all elements of the vector **h** are nonzero and have the same sign, i.e. if **h** > **0** or **h** < **0** holds true. Interestingly, the value of the applied wrench $\mathbf{w}_\mathrm{P}$ does not matter for checking force-closure. Whenever the kernel fulfills the criterion above, one can choose λ sufficiently large (small if **h** < **0**) to shift all cable forces into the positive region (see Fig. 3.16).

If the robot is in a wrench-feasible (acceptable) configuration, then there must be a range for the λ that leads to feasible cable force **f**. The conditions for wrench-feasibility can be written as follows

$$\underbrace{\max_{1\leq i\leq m} \frac{f_{\min} - f_{0,i}}{h_i}}_{\lambda_{\min}} \leq \lambda \leq \underbrace{\min_{1\leq i\leq m} \frac{f_{\max} - f_{0,i}}{h_i}}_{\lambda_{\max}} \,. \tag{3.31}$$

If the interval $[\lambda_{\min}; \lambda_{\max}]$ is empty, the pose is not wrench-feasible. Otherwise, all solutions can be computed from Eq. (3.30) by substituting the feasible value for λ.

For numerical computations, one has to determine an element of the kernel of $\mathbf{A}^\mathrm{T}$. This can easily be done using the singular value decomposition (SVD) that allows to directly compute the vector **h** as the direction associated to the singular value 0. Using SVD is a very robust but slow way to compute the kernel. Although SVD is available in many powerful numerical libraries and computer algebra systems, it might be tiresome to manually implement it for usage in other contexts such as in a real-time system. Basically, any linear algebra algorithm that allows to solve a homogeneous system can be applied to compute the nullspace.

3.6 Force Computation for RRPM

The determination of feasible force distribution for redundantly constrained cable robots is a major challenge for design and control of such robots. One has to distinguish two problems regarding force calculations for cable robots. The first problem is, if at least one solution $\mathbf{f}$ to Eq. (3.5) exist. Secondly, if there are many solutions how can one select one solution such that the cable force distributions are continuous along a trajectory, i.e. under continuous changes in the platform pose $(\mathbf{r}, \mathbf{R})$.

3.6.1 Force Computation as Optimization Problem

Verhoeven [473] and Gosselin [168] showed that one can find trajectories with continuous force distributions if one converts the under-determined linear system with constraints into an optimization problem using p-norms (Fig. 3.17) with $p > 1$. This leads to the following constrained optimization problem

$$\text{minimize } g(\mathbf{f}) = \|\mathbf{f} - \mathbf{f}_{\text{ref}}\|_p = \sqrt[p]{\sum_{i=1}^{m} (f_i - f_{\text{ref},i})^p} \qquad (3.32)$$

$$\text{subject to} \qquad f_{\min} \leq f_i \leq f_{\max} \qquad (3.33)$$

$$\text{linear constraints} \qquad \mathbf{w}_{\text{P},i} = -\sum_{j=1}^{m} A_{j,i}^{\text{T}} f_j \ , \qquad (3.34)$$

Fig. 3.17 Unit spheres for different p-norms with $p = \{1, 2, 4, 10, 40, 100\}$. It can be seen that for higher values of p the form smoothly approximates the enclosing box

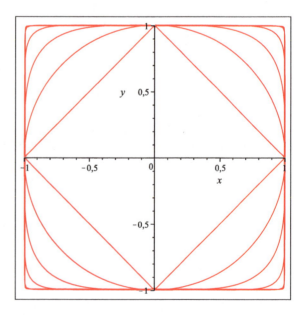

where $\mathbf{f}_{ref}$ is the desired cable force to be optimally approached. Using p-norms as objective function and the structure equation as constraints introduce a criterion to choose an appropriate solution amongst the infinitely many possible solutions. The p-norm can be understood a kind of potential field on the solution set that allows to compare different solutions based on their p-norm (Fig. 3.18). Due to the properties of the p-norm, the solution becomes unique and continuous along a trajectory. Contrary, using p-norms also introduces some trade-off: low values of p (especially using the Euclidian norm $p = 2$) allow for simple and numerically stable computation. Unfortunately, the potential field introduced by the p-norm sometimes favors infeasible solutions outside the allowed region over feasible solutions in the corners of the force space (Fig. 3.17). In contrast, high values of the p-norm provide a smooth mapping towards the corners of the box but introduce numerical problems with computing the high exponents and sometimes rapidly changed values. This can be seen from the comparison in Fig. 3.18 for p-norms between 2 and 100. In each of the six plots, the isolines for $\{0.2, 0.4, 0.6, 0.8, 1.0\}$ are shown for different p-norms. The figure illustrates that choosing the p-norms makes a compromise between numerical stability and coverage of the maximum volume of the hypercube $\mathcal{C}$. Furthermore, the chart for $p = 1$ unveils lack of differentiability where the isolines cross the coordinate axes.

Some iterative procedures to solve this general problem have been proposed, such as linear and quadratic programming, gradient methods, Dykstra method, and interval analysis. Especially for real-time control purposes, non-iterative methods with an acceptable amount of computation time were developed. A different approach for choosing a unique solution is based on computing the barycenter of the solution set $\mathcal{F}$. The computation time is driven by triangulations in the r-dimensional space [275, 277, 348] which become more complex for higher dimensions m. Currently, the barycentric algorithm has been implemented for redundancies up to $r = m - n = 2$.

3.6.2 Other Ways to Consider the Problem

Most authors address the minimization or maximization of the cable forces. Once these extremal values are found, one can easily choose a compromise between these distributions. Because of the convexity of the solution set, every weighted sum of the extremal solutions is also a valid solution. Tracking such a linear combination also leads to continuous force distributions.

Instead one may search for force distributions with minimal deviations amongst the cable forces, i.e. force distributions that are homogeneous. This presents a slightly

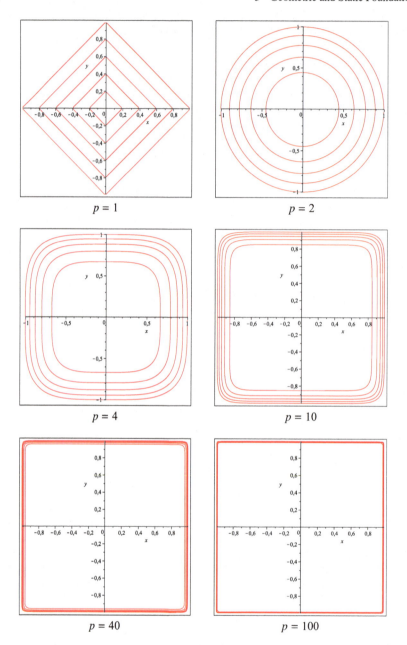

Fig. 3.18 Potential fields for different p-norms and values $v = \{0.2, 0.4, 0.6, 0.8, 1.0\}$. For high p values, the potential fields extend to the very corners of the box but the isolines of the potential field are compressed indicating an ill-conditioned numerical problem

more complex objective function that can be expressed by minimizing the standard deviations or higher moments. Minimizing the standard deviation can be achieved by the objective function

$$g(\lambda) = \sum_{i=1}^{m} \left[\overline{f}(\lambda) - [\mathbf{H}\lambda]_i \right]^2 \tag{3.35}$$

where

$$\overline{f}(\lambda) = \frac{1}{m} \sum_{i=1}^{m} [\mathbf{H}\lambda]_i \tag{3.36}$$

is the mean value of $\mathbf{H}\lambda$.

The following consideration shall highlight the connection between the general parameter form of the feasible force set $\mathcal{S}$ and the selection of a single solution. Starting from the parametric form Eq. (3.25) of the nullspace, we can convert this formula into an optimization problem. For this purpose, we take a closer look at the structure of the parameter form. The homogeneous solution $\mathbf{A}^{+T} \mathbf{w}_p$ is a constant vector where the spanning base of the nullspace $\mathbf{H}$ and the scale vector λ are the describing parameters of the optimization problem. Applying a p-norm for the objective function gives a multivariate polynomial in $\lambda_1, \ldots, \lambda_r$ which coefficients are linear combinations of the coefficients of $\mathbf{H}$, i.e. of the spanning base of the kernel of $\mathbf{A}^T$. This structure of the objective function must be kept in mind when looking for a suitable optimization algorithm.

3.6.3 On the Influence of Higher p-Norms

As discussed earlier, the p-norm of the force vector introduces a single measure on the cable force vector and in the context of optimization, we can interpret this measure as a selection criterion. One can find a value $||\mathbf{f}||_p \leq \tilde{f}$ for every p-norm such that all forces fulfilling the inequality are inside the cube $\mathcal{C}$, i.e. are feasible force distributions. A guaranteed region of existence and uniqueness for p-norms can be described by a unit superquadric with the implicit equation

$$\sum_{i=1}^{m} |f_i|^p = 1 \quad . \tag{3.37}$$

Now, the ratio between this unit superquadric and the volume of the unit hypercube $\mathcal{C}$ reveals how much of the theoretically possible force limits can be correctly mapped inside the hypercube by the respective p-norm. A closed-form formula was found to compute the volume of the generalized m-dimensional superquadric [485] which reads

Fig. 3.19 Ratio of the unit
m-dimensional superquadric
defined by the p-norm and
the unit m-dimensional
hypercube (p-axis in
logarithmic scale)

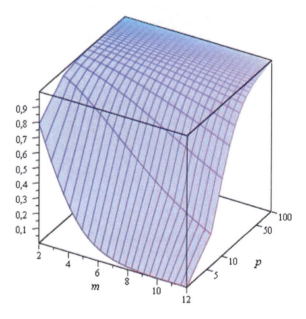

$$V_m^p = \frac{\Gamma(1 + \frac{1}{p})^m}{\Gamma(1 + \frac{m}{p})} \, 2^m \quad , \tag{3.38}$$

where m is the number of dimensions (cables), p is the chosen p-norm, and Γ is Euler's Gamma-function that generalizes the factorial function to real numbers $\mathbb{R}$. The ratio between the volume of the hypercube and the superquadric is shown in Fig. 3.19 for different p-norms and dimensions m. For the Euclidian norm with $p = 2$, the volume can be simplified

$$V_m^2 = \frac{\Gamma(1 + \frac{1}{2})^m}{\Gamma(1 + \frac{m}{2})} \, 2^m = \frac{\pi^{\frac{m}{2}}}{\Gamma(1 + \frac{m}{2})} \, 2^m \quad . \tag{3.39}$$

It can be seen that the superquadrics almost exploit the hypercube for higher p-norms of $p > 25$. Contrary for small p-norms (and especially for the 2-norm), the coverage of the hypercube is rather limited. Therefore, one has to abstain from guaranteed existence or analyze the respective algorithm more in detail. Although higher p-norms are highly beneficial in terms of region of guaranteed convergency, they also introduce severe numerical problems for implementation. This holds especially true for real-time systems with little support for arbitrary-precision arithmetic.

In the following section, we explain and discuss a number of algorithms to compute the force distributions.

3.7 Algorithms for Force Distribution

In this section, a wide variety of approaches for numerically solving the force distribution problem is presented and compared. Where possible, considerations on real-time capability, convergency, continuity, and computational complexity are added.

3.7.1 Linear Programming

Linear programming is a special form of convex optimization [438] where the cost function is linear in the parameters and the problem has linear constraints, i.e. equations and inequations in the independent parameters. A general form of a linear program is

$$\text{minimize} g(\mathbf{f}) = \mathbf{c}^{\mathrm{T}}\mathbf{f} \tag{3.40}$$
$$\text{subject to} \quad \mathbf{Bf} \leq \mathbf{l} \tag{3.41}$$
$$\text{with the linear constraints} \quad \mathbf{Qf} = \mathbf{d} \tag{3.42}$$

where we chose $\mathbf{c}^{\mathrm{T}} = [1, \ldots, 1]^{\mathrm{T}}$ to emulate the $p = 1$ norm.[4] The matrices and vectors $\mathbf{B}, \mathbf{Q}, \mathbf{l}$ and $\mathbf{d}$ are chosen such that the conditions Eqs. (3.33) and (3.34) match the structure of a linear program. Thus, one receives

$$\mathbf{B} = \begin{bmatrix} \mathbf{I} \\ -\mathbf{I} \end{bmatrix} \tag{3.43}$$

$$\mathbf{l} = \begin{bmatrix} \mathbf{f}_{max} \\ -\mathbf{f}_{min} \end{bmatrix} \tag{3.44}$$

$$\mathbf{Q} = \mathbf{A}^{\mathrm{T}} \tag{3.45}$$

$$\mathbf{d} = -\mathbf{w}_{\mathrm{P}} \quad . \tag{3.46}$$

Linear programs are mostly solved with the simplex algorithm that is robust, efficient, and well-understood. Anyway, also other algorithms are known to deal with linear programs.

3.7.1.1 Existence and Uniqueness

The simplex algorithm searches the optimum by visiting the vertices of the polytope defined by the constraints. The next vertex is selected such that the objective function is reduced. The procedure continues until all vertices have been visited (worst case)

---

[4]Note, that by choosing the 1-norm, one does not fulfill the assumptions required to receive a continuous solution along a trajectory. Instead, linear programs may generate steps in the force signal from time to time.

or no improvements can be found on a neighboring vertex. In general, it may happen that all vertices must be visited before finding the optimum. It is unknown if this can really happen for cable robots. Linear programs can be unbounded but this cannot happen for cable robots since $\mathbf{f}_{min}$ is always the smallest solution. Furthermore, it may happen that no solution is feasible at all which happens if the current pose does not belong to the wrench-feasible workspace.

If the optimization problem has solutions and is bounded, then the simplex algorithm will find the global optimum which is not necessarily unique. Note this lack of uniqueness is a source of discontinuity and it is inherent to linear programs. The simplex algorithm always provides one of the vertices of $\mathcal{F}$. Therefore, at least one force in the solution vector is equal to f_{min} or f_{max}. When following a trajectory, it may happen that the vertex changes that is optimal with respect to the linear program. In this case, a discontinuity appears in the force distributions. Clearly, this drawback hinders the usage of linear programming to compute set-point forces in most control applications. Linear programs are widespread tools in optimization and solving the linear program is sufficient to answer the question if a pose is wrench-feasible.

3.7.1.2 Computational Complexity

The computational time of the simplex algorithm can be strictly bounded. The worst case computation time depends on the number of vertices of the polytope which in turn is exponentially in the number of cables. Since the number of cables m is constant for a given robot, a real-time capable implementation is basically possible. Aside from solving the linear program, no further advanced linear algebra algorithms are needed. Linear programs can be employed for an arbitrarily large number of cables m.

3.7.2 Nonlinear Programming

The discontinuities generated by linear programming are a consequence of the implicit using the 1-norm as objective function. When extending the approach by using high p-norms such as the Euclidian norm, linear programming is no longer applicable. Instead, quadratic or other nonlinear methods have to be applied to solve the optimization problem. A typical form of quadratic programming is given by

$$\text{minimize} g(\mathbf{f}) = \frac{1}{2}\mathbf{f}^{\mathrm{T}}\,\mathbf{C}\mathbf{f} + \mathbf{c}^{\mathrm{T}}\,\mathbf{f} \tag{3.47}$$

$$\text{subject to} \quad \mathbf{Bf} \leq \mathbf{l} \tag{3.48}$$

$$\text{with the linear constraints} \quad \mathbf{Qf} = \mathbf{d} \ . \tag{3.49}$$

To emulate the 2-norm in the objective function, we choose $\mathbf{C} = \mathbf{I}$ for the quadratic part in $\mathbf{f}$ and neglect the linear part by setting $\mathbf{c} = \mathbf{0}$. The objective function becomes basically $g(\mathbf{f}) = ||\mathbf{f}||_2$ but also fits the formal specifications of a quadratic program. The values of $\mathbf{B}, \mathbf{Q}, \mathbf{l}$, and $\mathbf{d}$ can be chosen as described in Sect. 3.7.1.

3.7.3 Verhoeven's Gradient Method

Verhoeven [473] developed an optimization method highly specialized to cable robots. This algorithm is designed such that it can be applied for different values of p, especially for high values of p in order to approximate the infinite norm. The gradient method uses an iterative solver to determine two feasible solutions called $\mathbf{f}_{low}$ and $\mathbf{f}_{high}$ that are minimal/maximal with respect to the chosen p-norm (Fig. 3.20). Due to the convexity of the set $\mathcal{F}$, all points on the linear interpolation between the low and the high solutions are also feasible. Therefore, it is straightforward to adjust the level of internal tension in the robot by computing a weighted sum between $\mathbf{f}_{low}$ and $\mathbf{f}_{high}$. Since the method is designed for high values p (numerical examples are given for up to $p = 9$ [473]), it uses a sophisticated method to internally scale the magnitude of the intermediate results such that numerical errors due to canceling are avoided. Furthermore, a method is developed to control the step size.

3.7.3.1 Existence and Uniqueness

The gradient method may fail to find solutions although they exist. Due to its iterative structure, it is difficult to predict its worst case computational time. A major advantage of the method is that it can deal with large values for the p-norm as well as for highly redundant robots with many cables. Some examples are presented including $p = 9$

Fig. 3.20 A selection of force distributions is found by a lowest $\mathbf{f}_{low}$ and a highest $\mathbf{f}_{high}$ solution

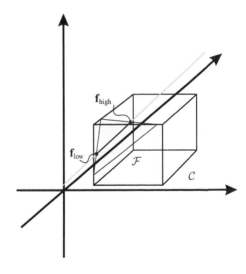

norm and a planar robot with $m = 25$ cables. For both cases, it receives a large region of convergency while maintaining the continuity of the computed force distributions. Therefore, it serves well for analysis and research purpose but is rather limited for application in force control.

3.7.3.2 Computational Complexity

A real-time capable implementation was not reported yet. From the simulation results, it can be expected to be fast enough for real-time purpose where the worst case computation time is not published or estimated.

3.7.4 Dykstra Method

Hassan uses the iterative Dykstra method to compute solutions of the structure equation and thus receives force distributions [199, 201, 203]. The idea is to perform an alternating series of projections P_S and P_C where beginning from an initial guess of the forces (e.g. $\mathbf{f} = \mathbf{0}$). These forces are firstly projected onto the solution space S. The resulting force distribution is projected back onto the cube C. Hassan showed that the algorithm converges to a force distribution with a minimal distance between S and C. If the sets are intersecting, this minimal distance is 0 and the desired solution is found. If no improvement on the force distribution can be made by this two-step projection, then the set $\mathcal{F}$ is empty and thus no solution exists. The alternating projects onto the cube C and the solution space S are illustrated in Fig. 3.21.

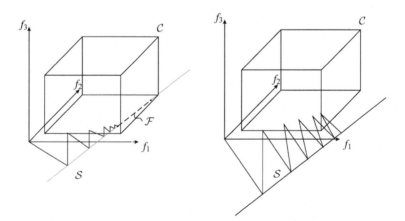

Fig. 3.21 Iterative projections computed by the Dykstra algorithm from C to S and back. Convergency towards a feasible solution in $\mathcal{F}$ (left). Situation outside the wrench-feasible workspace without convergency (right)

The first projection P_S from an arbitrary force $\mathbf{f}^{(i)}$ onto the solution space S is achieved by

$$P_S \quad : \quad \mathbf{f}^{(i+1)} = \left(\mathbf{I} - \mathbf{A}^{+T}\mathbf{A}^T\right)\mathbf{f}^{(i)} - \mathbf{A}^{+T}\mathbf{w}_P \quad . \tag{3.50}$$

The second projection of the Dykstra algorithm P_C onto the cube C is simply achieved by limiting each component f_j of the vector $\mathbf{f}^{(i)}$ to the force limits f_{min} and f_{max}, respectively, as follows

$$P_C \quad : \quad \mathbf{f}^{(i+1)} = [\widehat{f}_1, \dots, \widehat{f}_m]^T \tag{3.51}$$

with

$$\widehat{f}_j = \begin{cases} f_{min} & f_j < f_{min} \\ f_{max} & f_j > f_{max} \\ f_j & \text{otherwise} \end{cases} \tag{3.52}$$

Hassan shows that the algorithm converges in general but practical tests and also the figures in Hassan's paper indicate that the rate of convergency is very slow and also a test implementation indicated a high number of iterations until convergency was reached. Due to the projection steps, it seems that the convergency slows down when the minimum is approached. Therefore, a good initial guess can speed up the computation. In one implementation, the closed-form solution (see Sect. 3.7.5) was used as initial guess rather than starting with $\mathbf{f} = \mathbf{0}$. The general method proposed in the older work [199, 429] is only able to any one solution on the cube C. Note that for redundancy $r > 1$, there are in general still infinitely many solutions on the cube. Therefore, this method cannot be used to compute force distributions that are in general continuous along a trajectory.

3.7.5 Closed-Form Method

In the following, a method is presented to compute force distributions for cable robots with an arbitrarily large redundancy $r = m - n > 0$ of cables [396]. It is shown how to calculate a solution for this problem in closed-form for $p = 2$ with simple algebraic operations making an implementation in a real-time system straightforward. In order to find the solutions, the cable force vector $\mathbf{f}$ is split into

$$\mathbf{f} = \mathbf{f}_M + \mathbf{f}_V \quad , \tag{3.53}$$

where the i-th component of the vector $\mathbf{f}_M$ reads $f_{M,i} = (f_{min} + f_{max})/2$ which is the average feasible force and $\mathbf{f}_V$ is an arbitrary force vector. In other words, a coordinate transformation to the center of the cube C is performed. Thus, one can rewrite Eq. (3.5) to

$$\mathbf{A}^T\mathbf{f}_V = \underbrace{-\mathbf{w}_P - \mathbf{A}^T\mathbf{f}_M}_{\mathbf{b}} \quad . \tag{3.54}$$

It follows from Verhoeven's theorem [473] that a force distribution is continuous along a trajectory of the mobile platform if the distance to a reference vector is minimized. Here, we use the vector $\mathbf{f}_M$ in the center of the feasible forces $\mathcal{C}$ as reference and measure the distance using a p-norm with ($2 \leq p < \infty$). A similar definition of an optimal solution is [348]: In terms of reliability, a solution between the force limits f_{min} and f_{max} is desired to stay away from the critical force limits as far as possible, resulting in an average cable tension level. This assumption is a reasonable compromise between the minimum and maximum cable force level. The idea of the proposed method is to transform the problem using Eq. (3.53) such that one has to determine bounded, least-square solutions rather than searching for positive solutions, since the latter turned out to be a challenging problem.

Here, the Euclidian norm ($p = 2$) is used to determine the least-square solution of the equation, which fulfills Eq. (3.5) and has a minimum 2-norm (Euclidian norm) with respect to $\mathbf{f}_M$. This can be done by means of the Moore-Penrose generalized matrix inverse which is defined for matrices with more columns than rows as $\mathbf{A}^{+T} = \mathbf{A}(\mathbf{A}^T\mathbf{A})^{-1}$. Multiplying $\mathbf{A}^{+T}$ from the left hand side of Eq. (3.54) gives

$$\mathbf{f}_V = -\mathbf{A}^{+T}(\mathbf{w}_P + \mathbf{A}^T\mathbf{f}_M) \tag{3.55}$$

and substituting this into Eq. (3.53) yields

$$\mathbf{f} = \mathbf{f}_M - \mathbf{A}^{+T}(\mathbf{w}_P + \mathbf{A}^T\mathbf{f}_M) \quad . \tag{3.56}$$

Finally, the resulting force distribution $\mathbf{f}$ must be checked to be feasible, i.e. tested if it is consistent with the force limits. One can improve both the computation time and the numerical stability by solving a linear system instead of computing the pseudo-inverse matrix $\mathbf{A}^{+T}$ explicitly. To do so, one has to solve the linear system

$$\underbrace{\mathbf{A}^T\mathbf{A}}_{\mathbf{A}_S}\mathbf{f}_V = \underbrace{-\mathbf{w}_P - \mathbf{A}^T\mathbf{f}_M}_{\mathbf{b}_S} \tag{3.57}$$

with the symmetric system matrix $\mathbf{A}_S$, the given right hand side $\mathbf{b}_S$, and the sought force vector $\mathbf{f}_V$. This linear square system can be solved with known algorithms such as Cholesky decomposition which is very fast by exploiting the symmetry of $\mathbf{A}_S$ but insensitive to singular cases. The general purpose LU-decomposition (Gaussian elimination) is sensitive to singular matrices but slower that Cholesky decomposition. The Householder QR decomposition is sensitive to singular matrices and performs orthogonalization where finally the Jacobi singular value decomposition (SVD) is relatively slow but sensitive to singular cases and provides the highest numerical stability especially for poorly conditioned cases. The linear problem is relatively small and in general it holds true that dim $\mathbf{A}_S \leq n \leq 6$. Therefore, the numerical performance largely depends on how much the used numerical solver is optimized to deal with small problems where the scaling with the size of the linear system is second for performance. A comparison amongst the methods is shown in Fig. 3.22 indicating

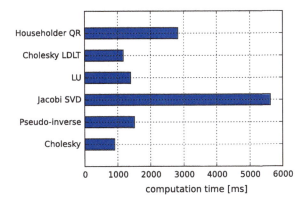

Fig. 3.22 Comparison of different numerical schemes to compute the force distribution. The benchmark was computed for the IPAnema 1 geometry at 229376 poses

a factor of six between the fastest method Cholesky and the slowest method Jacobi SVD. If not stated otherwise, the computations in the result section were executed using the Cholesky decomposition which showed sufficient numerical stability.

The closed-form method to compute the force distribution has the following properties:

- It satisfies exactly the structure equation (3.5).
- It satisfies Verhoeven's theorem, i.e. if $\mathbf{A}^{\mathrm{T}}$ and $\mathbf{w}_{\mathrm{P}}$ are continuous along a trajectory, then the computed forces $\mathbf{f}$ are also continuous.
- The force distribution $\mathbf{f}$ can be computed explicitly, where for the numerical computation only the operations matrix inverse, matrix transpose, and matrix multiplication are needed. Thus, computation time is strictly bounded allowing for use in real-time systems.
- Feasibility ($f_i \in \mathcal{C}$) can be checked straightforward simply by verifying that for each component f_i it holds true that $f_{\min} \leq f_i \leq f_{\max}$.
- The algorithm fails if $\mathbf{A}^{\mathrm{T}}\mathbf{A}$ is singular. This can be easily detected while inverting this matrix and such poses do not belong to the workspace due to rank deficit of the structure matrix $\mathbf{A}^{\mathrm{T}}$.

3.7.5.1 Existence and Uniqueness of the Force Distribution

Note, that this algorithm may also fail to find a feasible force distribution between the force limits although such a distribution exists. Three basic cases can be distinguished (Fig. 3.23) based on

$$\mathbf{f}_{\mathrm{V}} = \mathbf{f} - \mathbf{f}_{\mathrm{M}} = -\mathbf{A}^{+\mathrm{T}}(\mathbf{w}_{\mathrm{P}} + \mathbf{A}^{\mathrm{T}}\mathbf{f}_{\mathrm{M}}) \quad . \tag{3.58}$$

Using the radii r_i and r_o of the inner and outer sphere, respectively, the following cases must be distinguished:

Fig. 3.23 Distinction of cases for the closed-form determination of feasible force distributions

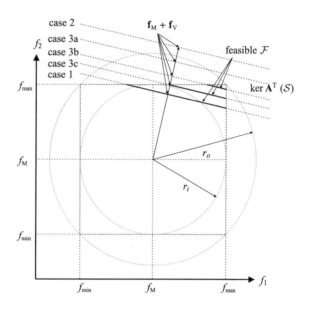

1. If $||\mathbf{f}_v||_2 < r_i = \frac{1}{2}(f_{\max} - f_{\min})$, then a feasible distribution exists and is bounded by the minimum and maximum forces in each cable. Furthermore, under this condition, the force distribution is additionally differentiable along a differentiable trajectory and applied wrench $\mathbf{w}_P$.
2. If $||\mathbf{f}_v||_2 > r_o = \frac{1}{2}\sqrt{m}(f_{\max} - f_{\min})$, then no feasible force distribution exists and the pose does not belong to the wrench-feasible workspace for the given applied wrench $\mathbf{w}_P$.
3. Elsewise the algorithm might or might not fail to find a feasible solution although it exists. In case 3a, no solution exists, thus it cannot be found. In case 3c, a feasible solution outside of sphere r_i is found. Although there are feasible solutions in case 3b (see Fig. 3.23), the algorithm determines a solution outside the feasible region. Currently, we have no simple criterion to distinguish between 3a, 3b, and 3c but the algorithm is robust in that way that if it rather misses an existing solution than supplying an invalid solution.

To study the unique domains where the algorithm guarantees to determine the solution, it is interesting to compare the volume of the inner m-hypersphere, the outer m-hypersphere, and the hypercubes given by the force limits. The results of this comparison are quite counter-intuitive for higher numbers of cables m. The volume V_m^2 of an m-dimensional sphere with radius r is given in Eq. (3.39). Figure 3.24 shows the relative volume of the enclosing sphere (case 2) and inner sphere (case 1) in logarithmic scale. One can see from the diagram that the distinction of cases between 3a, 3b, and 3c becomes an important issue when the number of cables increases. For eight cables, the inner sphere that represents the region of guaranteed convergency is only around 2.5% of the volume of the cube $\mathcal{C}$. It seems that the algorithm works out very well if the robot is in a fully-constrained position while

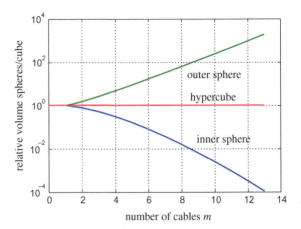

Fig. 3.24 Comparison of the volume of inner and outer m-spheres with the volume of the m-hypercube. The relative volume is presented on a logarithmic scale

the case 3b occurs more often for suspended configurations. This may be explained when considering that wrench-feasible poses of a fully-constrained robot are located in a similar region as the force-closure poses. For force-closure poses, a subset of the kernel of the structure matrix lie in the positive half-space of $\mathbb{R}^m$.

As discussed in [473], this problem can be overcome when using higher p-norms. The boundary of the hypercube is more accurately approximated for larger p leading to a larger region of convergency. For $p \rightarrow \infty$, convergency is achieved for the full hypercube. Contrary only for $p = 2$, a closed-form solution is known. When setting $p > 2$, one has to rely on an iterative algorithm and also numerical stability becomes an issue.

3.7.5.2 Implementation Issues and Computational Complexity

Using the cable force $\mathbf{f}$ determined from Eq. (3.56), one can answer the important questions of existence and value of the optimal solution by simply checking if it is feasible, i.e. $\mathbf{f} \in [f_{\min}; f_{\max}]$. Thus, Eq. (3.56) renders a closed-form solution to the problem which only involves matrix multiplications, matrix transpose, and one matrix inversion. Thus, the computation time is well-defined and strictly bounded. The latter is very important for applications in real-time control. Especially for a high number of cables, the computation is relatively cheap.

For implementation in an algorithm, one has to carefully check, if the inverse $(\mathbf{A}^T\mathbf{A})^{-1}$ exists. Physically speaking, this inverse exists if the robot is in a nonsingular configuration. Mathematically speaking, this can be checked if $\mathbf{A}^T$ has full row-rank.

A major advantage of the presented algorithm is its efficiency for higher degrees of redundancy $r = m - n$. Since Eq. (3.56) presents the solution for any $r > 0$, one can investigate the computational efforts for different r. The number of operations to evaluate Eq. (3.56) is driven by the matrix multiplications and the matrix inversion. Thus, the overall effort is $\mathcal{O}(n^2 m) + \mathcal{O}(n^3)$ (only $\mathcal{O}(n^2 m)$ if linear solving is used instead of the pseudo-inverse) and since in general the degree-of-freedom for robots is bounded by $n \leq 6$, the complexity for the algorithm is only $\mathcal{O}(m)$. This reveals that the runtime for force calculation is acceptable also for highly redundant robots.

For efficient numerical computation, one can avoid the computation of the pseudo-inverse matrix $\mathbf{A}^{+T}$ as outlined in Eq. (3.57) by rewriting the closed-form equation to a linear system. This system is in turn solved by Cholesky decomposition or Gauss elimination which is of order $\mathcal{O}(n^2 m)$ and also for small m faster than inverting the matrix.

3.7.6 Improved Closed-Form Solution

The method presented in the previous section can be improved by a recursive procedure in order to overcome its limitations in terms of coverage of the wrench-feasible workspace [392]. We present this algorithm in a new section since the closed-form method above is also of use when applying it as initial guess for iterative schemes such as the Dykstra method (see Sect. 3.7.4) or the puncture method (Sect. 3.7.9).

Based on our results in the previous section, we developed a formula to compute a solution for the force distribution problem in closed-form [396]. The basic idea of this method is maintained and we perform a coordinate transformation to the medium feasible cable force $\mathbf{f}_M = \frac{1}{2}(\mathbf{f}_{min} + \mathbf{f}_{max})$. This also changes parts of the optimization problem from constrained optimization to pure minimization. The cable forces $\mathbf{f}$ can be computed as

$$\mathbf{f} = \mathbf{f}_M + \mathbf{f}_v = \mathbf{f}_M - \mathbf{A}^{+T}(\mathbf{w}_P + \mathbf{A}^T \mathbf{f}_M) \quad , \tag{3.59}$$

As discussed above, this formula might fail to provide a feasible solution although such a solution exists, if the magnitude of the variable part $\mathbf{f}_v$ of the force distribution is in the range

$$\frac{1}{2}(f_{max} - f_{min}) \leq ||\mathbf{f}_v||_2 \leq \frac{1}{2}\sqrt{m}(f_{max} - f_{min}) \quad . \tag{3.60}$$

If $||\mathbf{f}_v||_2$ violates the upper limit, no solution exists and if it is below the lower limit the distribution is feasible. This undefined case occurs amongst others close to the boundary of the wrench-feasible workspace, for robots with many redundant cables, and for redundant robots in suspended configuration.

In the following, we present an extension of the method such that feasible force distributions are found in almost all cases where the original method fails. The closed-form solution is guaranteed to fulfill the force equilibrium but may violate the force limits. The following approach is proposed:

1. Equation (3.59) is used to compute a force distribution. If this initial guess already fulfills the cable force conditions, we have the sought solution and stop the algorithm.

2. Otherwise, let i be the cable with the largest force over (under) the maximum (minimum) feasible cable force. If one moves from this distribution along the spanning base of the structure matrix kernel, one must cross the value where f_i reaches its maximum (minimum) feasible value.

3. Therefore, it is assumed[5] that a feasible force distribution minimizing the 2-norm can only be found, if this cable force is fixed to its maximum (minimum) value f_{max} (f_{min}). Using a constant value for cable force f_i simplifies the force distribution problem as follows

$$\mathbf{A}'^{T}\mathbf{f}' + \mathbf{w}'_{P} = \mathbf{0} \quad \text{with} \quad \mathbf{w}'_{P} = f_{max}\left[\mathbf{A}^{T}\right]_{i} + \mathbf{w}_{P}, \qquad \left(\mathbf{w}'_{P} = f_{min}\left[\mathbf{A}^{T}\right]_{i} + \mathbf{w}_{P}\right),$$
$$(3.61)$$

where $\mathbf{A}'^{T}$ is the structure matrix and $\mathbf{f}'$ the cable forces vector with the i-th column/element dropped, respectively. $[\mathbf{A}^{T}]_{i}$ denotes the i-th column of the matrix $\mathbf{A}^{T}$. Thus, we have reduced the actuator redundancy r by 1.

4. Now, we compute the solution by recursively reducing the order and computing the closed-form solution by going to step 1 until:

   a. We find a feasible distribution,
   b. The degree-of-redundancy becomes $r = 0$, or
   c. Equation (3.60) proofs that no feasible solution exists because the computed force violate Eq. (3.60).

Therefore, we find the desired cable force distribution (if it exists) with at most r evaluations of the closed-form formula. This is to the best of the author's knowledge the fastest way to compute force distributions for robots with high redundancy, e.g. robots with $m = 12$ cables.

3.7.7 Barycentric Force Distribution Method

The barycentric approach was developed by Mikelsons et al. [348]. Mikelsons showed for the barycentric approach that the convex solution set $\mathcal{F}$ is continuous along a trajectory. Therefore, its barycenter $\mathbf{f}_{B}$ must also be continuous along a trajectory and it must be an element of that set because the set $\mathcal{F}$ is convex. Furthermore, $\mathcal{F}$ is a polyhedron that can be decomposed by triangulation for any degree-of-redundancy $r > 0$ through simplices. An algorithm for the case $r = 2$ was presented and its real-time capability was demonstrated [348]. Later, Lamaury [275, 277] proposed an improved algorithm that speeds up the computation by reducing the number of considered vertices. Although the approach can be applied for arbitrary $r > 0$, only implementations for $r = 2$ were reported in the literature.[6] The challenge for

---

[5]Unfortunately, we have no formal proof that this holds true in general.

[6]Some algorithms for CRPM ($r = 1$) can be interpreted as special versions of barycentric method but the triangulation degenerated to a trivial case for the one-dimensional kernel because only one line segment needs to be considered and the sought solution is just the middle between the two ends.

higher redundancy $r > 2$ is the computation of the triangulation that becomes more involved when simplices of general dimension must be considered.

3.7.7.1 Vertex Computation

There are two ways to compute the vertices of the solution set $\mathcal{F}$. Firstly, one can intersect the planes that generate the half-spaces of $\mathcal{C}$ with the kernel of the matrix, e.g. fix r components in the structure equation to compute the solution of the thereby fully defined linear system. Only if the remaining n computed forces are feasible, it is a vertex of $\mathcal{F}$.

Secondly, one can project the $2m$ hyperplanes of $\mathcal{C}$ into the r-dimensional kernel of the structure matrix where the m-dimensional hyperplanes degenerate to r-dimensional hyperplanes.[7] Then, one computes all intersections of every permutation of r lower dimensional hyperplanes.

Once the vertices of $\mathcal{F}$ are determined, one has to triangulate the set $\mathcal{F}$. For $r = 2$, one can compute the mean value of all vertices which must lie inside the convex polygon $\mathcal{F}$ and connect two neighboring vertices with this central point. Firstly, the area and barycenter is computed for each of these triangles and secondly the common barycenter of all the triangles is computed.

3.7.7.2 Computational Complexity

A time consuming step in the computation of the triangulation is the determination of the vertices of $\mathcal{F}$. The vertices can be determined choosing r cable forces to be minimal or maximal and solve the resulting linear $n \times n$ system. There are m over r possible permutations to choose the constant cable forces and for each of these choices there are 2^r combinations to assign the minimum and maximum cable force to the cables. Thus, the number of linear systems to solve is

$$n_{\text{Systems}} = \binom{m}{r} 2^r = \frac{m!}{r!\,(m-r)!} 2^r = \frac{m!\,2^r}{r!\,n!} = \frac{m!\,2^{m-n}}{(m-n)!\,n!} \quad . \tag{3.62}$$

The computational costs to solve a linear system with n equations is $\mathcal{O}(n^2)$ and can be considered constant since n is bounded by six in general.

We have to firstly project $2m$ planes into the kernel space where we receive again $2m$ hyperplanes. Secondly, we have to compute all intersections of r from $2m$ hyperplanes leading to

$$n_{\text{kernel}} = \binom{2m}{r} = \frac{2m!}{r!\,(2m-r)!} \tag{3.63}$$

---

[7]The projected hyperplanes have a dimension less than r in special degenerated cases.

Table 3.3 Comparison of the number of linear systems to check when computing all vertices of the solution space $\mathcal{F}$ for robots with $n = 6$ degrees-of-freedom

m	r	m-dimensional force space	r-dimensional kernel space
7	1	14	14
8	2	112	120
9	3	672	816
10	4	3360	4845
11	5	14784	26334
12	6	59136	134596

linear systems to be solved, where the dimension of the linear systems to solve is $r \times r$. Some reductions can be achieved when we can skip the intersection tests for some combinations that involve the intersections between the pairwise parallel planes.

In both cases, the computation time can be strictly bounded. Anyway, the actual number of systems to check becomes quite large as can be seen from Table 3.3. When computing the intersections in kernel space, one has to check moderately more systems.

3.7.8 Weighted Sums of Vertices

In order to receive a low tension level for the force distribution, Bruckmann [68] proposed a weighted sum of the vertex points of the polytope $\mathcal{F}$. The force distributions received this way are almost continuous, because the volume of $\mathcal{F}$ is continuous and the distance between the vertices is introduced as additional weight that allows for some smoothing whenever the number of vertices of $\mathcal{F}$ changes along a trajectory. The algorithm is a four step procedure:

1. Compute all vertex points $\lambda_{v,i}$ of the set $\mathcal{F}$ (see also Sect. 3.7.7 above) and let k be the number of vertices $\lambda_{v,i}$.
2. Compute the weights

$$a_i = \frac{\sum_{j=1}^{r} ||\lambda_{v,i} - \lambda_{v,j}||_p}{||\lambda_{v,i}||_p} \quad \text{for} \quad 1 \le i \le k \ . \tag{3.64}$$

3. Compute the weighted sum

$$\lambda_a = \frac{\sum_{i=1}^{k} \lambda_{v,i} a_i}{\sum_{i=1}^{k} a_i} \ . \tag{3.65}$$

4. Compute the force distribution $\mathbf{f} = \mathbf{H}\,\lambda$ where the matrix $\mathbf{H}$ is the spanning basis of the kernel and the vector $\lambda = [\lambda_1, \ldots, \lambda_r]^\mathrm{T}$ collects the weights computed above.

Based on its design principle, the weighted sum method guarantees to find a solution whenever such a solution exists since it relies on the vertices of the set $\mathcal{F}$. However, as explained above, the computation of all vertices is time consuming and the weighted sum method suffers from this problem as well, especially if the degree-of-redundancy is higher.

3.7.9   Puncture Method

The puncture method aims at computing a continuous series of force distributions under real-time constraints where the forces in the cables are minimized [355, 356]. The basic idea of the puncture method is to take an initial guess $\mathbf{f}_1$ inside the solution set $\mathcal{F}$ which is computed by the closed-form solution (or possibly another straightforward method). This solution is unique and continuous along trajectories. Secondly, a solution $\mathbf{f}_\mathrm{o}$ of the structure equation in the kernel of the structure matrix $\mathbf{A}^\mathrm{T}$ is computed which is close to the origin. Also, this estimation shall be unique and continuous. Since the kernel $\mathcal{S}$ is a linear space, each point on the connecting line between the $\mathbf{f}_1$ and $\mathbf{f}_\mathrm{o}$ are solutions of the structure matrix. The idea of the puncture method is to approach the feasible force distributions $\mathcal{C}$ from a point close to the origin and thus to identify points on $\mathcal{S}$ with small forces. Using a connecting line between a feasible solution in $\mathcal{F}$ and a reference point close to the origin, the minimum force is computed.

Using the improved closed-form method as initial guess for the puncture procedure above, it was shown that the region of convergency of the puncture method can be increased [355] where the computational time is slightly increased due to the dimension reduction technique. The computational time of the puncture method mainly depends on the time for computing the (improved) closed-form solution; therefore, its use in real-time applications is mostly the same as for the closed-form method.

3.7.10   Comparison of the Methods

In the previous sections, a couple of methods were discussed to compute force distribution with different approaches and numerical techniques. In the following, we compare these approaches and elaborate the differences and properties. A comparison of some force distribution methods and their properties are given in Table 3.4. We briefly discuss the properties listed in the table. An algorithm is said to be *real-time capable* if the computation time is reasonably short (in the scale of milliseconds on

Table 3.4 Comparison of the different methods to compute force distributions

Method	Real-time capable	Force level	Workspace coverage	Continuity	Max. redundancy	Computation time
Linear programming	No	Any	Yes	No	Any	Fast
Quadratic programming	Yes	hi, lo	N/A	Yes	Any	Medium
Nonlinear programming	Part.	hi, lo	N/A	Yes	Any	Medium
Gradient method	No	Param	No	Yes	Any	Medium
Projected gradient	N/A	N/A	N/A	Yes	N/A	N/A
Dykstra method	No	Any	Yes	No	Any	Slow
Closed-form	Yes	Any	No	Yes	Any	Fast
Improved closed-form	Yes	Any	Yes	Yes	Any	Fast
Barycentric	Yes	mi	Yes	Yes	$r = 2$	Fast
Weighted sum	Yes	mi	Yes	Mostly	Any	Medium
Kernel translation	Yes	hi, mi, lo	Yes	Yes	$r = 1$	Fast
Available wrench set	No	hi, mi, lo	Yes	No	Any	Slow
Puncture method	Yes	lo	No	Yes	Any	Fast

an industrial PC). The worst case computation time can be strictly bounded and a real-time implementation was reported in the literature. Some iterative methods were successfully used for computation in real-time although their worst case computational time was not determined. The *force level* may be chosen, e.g. the algorithm may aim at finding minimal (lo), maximal (hi), average (mi), or any solution (any). Furthermore, there might be a parameter (param) that allows to smoothly adjust the force level of tension between low and high. The property *workspace coverage* indicates that the respective algorithms determine a force distribution if it exists. A couple of authors [396, 473] reported approaches that may fail to find force distributions for special poses of the wrench-feasible workspace. An algorithm is said to provide *continuity*, if continuous trajectories in the pose $(\mathbf{r}, \mathbf{R})$ as well as in the applied wrench $\mathbf{w}_P$ produce continuous trajectories in the cable forces $\mathbf{f}$, except for crossing a singularity. Some methods are limited to a certain degree-of-redundancy r either because they are specific or because their implementation can hardly be generalized to arbitrary r. The evaluation of the *computation time* is problematic because it requires comparable implementations which are not available for all methods reported in the literature. Anyway, it was tried to set up a basic ranking taking into account how complex the underlying numerical methods are. For example, linear system solving is considered to be faster than inverting a matrix, which in turn is faster than computing a singular value decomposition or solving an optimization problem. Designing a real-time system might become involved if an advanced numerical algorithm such as advanced optimization or singular value decomposition shall be used. This is due to lack of appropriate real-time capable implementations of the algorithm although the algorithm is part of every state of the art numerical toolbox. The computational speed depends on the degree-of-redundancy in addition to the algorithm's complexity. For this assessment, a low degree-of-redundancy was assumed.

3.7.11 Simulation Results

In the following, some of the presented algorithms are evaluated and numerical results are compared to other known approaches. As a reference, this provides some guidelines for selecting a force calculation method for a specific purpose.

3.7.11.1 Barycentric, Optimizer and Closed-Form Results on Segesta

Here, the barycentric approach [348] as well as a nonlinear general purpose optimization algorithm are compared to the closed-form solution. It is reasonable to evaluate the methods by applying them to the real-time control system of the prototype Segesta.

Some of the methods listed above are profiled with respect to their application on a real-time system. The prototype Segesta was developed at the Chair for Mechatronics, University of Duisburg-Essen, during the last decade [139, 210]. For this case study, it was equipped with $m = 8$ cables and the winches were directly driven by electronic commutated motors. A real-time control system DS1005 from dSPACE (PowerPC 705, 480 MHz) is used which was programmed in the MATLAB/Simulink language. The geometrical properties of the Segesta prototype (Fig. 9.16) are given in Table 9.12. The feasible forces were set to $f_{min} = 10$ N and $f_{max} = 1000$ N, respectively.

To compare the three methods regarding the performance of the force calculation a screw-shaped trajectory pictured in Fig. 3.25 was used. As shown in Fig. 3.26, the optimizer approach delivers continuous solutions along the whole trajectory. Furthermore, these force distributions lead to a relatively low tension level in the cables, i.e. cable forces are optimal in the sense of minimality. The barycentric approach also (Fig. 3.26) delivers feasible continuous force distributions along the whole trajectory,

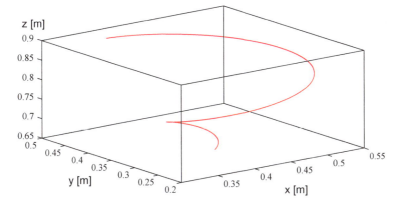

Fig. 3.25 Screw-shaped trajectory in the workspace of Segesta

but the computed force distributions lead to a much higher tension level in the cables, i.e. the force distributions are optimal in the sense of maximizing the distance to the force limits. The closed-form approach delivers force distributions (Fig. 3.26) that look very similar to the tensions calculated by the barycentric approach, but it fails where the path lies very close to the workspace boundary. This is an example for the case 3b (see Sect. 3.7.5), i.e. a feasible solution exists but is not found since it lies outside the inner sphere. The calculation times measured on the real-time system are given in Fig. 3.27, where for the closed-form method the slower implementation with matrix inverse is used instead of the slightly faster implementation based on Cholesky decomposition. Therefore, the closed-form force calculation is suitable for real-time control as long as the end-effector remains inside the (restricted) workspace, which is covered by the closed-form approach. In this case, the closed-form approach is superior to the others in terms of calculation time. This holds especially for cases with higher actuator redundancies $r > 2$, where no practical application of the barycentric approach was reported and lower performance is expected due to difficult triangulation in high dimensions. The usage of the optimizer approach is dangerous in real-time control for all redundancies due to the in general non predictable worst case runtime.

3.7.11.2   Comparison for IPAnema 1 Geometry

In the following, we present a case study for another robot geometry. The behavior of different algorithms for force distribution is compared using a sample trajectory that is depicted in Fig. 3.28. The way-points of the trajectory are indicated by numbers 0 to 8 and the following plots with cable forces against time t have additional marks above the x-axis indicating the way-points for better reference. The positions were linearly interpolated between the way-points. The trajectory was chosen so that the robot moves in different regions of the workspace and finally crosses the boundary

Fig. 3.26 Comparison of the
force distributions along the
shown screw-shaped
trajectory using three
different force calculation
algorithms

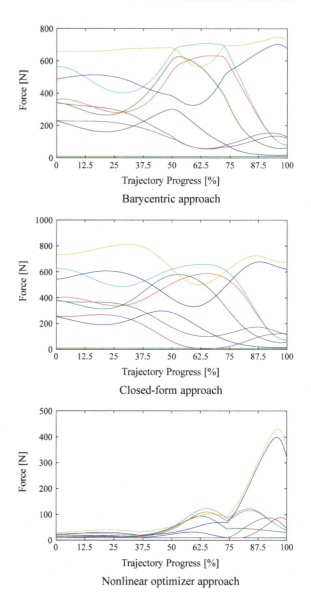

of the wrench-feasible workspace between way-point 7 and 8. The force limits were
set to reference value $f_{min} = 1$ and $f_{max} = 10$ N for the sake of simplicity. Inertia
effects of the platform were neglected.

Figure 3.29 illustrates the proposed improved algorithms based on closed-form
estimation and correction for the remaining cables. From the diagram, one can see
that the force distribution is continuous along the trajectory and the magnitude of the
forces are on a medium level. When approaching the boundary of the workspace (e.g.

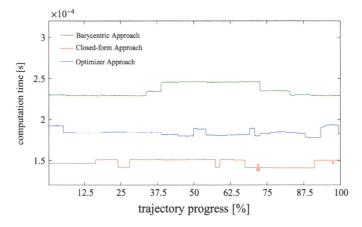

Fig. 3.27 Computation time for different force distribution methods using a dSPACE system for real-time control

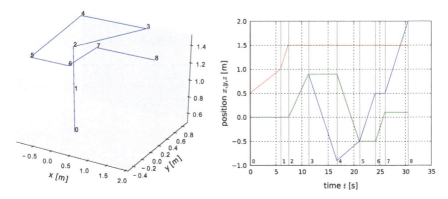

Fig. 3.28 Test trajectory used for the evaluation. Left: spatial curve of the trajectory. Right: $[x, y, z]$ coordinates of the motion over time t

between $t = 10.0$ s and $t = 17.0$ s), one or two cable forces reach the minimum cable force and remain constantly at the limit. It can be seen from the shape of the diagram that the cable forces quickly increase after leaving the workspace. Anyway, the force distributions remain continuous after crossing the boundary of the workspace.

In Fig. 3.30, the force distribution is shown for the original closed-form method for comparison. When the platform remains in the inner region of the workspace, the results match exactly with the force distributions computed with the improved closed-form method. Close to the boundary of the workspace between $t = 10$ s and $t = 17$ s, the closed-form formula fails to compute force distributions although such distributions exist as it can be seen between way-point 2 and 5 and also between 7 and 8, where the closed-form solution is not able to prevent some cables from violating the lower force limits. One can conclude that for such poses fixing one or two cable

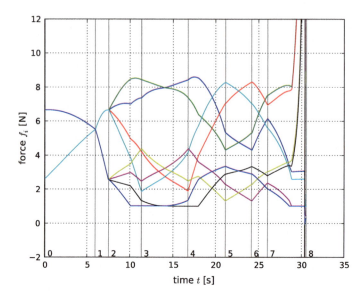

Fig. 3.29 Force distributions computed with the improved closed-form solution

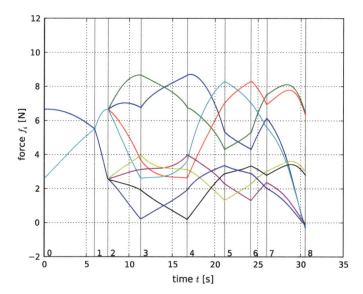

Fig. 3.30 Force distributions computed with the closed-form method

forces to the lower limit f_{min} is effective to compute feasible force distributions where the conventional closed-form method fails.

Cable forces computed with the Dykstra method are presented in Fig. 3.31. For the Dykstra method, it can be observed that some cable forces get limited to the minimal values when the boundary of the workspace is approached. After crossing

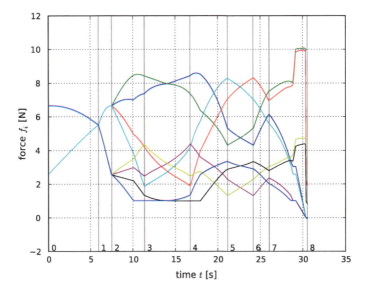

Fig. 3.31 Force distributions computed with Dykstra method

the workspace border between way-point 7 and 8, force distributions computed with
Dykstra show a different behavior compared to the proposed scheme.

In Fig. 3.32, the computed forces for the first cable f_1 are compared for different methods (closed-form, improved closed-form, Dykstra, and the weighted sums
method). Some methods like the uncorrected weighted sum method do not even provide continuous shapes for the forces which becomes evident between way-point 0
and 2. Other methods show discrete steps at certain points on the trajectory.

It is well-known that the solution set of the force distribution problem is a convex
polytope. The number of the vertices of the convex hull of the solution is determined
along the trajectory. It turns out that typically the number of vertices varies between
5 and 13 but for special poses on the trajectory the number of vertices may reach
high values above 30 (Fig. 3.33). Although the structure of the solution set undergoes
massive changes, no effects on the continuity of the presented method were found.

3.7.11.3 Computation of the Number of Vertices of $\mathcal{F}$

The number of vertices of the solutions set $\mathcal{F}$ is further studied in the following. The
case study is exemplified with the IPAnema 1 geometry, the force limits were assumed
to be $f \in [1; 10]\,\mathrm{N}$, and some xy cross sections of the workspace are considered
for $z = 1\,\mathrm{m}$. Different wrenches $\mathbf{w}_\mathrm{P}$ are applied to the platform, where the forces
are between zero and half of the maximum cable force $f_{\max}$, i.e. $||\mathbf{w}_\mathrm{P}||_\infty \leq \frac{1}{2} f_{\max}$.
Results are shown in Fig. 3.34. The outer blue contour indicates that no vertex was
found at all which is equivalent to the boundary of the wrench-feasible workspace.

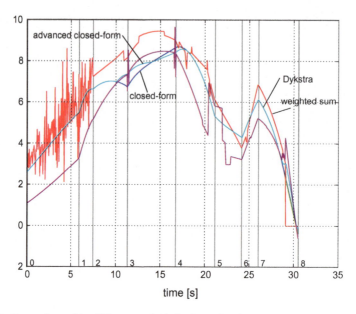

Fig. 3.32 Comparison of the different methods for force distribution

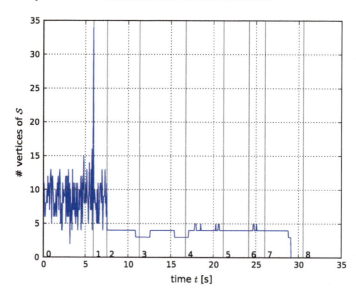

Fig. 3.33 Number of vertices of the solution set $\mathcal{F}$ computed by a brute force approach, i.e. solving for the homogeneous linear system where all permutations of two forces have been set to f_{min} and f_{max}, respectively

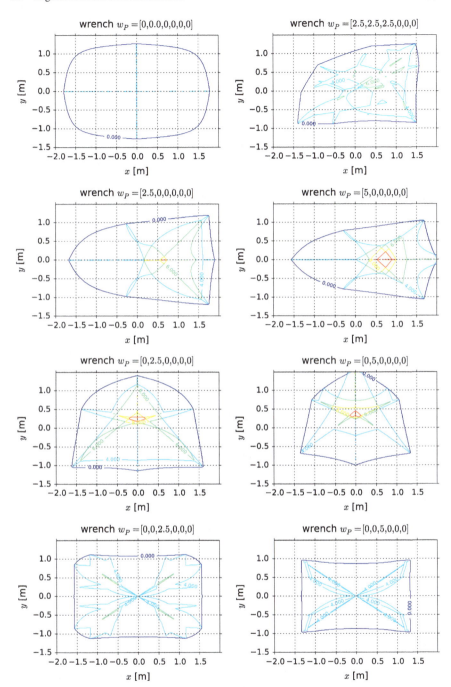

Fig. 3.34 The available cable force distributions $\mathcal{F}$ are computed from their vertices for different applied wrenches $\mathbf{w}_P$. The plots show the number of vertices of $\mathcal{F}$ per pose for the IPAnema 1 robot design in xy cross sections for $z = 1\,\text{m}$

Table 3.5 Comparison of computation time on an Intel Core i5-3320M 2.6 GHz, Visual C++ 2010 for 344 000 evaluations of the force distribution

Algorithm	Calculation time [ms]	Relative time (%)	Evaluations per ms
Closed-form	1173	100	293
Improved closed-form	3359	286	102
Dykstra	71612	6103	5
Weighted sum	48512	4134	7

The regions where multiple vertices can be found inside the workspace have a quite complex structure. The boundary of the workspace is mostly a smooth curve where the inner regions frequently have cusp points.

3.7.12 Computation Time

A comparison of the computation time is difficult because the computation time is influenced by the maturity of the implementation as well as by the underlying numerical algorithms, the used compiler, the CPU of the real-time system, and the operating system. In the performance tests presented here, four algorithms were used for workspace computation with around 344 000 evaluations on an Intel Core i5-3320M. As reference, some numbers are given in Table 3.5. The table lists both absolute and relative computation time to allow for comparison amongst the algorithms as well as to present an estimate on the usability in a real-time controller. As expected, the closed-form solution works faster[8] than its improved version but the difference is comparably small. Both methods allow for many evaluations based on a controller cycle time of 1 ms and their implementations only require matrix multiplication and solving of a linear system.

3.8 Stiffness

After the intensive study of methods for computing feasible distributions in the cable force, the following section is dedicated to investigate the effect on forces in cables on the robot. Clearly, forces lead to elastic reaction in the cables and thus the platform will undergo some motion if loaded with a force. This effect is pose-dependent since the geometrical structure of the robots largely effects how much displacement is contributed by each cable. The stiffness of a cable robot represents its ability

---

[8]In contrast to the numerical study using the dSpace system, the implementation used in this test makes use of the linear solving with matrix decomposition rather than the slower computation of the pseudo-inverse matrix $\mathbf{A}^{+T}$.

to withstand forces and torques in any direction. Indeed, when cable robots were developed as improved cranes, their unique property over cranes is their nonvanishing stiffness [106]. Verhoeven describes a simple linear elastic stiffness model where the cables are modeled to be linear springs [473]. Therefore, stiffness of cable robots is closely connected to the concept of wrench-closure or wrench-feasibility [33] and there is a connection between stiffness and stability. The cable tension contributes asymmetrically to the stiffness matrix. Thus, the stability of a pose depends also on the cable tension distribution. Behzadipour names conditions to analyze stability of the cable robot. Williams [494] tackles the problem of computing the stiffness matrix for a cable robot with twelve cables. Hassan [202] analyzes the stiffness of a flat but large cable robot for storage retrieval and proposed to compute the stiffness matrix from the cables and from the drive-trains. Furthermore, the lowest eigenfrequency was expected be around 0.3 Hz. Yu addressed the problem of actively adjusting the stiffness of a cable robot through a feedback control [512]. Nguyen [361, 362] discusses the homogenization of the stiffness matrix and the effect of sagging cables and its interconnection with the stiffness of the robots is derived in these contributions. Schmidt [432] shows experimentally that for a planar robot the stiffness only depends on the geometric stiffness. The concept of *geometric stiffness* [33] is lately investigated by Surdilovic [450] and Kraus [259]. An interesting finding of the latter work was that force control reduces the stiffness of an over-constrained cable robot.

The stiffness of a cable robot is characterized by the infinitesimal displacements $\delta\mathbf{y}$ of the mobile platform that are generated by infinitesimal wrenches $\delta\mathbf{w}_P$ applied to it (Fig. 3.35). Here, we restrict ourselves to the consideration of linear dependencies leading to the mapping

$$\delta\mathbf{w}_P = \mathbf{K}_{os}(\mathbf{y})\delta\mathbf{y} \quad , \tag{3.66}$$

where $\mathbf{K}_{os}(\mathbf{y})$ is the pose-dependent stiffness matrix in the operational space. Some considerations on nonlinear cable behavior are discussed in the literature [259]. If the inverse mapping $\mathbf{C}(\mathbf{y})$ exists, it is called *compliance matrix* and it relates wrenches $\delta\mathbf{w}_P$ to displacements of the platform $\delta\mathbf{y}$.

Fig. 3.35 Model of the stiffness of a 3R3T robot

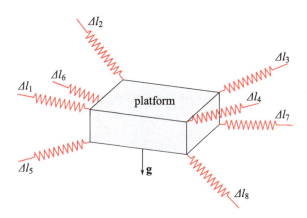

3.8.1 Cable Stiffness

The stiffness of cable robots is a primary result of elastic deformations of its components, especially of the cables, the winches, the platform, the actuators, and the machine frame. Firstly, the cables itself exhibit an elastic behavior. This holds true especially for large-scale cable robots since the compliance of the cables is roughly proportional to their length. If winches are used to move the robot, the stiffness depends on the free length of the cables and therefore the stiffness of each cable depends on the pose. For cable robots with linear actuators, the effective length of the cables is constant and so is the stiffness of each cable. Let k_i be the stiffness coefficient of the i-th cable, then the stiffness in configuration space is [473]

$$
\delta \mathbf{f} = \begin{bmatrix} k_1 & & 0 \\ & \ddots & \\ 0 & & k_m \end{bmatrix} \delta \mathbf{l} = \mathbf{K}_c \, \delta \mathbf{l} \quad , \tag{3.67}
$$

where $\mathbf{K}_c$ is the diagonal actuator stiffness matrix in the generalized coordinates in the direction of the cables. Even if the cables are considered to be linear springs, we have to take into account, that the effective cable length and thus also the spring constant of the cable may be variable since the coiled part of the cables does not contribute to the stiffness. The spring constant k_i for cable i becomes

$$
k_i = \frac{k'}{l_i + l_i^0} \tag{3.68}
$$

where k' is a material constant characterizing the stiffness of the cable material per length unit. Furthermore, l_i is the variable length as determined from the inverse kinematics and l_i^0 is a cable robot design specific constant length of each cable, e.g. inside the winches between the guiding pulleys, or for robots with linear drives the total length of the cables in the pulley tackle. In the latter case, there is no variable part l_i effective for the stiffness since the effective length for the stiffness does not changes if the cable is in a pulley tackle. In general, the stiffness constant depends on many technical details of the cable. As an estimation, the cable is assumed to be homogeneous cylindrical strut. Then, we use

$$
k' = E_c A_c \quad , \tag{3.69}
$$

where E_c is Young's modulus for the cable material and A_c is the cross section of the cable. To compute the cross section $A_c = \pi \nu_c r_c^2$, we need the radius r_c of the cables and a constant $\nu_c \in [0; 1]$ that is a cable specific constant characterizing the amount effective fibers in the cable cross section. In practice, the stiffness of a real cable is therefore lower than the full cylinder computed from the radius and we come back to this problem in Sect. 3.8.5.

Using the structure matrix $\mathbf{A}^{\mathrm{T}}$ and its transpose, Eq. (3.67) is transformed into operational space as follows

$$\delta\mathbf{f} = \mathbf{A}^{\mathrm{T}} \begin{bmatrix} k_1 & & 0 \\ & \ddots & \\ 0 & & k_m \end{bmatrix} \mathbf{A}\,\delta\mathbf{l} \tag{3.70}$$

$$\delta\mathbf{f} = \underbrace{\qquad\qquad\qquad}\ \ \mathbf{K}_{\mathrm{o}}\qquad\quad \delta\mathbf{l} \tag{3.71}$$

revealing the pose-dependent stiffness mapping with the stiffness matrix $\mathbf{K}_{\mathrm{o}}$ in Cartesian coordinates in operational space.

The drive-trains with their actuators and gears are a second source of compliance. They show an elastic behavior due to their mechanical design and also as a response of the control system to applied forces. The latter effect strongly depends on the control algorithms and there is a strong dynamic influence. The linear elastic deformation of the drive-train can be easily integrated into the stiffness mapping. Let k_{A} be the stiffness of the drive-train. Then, we can summarize the stiffness mapping for the i-th cable as follows

$$\frac{1}{k_i} = \frac{1}{k_{\mathrm{A}}} + \frac{l_i + l_i^0}{k'} \quad. \tag{3.72}$$

Using the modified coefficients, one computes the stiffness mapping from Eq. (3.66).

The frame of the robot may also be source of compliance in the system. For small robots, the frame can be designed to be very stiff and react with negligible displacements even for high static forces in the cables. Especially for larger robots without a closed framework structure for attaching the proximal anchor points, the stiffness and vibration in the frame became a severe source of disturbance since cable robots apply very high impetus on the frame when accelerating and decelerating. The stiffness of the frame can be computed with conventional finite element analysis (FEA) techniques which depend on the mechanical design of the frame rather than on something special for cable robots.

3.8.2 Geometric Stiffness

The following introduction to the geometric stiffness of cable robots is based on the work by Surdilovic [450] and Kraus [259]. It takes into account the influence of the geometric deviations in the displacement of the platform under external load. The linear consideration of stiffness effects disregards the influence of a changing structure matrix $\mathbf{A}^{\mathrm{T}}$ causing nonlinear effects even when considering a linear spring model for the cables. The second order terms are well-known from elasticity theory where these terms cause e.g. buckling effects for thin beams under axial stress. Taking into account the influence of a changing structure matrix, the stiffness for a given pose is characterized by

$$\delta\mathbf{w}_{\mathrm{P}} = -\mathbf{A}^{\mathrm{T}}\delta\mathbf{f} + -\frac{\partial\mathbf{A}^{\mathrm{T}}}{\partial\mathbf{y}}\,\mathbf{f}\delta\mathbf{y} \tag{3.73}$$

where the derivatives of the structure matrix $\mathbf{A}^{\mathrm{T}}$ have to be taken with respect to the world coordinate $\mathbf{y}$. As shown by Kraus [259], the equation for the stiffness matrix becomes

$$\mathbf{K}_{\mathrm{os}}\,\delta\mathbf{y} = \underbrace{-\frac{\partial\mathbf{A}^{\mathrm{T}}}{\partial\mathbf{y}}\mathbf{f}\,\delta\mathbf{y}}_{\mathbf{K}_{\mathrm{G}}} + \underbrace{\mathbf{A}^{\mathrm{T}}\mathbf{K}_{\mathrm{c}}\mathbf{A}\,\delta\mathbf{y}}_{\mathbf{K}_{\mathrm{O}}} \tag{3.74}$$

and the stiffness of the robot is depending on the actual tension $\mathbf{f}$ in the cables. The connection between the stiffness and thus the eigenfrequencies with the tension can be observed at every string instrument such as guitars where an increase in the tension leads to a higher tone. Thus, the stiffness becomes a function of the cable forces and consists of two parts. Firstly, the pose-dependent cable stiffness matrix $\mathbf{K}_{\mathrm{c}}$ is the structure matrix described above and represents the linear elongation of the cables. If the cables have a nonlinear characteristic, the cable stiffness can also depend on the tension [259]. Secondly, the geometric stiffness matrix $\mathbf{K}_{\mathrm{G}}(\mathbf{f})$ represents the second order terms and depends both on the pose and on the tension in the cables. The cable stiffness matrix is in general symmetric which becomes clear when considering the equation $\mathbf{K}_{\mathrm{O}} = \mathbf{A}^{\mathrm{T}}\mathbf{K}_{\mathrm{c}}\mathbf{A}$. The geometric stiffness matrix adds in general an asymmetric part to the overall stiffness $\mathbf{K}_{\mathrm{os}}$ of the robot. Thus, static stability of the robot depends in general on the tension in the cables.

3.8.3 Stability

The problem of stability is discussed by Behzadipour [33]. Within this contribution, a theorem is set up stating that a cable-driven parallel robot can be stabilized in the absence of external load if the structure matrix is regular and the active stiffness matrix is positive definite. Following the derivation presented in [33] with the symbols introduced in this book, one can analyze the stabilization if no external loads $\mathbf{w}_{\mathrm{P}}$ are applied from the matrix $\mathbf{Z}$

$$\mathbf{Z} = -\sum_{i=1}^{m} f_i(\mathbf{u}_i \cdot \mathbf{b}_i)\mathbf{I} + \sum_{i=1}^{m} f_i(\mathbf{b}_i \, \mathbf{u}_i^{\mathrm{T}}) \quad . \tag{3.75}$$

Then, one sets up the characteristic polynomial for the matrix $\mathbf{Z}$ as

$$\det(\mathbf{Z} - \lambda\mathbf{I}) = \lambda^3 - z_2\lambda^2 + z_1\lambda - z_0 = 0 \tag{3.76}$$

where the coefficients z_2, z_1, z_0 are given by

$$z_2 = \sum_{i=1}^{m} f_i(\mathbf{u}_i \ \mathbf{b}_i) \tag{3.77}$$

$$z_1 = \sum_{i=1}^{m} \sum_{j=i+1}^{m} f_j f_i(\mathbf{b}_i \times \mathbf{b}_j)(\mathbf{u}_i \times \mathbf{u}_j) \tag{3.78}$$

$$z_0 = \sum_{i=1}^{m} \sum_{j=i+1}^{m} \sum_{k=j+1}^{m} f_j f_i f_k \left(\mathbf{b}_i(\mathbf{b}_j \times \mathbf{b}_k)\right) \left(\mathbf{u}_i(\mathbf{u}_j \times \mathbf{u}_k)\right) \ . \tag{3.79}$$

With these expressions, one can check stability with the simple conditions [33]

$$z_2 > 0 \tag{3.80}$$
$$z_2^2 + z_1 > 0 \tag{3.81}$$
$$z_0 - z_2 z_1 > 0 \ . \tag{3.82}$$

In contrast, the matter of stability in the presence of a nonvanishing platform wrench $\mathbf{w}_\mathrm{P}$ seems to be an open issue.

3.8.4 Stiffness Evaluation

The stiffness of a cable robot is both a pose- and direction-dependent phenomena. Therefore, the computation of bounds for the stiffness is relevant. This is basically possible by considering the eigenvalues $\lambda_{\mathrm{K},i}$ of $\mathbf{K}_\mathrm{os}$ where the smallest (largest) eigenvalue corresponds to the smallest (largest) stiffness and its eigenvectors unveil the direction where the extremal stiffness appears. If the smallest eigenvalue vanishes, the robot is in a singular configuration and the robot becomes uncontrollable in the respective direction.

For robots with the motion patterns 1R2T, 2R3T, and 3R3T, the stiffness matrix consists of translation and rotation (force and torque). From the extensive study of dexterity, it is known that measures involving rotation and translation are difficult to compare [442]. To deal with this problem, it was proposed (see e.g. [322, 473]) to introduce a normalizing length. Also, Nguyen [361, 362] discusses the homogenization of the stiffness matrix. Accordingly, the singular values of the stiffness matrix can be used to characterize the stiffness of the robot and the following relation holds true [362]

$$\sigma_\mathrm{min} = \frac{1}{||\mathbf{K}^{-1}||_2} \leq \frac{||d\mathbf{w}_\mathrm{P}||_2}{||d\mathbf{y}||_2} \leq ||\mathbf{K}||_2 = \sigma_\mathrm{max} \ \forall \ d\mathbf{y} \neq \mathbf{0} \ , \tag{3.83}$$

where $\sigma_\mathrm{min}, \sigma_\mathrm{max}$ are the smallest and largest singular values of $\mathbf{K}$, respectively. In order to derive a dimensional homogenization $\mathbf{K}_\mathrm{H}$, it is proposed to transform the stiffness matrix by

$$\mathbf{K}_\mathrm{H} = \mathbf{S}_\mathrm{H}^{-1} \mathbf{K} \mathbf{S}_\mathrm{H}^{-1} \quad , \tag{3.84}$$

where the transformation matrix $\mathbf{S}_\mathrm{H} = \mathrm{diag}(1, 1, 1, ||\mathbf{r}_\mathrm{M}||_2, ||\mathbf{r}_\mathrm{M}||_2, ||\mathbf{r}_\mathrm{M}||_2)$ introduces the normalizing length $||\mathbf{r}_\mathrm{M}||_2$ which represents a reference point $\mathbf{r}_\mathrm{M}$ of the platform relative to the frame $\mathcal{K}_\mathrm{P}$. It seems reasonable to choose this point such that the stiffness is mapped to the hull of the platform, i.e. the length should be the maximum distance from $\mathcal{K}_\mathrm{P}$ to the relevant operation point of the platform.

3.8.5 Cable Parameters

In general, the elastic behavior of cables is a highly complex phenomena: The book from Feyrer treated the basic behavior of steel cables [149]. Even for quite simple operation conditions like cables in static construction, one has to deal with complex empirical data to predict the cable behavior. For dynamically stressed steel cables like they are used in cable robots, the situation is even much more complex. Some first results related to cable robots are presented by Weis et al. [489]. Many cables used for cable robots are made of some kind of synthetic fibers such as polyethylene and polyamide. Compared to steel cables, little is known about the behavior of these materials under practical conditions. But even for steel cables, the typical operation conditions of highly dynamic cable robots have been hardly investigated. In contrast to common applications of running cables, bending with high velocities in different and frequently changing directions are typical operation conditions for a cable robot. Research on steel cables is still a mostly empiric science and little attention has been paid to the conditions relevant for cable robots due to lack of applicability in other fields.

One could expect that mechanical performance figures such as Young's modulus and the breaking load can easily be specified but unfortunately this is not true. Young's modulus E_c of a steel cable depends amongst others on the number of strands in the cable, the cable's and the strand's geometry, and the pretension of the cable. The same holds true for durability, breaking load, and bending fatigue strength. In practice, one has to rely on more detailed computations, experimental results, as well as sufficiently large safety factors.

Therefore, Table 3.6 can only provide some basic data rather than dependable parameters. It must be emphasized that the presented data *must not* be used for robot for winch design without intensive experimental verification. Reviewing literature on the topic reveals a broad width for the parameters and there seems to be significant inconsistencies in definitions and parameter values. The values presented in the tables are taken from Feyrer [149], Hearle [204], Mammitzsch [311], and Michael [341]. The parameters are cross-checked with public sources in the internet (such as Wikipedia) as well as parameters given by companies producing or processing syn-

Table 3.6 Cable materials and related properties: Young's modulus E, Young's modulus for a cable E_C, tensile strength σ, density ϱ_C, and specific strength σ_S. Units: [MPa] = [Nmm^{-2}]. The overview is compiled from a number of sources: Mammitzsch [311], Hearle [204, p. 72, p. 96, p. 105], Michael [341, p. 32], Feyrer [149, p. 92–94], Wikipedia: http://en.wikipedia.org/wiki/specific_strength, kuraray: http://www.vectranfiber.com

Material	E [MPa]	E_C [MPa]	σ [MPa]	ϱ_C [g/cm^3]	σ_S [km]
Steel	210000	110000	500	7.8	25–51
Copper	100000–130000		220	8.92	
High-modulus polyethylene (Dyneema)^a	95000		3000	0.95–0.97	300–400
Aramid (Kevlar)	59000–127000		2800	1.45	235
Polyamide (PA, Perlon/Nylon)	2300		78	1.14	7.04
Polyester (PES)	1000–5000		50–100	1.38	85
Polypropylene (PP)	15000–18000		600-650	0.91	9.06
Polyethylene (PE)	1000		20–30	0.96	
LCP (Vectran)	65000		2900	1.41	79–215
PBO (Zylon)	270000		5800	1.52	384
Carbon	235000		3400	1.78	195
Silk	8000–15000	8000–12000	350	1.25–1.37	50
Hemp	69000		310–390	1.48	25–52

^{a}also known as HPPE: high-performance polyethylene, HMPE: high-modulus polyethylene, Dyneema SK65

thetic fibers (LIROS,[9] Kuraray,[10] Suter[11]). Furthermore, in Table 3.7, some reference values can be found for friction according to Hearle [204].

3.8.6 Examples

In the following, some experimental data for the cable robot IPAnema 1 is presented, which parameters are listed in Table 9.1. In the experimental setup, cables of type LIROS D-PRO made from Dyneema were used. The diameter of the cables was

---

[9]LIROS GmbH, Berg, Germany. http://www.liros.com.

[10]Kuraray Co. Ltd., Chiyoda, Japan. http://www.vectranfiber.com.

[11]Suter Kunststoffe AG, Fraubrunnen, Switzerland, http://www.swiss-composite.ch/.

Table 3.7 Friction coefficients for different synthetic fiber ropes on steel according to Hearle [204]

Material	Dry friction coefficient μ_C	Lubricated friction μ_L
Polyamide (PA)	0.10–0.12	0.12–0.15
Polyester (PES)	0.12–0.15	0.15–0.17
Polypropylene (PP)	0.08–0.11	0.08–0.11
Aramid (Kevlar)	0.12–0.15	0.15–0.17
HMPE (Dyneema)	0.08–0.11	0.08–0.11
Steel cable	0.12–0.15	0.10–0.12

Table 3.8 Experimental determination of the stiffness and compliance matrix for the IPAnema 1 prototype. For seven poses of the platform, the translational stiffness was determined by applying a load to the platform. Measurements were taken with and without the load. The static load was applied through an attached cable with a mass element as counter weight. The Euclidian displacement of the platform was measured with a laser tracker with high accuracy

Position	r [mm]	Compliance matrix **C** [μm/N]	Stiffness matrix **K** [N/μm]
1	$\begin{bmatrix} 0 \\ 0 \\ 1500 \end{bmatrix}$	$\begin{bmatrix} 7.28 & 1.08 & -0.69 \\ -0.07 & 9.33 & 0.00 \\ 0.37 & 0.00 & 24.70 \end{bmatrix}$	$\begin{bmatrix} 0.137 & -0.015 & 0.003 \\ 0.001 & 0.107 & 0.000 \\ -0.002 & 0.000 & 0.040 \end{bmatrix}$
2	$\begin{bmatrix} 0 \\ 0 \\ 500 \end{bmatrix}$	$\begin{bmatrix} 6.88 & 1.17 & -0.62 \\ -0.08 & 12.05 & 0.16 \\ 0.70 & 1.42 & 22.10 \end{bmatrix}$	$\begin{bmatrix} 0.144 & -0.014 & 0.004 \\ 0.001 & 0.082 & -0.000 \\ -0.004 & -0.004 & 0.045 \end{bmatrix}$
3	$\begin{bmatrix} 0 \\ 0 \\ 1000 \end{bmatrix}$	$\begin{bmatrix} 6.57 & 1.19 & -1.03 \\ -0.05 & 9.81 & -0.11 \\ -0.28 & 1.70 & 22.31 \end{bmatrix}$	$\begin{bmatrix} 0.152 & -0.019 & 0.006 \\ 0.000 & 0.101 & 0.000 \\ 0.001 & -0.008 & 0.044 \end{bmatrix}$
4	$\begin{bmatrix} -1200 \\ 0 \\ 1000 \end{bmatrix}$	$\begin{bmatrix} 10.89 & 7.92 & -1.71 \\ -0.55 & 15.91 & -0.36 \\ -4.30 & -0.26 & 20.49 \end{bmatrix}$	$\begin{bmatrix} 0.092 & -0.045 & 0.006 \\ 0.003 & 0.061 & 0.001 \\ 0.019 & -0.008 & 0.050 \end{bmatrix}$
5	$\begin{bmatrix} -800 \\ -300 \\ 600 \end{bmatrix}$	$\begin{bmatrix} 7.04 & 2.17 & -0.66 \\ -2.38 & 12.59 & 0.29 \\ -1.27 & 0.57 & 21.92 \end{bmatrix}$	$\begin{bmatrix} 0.134 & -0.023 & 0.004 \\ 0.025 & 0.075 & -0.000 \\ 0.007 & -0.003 & 0.045 \end{bmatrix}$
6	$\begin{bmatrix} -800 \\ -600 \\ 1400 \end{bmatrix}$	$\begin{bmatrix} 7.56 & 0.15 & -0.74 \\ -4.52 & 11.24 & 2.04 \\ -2.00 & 2.14 & 21.77 \end{bmatrix}$	$\begin{bmatrix} 0.131 & -0.002 & 0.004 \\ 0.051 & 0.089 & -0.006 \\ 0.007 & -0.009 & 0.047 \end{bmatrix}$
7	$\begin{bmatrix} 0 \\ -900 \\ 1000 \end{bmatrix}$	$\begin{bmatrix} 6.43 & 0.47 & 0.36 \\ 0.64 & 13.79 & 3.03 \\ 0.19 & 2.78 & 20.06 \end{bmatrix}$	$\begin{bmatrix} 0.156 & -0.004 & -0.002 \\ -0.007 & 0.075 & -0.011 \\ -0.000 & -0.010 & 0.051 \end{bmatrix}$

$d_c = 1.5$ mm. The platform was moved to the measurement positions **r**. Then, an additional cable was connected to a mass element of $m_L = 10$ kg by means of a pulley. Therefore, a static force $f = m_L g$ was exerted on the mobile platform. For each measurement position **r**, the pulley was located such, that the force acts along each coordinate axis, both in positive and negative direction. For the measurement positions 4–7, it was for practical reasons not possible to fix the pulley above the platform. Therefore, in these cases only five rather than six forces were exerted on the platform. The Cartesian displacement Δ**r** of the platform was measured using a Leica Absolute Tracker AT901-MR with a certified absolute accuracy of less than $10\,\mu$m. For the magnitude of the force, the direction of the forces and the displacements measured by the laser tracker, the stiffness matrix was computed. The results of the stiffness determination are shown in Table 3.8.

3.9 Conclusion

Based on the assumption of massless, straight, and perfectly stiff cables, this leads to a purely geometric model for kinematics and statics. There exists a rich collection of tools to deal the robot under these assumptions. The conditions for defining the minimum and maximum cable force outline the limitations of the standard model and characterizes the transition to the advanced kinematic model that deal with the effect of pulleys, elastic deformations, and hefty cables. These effects are subject in Chap. 7. Since cable robots with more cables than degrees-of-freedom are kinematically over-constrained and statically under-constrained, special methods are required to deal with such robots. A broad collection of methods to test for existence and to compute possible force distributions is discussed and compared in this chapter. Summing up the findings of this chapter on statics, one can state that there is not one optimal method for force distribution but the method of choice depends on the main requirement of the application. Some methods are efficient in terms of computation time especially when considering the application in a real-time system. Other methods put emphasis on simplicity, workspace coverage, and applicability for robot with higher degrees-of-redundancy.

The stiffness of cable robots can be estimated based on the structure matrix. Three main effects influence the actual stiffness of a cable robot. Firstly, the elastic stiffness of the cables. Secondly, the effective stiffness of the winches including motors, gears, and clutches has a direct influence on the stiffness. This also includes the closed-loop servo control in the drives. Finally, the so-called geometric stiffness has an effect on the cable robot's stiffness which become particularly apparent for planar robots and close to singular configurations. The sensitivity of a cable robot's stiffness to uncertainties in the cable's physical parameters remains an open issue.

Chapter 4
Kinematic Codes

Abstract This chapter deals with the kinematic transformation and its derivatives for the standard model. Then, singularities are introduced for cable robots. An overview of kinematic codes is given and algorithms for real-time capable codes are proposed.

4.1 Introduction

In the last chapter, we introduced the kinematic basics for cable robots where no details on the actual computation on the kinematic functions were presented. Since the algorithm that is used to solve the problem may be quite involved, it is worthwhile to discuss the issue within this chapter at length. In this setting, a *kinematic code* is an algorithm to provide a numerical solution for the mathematical problem described by the kinematic model. Therefore, kinematic codes have a clear focus on how to compute the solution accurately and efficiently. Theoretical kinematics aims at revealing and understanding the mathematical structure of the underlying problem. Clearly, the theoretical analysis of the kinematic problems provides essential properties such as the number of solutions that can be exploited to find algorithms for their solutions. The kinematic codes are a building block in the robot motion controller or are parts of the design procedure. Both applications demand for high computational efficiency where the control system additionally requires a deterministic runtime to fulfill real-time constraints in the controller. As it is carried out in this chapter, the latter is challenging since the mathematical problem to be solved is involved. Providing deterministic upper bounds on the computational time is possible but needs special consideration for the worst case computational time. In contrast, if kinematic codes are subjected to design and analysis procedures, one is usually only interested in average computational time where worst case behavior is negligible.

When considering a cable robot where the number of cables equals the degrees-of-freedom, i.e. $m = n$, and all cables are under tension, we can resort to a rich literature on forward and inverse kinematics of rigid parallel robots, see Merlet [322] for an overview of this subject. Since there are good reasons to build both kinematically

© Springer International Publishing AG, part of Springer Nature 2018   119
A. Pott, *Cable-Driven Parallel Robots*, Springer Tracts in Advanced
Robotics 120, https://doi.org/10.1007/978-3-319-76138-1_4

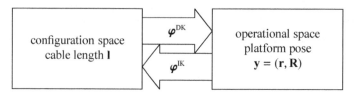

Fig. 4.1 Relation between forward and inverse kinematics functions

over-constrained and under-constrained cable robots it is worthwhile to study their
kinematics in depth.

4.2 Kinematic Transmission Functions

The *kinematic transmission functions* or *kinematic transformation* relate the platform pose **y** in operational space defined by the position of the platform **r** and the
orientation of the platform **R** to the length of the cables **l** in configuration space and
vice versa (Fig. 4.1). The function φ^{IK} calculating the length of the cables from a
given pose of the platform is called *inverse kinematics* or *backward transformation*.
Contrary, the *forward kinematic* function φ^{DK} determines the pose of the platform
from given cable length. For some problems such as singularities and stiffness, it is
important to investigate the infinitesimal relations between platform pose and cable
length and thus, one has to study differential kinematics.

4.2.1 Inverse Kinematics

For *inverse kinematics* or *backward kinematics transformation*, the geometry of the
machine frame a_i and mobile platform b_i are known, the position **r** and the orientation
R of the mobile platform are given, and the respective cable lengths **l** are sought.
This is the typical situation in open loop position control, where the motion planner
provided the desired motion of the mobile platform and, in order to generate the
set-point values for the actuators, we have to compute the required length of the
cables. In the following, we will consider the pure kinematic relations for a given
pose disregarding their static or dynamic stability. Thus, one searches for a formula
to calculate the inverse kinematics function $\varphi^{IK} : \mathbb{R}^n \to \mathbb{R}^m$ for a parameterization
of the pose **y** in the following form

$$\mathbf{l} = \varphi^{IK}(\mathbf{y}) \ . \tag{4.1}$$

In the context of inverse kinematics, the choice of the parameterization of the pose **y**
depends on the motion pattern. If orientations are involved, the parameterization of

the rotation matrix $\mathbf{R}$ is of lesser importance but it is a matter of convenience which convention is preferred by the user. For the standard model, it is trivial to derive a relation in closed-form. Calculating the Euclidian norm from Eq. (3.2) yields

$$l_i = |\mathbf{l}_i| = |\mathbf{a}_i - \mathbf{r} - \mathbf{R}\mathbf{b}_i| \quad \text{for} \quad i = 1, \ldots, m \tag{4.2}$$

where the desired cable length l_i is a function of the pose $(\mathbf{r}, \mathbf{R})$ of the platform and of the geometrical parameter $(\mathbf{a}_i, \mathbf{b}_i)$. As it can be seen from the equations, the solution of the inverse kinematics function always exists, is always unique,[1] and it exists for arbitrary poses $(\mathbf{r}, \mathbf{R})$. The inverse kinematics function can be calculated for any number of cables. Note that the existence of the inverse kinematics solution does not guarantee that the determined configuration is mechanically stable or even belongs to the wrench-closure workspace. In Chap. 5, different methods are discussed to characterize a pose in order to define the workspace.

The numerical efforts to calculate the inverse kinematics are low since the number of operations is strictly limited and relatively small. Even little-endian hardware such as micro-controllers can easily implement real-time capable inverse kinematics codes. Using automatic code generation with a computer algebra system, one can generate an efficient subroutine for the inverse kinematics. Using an Euler angle parameterization for the rotation matrix $\mathbf{R}$ and the general geometry of an eight cable robot, it requires only six trigonometric function evaluations and 242 arithmetic floating point operations to compute the inverse kinematics. The computation time for a general purpose implementation was determined[2] to be around $0.067\,\mu s$ for a cable robot with eight cables which allows for almost unrestricted use.

4.2.2 Forward Kinematics

The terms *forward kinematics*, *direct kinematics*, or *direct geometric problem* are synonyms and aim at estimating the platform pose from the given length of the cables. The main application of *forward kinematics codes* or the *forward kinematics transformation* lies in the control system where the sensors of the robot deliver a measurement for the cable length. In most systems, no direct measurement of the platform pose is possible either because of high costs or because such measurement systems have a significant lag in processing the measurement. In contrast, encoders and resolvers provide the cable length without relevant delay. For the forward kinematics function or forward position problem, a mapping $\varphi^{\text{DK}} : \mathbb{R}^m \to \text{SE}_3$ in the following form

$$\mathbf{y} = (\mathbf{r}, \mathbf{R}) = \varphi^{\text{DK}}(\mathbf{l}) \tag{4.3}$$

---

[1] More precisely, the solution is only unique if we ignore the solutions that provide a negative length for the cable which is considered to be infeasible from a physical point of view.

[2] Computation time determined on an Intel Core i5-3320M 2.6 GHz, Microsoft Visual C++ 2010.

is sought. Computing this mapping is in general much more complicated than the inverse kinematics problem we have dealt with in the previous sections. Depending on the cable length $\mathbf{l}$ and the robot configurations, there might be no solutions, one, multiple solutions, and even infinitely many solutions, possibly located in distinct sets. In the setting of this work, we distinguish between the forward kinematics as the general mathematical problem that studies the underlying structure of the kinematic code or the transformation function which can be understood as computer programs that provide numerical values. The former aims at mathematical insight. Therefore, one is interested in the theoretical number of solutions, the actual number of real solutions of a given robot family, and a classification of the equations. In contrast, the kinematic code fulfills the needs of the control engineer and its application perspective by providing a dumb but efficient numerical solution. A couple of different codes are addressed in Sect. 4.3.

For spatial cable robots with six cables and six degrees-of-freedom, the results for conventional parallel robots hold true: In general, up to 40 poses may exist that match given cable length for the standard model. If the robot is a generic design, i.e. certain geometric relations hold for the mobile platform or machine frame geometry, the maximum number of real solutions is typically reduced. However, for some non-generic designs the robot may maintain its maximum forward kinematics solution set [216, 217].

4.2.3 First-Order Differential Kinematics

Given the forward kinematics function $\mathbf{y} = \boldsymbol{\varphi}^{\mathrm{DK}}(\mathbf{l})$, one formally derives the forward kinematics function with respect to time t

$$\dot{\mathbf{y}} = \dot{\boldsymbol{\varphi}}^{\mathrm{DK}}(\mathbf{l}, \dot{\mathbf{l}}) = \frac{\partial \boldsymbol{\varphi}^{\mathrm{DK}}(\mathbf{l})}{\partial t} = \frac{\partial \boldsymbol{\varphi}^{\mathrm{DK}}(\mathbf{l})}{\partial \mathbf{l}} \frac{\partial \mathbf{l}}{\partial t} = \mathbf{J}^{\mathrm{DK}} \dot{\mathbf{l}} \quad , \tag{4.4}$$

where $\mathbf{J}^{\mathrm{DK}}$ is the Jacobian matrix of the kinematics function $\boldsymbol{\varphi}^{\mathrm{DK}}$ and $\dot{\mathbf{y}} = [\mathbf{v}^{\mathrm{T}} \ \boldsymbol{\omega}^{\mathrm{T}}]^{\mathrm{T}}$ is the twist of the platform collecting the linear velocity $\mathbf{v}$ and the angular velocity $\boldsymbol{\omega}$ (Fig. 4.2). The equation presents a pose-dependent linear mapping between the velocities of the mobile platform $\dot{\mathbf{y}}$ and the velocities $\dot{\mathbf{l}}$ of the cables. Due to kinetostatic duality, the transposed kinematic Jacobian matrix is identical to the structure matrix $\mathbf{A}^{\mathrm{T}}$ that relates cable forces to the platform wrench and it holds true

$$\mathbf{J}^{\mathrm{DK}} = -\mathbf{A} \quad . \tag{4.5}$$

The minus in the equation above is due to the conventions used to define the cable forces: By definition, the force vector for each cable points from the platform towards the winches. This seems to be clear by intuition and leads to positive forces in the cables. However, such positive forces lead to a contraction of the cables. Therefore, positive tension is coupled to negative velocities in the cables and vice versa.

Fig. 4.2 First-order differential kinematics of one cable

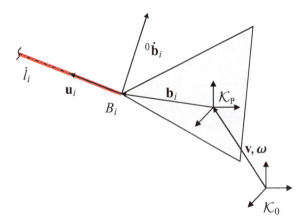

When comparing and converting between Jacobian matrix $\mathbf{J}$ and structure matrix $\mathbf{A}^\mathrm{T}$, one must take into account that different parameterizations used to compute the matrices may be incompatible. More precisely speaking, one can set up the matrices in different coordinates. The simple relation expressed through Eq. (4.5) holds true if, and only if, the wrench $\mathbf{w}_\mathrm{P}$ composed of forces and torques applied exactly in the directions of the respective twist $\dot{\mathbf{y}}$, i.e. the velocities of the generalized coordinates. If rotations are involved in the computation, special care must be taken. The angular velocity $\boldsymbol{\omega}$ can be understood as the natural derivation of the orientation matrix $\mathbf{R}$ with respect to time. However, if an angular model is used for the parameterization of the orientation matrix, the angular velocity $\boldsymbol{\omega}$ must be computed with a special formula that maps the derivatives of the parameters. To highlight this problem, we consider for example Euler angles φ, ψ, θ with $\mathbf{R} = \mathbf{R}_\mathrm{z}(\varphi)\mathbf{R}_\mathrm{x}(\psi)\mathbf{R}_\mathrm{z}(\theta)$ and their respective velocities $\dot{\varphi}, \dot{\psi}, \dot{\theta}$ in the parameter model. Note that the effective angular velocities of the parameter rotation model relate to their respective local coordinate frame. Now, the angular velocity $\boldsymbol{\omega}$ has to be computed using the kinematic Euler equation and the angular velocity becomes $\boldsymbol{\omega} = \dot{\boldsymbol{\varphi}} + \dot{\boldsymbol{\psi}} + \dot{\boldsymbol{\theta}}$ which expands to

$$\boldsymbol{\omega} = \dot{\varphi}\begin{bmatrix} 0 \\ 0 \\ 1 \end{bmatrix} + \dot{\psi}\begin{bmatrix} \cos\varphi \\ \sin\varphi \\ 0 \end{bmatrix} + \dot{\theta}\begin{bmatrix} \sin\varphi\sin\psi \\ -\cos\varphi\sin\psi \\ \cos\psi \end{bmatrix} . \tag{4.6}$$

Using the dualism for statics, the generalized forces relate to the instantaneous direction of the generalized forces. Therefore, the torques applied to the platform need to be transformed into the respective direction.

For robots that do not have a closed-form forward kinematics the derivation of the Jacobian matrix from the nonlinear forward kinematics mapping is not straightforward due to the lack of an equation to be differentiated. If no symbolic expressions are available, one can approximate the Jacobian matrix through finite differences. However, round-off errors from solving the nonlinear system make it difficult to compute accurate estimates. Furthermore, the ambiguity of the forward kinematics mapping requires careful considerations to use numerical values from the same solution branch in the finite differences. For practical purpose, the Jacobian matrix $\mathbf{J}$ can often be calculated from Eq. (3.5) rather than deriving the differentials in Eq. (4.4).

In the same way, in which one defines the Jacobian matrix $\mathbf{J}^{DK}$ for the forward kinematics mapping in Eq. (4.4), the derivation with respect to time t of the inverse kinematics are taken

$$\dot{\mathbf{l}} = \dot{\boldsymbol{\varphi}}^{IK}(\mathbf{y}, \dot{\mathbf{y}}) = \frac{\partial \boldsymbol{\varphi}^{IK}(\mathbf{y})}{\partial t} = \frac{\partial \boldsymbol{\varphi}^{IK}(\mathbf{y})}{\partial \mathbf{y}} \frac{\partial \mathbf{y}}{\partial t} = \mathbf{J}^{IK} \dot{\mathbf{y}} \quad, \tag{4.7}$$

where the Jacobian matrix $\mathbf{J}^{IK}$ maps the platform twist to the velocities in the cables. Following the discussion on kinetostatic duality, the transpose of $\mathbf{J}^{IK}$ maps cable forces to generate platform forces. If $m = n$, then both Jacobian matrices are square and in a kinematically regular pose

$$\mathbf{J}^{DK} = \mathbf{J}^{IK-1} \tag{4.8}$$

holds true. An interesting effect in this formula is that one derives the forward Jacobian matrix as a function of the cable length $\mathbf{l}$ where the inverse Jacobian matrix is usually a function of the pose $\mathbf{y}$.

According to Verhoeven [473], the velocity mapping per cable is given by

$$\dot{l}_i = - \begin{bmatrix} \mathbf{u}_i^T & (\mathbf{b}_i \times \mathbf{u}_i)^T \end{bmatrix} \begin{bmatrix} \mathbf{v} \\ \boldsymbol{\omega} \end{bmatrix} \tag{4.9}$$

$$= -\mathbf{u}_i^T (\mathbf{v} + \boldsymbol{\omega} \times \mathbf{b}_i) \tag{4.10}$$

$$= -\mathbf{u}_i^{T} \, {}^0\dot{\mathbf{b}}_i \quad, \tag{4.11}$$

where $\mathbf{v}$ and $\boldsymbol{\omega}$ are the platform's linear and angular velocities, respectively. This formula is interesting since it depends only on the instantaneous geometric values and thus allows for a geometric interpretation of the velocity mapping. Recall that ${}^0\dot{\mathbf{b}}_i$ is the absolute velocity of point B_i decomposed in the world coordinate frame $\mathcal{K}_0$. Since the axis vector $\mathbf{u}_i$ is a unit vector, we can apply the Euclidian norm in order to estimate the maximum cable velocities

$$\dot{l}_{max} = |\dot{l}_i| \le |{}^0\dot{\mathbf{b}}_i| \quad, \tag{4.12}$$

i.e. the maximum cable velocity $\dot{l}_{max}$ is in general equal or smaller than the maximum absolute velocity of its respective distal anchor point. When considering only translational motion, the maximum velocity of the cable is in general smaller or equal to the maximum translational velocity of the platform. The maximum velocity of the cable is actually needed if (and only if) the platform is moving exactly towards or away from the winch, i.e. when the platform moves in the direction $\mathbf{u}_i$ of the cable. When executing a realistic trajectory, this situation is rarely reached.

4.2.4 Singularities

For fully-constrained parallel robots, the general approach to analyze singularities was introduced by Gosselin and Angeles [171]. The implicit nonlinear loop closure condition Eq. (3.1) can be written as a function of the pose parameters $\mathbf{y}$ and the cable length $\mathbf{l}$ as

$$\mathbf{v}(\mathbf{y},\mathbf{l}) = \mathbf{0} \quad , \tag{4.13}$$

and differentiation yields

$$\underbrace{\frac{\partial\mathbf{v}(\mathbf{y},\mathbf{l})}{\partial\mathbf{y}}}_{\mathbf{J}_A}\delta\mathbf{y} + \underbrace{\frac{\partial\mathbf{v}(\mathbf{y},\mathbf{l})}{\partial\mathbf{l}}}_{\mathbf{J}_B}\delta\mathbf{l} = \mathbf{0} \quad . \tag{4.14}$$

The dimension of the matrix $\mathbf{J}_A$ is $\mathbb{R}^{m\times n}$, whereas the matrix $\mathbf{J}_B$ is in general square $\mathbb{R}^{m\times m}$. Substituting the known expressions of the standard model into the closure expression yields the ith column as block matrix

$$[\mathbf{J}_A]_i = -\begin{bmatrix} 2(\mathbf{a}_i - \mathbf{R}\mathbf{b}_i - \mathbf{r}) \\ 2\mathbf{R}\mathbf{b}_i \times (\mathbf{a}_i - \mathbf{R}\mathbf{b}_i - \mathbf{r}) \end{bmatrix} \tag{4.15}$$

and the diagonal matrix

$$\mathbf{J}_B = \begin{bmatrix} 2l_1 & 0 & 0 \\ 0 & \ddots & 0 \\ 0 & 0 & 2l_m \end{bmatrix} = 2\mathbf{L} \quad , \tag{4.16}$$

where the Jacobian matrix $\mathbf{J}_B$ is proportional to the already known matrix $\mathbf{L}$ that contains the cable lengths as diagonal elements and $\mathbf{J}_A$ is proportional to the

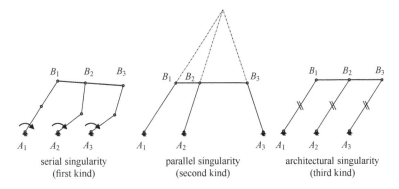

Fig. 4.3 Examples of the three kinds of singular configurations for conventional parallel cable robots

non-normalized structure matrix $\widehat{\mathbf{A}}^{\mathsf{T}}$. Thus, the analysis reveals that singularities depend on well-known matrices (Fig. 4.3).

If $\mathbf{J}_A$ or $\mathbf{J}_B$ is singular, then the robot is in a singular configuration. One can classify three types [171]:

1. $\det \mathbf{J}_B = 0$: In this case, the robot looses one or more degrees-of-freedom (under-mobility or serial singularity). Theoretically, the length l_i of one cable could be zero which means that in this pose the anchor point of the platform B_i coincides with the base A_i. From a practical point of view, such configurations are infeasible anyway because a collision between the distal and proximal anchor points occurs before the kinematic degeneration can be reached. Such configurations can be easily detected and characterized due to their simple geometric interpretation.
2. $\det \mathbf{J}_A = 0$ or rank $\mathbf{J}_A < n$ if $m \neq n$: In this cases, the robot gains one or more degrees-of-freedom (over-mobility or parallel singularity). This is the typical case for parallel robots and there is no counterpart for this effect in serial robots. An infinite motion of the platform is possible without moving the actuators. In such a singular posture, the robot cannot withstand certain wrenches and thus undergoes infinite motion even if all cable lengths **l** are fixed and all cables are under tension. Furthermore, the stiffness of the robot arising from the stiffness of the cables drops to zero in one or more directions.
3. $\det \mathbf{J}_A = \det \mathbf{J}_B = 0$: This is only possible with special (non-generic) robot design parameters. If it holds true for any pose of the robot, the robot is called architecturally singular. A well-designed robot should be free of architectural singularities. Especially for cable robots, where we have seen that the first kind of singularity is rather specific, architectural singularity has no impact in practice.

In case of $m = n \mathbf{J}_A$ is quadratic. If $\mathbf{J}_A$ is also regular, one derives

$$\delta \mathbf{y} = -\mathbf{J}_A^{-1} \mathbf{J}_B \, \delta \mathbf{l} \tag{4.17}$$

and the Jacobian matrix becomes $\mathbf{J}^{DK} = -\mathbf{J}_A^{-1} \mathbf{J}_B$. For cable-driven parallel robots, the matrix $\mathbf{J}_B$ is always a diagonal matrix since each winch-cable system (leg of the robot) has exactly one generalized coordinate l_i if the robot has no passive cables. As pointed out by Ma and Angeles [303], one can also distinguish between three sources of singularity. Firstly, *configuration singularities* relate to the type 1 and 2 singularities described above where lower-dimensional regions occur in either the configuration or operational space. Secondly, *architectural singularities* are an effect of inappropriate geometrical parameters and are inherent to a cable robot design. Finally, *parameter singularities* are artificially introduced by the mathematical description of the robot motion. The well-known singular configurations of Euler and Bryant angles are typical examples for such singularities. The local of such singularities can be changed by using another mathematical model for the robot. In the case of the orientation, one can use e.g. quaternion for the parameterization to eliminate parameter singularities.

For IRPM, CRPM, and RRPM typed robots, the Jacobian matrix $\mathbf{J}_A$ is not quadratic and thus cannot be inverted. For fully-parallel robots, we always have an equal number of legs and generalized coordinates **l**. Furthermore, the matrix $\mathbf{J}_B$ can easily be inverted and one gets the mapping

$$\delta \mathbf{l} = -\mathbf{J}_{\text{B}}^{-1} \mathbf{J}_{\text{A}} \, \delta \mathbf{y} \quad . \tag{4.18}$$

For cable robots, it is useful to generalize the concept since one usually often has to deal with robots where the Jacobian matrix $\mathbf{J}_{\text{A}}$ is rectangular and thus we cannot refer to the classification based on the determinant. Contrary, for cable robots, only one type of singularity is of practical importance, i.e. over-mobility [473]. A necessary and sufficient condition for a pose $\mathbf{y}$ to be singular is that the structure matrix is rank deficient which implies

$$\text{rank } \mathbf{A}^{\text{T}}(\mathbf{y}) < n \quad . \tag{4.19}$$

Based on the considerations above, this is equivalent to a rank deficit in the matrix $\mathbf{J}_{\text{A}}$. Verhoeven [473] gives an analysis of the conditions for singularities based on the motion pattern of the platform. The purely translational systems 2T and 3T are free of singularities if the design is not architecturally singular or in other words: Such system cannot have a limited number of singular poses. Either all poses are singular or they are free of singularities.

For complex motion patterns such as 1R2T, 2R3T, and 3R3T, a general classification becomes more involved. There are some general aspects that seem clear. Generic designs that share a proximal or distal anchor point tend to reduce the number of singularities. Furthermore, singularities can be removed by adding redundant cables to the system. The latter can be easily seen from Eq. (4.19): Adding a cable and thus a column to the structure matrix $\mathbf{A}^{\text{T}}$ gives additional opportunities to increase the rank of the matrix to maximum of n.

A couple of other methods for finding singular configurations beside this algebraic approach exist. An exhaustive classification of singularities for parallel robots was given by Merlet [317] through geometric considerations. In this approach, the legs of the robot are represented by line coordinates (Plücker coordinates) and Merlet used Grassmann geometry to characterize different singular configurations. It was shown that singularities can be mapped to special geometric conditions of line coordinates [317]. These geometric conditions can be used to geometrically construct the singular locus of parallel robots.

As stated previously, for a fully-constrained cable robot, the structure matrix is rectangular and therefore it is not possible to compute the determinant. However, we can compute the singular values of the matrix. If the smallest singular value is 0, the respective matrix is rank deficient and thus the robot is in a singular configuration. In Figs. 4.4 and 4.5, the singular values of the structure matrix have been computed for cross sections through the workspace of the IPAnema 1 robot (Table 9.1). Note that the singular values of the structure matrix $\mathbf{A}^{\text{T}}$ can be easily and robustly computed with singular value decomposition (SVD) without considering whether a pose is wrench-closed or wrench-feasible. The minimum singular value is mostly uniformly distributed around the central axis of the robot where the largest singular value has a clear polarization, i.e. it is almost constant along the y-axis and strongly varying along the x-axis.

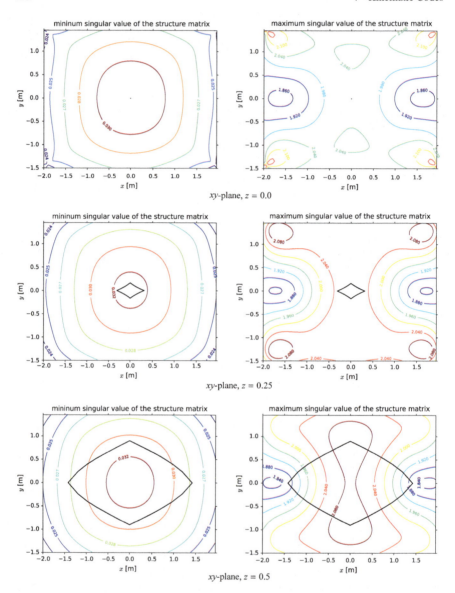

Fig. 4.4 Contour plots for the minimum (left column) and maximum (right column) singular value of the structure matrix for horizontal cross section in the xy-plane of the IPAnema 1 robot. Cross sections have been computed for the following values for $z = \{0.0, 0.25, 0.5\}$ m and the wrench-feasible workspaces are overlaid in black as reference

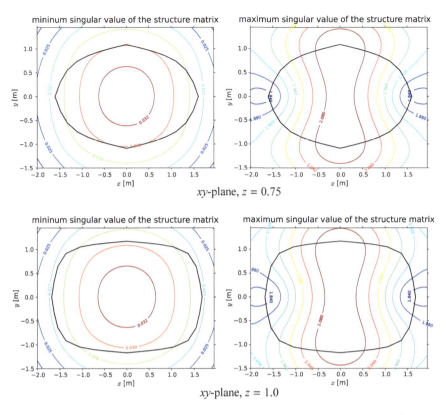

Fig. 4.5 Contour plots for the minimum (left column) and maximum (right column) singular value of the structure matrix for horizontal cross section in the xy-plane of the IPAnema 1 robot. Cross sections have been computed for the following values for $z = \{0.75, 1.0\}$ m and the wrench-feasible workspaces are overlaid in black as reference

4.2.5 Second and Higher Order Differential Kinematics

Second order kinematics is used for dynamics since Newton's axiom relates forces to accelerations and inertia. We can drive the acceleration transmission function from Eq. (4.9) by derivation with respect to time

$$\ddot{\mathbf{l}} = \mathbf{J}\ddot{\mathbf{y}} + \dot{\mathbf{J}}\dot{\mathbf{y}} \quad . \tag{4.20}$$

Rewriting this component-wise yields

$$\ddot{l}_i = -\mathbf{u}_i^{\mathrm{T}} {}^{0}\ddot{\mathbf{b}}_i - \dot{\mathbf{u}}_i^{\mathrm{T}} {}^{0}\dot{\mathbf{b}}_i \quad . \tag{4.21}$$

Third order differentials become even more involved but it seems that third and higher order derivatives are not needed for practical or theoretical computations for cable robots.[3] Anyway, if such equations are required it is straightforward to compute these equations from Eq. (4.20) by simple differentiation.

4.2.6 Kinematics for Under-Constrained Robots

According to the classification of cable robots by Ming and Higuchi [350], robots with less cables m than degrees-of-freedom n are called *incompletely restrained positioning mechanisms* (IRPM) or *cable-suspended parallel mechanisms*. Such robots are always operated in a *suspended* configuration[4] where the winches are located above the mobile platform and gravity is additionally employed to keep the cables under tension. The NIST RoboCrane [7] was based on this approach and a couple of later works followed this design approach [173, 406]. Such suspended cable robots are sometimes said to be in *crane-configuration* since applied wrenches are usually considered second to the gravity force caused by the mass of the platform and its possible payload. To model such robots, it was proposed to consider gravity as an additional cable that is pointing in the direction of gravity, independent from the current position of the platform [473]. The wrench $\mathbf{w}_p$ produced by the robot is limited both in its directions and in its magnitude. Therefore, the possible accelerations are also strictly bounded. Compared to fully-constrained robots, the wrench-closure workspace of suspended robots may be empty.

If the cable robot is under-actuated $m < 6$ in addition to being suspended, a number of properties differ from the fully-constrained case. As for any other under-actuated manipulator, only m independent directions of motion can be generated through the actuators. If the number of cables m is less than six, $6 - m$ linearly independent directions exist in SE_3 where no infinitesimal motion can be generated although such motion may be consistent with the constrains imposed by the cables [105, 175, 284, 516, 534]. New theoretical problems arise for such crane-like robots since the determination of the static equilibrium poses requires a distinct approach for modeling compared to the fully-constrained cable robots discussed so far. The so-called *geometrico-static modeling* [88] involves both kinematic (geometric) constraints described by the position of the proximal and distal anchor points as well as static equations arising for force equilibrium of the mobile platform. A simple planar case was discussed by Jiang [228]. The general base for three cables is handled in detail by Carricato and Merlet [89]. The underlying mathematical problem turns out to be more complex as one would expect and the problem can be transformed to finding the roots of a polynomial of degree 156. Using four cables is even more involved and requires an univariate polynomial of degree 216 [86]. Since such equations are difficult to handle with conventional numeric methods, Berti [34, 35, 37] approached

---

[3] At least, the author is not aware of a use-case.

[4] More precisely speaking, such robots are always in operation with a constantly applied external wrench that is independent from the platform pose. Similarly, one can add springs to the robot in order to create a pose-dependent artificial potential field as studied by Duan [131].

Table 4.1 Maximum number of solutions for the forward kinematics of the standard model

Number of cables m	Number of solution
1	1
2	24
3	156
4	216
5	140
6	40
>6	40

the numerical solving with interval analysis to rigorously bound the numerical errors. Collard [99] formulated the kinematic problem of under-constrained cable robots as optimization problem and presented numerical examples for over-actuated but suspended robots with a huge number of cables. An overview is shown in Table 4.1.

For IRPM, one can evaluate the inverse kinematics given by Eq. (4.2) to compute cable length **l** for a given pose $(\mathbf{r}, \mathbf{R})$ disregarding the insufficient number of cables. However, for such under-constrained cable robots, the configuration space is a lower-dimensional manifold compared to the operational space of the platform poses. This becomes clear if one considers the Jacobian matrix $\mathbf{J}^{\text{DK}} \in \mathbb{R}^{m \times n}$ from Eq. (4.4) that describes the instantaneous mobility of the platform. At a regular pose, the rank of the Jacobian matrix

$$\text{rank } \mathbf{J}^{\text{DK}} = m \quad , \tag{4.22}$$

since $m < n$ and therefore, only m independent displacements can be generated by the robot. Thus, there exist $(n - \text{rank } \mathbf{J}^{\text{DK}}) > 0$ linearly independent infinitesimal displacements $\delta\mathbf{t}_i$ with

$$\mathbf{J}^{\text{DK}}\delta\mathbf{t}_i = \mathbf{0}, \quad i = 1, \dots, (n - \text{rank } \mathbf{J}^{\text{DK}}) \tag{4.23}$$

that are instantaneously inaccessible through changes $\delta\mathbf{l}$ of the cable length at the current pose (Fig. 4.6). Although the number of these independent displacements is constant (except for singularities of second kind), the spanned subspace ker $\mathbf{J}^{\text{DK}}$ is pose-dependent. These displacements can only be generated if an additional wrench is applied to the platform. Thus, an $(n - m)$-dimensional subspace of SE$_3$ cannot be reached with the robot being in force equilibrium.

To evaluate the inverse kinematics, one has to check if the platform is in static equilibrium of the cable forces and external wrench. If the desired pose $(\mathbf{r}, \mathbf{R})$ is statically stable, one can apply the simple procedure for inverse kinematics as described above. Since the workspace of such IRPM is a lower-dimensional manifold embedded into the Euclidian motion group SE$_3$, it is, in the presence of the unavoidable uncertainties, rather unlikely that an arbitrarily chosen pose for the mobile platform is stable and can be achieved with positive cable tensions. Therefore, the question of

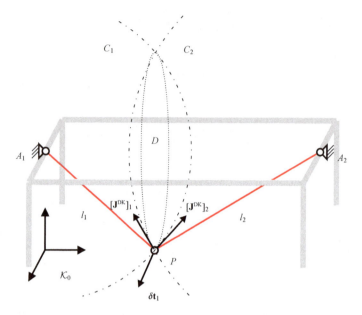

Fig. 4.6 Mobility of under-constrained cable robots with two cables. Only some displacements $[\mathbf{J}^{DK}]_1$ and $[\mathbf{J}^{DK}]_2$ can be generated by changing the length of the cables while the displacement $\delta\mathbf{t}_1$ is inaccessible through the actuators

how to deal with the situation when a desired pose relates to an unstable configuration arises.

Inverse kinematics gets an involved problem if one expects the inverse kinematics code to compute cable length for a stable pose $\mathbf{y}^*$ that is close to the desired pose $\mathbf{y}$. To do so, one also has to take the effect of external wrenches into account including gravity on the platform. Basically, one searches for a pose $\mathbf{y}^*$ in the wrench-feasible workspace (see Sect. 5.2.2) that is close to the given pose $\mathbf{y}$. Here, we touch the well-known problem that there is no unique way to measure proximity in the Euclidian motion group SE_3. One usually assumes that the external wrench is caused by gravity. Therefore, inverse kinematics of under-constrained cable robots depends on both the geometric and the static properties of the robot. As a result, the behavior of the robot depends both on its location of the anchor points given by $\mathbf{a}_i$ and $\mathbf{b}_i$ as well as on its mass m_P and its center of gravity $\mathbf{r}_M$. In the literature, the approach is referred to as *geometrico-static kinematics* [35, 36, 87, 88, 90] and combines kinematic (geometric) equations with static (force equilibrium) equations. In other works, such approaches are called *kinetostatic* models. Another approach to this problem is to find a trajectory where dynamic inertia forces can be used to balance the robot also at poses where no static equilibrium can be reached [308]. In this case, the kinematics equations have to be solved along with the dynamic equations and kinematics becomes closely coupled to motion planning.

The computation of forward kinematics remains challenging for IRPM. For point-shaped platforms, the problem is rather simple. The planar case (2T) is trivial since $m < n$ relates to a cable robot with only one cable which kinematically restrict the platform to a circle that has exactly one minimum point. In the spatial case (3T), we have two cables constraining the platform onto a circle which is again easy to deal with. A more involving problem occurs for finite platforms in the plane and especially in space.

For a cable robot with two cables and 1R2T motion pattern, the center of gravity is kinematically constrained to a curve that is equivalent to the coupler curve of a four-bar mechanism [88]. To model the behavior of such a cable robot under gravity, it was proposed to search for the minima of such a curve. Furthermore, it is shown that this condition is equivalent to the results received for the minima of a constrained optimization problem.

For the modeling of the geometrico-static model, Carricato et al. [35, 87, 88, 90] used line coordinates. This modeling approach is equivalent to the notation used in this work and the basic geometrico-static conditions reads

$$l_i = ||\mathbf{a}_i - \mathbf{r} - \mathbf{R}\mathbf{b}_i||_2 \quad \text{for} \quad i = 1, \ldots, m \tag{4.24}$$

$$\mathbf{0} = \mathbf{A}^{\mathrm{T}}\mathbf{f} + \mathbf{w}_{\mathrm{P}} \ , \tag{4.25}$$

which relates to $m + 6$ scalar equations that involve $6 + 2m$ variables which are the six pose parameters used to parameterize $\mathbf{y} = (\mathbf{r}, \mathbf{R})$, the m cable lengths l_i, and the m cable forces f_i. To compute the inverse kinematics, the six pose parameters are given and one receives $m + 6$ equations for $2m$ unknown parameters. Thus, for IRPM with $m < n$, one gets in general an over-constrained nonlinear system.

There is an interesting connection between the forward kinematics of IRPM and the dynamic simulation presented in Chap. 6. From a kinematic point of view, the coupled geometrico-static equations present an involved mathematical problem. Interestingly, from a dynamic point of view, there is little difference in modeling IRPM, CRPM, and RRPM robots. Thus, using the dynamics formulation, one can integrate the differential equations over a sufficiently long time period to find the pose $(\mathbf{r}, \mathbf{R})$ where the dynamic system is at rest in equilibrium. Hence, the pose found in this way also solves the forward kinematics problem. The connection between forward kinematics and dynamics is further discussed in Sect. 6.3.5 in terms of energy.

4.3 Forward Kinematics Codes

A fundamental problem for parallel mechanisms including cable robots is the solution of the forward kinematics. For serial robots, it is straightforward to calculate the pose of the end-effector from given joint coordinates [114] and the result is always unique, even if the robot has kinematically redundant articulated joints. For parallel robots, this problem can be very complex and there exist quite simple examples for robots having more than one pose which can be related to one set of joint coordinates. As late as 1992, it was shown that the maximum number of poses of spatial parallel robots of the Stewart–Gough-type are bounded by 40 [283, 420], where some of these poses may be in the imaginary plane. Later, Dietmaier found an example where all 40 poses are real [122]. An algorithm to solve the general kinematics was provided by Husty [214], who applied kinematic mapping [58] to transform the kinematic equations into a projective space. In this space, the equations are algebraic leading to a polynomial of degree 40. Unfortunately, this algebraic approach seems inadequate for real-time implementation. Furthermore, highly specialized numerical methods are needed to cope with 40th order polynomial and numerical stability becomes an important issue. Adding more constraints does not generally reduce the number of solutions [216, 217] if some geometrical relations are satisfied. Lately, it was shown that for robots with less than six legs, the equations may lead to some hundreds of solutions where dozens of real solutions were found ad-hoc [314].

For less than six cables, the kinematic problem is surprisingly more complex since the geometric relations and the static equations get coupled making the underlying mathematical problem more involved. The number of complex solutions for robots with five cables is 140. For four cables, the number of solutions was found to be 216 [319] where some of these can be complex. For three cables, the maximum number of solutions was found to be 156. For two cables, in total 24 solutions exist in two sets with 12 solutions each. Only for robots with a single cable, the situation is trivial and such crane-like robots have, as expected, a unique solution. An overview is given in Table 4.1 according to [34].

As discussed earlier, at least $m = n + 1$ cables are required to fully control the motion for a mobile platform with n degrees-of-freedom [349]. Therefore, many cable robots are under-determined or over-actuated with respect to the distribution of forces in the cables and over-determined with respect to forward kinematics (Fig. 6.10a). For over-constrained cable robots, one could expect that the number of solutions is reduced to at most one due to the additional constraints but, unfortunately, in general this is not the case [216, 217]. Special non-generic geometries maintain this maximum solution set. At first sight, these results were considered to be only relevant for theoretical considerations while the geometry of practically built robots does not have this non-generic property. To solve the over-constrained equations, Merlet used interval analysis to compute all solutions of the forward kinematics of parallel robots [321] in a guaranteed way. A more specialized method was presented for cable robots with linear drives and elastic deformation in the cables [324]. A closed-form kinematic code for the so-called 3-2-1 configuration is well suitable for real-time applications [389, 465] but relies on a special non-generic geometry where at least three cables share a common distal anchor point. Bruckmann [72] presented a method to cope with winches using pulley mechanisms to guide the cables. A real-time capable code for generic over-constrained geometries was presented [390] and extended by Schmidt [434]. Liwen recently used the same approach of minimizing the potential energy but employed sequential quadratic programming [297].

For practical computations in the controller of the robot, one is mostly interested in tracing one of these solutions inside the workspace. The solution at hand is the configuration that arises from the initial geometric conditions when firstly assembling the robot. Up to now, no method that guarantees to distinguish between the different assembly modes is known. A common work-around is to start with one known pose and use an iterative scheme to track that solution [68, 139]. In practice, forward kinematics has to be used inside the controller to monitor the pose of the platform or to determine the initial platform position after power on. Here, one has to deal with small errors in the kinematic parameters due to disturbances in the measurement and other uncertainties. This can result in finding no solutions which is related to measuring too short cable length or infinitely many solutions which may result from having too long cables. In both cases, additional assumptions have to be made to keep the control system stable. The latter problem becomes really involved if first and higher order derivatives are also needed.

4.3.1 Classification and Approaches

Firstly, one can classify the algorithms based on the number of cables m and degrees-of-freedom n:

- IRPM with more degrees-of-freedom n than cables m: Forward kinematics equations are under-constrained and we have most probably to deal with infinitely many solutions. Furthermore, the set of solutions depends on applied forces such as gravity.
- IRPM where the degrees-of-freedom n equal the number of cables m: Here, we can apply some algorithms from conventional parallel robots. Still, we have to check if the determined solution is stable.
- CRPM and RRPM: The robot is kinematically over-constrained and in the presence of uncertainties no exact solution exists.

Secondly, the algorithms are classified with respect to the motion pattern of the platform (see Sect. 2.2.2). We have to distinguish especially between point-platforms with motion pattern 2T and 3T and bodies with 1R2T or 3R3T motion pattern.

Thirdly, specialized algorithms can be found for non-generic geometrical configurations of the robot, such as multiple cables that share a common point on the frame or on the platform. Another typical type of constraints used to simplify the kinematic equations are relations in the geometry such as all anchor points lying on a common line or in a common plane. More exotic kinematic constraints on the geometry are analyzed [217].

Another important criterion for a kinematic code is the computation time. Due to continuously raising computation power, we will distinguish *real-time* codes that determine the solution within a strictly bounded and relatively short time and general methods, where it is not possible to give an upper bound of the computation time. Some general methods can be used in real-time context if it is possible to configure the algorithm in so far as that the computation time is strictly bounded, e.g. by providing an upper limit on the number of iterations. In this setting, we may derive modified kinematic codes that allow to approximate the solution within a reasonable time and with acceptable accuracy. Anyway, such trade-offs need intensive testing before they can be applied in a controller.

Finally, we classify the algorithms based on the number of solutions they are able to find. *Local methods* mostly use an initial estimate to determine a solution that is close to that estimate, whereas *global methods* determine many or even all solutions.

The equations characterizing the forward kinematics for CRPM and RRPM form an over-constrained system and for arbitrary cable length **l** one cannot find solutions **y** that fulfill the constraint Eq. (3.1). In practice, uncertainties such as noise, systematic errors from the measurement system, inaccuracies in the geometric parameters, and simplifications in modeling lead to the situation that no exact solution can be found. There are different approaches to deal with this problem:

- One can solve the forward kinematics using a *general solver* for six cables to obtain one or more candidates for the pose $\mathbf{y}_i$. Then, it can be checked which of these solutions are consistent with the redundant constraints. A couple of different solvers have been presented for the forward kinematics of Stewart–Gough platforms including gradient methods, interval analysis, kinematic mapping, optimization methods, and some closed-form solutions for non-generic geometries. An overview can be found in [322].
- In the presence of actuator redundancy, the problem of forward kinematics can be interpreted as an *optimization problem*, where the roots of Eq. (3.1) need to be approximate best possible. The constraints are distance equations and can be interpreted as the squared distance from an exact solution. Thus, summing up the errors from each constraint leads to a least-square problem.
- Interpreting the constraints imposed by the cables as unilateral constraints, the resulting problem of forward kinematics is a *constraint satisfaction problem* (CSP) [196, 388] which has in general infinite solutions where a couple of disconnected compact sets may exist. Interval analysis can be used to find all these regions in a guaranteed way. This kind of solver is very slow and the approximations of the sets are complicated to use for applications like controller design. Nevertheless, this approach calculates the solutions in a very general way and can serve as reference.
- Using techniques like *kinematic mapping* [58], the constraints are transformed into a *univariate polynomial*. This kind of equation may be inappropriate for solving with numerical methods but provides insight into the structure of the underlying mathematical problem. The order of the polynomial is an upper bound on the number of solutions where in general the number of real solutions might be even smaller. Furthermore, the coefficients of the polynomial provide information on constraints for non-generic designs that may have a smaller solution set.
- One can formulate the forward kinematics in terms of *quadratic distance equations* between the distal anchor points with limits on the distance between $\overline{A_i B_i}$, $1 \leq i \leq m$ and exact distances between the platform anchor points $\overline{B_i B_j}$, $1 \leq i, j \leq m, i \neq j$. This approach can be compared to the set of equations analyzed with interval analysis for Stewart–Gough platforms [321] or with other numerical algorithms

specialized on this specific type of equations. For example, such a solver was proposed in [464] for a measurement device.

• An approach based on neural networks and interval analysis was presented by Schmidt [433].

4.3.2 General Challenges with Forward Kinematics

One can use numerical schemes like Newton–Raphson or Levenberg–Marquardt algorithms to numerically solve the closure constraints (3.1). Iterative solvers have the drawback that their region of convergency is limited and difficult to predict. If multiple solutions exist, it is difficult but possible [321] to find all of them, e.g. by using interval analysis. Furthermore, we consider the set of solutions $\mathcal{Y}_A = \{\mathbf{y}_1, \ldots, \mathbf{y}_k\}_A$ that is calculated for a configuration A for a vector of cable length $\mathbf{l}_A$ and the set of solutions for a configuration B with solutions $\mathcal{Y}_B = \{\mathbf{y}_1, \ldots, \mathbf{y}_k\}_B$ that follows from another vector of cable length $\mathbf{l}_B$. Now, there seems to be no simple way or it might even be generally impossible to find a one-to-one correspondence between the elements of $\mathcal{Y}_A$ and $\mathcal{Y}_B$. Practically, the robot is in a particular configuration $\mathbf{y}_{A,i}$ and one wants to receive only the corresponding solution $\mathbf{y}_{B,i}$ from $\mathcal{Y}_B$ that can be physically reached from that particular pose $\mathbf{y}_{A,i}$ without disconnecting cables or crossing singularities. Some recent results [85, 215, 519] for conventional parallel robots even suggest that corresponding configuration in $\mathcal{Y}_B$ depends on the path between $\mathbf{l}_A$ and $\mathbf{l}_B$ since parallel robots are able to change their assembly mode without crossing a singularity. It might even happen that the numbers of (real) elements in the sets $\mathcal{Y}_A$ and $\mathcal{Y}_B$ do not match. To the best of the author's knowledge, there is no solution to this problem in the literature.

4.3.3 The 3-2-1 Configuration

Some cable robots have a generic geometry with only three distinct points on the platform or, in other words, multi cables share a common anchor point on the platform. Such generic designs have special kinematic properties and it is possible to derive specialized but efficient algorithms in this case. Explicit formulas are desired and the derived formulas are based on the work by Thomas [465] which takes robots with a so-called 3-2-1 or, as discussed below, in a 3-2-2 configuration into account. For parallel cable robots with seven cables, the distal attachment points $B_1, \ldots, B_7$ can be written as three distinct attachment points on the platform which are called B_A, B_B, B_C. Furthermore, for the configuration it is assumed that either two or three cables share a common distal anchor point on the mobile platform. This configuration is shown in Fig. 4.7 and was used for a couple of recently built cable robots, since it minimizes the restrictions caused by cable interference elsewhere in the workspace, see also [473].

Fig. 4.7 Six
degrees-of-freedom parallel
cable robot with three
distinct attachment points on
the platform (the so-called
3-2-2 configuration)

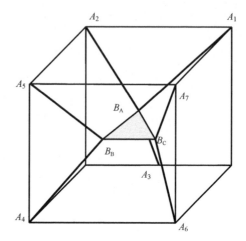

Fig. 4.8 View on the
intersection of three spheres
S_i in the plane E_3 containing
the three center points

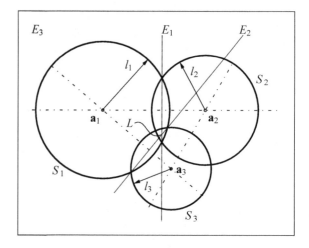

The presented calculation derives the equations in closed-form as they are used
for the controller. It is entirely based on a geometric procedure. Firstly, it is recalled
that the distal attachment points B_i have to be located on spheres S_i with radii l_i that
center around the proximal attachment points A_i. If three attachment points on the
platform coincide as shown for B_C in Fig. 4.7, their location in space can be calculated
from the intersection of these three spheres. Once B_C on the platform is known, the
distances between B_C and B_A, B_B can be used as distance constraint. Thus, one can
apply the same procedure to determine B_A, B_B as well. This provides a closed-form
solution for the forward kinematics of the robot. Therefore, the determination of the
intersection of three spheres is described in the next section.

An important step in solving this special forward kinematics problem is to compute the intersection (Fig. 4.8) of the three given spheres S_1, S_2, S_3 defined by their

centers $\mathbf{a}_1$, $\mathbf{a}_2$, $\mathbf{a}_3$ and the radii l_1, l_2, l_3, respectively.[5] Generally, there are four possibilities for the number of intersections. There might be no solution, one solution, two symmetric solutions, or an infinite number of solutions. The latter case only occurs if $\mathbf{a}_1$, $\mathbf{a}_2$, $\mathbf{a}_3$ define a line. From a kinematic point of view, one of the three constraints degenerates and this happens only in a singular configuration. Firstly, intersection occurs between the spheres only if

$$|l_i - l_j| < ||\mathbf{a}_i - \mathbf{a}_j||_2 < l_i + l_j \quad \text{for} \quad i, j \in 1, 2, 3, i \neq j \tag{4.26}$$

holds true. Otherwise, there is no solution because the distance between the spheres is either too big or too small to allow for intersection. Now, let the plane E_3 be defined by the points $\mathbf{a}_1$, $\mathbf{a}_2$, $\mathbf{a}_3$. Then, its unit normal vector $\mathbf{n}_3^0$ is given by

$$\mathbf{n}_3^0 = \frac{(\mathbf{a}_1 - \mathbf{a}_2) \times (\mathbf{a}_1 - \mathbf{a}_3)}{||(\mathbf{a}_1 - \mathbf{a}_2) \times (\mathbf{a}_1 - \mathbf{a}_3)||_2} \ . \tag{4.27}$$

Furthermore, the planes E_1 and E_2 defined by the intersection of S_1, S_2 and S_1, S_3, respectively, are given by

$$E_1 : \mathbf{x} \cdot \mathbf{n}_1 = \frac{\mathbf{a}_2^2 - \mathbf{a}_1^2 - l_2^2 + l_1^2}{2} \tag{4.28}$$

$$E_2 : \mathbf{x} \cdot \mathbf{n}_2 = \frac{\mathbf{a}_3^2 - \mathbf{a}_1^2 - l_3^2 + l_1^2}{2} \tag{4.29}$$

where the normal vectors $\mathbf{n}_1$, $\mathbf{n}_2$ are constructed from $\mathbf{n}_1 = \mathbf{a}_2 - \mathbf{a}_1$, $\mathbf{n}_2 = \mathbf{a}_3 - \mathbf{a}_1$. The sought intersection(s) $\mathbf{x}_i$ of all three spheres is located on the line

$$L : \mathbf{x} = \mathbf{x}_0 + \lambda \mathbf{n}_3^0 \ , \tag{4.30}$$

where $\mathbf{x}_0$ is the intersection of the planes E_1, E_2, E_3. Therefore, one can compute $\mathbf{x}_0$ from the linear system

$$\begin{bmatrix} \mathbf{n}_1 & \mathbf{n}_2 & \mathbf{n}_3^0 \end{bmatrix}^T \mathbf{x}_0 = \begin{bmatrix} \dfrac{\mathbf{a}_2^2 - \mathbf{a}_1^2 - l_2^2 + l_1^2}{2} \\[2mm] \dfrac{\mathbf{a}_3^2 - \mathbf{a}_1^2 - l_3^2 + l_1^2}{2} \\[2mm] \mathbf{a}_1 \cdot \mathbf{n}_3^0 \end{bmatrix} \ . \tag{4.31}$$

The parameter λ is then determined from

$$\lambda = \pm \sqrt{l_1^2 - (\mathbf{x}_0 - \mathbf{a}_1)^2} \ . \tag{4.32}$$

---

[5]Note that the indices 1, 2, 3 in this section relate to the three winches chosen with the common point B on the platform. Without loss of generality, one can renumber the winches so that the first three have the common point on the platform.

Substituting λ into the parameter form of the line L, Eq. (4.30) yields the two points of intersections, as expected.

Since the intersection of three spheres yields two solutions in regular configurations of the robot, up to eight different sets of solutions B_A, B_B, B_C exist. Finally, one has to check which of these solutions is the sought-for by testing a redundant constraint. This is done by testing if the distance between the points B_A, B_B matches the known constant distance

$$||\mathbf{b}_A - \mathbf{b}_B||_2 = \text{const} \tag{4.33}$$

between these points. In the presence of uncertainties such as measurement and tracking errors of the cable length, one has to use a threshold for the error of the redundant constraint.

4.3.4 Numerical Methods for Redundantly Restrained Robots

In this section, we consider numerical methods to solve the forward kinematics of CRPM and RRPM type robots. The algorithm is designed for the use in the controller. Therefore, the geometry shall be generic and the algorithm must work under real-time conditions. An algorithm in two steps for solving the forward kinematics of redundant cable robots in general is presented. In the first step, interval techniques are used to estimate guaranteed bounds on the pose. In the second step, the bounds are used to calculate an estimate of the pose which is the initial value for an iterative Levenberg–Marquardt solver.

4.3.4.1 Estimating an Initial Pose for Iterative Schemes

The algorithm assumes that the mobile platform is small compared to the measured cable length l_i. In general, it holds true that for any orientation of the mobile platform the coordinates of the TCP are inside the spheres around the anchor points of the frame $\mathbf{a}_i$ and with a radius equal to $||\mathbf{l}_i||_2 + ||\mathbf{b}_i||_2$. For practical calculations, a transformation of the platform vectors $\mathbf{b}_i$ that minimizes their lengths is useful to increase the quality of the estimate, e.g. to transform the platform in a reference point computed from the average of the points $\mathbf{b}_i$. It is straightforward to calculate axis-aligned bounding boxes for each of the m spheres. Using only the intersection of those m boxes yields guaranteed bounds $\mathbf{r}_{\min} = [x, y, z]_{\min}^T$ and $\mathbf{r}_{\max} = [x, y, z]_{\max}^T$ for the position of the platform. Then, any point in the box can be used as starting point for the iteration, whereas the center $\mathbf{r}_M = \frac{1}{2}(r_{\min} + r_{\max})$ of the box is used as initial value of the iterative Levenberg–Marquardt algorithm.

This initial estimate can be improved if one computes the exact intersection of the spheres that leads to a smaller region and thus to a better estimate. However, the possible gains from such a procedure seem to be little compared to the significant efforts. Note that such improvements of the initial guess are essentially required only to enter the region of convergency of the following iterative scheme.

4.3.4.2 Assumptions for Forward Kinematics

Here, we present an algorithm for forward kinematics to be used rather for real-time control of a well-designed robot than for analysis of possibly ill-conditioned or architecturally singular robots. The following assumptions were made taking into account practical needs:

- The sought-for pose $\mathbf{y}$ of the mobile platform to be estimated belongs to the workspace (positive tension in the cables) and the control error measured by the length sensors is moderate. If the cable lengths are too short, it may cause either overloading the motors or breaking the cables. If the cable lengths are too long, we lose the control on the platform. In both cases, the control system must perform an emergency stop rather than computing a theoretical solution that cannot be generated by a real robot.
- The cables of the robot are elastic allowing for small changes in length around the given length l_i. Nevertheless, the presented algorithm does not take into account changes in the length due to the actual tension.
- The geometry of the mobile platform $\mathbf{b}_i$ was chosen so that the rotation matrix $\mathbf{R} = \mathbf{I}_3$ is in the workspace or close to the workspace. This is a minor restriction since cable robots only allow for relatively small orientation workspace and we choose a pre-orientation of all $\mathbf{a}_i$ and $\mathbf{b}_i$ such that the workspace is somewhat centered around $\mathbf{R} = \mathbf{I}$.
- The cable robot has more cables m than degrees-of-freedom n, i.e. it is kinematically over-constrained.
- The size of the mobile platform is small compared to the machine frame, i.e. $||\mathbf{b}_i - \mathbf{b}_j||_2 \ll ||\mathbf{a}_i - \mathbf{a}_j||_2$ for $i, j = 1, \ldots, m$ $i \neq j$. This is fulfilled for most cable robots.
- The robot is in a fully-constrained pose. For suspended robots at least, the initial pose estimator needs tuning to exploit the minimum energy condition.

The algorithm should satisfy the following requirements:

- Real-time capability: the computation time of the algorithm must be strictly bounded and in the range of milliseconds on available real-time hardware.
- The geometry of the robot is generic, i.e. no special constraints like linearity, planarity, etc. are assumed for the mobile platform $\mathbf{b}_i$ or the machine frame $\mathbf{a}_i$. Nevertheless, it is assumed that the robot geometry is designed to avoid architectural singularities and the like.
- Errors have to be reported reliably, e.g. if no solution is found because it does not exist. Note that for the control system such conditions are exceptions requiring an emergency stop of the robot and we must reliably detect such situations.

From Eq. (3.1), we receive m nonlinear equations for forward kinematics

$$v_i(\mathbf{l}, \mathbf{r}, \mathbf{R}) = ||\mathbf{a}_i - \mathbf{r} - \mathbf{R}\mathbf{b}_i||_2^2 - l_i^2 = 0 \quad \text{for} \quad i = 1, \ldots, m \tag{4.34}$$

that form an over-constrained system. Here, we consider the cables to be linear springs. In general, we cannot expect to solve the above-mentioned equations exactly, but we can minimize the error which can be interpreted as minimizing the potential energy in pretensed cables. Let $U = \sum_i U_i$ be the potential energy of the system and the contribution from each cable reads

$$U_i = \frac{1}{2} k_i v_i^2 \quad , \tag{4.35}$$

where k_i is the stiffness of the ith cables. We assume all cables to have the same stiffness k_i. Then, the minimum of the potential energy U of the system does not depend on the specific value of the stiffness k' and the function for forward kinematics yields

$$\varphi^{\mathrm{DK}}(\mathbf{l}) = \min_{\mathbf{r}, \mathbf{R}} \sum_i^m v_i^2(\mathbf{l}, \mathbf{r}, \mathbf{R}) \quad , \tag{4.36}$$

where the vector $\mathbf{l} = [l_1, \ldots, l_m]^{\mathrm{T}}$ contains the given cable lengths. The function $\varphi^{\mathrm{DK}}(\mathbf{l})$ yields the values $\mathbf{r}^*, \mathbf{R}^*$ that minimize the right side of Eq. (4.36). The idea of finding the pose that minimizes the potential energy can be extended by weighting the energy contribution of each cable to the length of the cable in order to reflect the different spring constants that are reciprocally proportional to the cable's length. We have already discussed this issue in the context of stiffness in Sect. 3.8. Thus, some further fine tuning can be achieved by adding offsets to the effective cable length since there is usually also a certain part of the cable inside the winches or between the winch and the last pulley that contributes to the spring constant but is not kinematically effective. Furthermore, the minimization of the potential energy also reveals an interesting connection between kinematics and dynamics as carried out in Chap. 6.

4.3.4.3   Real-Time Algorithm

In the literature, iterative schemes [72, 139] as well as interval methods [321, 324] were proposed for forward kinematics where the first methods may suffer from not converging, while the latter does not fulfill real-time constraints. Although the computation time of interval methods can be strictly bounded, the worst case computation time is typically so large that they are not applicable for real-time computation. In practice, efficient heuristics are applied to the interval schemes where no specific time bound can be guaranteed. In the following, both approaches are combined in two process steps. Firstly, an initial solution for the pose of the platform $\mathbf{y}_0$ is estimated together with guaranteed bounds through an interval analysis inspired technique. Secondly, a Levenberg–Marquardt algorithm is used to iterate the platform pose from this initial estimate through a least square approach of the over-constrained nonlinear equations.

In the first step of the proposed algorithm, an estimate of the pose $\mathbf{y}_0$ is determined. To estimate the initial position of the platform, an interval analysis inspired approach is adopted. Although inspired by interval analysis, the implementation performs the computation with standard arithmetics since interval libraries are not available on all real-time systems. The basic idea is to strictly bound the position of the TCP. This can be done by axis-aligned bounding boxes that are placed around the winches. For fully-constrained cable-driven parallel robots, the anchor points $\mathbf{a}_i$ are distributed around the workspace. Therefore, the region of intersection of these boxes is relatively small and can be used as an initial estimate for the position.

For the pose estimation, we proceed as follows: The vector loop expressed in Eq. (3.1) can be rewritten to

$$\mathbf{a}_i - \mathbf{r} = \mathbf{l}_i + \mathbf{R}\mathbf{b}_i \quad . \tag{4.37}$$

Since rotation with the matrix $\mathbf{R}$ is conserving the length of an arbitrary vector $\mathbf{b}_i$, it generally it holds true that

$$||\mathbf{R}\mathbf{b}_i||_2 = ||\mathbf{b}_i||_2 \quad . \tag{4.38}$$

Applying the triangle inequality and removing the rotation matrix yields

$$||\mathbf{a}_i - \mathbf{r}||_2 \le l_i + ||\mathbf{b}_i||_2 \quad , \tag{4.39}$$

where the vector $\mathbf{l}_i$ is replaced by its length l_i. Thus, the TCP lies inside a sphere with radius $l_i + ||\mathbf{b}_i||_2$ around the anchor point $\mathbf{a}_i$. Using an interval estimation for this sphere by enclosing the sphere with an axis-aligned box, one receives the bounds

$$\left. \begin{array}{l} \mathbf{r}_i^{\text{low}} = \mathbf{a}_i - (l_i + ||\mathbf{b}_i||_2)[1, 1, 1]^{\text{T}} \\ \mathbf{r}_i^{\text{high}} = \mathbf{a}_i + (l_i + ||\mathbf{b}_i||_2)[1, 1, 1]^{\text{T}} \end{array} \right\} \quad . \tag{4.40}$$

Then, the intersection of all m bounding boxes is calculated from

$$\mathbf{r}^{\text{low}} = \max_i \mathbf{r}_i^{\text{low}} \quad \text{and} \quad \mathbf{r}^{\text{high}} = \min_i \mathbf{r}_i^{\text{high}} \quad . \tag{4.41}$$

If $\mathbf{r}^{\text{low}} > \mathbf{r}^{\text{high}}$, then we have strictly proven that the forward kinematics function has no solutions for the given cable length $\mathbf{l}$. Otherwise, the center $\mathbf{r}_0 = \frac{1}{2}(\mathbf{r}^{\text{low}} + \mathbf{r}^{\text{high}})$ of this bounding box is used as initial estimate. Note that the equations above are trivial to implement on a computer and highly efficient for any number of cables m. If the forward kinematics has multiple solutions, all solutions are guaranteed to be contained in the box.

In Fig. 4.9, a simplified example of a planar 1R2T cable robot with three winches illustrates the geometrical interpretation of the procedure. The dashed circles are centered around the proximal anchor points $\mathbf{a}_i$ and have a radius of $l_i + ||\mathbf{b}_i||_2$. The enclosing rectangles around the circle represent the interval estimate of these circles. The light gray rectangle is the intersection of the boxes and is guaranteed to contain

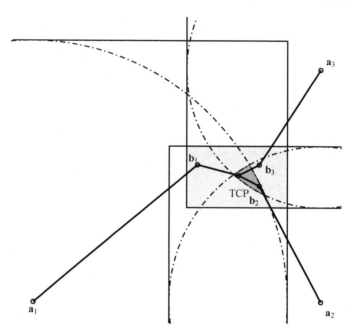

Fig. 4.9 Bounding of solution with axis-aligned boxes

the TCP. The dark gray region can be computed from the intersection of circles which typically gives a better estimate but is also more complex to determine. For example, using interval consistency methods, one can shrink the box to contain only the dark gray region [196]. However, here we simply use the center of the light gray box to start the iteration with the Levenberg–Marquardt method. Note that this bounding technique exploits the over-constrained equations since more equations impose more restrictions on the box and thus produce more accurate estimates. Further improvements can be achieved by additionally considering $l_i - ||\mathbf{b}_i||_2$ as minimum radius, where the computations become more involved in this case. In particular, the computation of the intersection of spherical shells can lead to disconnected regions with possible solutions.

To determine the sought pose $\mathbf{y}$ of the cable robot, a Levenberg–Marquardt method is applied [304]. Given a function $\boldsymbol{\varphi} : \mathbb{R}^n \rightarrow \mathbb{R}^m$ with $m \geq n$, the Levenberg–Marquardt algorithm can be used for obtaining the argument $\mathbf{y}$ that minimizes $||\boldsymbol{\varphi}(\mathbf{y})||_2$. This is done by an iterative procedure $\mathbf{y}_{i+1} = \mathbf{y}_i + \mathbf{h}_i$ where a step $\mathbf{h}_i$ of the Levenberg–Marquardt algorithm is determined by solving the linear system

$$\left[\mathbf{J}_v(\mathbf{y}_i)\mathbf{J}_v^{\mathrm{T}}(\mathbf{y}_i) + \mu\mathbf{I}\right]\mathbf{h}_i = \mathbf{J}_v^{\mathrm{T}}(\mathbf{y}_i)\boldsymbol{\varphi}(\mathbf{y}_i) \quad , \tag{4.42}$$

where μ is the damping parameter and $\mathbf{J}_v$ is the Jacobian matrix of constraints $\boldsymbol{v}$ defined as

$$\mathbf{J}_v = \begin{bmatrix} \dfrac{\partial v_1^2}{\partial y_1} & & \dfrac{\partial v_1^2}{\partial y_n} \\ & \ddots & \\ \dfrac{\partial v_m^2}{\partial y_1} & & \dfrac{\partial v_m^2}{\partial y_n} \end{bmatrix}. \tag{4.43}$$

For the numerical scheme, two threshold parameters ε_1 and ε_2 are proposed [304]. The procedure is terminated if one of the following conditions is reached

$$||\mathbf{h}_i||_2 < \varepsilon_2(||\mathbf{y}_i||_2 + \varepsilon_2) \tag{4.44}$$

$$||\mathbf{J}_v(\mathbf{y}_i)\boldsymbol{\varphi}(\mathbf{y}_i)||_2 < \varepsilon_1. \tag{4.45}$$

The first condition relates to little improvements in the iteration due to a small step size $\mathbf{h}_i$ and the second condition relates to the derivative approaching a stationary point which is hopefully a local minimum.

The computational effort for the bounding procedure is constant and negligibly small. The effort for each iteration step is constant and a maximum number of iterative steps is defined. Thus, the algorithm can be integrated into a real-time environment, given that a reasonably small number of steps is needed. In Sect. 4.3.4.5, the convergency of the algorithm is numerically investigated.

An interesting extension to the basic scheme is to use a box-constrained Levenberg–Marquardt solver instead of the unconstrained version described above. It is possible to assign strict bounds on the pose parameters $\mathbf{y}$ during the optimization process. Since the pose estimation procedure provides guaranteed bounds on the position through $\mathbf{r}^{\text{low}}$ and $\mathbf{r}^{\text{high}}$ and the angular model is restricted to the interval of $[-\pi; \pi]$ anyway, the application of box constraints is straightforward. The LevMar implementation [301] provides both constrained as well as unconstrained versions of the Levenberg–Marquardt methods. The application of the constrained version is slower but bounds the pose to the pre-computed box, fully exploiting the pose estimation procedure.

The typical application of these numerical methods is to compute the kinematic transformation in the real-time controller of a cable robot. Thus, checking for appropriate geometry and tuning of algorithm parameters are acceptable if the numerical performance is improved. The efficiency of the pose estimation procedure can be improved by performing a rigid body transformation to the platform anchor points $\mathbf{b}_i$. Firstly, the platform frame $\mathcal{K}_\text{P}$ needs not to match the task-related TCP but the geometric center of the platform. Thus, the platform frame $\mathcal{K}_\text{P}$ is virtually translated for the kinematic code to the geometric center of platform $\mathbf{b}_c = \frac{1}{m} \sum_i^m \mathbf{b}_i$. Then, the platform anchor points are transformed by $\mathbf{b}_i' = \mathbf{b}_i - \mathbf{b}_c$. Secondly, numerical convergency is improved if the final orientation matrix $\mathbf{R}$ is as close to the identity matrix as possible. Thus, a rotation $\mathbf{R}_c$ is applied to the anchor points $\mathbf{b}_i'' = \mathbf{R}_c \mathbf{b}_i'$ so that the orientation $\mathbf{R} = \mathbf{I}_3$ is inside the workspace. Optimal values for $\mathbf{R}_c$ can be determined from a detailed analysis of the workspace. However, reasonable improvements can be achieved when determining $\mathbf{R}_c$ from intuition such that the home pose of the robot has an orientation equal to the identity matrix.

Clearly, the determined pose computed with the kinematic code has to be transformed back to the original coordinate system by rotating with $\mathbf{R}_C^\mathsf{T}$ and translating along $\mathbf{b}_c$.

4.3.4.4   Implication for Convergency

To further characterize the optimization problem at hand, we consider the objective function $g : \mathbb{R}^m \to \mathbb{R}$ as follows

$$g(\mathbf{l}, \mathbf{r}, \mathbf{R}) = \sum_i^m v_i^2 = \sum_i^m \left(||\mathbf{a}_i - \mathbf{r} - \mathbf{R}\mathbf{b}_i||_2^2 - l_i^2 \right)^2 \quad . \tag{4.46}$$

In order to compute derivatives, we introduce a parameterization of the rotation matrix $\mathbf{R}$ through an angular model with the angles a, b, c. These can be chosen to be e.g. Euler angles or Bryant angles. The pose is thus denoted by $\mathbf{y} = [x, y, z, a, b, c]^\mathsf{T}$. Computing the gradient $\mathbf{G}$ of g yields

$$\mathbf{G} = \nabla g(\mathbf{y}) = \begin{bmatrix} \dfrac{\partial g}{\partial x} \\ \dots \\ \dfrac{\partial g}{\partial c} \end{bmatrix}, \tag{4.47}$$

containing six partial derivatives of the objective function. Since the objective function g is differentiable, the sought optimum corresponds to the pose where $\nabla g = \mathbf{0}$, given exact cable lengths. Furthermore, we will consider the Hessian matrix $\mathbf{H}_v$ of the function g in order to characterize the number and type of extremal values of g. The Hessian matrix of g is given by

$$\mathbf{H}_v = \frac{\partial^2 g}{\partial \mathbf{y}^2} = \begin{bmatrix} \dfrac{\partial^2 g}{\partial x\, \partial x} & \cdots & \dfrac{\partial^2 g}{\partial x\, \partial c} \\ \vdots & \ddots & \\ \dfrac{\partial^2 g}{\partial c\, \partial x} & & \dfrac{\partial^2 g}{\partial c\, \partial c} \end{bmatrix}, \tag{4.48}$$

where the Hessian matrix is symmetric according to the theorem of Schwarz because the function g is continuously differentiable in $\mathbf{y}$.

Numerical studies as well as experimental results from several years of operation of the robot controller indicate that the kinematic code built from Levenberg–Marquardt optimization of g performs both stable and reliable in practice. However, little analysis has been made so far to elaborate a theoretical foundation. In the following section, we present some case studies for point-shaped platforms as well as for planar robots with one rotational degree-of-freedom and two translational degrees-of-freedom (1R2T).

The 2T and 3T Cases

In this section, we analyze the procedure for the generalized robot geometry. We consider the objective function g for robots of the 2T and 3T type. The geometric condition for robots with these two motion patterns is that all cables are connected to the same point on the platform and thus, all vectors $\mathbf{b}_i$ are equal. Without loss of generality, we can therefore assume $\mathbf{b}_i = \mathbf{0}$. Consequently, the equations of the objective function are greatly simplified. To further characterize the optimization problem at hand, we consider the objective function $g : \mathbb{R}^m \to \mathbb{R}$ in 2T case as follows

$$g(\mathbf{l}, \mathbf{r}) = \sum_{i}^{m} \left(\|\mathbf{r} - \mathbf{a}_i\|_2^2 - l_i^2 \right)^2 \tag{4.49}$$

and substituting the parameters of the position $[x, y]^\mathrm{T}$ for the vectors into the expression gives

$$g = \sum_{i}^{m} \left((x - a_{ix})^2 + (y - a_{iy})^2 - l_i^2 \right)^2 . \tag{4.50}$$

Thus, the gradient $\mathbf{G}$ with respect to x and y is computed as follows

$$\mathbf{G} = \sum_{i}^{m} \begin{bmatrix} 4 \left((x - a_{ix})^2 + (y - a_{iy})^2 - l_i^2 \right) (x - a_{ix}) \\ 4 \left((x - a_{ix})^2 + (y - a_{iy})^2 - l_i^2 \right) (y - a_{iy}) \end{bmatrix} \tag{4.51}$$

and the Hessian matrix $\mathbf{H}_\nu$ becomes

$$\mathbf{H}_\nu = \begin{bmatrix} H_{xx} & H_{xy} \\ H_{xy} & H_{yy} \end{bmatrix} , \text{ with} \tag{4.52}$$

$$H_{xx} = \sum_{i}^{m} 12(x - a_{ix})^2 + 4(y - a_{iy})^2 - 4l_i^2 \tag{4.53}$$

$$H_{xy} = \sum_{i}^{m} 8(x - a_{ix}) + 8(y - a_{iy}) \tag{4.54}$$

$$H_{yy} = \sum_{i}^{m} 4(x - a_{ix})^2 + 12(y - a_{iy})^2 - 4l_i^2 \tag{4.55}$$

where for the 3T case the gradient G is extended with the respective terms for the z-coordinate and the Hessian matrix consists of some additional trivial derivatives. A sufficient condition for the optimum of the function g to be unique is that the gradient $G = 0$ and the Hessian matrix $\mathbf{H}_\nu$ is positive definite. For a symmetric 2×2 matrix, this check can be done by testing if the determinant is positive. The eigenvalues of a symmetric matrix are real, therefore, both eigenvalues are positive if the

Table 4.2 A example planar robot with 1R2T motion pattern: platform vectors $\mathbf{b}_i$ and base vectors $\mathbf{a}_i$

Cable i	Base vector $\mathbf{a}_i$	Platform vector $\mathbf{b}_i$
1	$[-2.0, 2.0]^{\mathrm{T}}$	$[-0.05, 0.1]^{\mathrm{T}}$
2	$[2.0, 2.0]^{\mathrm{T}}$	$[0.05, 0.1]^{\mathrm{T}}$
3	$[2.0, 0]^{\mathrm{T}}$	$[0.05, -0.0]^{\mathrm{T}}$
4	$[-2.0, 0]^{\mathrm{T}}$	$[-0.05, -0.0]^{\mathrm{T}}$

determinant is positive. To demonstrate the procedure, we use the geometric parameters for $\mathbf{a}_i$ given in Table 4.2. With actual numbers for the geometry, the determinant of $\mathbf{H}_\nu$ becomes a multivariate polynomial in the position $[x, y]^{\mathrm{T}}$ and the cable length $[l_1, \ldots, l_m]^{\mathrm{T}}$. This polynomial allows to consider the general relation for arbitrary cable length. To remove the dependency from the cable length, the inverse kinematics equation is inserted into $\mathbf{H}_\nu$. This corresponds to the ideal situation without measurement errors or disturbances in the cable lengths. Executing the substitution with computer algebra gives a surprisingly simple expression

$$\det \mathbf{H}_{\nu,\,\text{Ideal}} = 1024(x^2 + 4(y-1)^2 + 4) \quad, \tag{4.56}$$

where for the determinant of $\mathbf{H}_{\nu,\,\text{Ideal}}$ the geometric parameters $\mathbf{a}_i$ are listed in Table 4.2. This expression is obviously always positive for all x and y. Therefore, we have shown for the example robot that the solution is always unique for the forward kinematics by the energy method. The result is also illustrated in Fig. 4.10a that shows the eigenvalues of the matrix $\mathbf{H}_{\nu,\,\text{Ideal}}$ over the area covered by the frame and in Fig. 4.10b we plot the determinant in the same region. From the positive definiteness of the Hessian matrix, we conclude that the objective function is convex which means that we can find a unique solution in the optimization problem to solve the forward kinematics.

The Planar Case 1R2T

We apply the same approach to the 1R2T case where the equations are slightly more complex. Again, we express the position of the platform with the coordinates $\mathbf{r} = [x, y]^{\mathrm{T}}$ and the rotation is given by the rotation matrix $\mathbf{R}$ which is parameterized by the angle φ. Thus, for the 1R2T case, the geometry $\mathbf{b}_i$ of the mobile platform cannot be removed from the equation and we deal with the general case of having arbitrary vectors $\mathbf{b}_i$. Substituting the known quantities into the general over-constrained objective function g (4.46) yields

$$g = \sum_{i}^{m} ((x + \cos(\varphi)b_{ix} - \sin(\varphi)b_{iy} - a_{ix})^2$$
$$+ (y + \sin(\varphi)b_{ix} + \cos(\varphi)b_{iy} - a_{iy})^2 - l_i^2)^2 \quad. \tag{4.57}$$

We compute the gradient $\mathbf{G} = [G_x, G_y, G_\varphi]^{\mathrm{T}}$ as follows

Fig. 4.10 Evaluation of the
smallest eigenvalues and the
determinant of the Hessian
matrix $\mathbf{H}_{\nu,\,\text{Ideal}}$ within the
frame of the 2T robot

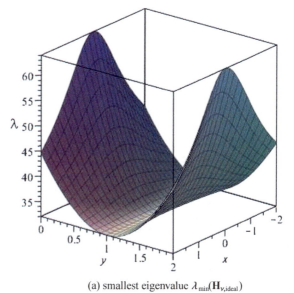

(a) smallest eigenvalue $\lambda_{\min}(\mathbf{H}_{\nu,\text{ideal}})$

(b) determinant $\det \mathbf{H}_{\nu,\text{ideal}}$

$$
\begin{aligned}
G_x = \sum_i^m & 4((x + \cos(\varphi)b_{ix} - \sin(\varphi)b_{iy} - a_{ix})^2 \\
& (y + \sin(\varphi)b_{ix} + \cos(\varphi)b_{iy} - a_{iy})^2 - l_i^2) \\
& (x + \cos(\varphi)b_{ix} - \sin(\varphi)b_{iy} - a_{ix})
\end{aligned} \tag{4.58}
$$

$$G_y = \sum_i^m 4((x + \cos(\varphi)b_{ix} - \sin(\varphi)b_{iy} - a_{ix})^2$$

$$(y + \sin(\varphi)b_{ix} + \cos(\varphi)b_{iy} - a_{iy})^2 - l_i^2)$$

$$(y + \sin(\varphi)b_{ix} + \cos(\varphi)b_{iy} - a_{iy}) \tag{4.59}$$

$$G_\varphi = \sum_i^m 4[(x + \cos(\varphi)b_{ix} - \sin(\varphi)b_{iy} - a_{ix})^2$$

$$+ (y + \sin(\varphi)b_{ix} + \cos(\varphi)b_{iy} - a_{iy})^2 - l_i^2]$$

$$[(x + \cos(\varphi)b_{ix} - \sin(\varphi)b_{iy} - a_{ix})$$

$$(-\sin(\varphi)b_{ix} - \cos(\varphi)b_{iy})$$

$$+ (y + \sin(\varphi)b_{ix} + \cos(\varphi)b_{iy} - a_{iy})$$

$$(\cos(\varphi)b_{ix} - \sin(\varphi)b_{iy})] \tag{4.60}$$

Evaluating the Hessian matrix is possible by repeating the procedure of the case study for the 2T type; however, we do not reproduce the coefficients of the matrix here due to space limitation. To study the expected convergency of the optimization problem, we apply the procedure outlined above. Substituting both a geometry given by Table 4.2 and the ideal cable length into the Hessian matrix provides the desired equations for the determinant of the Hessian matrix. The evaluation with computer algebra provides an expression with around 250 operations to compute the determinant for a pose $\mathbf{y} = [x, y, \varphi]^T$. Results from the computation of the determinant are shown in Fig. 4.11. As one can see in the figures, the determinant is positive for two coordinate planes. A numerical search also shows zero crossings within the workspace. Therefore, we expect the solution to be unique inside the robot machine frame.

4.3.4.5   Implementation and Experimental Results

For this case study, the cable-driven parallel robot IPAnema 1 (Fig. 9.5, see Sect. 9.3.1) is applied. On an industrial PC, the interpolation cycle time of the trajectory generator is between 1 and 4 ms depending on the robot and the version of the control system. The kinematic code for inverse and forward kinematics described in Sect. 4.3.4.3 is implemented into the control system in C language, where the implementation of the Levenberg–Marquardt algorithm is based on the open implementation by Lourakis [301] and lately also on cminpack.

The computation time was determined both on a desktop PC (Intel Core 2 Duo, 2.26 GHz) and on a PC-based real-time controller system (Intel Core 2 Duo, 2.4 GHz). For the following numerical study, the geometry of the IPAnema 1 system was applied (Table 9.1). In total, 5000 randomly chosen poses $\mathbf{y}_i$ within the workspace of the robot were tested and different magnitudes of noise were added to the cable length, simulating measurement and control errors. The thresholds for termination of the Levenberg–Marquardt algorithm were chosen to be $\varepsilon_1 = \varepsilon_2 = 10^{-17}$ and the

Fig. 4.11 Evaluation of the determinant of the Hessian matrix $\mathbf{H}_{\text{Ideal}}$ within the frame of the 1R2T robot. The right plot shows the value of the determinant for $\varphi = 0$ in the xy-plane where the left plot shows the value of the determinant for $y = 0.5$ in the $x\varphi$-plane

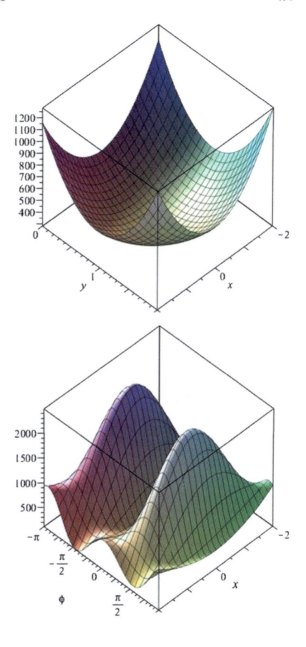

initial damping parameter $\mu = 10^{-3}$. As a parameterization for the pose $\mathbf{y}$, Cartesian position $\mathbf{r} = [x, y, z]^{\mathsf{T}}$ and Bryant angles $\mathbf{R} = \mathbf{R}_Z(c)\mathbf{R}_Y(b)\mathbf{R}_X(a)$ were used. The maximum number of iterations was set to 100 which was never reached in practice. On the desktop PC, the average computation time per evaluation was determined to be

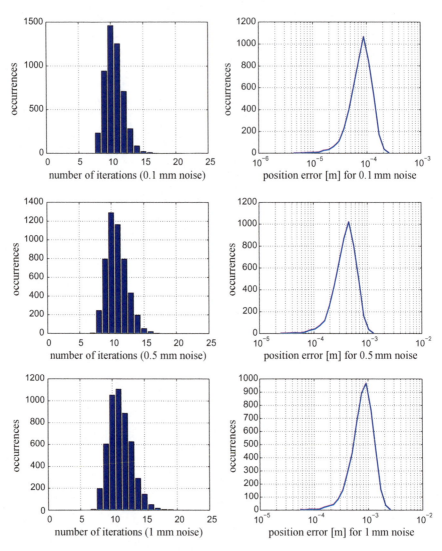

Fig. 4.12 Numerical results for the forward kinematics using Levenberg–Marquardt algorithm. Left column: Histograms of number of iterations for noise 0.1, 0.5, 1 mm on the cable length. Right column: Histograms of position error for noise 0.1, 0.5, 1 mm on the cable length

97 µs.[6] Since the used Windows operating system lacks a high-precision timer with a resolution less than one millisecond, the worst case computation time could not be determined accurately. However, the measured average time and the distribution of required iterations until convergency (Fig. 4.12 left) were encouraging. The number

---

[6]For comparison, a computation time of 47 µs was determined on a more recent Core i5-3320M with 2.6 GHz.

Table 4.3 Evaluation of the forward kinematics for the IPAnema 1 geometry using different implementations of the Levenberg–Marquardt algorithm. The resulting figures are average values for 52250 poses in the robot's workspace. More details on the subject can be found in the literature [431]

Algorithm/setting	Cable length error [mm]	Computation time [μs]	Number of iterations [average (min–max)]	
Analytic Jacobian	N/A	19.5	6.86	(3–8)
Analytic Jacobian	0.1	42.1	8.61	(5–17)
Analytic Jacobian	0.5	42.8	9.02	(5–18)
Analytic Jacobian	1.0	49.1	9.60	(5–18)
Analytic Jacobian, angle preconditioning	N/A	17.2	5.53	(3–6)
Without Jacobian	N/A	29.7	7.94	(3–12)
Bounded, analytic Jacobian	N/A	15.0	4.83	(2–6)
Bounded, without Jacobian	N/A	59.6	4.83	(2–6)

of iterations was between seven and twenty for all poses tested where a typical number of ten iterations is needed. The determined error between the nominal pose $\mathbf{y}_i$ and the determined pose were correlated to the noise in the cable length (Fig. 4.12 right). One can see that the average error of the poses is almost equal to the errors in cable length.

Different variants of the Levenberg–Marquardt algorithm are compared in Table 4.3 with respect to computation time and number of iterations. All computation times are measured on an Intel Core i5-3320M 2.6 GHz, Visual C++ 2010 using the so-called LevMar implementation [301] as solver. A regular grid with values in the range

$$\left\{\mathbf{r} \in \mathbb{R}^3 \mid [-1.5, -1, 0.5]^{\mathrm{T}} \le \mathbf{r} \le [1.5, 1, 1.5]^{\mathrm{T}}\right\} \tag{4.61}$$

was used for the evaluation. For each position of the grid, the nominal cable length $\mathbf{l} = \boldsymbol{\Phi}^{\mathrm{IK}}(\mathbf{r}, \mathbf{I})$ was determined and the resulting cable length $\mathbf{l}$ was fed into the forward kinematics code computing the pose $(\mathbf{r}, \mathbf{R}) = \boldsymbol{\Phi}^{\mathrm{DK}}(\mathbf{l})$ with different algorithms. From the table, it can be seen that one receives rapid convergency with all algorithms and that the use of the analytic Jacobian matrix can largely speed-up the computation. Introducing some disturbance in the scenario slightly decreases computational performance and increases the required average number of iterations from around 6.9 up to 9.6. However, all computation times are in a range that can be achieved with a PC-based real-time control system.

On the real-time controller system, it was not possible to measure the exact time that was consumed for the kinematic transformation but one can only measure the overall time consumed by the transformation and all other controller codes for each cycle. During some hours of operation, no violations of the cycle-time were reported by the real-time control system and the computation time for the whole CNC-kernel, including the kinematic transformation while moving along smooth trajectories, was always less than 1 ms.

4.3.5 Force-Based Forward Kinematics

An operational space formulation of the forward kinematics was presented in [342]. In this approach, an estimate of the stiffness matrix is used as basis for the forward kinematics

$$\widehat{\mathbf{f}} = \mathbf{K}_c \, \Delta \mathbf{l} = k' \mathbf{L}^{-1} (\boldsymbol{\varphi}^{\mathrm{IK}}(\mathbf{y}) - \mathbf{l}) \quad \text{for} \quad i = 1, \ldots m, \tag{4.62}$$

where $\Delta \mathbf{l}$ is the deviation in the cable length, $\mathbf{K}_c$ represents the stiffness model of the cables as introduced in Sect. 3.8, $\mathbf{L}^{-1}$ is the diagonal matrix with the reciprocal cable length, and k' is the stiffness coefficient of the cables (see also Sect. 3.8). The n-dimensional objective function $\mathbf{g}_{os}$ to be minimized becomes

$$\mathbf{g}_{os}(\mathbf{l}, \mathbf{y}) = \mathbf{W}(\mathbf{A}^{\mathrm{T}}\widehat{\mathbf{f}} - \mathbf{w}_0) \quad, \tag{4.63}$$

where $\mathbf{w}_0$ is the applied wrench resulting from the mass of the platform and $\mathbf{W}$ is a weighting matrix to relate rotational to translational displacement. The operational space formulation of the forward kinematics yields the following optimization problem

$$\boldsymbol{\varphi}_{os}^{\mathrm{DK}}(\mathbf{l}) = \min_{\mathbf{y}} \sum_{i=1}^{n} (\mathbf{g}_{os}(\mathbf{l}, \mathbf{y}))^2 \quad, \tag{4.64}$$

which can be solved again with a Levenberg–Marquardt solver.

4.4 Conclusion

For fully-constrained cable robots, the inverse kinematics transformation is straightforward and can be easily implemented even on low-endian computers such as micro controllers. In turn, forward kinematics is a challenging task and different special purpose codes were proposed to solve it. For some non-generic cases, like the 3-2-1 configuration, one can find closed-form solutions. However, for over-constrained robots, one has to deal with the problem of an over-constrained system of equations. In this setting, one can ease the problem by considering elastic cables which leads to an optimization problem that minimizes the energy in the cables and thus the deviations from the ideal cable length. An approach for rapid estimation of a guaranteed trust region for starting an iterative search is proposed. This method is simple to use and rigorous in its estimation. Different formulations are proposed and also approaches to solve the optimization problem under real-time constrains are tackled. The implementation shows a reasonable numerical performance and proves useful in real-world application. Experimental tests suggest solid convergency and a method is proposed to test for stability of the convergency by considering the Hessian matrix of the energy function. Open issues remain for redundant robots in crane configuration. Here, both pose estimation and iteration are less stable since the heuristics proposed here are less efficient.

Chapter 5
Workspace

Abstract This chapter deals with different types of workspace, the criteria used for workspace determination, as well as with algorithms to actually calculate the workspace. In the last part, the influence of the different criteria is compared.

5.1 Introduction

The workspace of a robot is an important property and its characterization is crucial for planning the robot's application. The workspace of some types of robots, e.g. Cartesian gantry robots, can be described with simple geometric primitives such as boxes or cylinders. For these robots, one can give simple but meaningful parameters such as length, height, width, or radius to characterize the dimensions of interest. The workspace description gets more involved when considering translation and orientation. For ease of understanding, the workspace model should be simple, however, the topology of the rotation group is not isomorphic to the Cartesian space. Unfortunately, the motion is more complex for parallel robots and especially for cable robots where translation and orientation are strongly coupled. But since most application engineers have a simple representation of the workspace in mind, one may have to finally reduce the used workspace of a robot to a subset that can be represented by a geometric primitive such as a box, a cylinder, or a sphere (see Fig. 5.1).

5.1.1 Literature Overview

Early studies of suspended cable robots including the consideration of the workspace are presented by Albus [6]. A mathematical sound definition of force-closure, workspace, stiffness, and intersection is presented by Verhoeven [476, 477].

© Springer International Publishing AG, part of Springer Nature 2018
A. Pott, *Cable-Driven Parallel Robots*, Springer Tracts in Advanced
Robotics 120, https://doi.org/10.1007/978-3-319-76138-1_5

158   5   Workspace

Fig. 5.1 The boundary of
the workspace of a parallel
robot is generally curved and
the largest axis-aligned box
may be significantly smaller

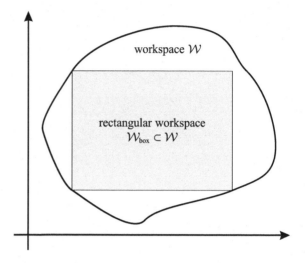

Fattah [141, 142] studies the workspace of suspended planar robots and derived
analytic formulations for the workspace [143]. Riechel [414] presents the structure
and force distributions for a point-mass cable robot in a suspended configuration.
Based on these ideas, analytic formulations for the workspace are derived. Verhoeven
[473] shows that the wrench-closure workspace of CRPM and RRPM robots is
in general bounded by polynomial surfaces and also provides an explicit formula
to compute the polynomials. Rezazadeh [412] proposes a method to compute the
boundary of the workspace of cable-driven multi-body systems.

Gouttefarde [183, 185] shows that the wrench-closure workspace of planar robots
consists of conic sections and elaborated on techniques to determine the boundaries
of that workspace. Later, a technique to compute cross sections of the wrench-closure
workspace of cable robots is proposed and it is shown that the constant orientation
workspace is bounded by cubic surfaces [187]. Gouttefarde [184] provides some
theorems to characterize the boundary of the wrench-closure workspace for six
degrees-of-freedom robots with seven cables. Using other arguments, the results
from Gouttefarde are generalized to spatial robots by Stump [443, 444]. Hadian
[192] studies the wrench-feasible workspace of a specific 6-6 suspended cable robot
and derives explicit formulas for cross section of the translation workspace. Azizian
[23, 24] determines the boundaries of the wrench-feasible workspace for planar
robots. Afshari proposes a method to determine the Jacobian matrix in order to speed
up workspace computation [3]. The workspace is computed through cross sections.
Hassan [202] presents an analytical expression for the wrench-closure workspace for
the example of a storage retrieval machine. Therefore, the author exploits the symmetric geometry of six degrees-of-freedom robots with eight cables by essentially
considering a cross section of the workspace to simplify the statics consideration to
an equivalent of a planar root. Then, the separating hyperplane approach is used to

compute the workspace. Recently, an analytic determination of the boundary of the wrench-closure workspace is reported by Sheng [436].

Bosscher [45, 47, 48] introduces the concept of the available net wrench. The boundaries of the wrench-feasible workspace are determined by the geometric properties of the available net wrench. Bouchard [61] studied the workspace of the Large Adaptive Reflector (LAR), a concept for a very large spatial cable robot. As an extension, the concept of the available wrench set is developed and a test is described to check if a set of applied wrenches can be generated for a specific pose [62].

Brau [65, 66] investigates the so-called tension capable workspace whose definition matches the wrench-feasible workspace for a four cable three degrees-of-freedom robot. Amongst others, the largest sphere which is fully enclosed in the workspace is determined.

The author proposes a constraint satisfaction problem (CSP) formulation for workspace determination of parallel robots that allows to take into account leg length, limits on the passive joints, dexterity criterion, and leg interfaces [395, 397]. Based on this approach, Bruckmann [78] developed an interval test for wrench-feasibility allowing for a guaranteed and continuous workspace computation based on interval analysis. Furthermore, a method for workspace analysis based on a CSP formulation is presented [72]. Gouttefarde [182, 188] uses interval analysis to determine the wrench-feasible workspace where a new test for wrench-feasibility based on Rohn's theorem [150] is used.

A performance index to measure the proximity to the boundary of the workspace is proposed by Verhoeven [473]. Pusey [405, 406] presents workspace studies based on the so-called global condition index, which is the average of the condition number of the structure matrix for the computation of total orientation workspace as well as for parameter design studies. Hadian [192] also uses the global condition index for a 6-6 suspended IRPM robot. Guilin and Yang [190, 503] use the tension factor, which is the pose-dependent ratio between the smallest and largest actual tension in the cables, as performance index. The authors argue that for testing the total orientation workspace it is sufficient to test if the upper bound and the lower bound of the orientation range for each position. Lin [289] also uses the global condition index for a suspended 6-6 robot and presents some numerical results for design studies with different proportions of the robot geometry for a simplified symmetric machine (SSM) geometry. Tang [458] proposes a quality index for the robot pose which is essentially the standard deviation of the cable forces and, based on this index, also a quality index for the workspace by averaging the pose index. Furthermore, the volume for some example robots is estimated.

A couple of works developed special kinds of evaluation procedures to test for the workspace. Ebert-Uphoff [134, 480] discovered the connection between the multi-fingered grasping and cable-driven parallel robots and used the antipodal theorem as tool for cable robot workspace computation. McColl [316] followed this approach to test for wrench-feasibility and thus for workspace computation. Loloei [298] uses a linear matrix inequalities formulation for the structure equation and applied the projective method to test for wrench-closure of poses. For the proposed LMI method, a computation time per pose of around 15 ms is reported. Another approach [299,

300] based on the so-called set of fundamental wrenches is used to compute the workspace as intersection of sub-workspaces. A unifying approach for workspace determination is addressed by Liu [296] who proposes a common approach to deal with the workspace of IRPM, CRPM, and RRPM. Alp presents some workspace studies as part of his control considerations [4]. Oh [365] also deals with the feedback control of a suspended robot with six cables and derives a workspace condition from the stability of his controller. This approach is further extended to compute the admissible workspace for the set-point controller [367].

The workspace of suspended robots is also analyzed by Hamedi [193] who studied the constant orientation workspace and total orientation workspace of suspended spatial cable robots with six cables.

Pham [384] presents geometric parameter studies taking into account wrench-feasibility and stiffness-based workspace computation for planar robots. Later, a method for testing wrench-closure is proposed that is based on a dimension reduction by projecting the spanning vectors of the structure matrix in order to test if the origin is fully enclosed by the columns of the structure matrix [385].

Barrette [28] defines the concept of the dynamic workspace for planar robots and extents the concept also to spatial workspace with six cables.

Ghasemi [161–163] computes the wrench-closure workspace of planar CRPM and RRPM robots and given computation times for the workspace of around 10 seconds. The performance of workspace determination is especially important when using workspace evaluations as parts of a design procedure. Gouttefarde [186] performs workspace computing as part of the parameter synthesis procedure to create a cable robot with a given workspace. An idea using interval analysis is proposed by Bruckmann [68]. Arsenault [17] analyzed workspace of planar robots taking into account the prestress of the cables for use in design.

A couple of works have been published on special robot configurations. Ferraresi [146–148] analyzed the workspace of a six degrees-of-freedom robot with nine cables using numerical methods where a specific method is proposed for testing if a pose belongs to the wrench-closure workspace. Williams [494] proposes a concept for a cable robot with twelve cables and linearly movable proximal attachment points. The numerical studies of the workspace are encouraging in terms of size, shape, and cable environment collisions at the price of a huge number of actuators. Zhang [521, 523] presents an analysis for wrench-closure workspace for a cable robot in 3-3-1-1 configuration.

The determination of the workspace becomes more involved if extensions to the standard model are taken into account. Korayem [252] analyzed the workspace of suspended robots with elastic and hefty cables, thus taking the effect of sagging of the cables into account. Furthermore, a comparison between the workspace of a planar IRPM for ideal cables and hefty cables is presented. Riehl [416] presents a workspace analysis for a huge robot taking the effect of sagging cables into account.

Alikhani [11] presents the workspace of a special robot called BetaBot. The authors state that the robot has a pure translation workspace. Diao and Ma [117, 118] propose a wrench-closure test for CRPM cable robots and compute the wrench-closure workspace by testing a discrete six-dimensional grid of poses.

5.1.2 Workspace Definitions

In general, the workspace of a robot is subset $\mathcal{W}$ of the Euclidian motion group SE_3 that can be generated by moving the robot's end-effector frame and which is measured for a characteristic frame $\mathcal{K}_P$ fixed to the end-effector with respect to an inertial world frame $\mathcal{K}_0$. The SE_3 is represented as a pair $(\mathbf{r}, \mathbf{R}) \in \mathbb{R}^3 \times SO_3$ as it is introduced in the previous sections. Some robots do not realize the full six-dimensional motion pattern but nevertheless their workspace is embedded in the most general Euclidian motion group. For some simple cases, e.g. pure translational motion, one can easily separate the *used* from the *unused* degrees-of-freedom of the motion group. This is in general not possible. Since this most general definition of the workspace is quite abstract, there exist a couple of simplified definitions of the workspace that address mostly application-driven restrictions to describe a subset of the generally six-dimensional space. Many of the definitions reduce or project the six-dimensional workspace onto a two or three-dimensional Cartesian space since this workspace can be easier imagined and visualized. It is important to recall that each of these workspace definitions imply certain assumptions which will be explained in the following sections.

5.1.3 Geometric Descriptions

In the following, we give an enumeration of different commonly used subsets of the generally six-dimensional workspace $\mathcal{W}$ (see also [322]):

- *Translation Workspace* (or constant orientation workspace) $\mathcal{W}_{CO}$: The translation workspace is a slice of the general workspace for one fixed orientation $\mathbf{R}_0$ of the platform defined by

$$\mathcal{W}_{co}(\mathbf{R}_0) = \left\{ \mathbf{r} \in \mathbb{R}^3 \mid \mathbf{y} = (\mathbf{r}, \mathbf{R}), \mathbf{R} = \mathbf{R}_0 \right\} \quad . \tag{5.1}$$

The dimension of this workspace is two for planar robots with motion pattern 2T and 1R2T or three for spatial robots with motion pattern 3T, 2R3T, and 3R3T. Therefore, the translation workspace can be easily visualized.

- *Orientation Workspace* $\mathcal{W}_O$: Contrary to the translation workspace, the orientation workspace (or rotational workspace) is a slice of the general workspace for one given position $\mathbf{r}_0$

$$\mathcal{W}_o(\mathbf{r}_0) = \left\{ \mathbf{R} \in SO_3 \mid \mathbf{y} = (\mathbf{r}, \mathbf{R}), \mathbf{r} = \mathbf{r}_0 \right\} \quad . \tag{5.2}$$

The dimension of this workspace is one for 1R2T robots, two for 2R3T robots and three for 3R3T robots. In the latter cases, visualization might present some problems since the orientation workspace of a 3R3T robot has the topology of the special orthogonal group SO_3 which is not an Euclidian space. Commonly used mappings like Euler angles always introduce some distortion and singularities to the plots. Singularity-free parameterizations like quaternion are not intuitive to understand and cannot be easily visualized in a three-dimensional plot.

- *Inclusion Orientation Workspace* $\mathcal{W}_{IO}$: at least one orientation $\mathbf{R}$ from a set of orientations $\mathcal{R}_0 \subset SO_3$ belongs to the workspace, i.e.

$$\mathcal{W}_{IO}(\mathcal{R}_0) = \left\{ \mathbf{r} \in \mathbb{R}^3 \mid \mathbf{y} = (\mathbf{r}, \mathbf{R}), \mathbf{R} \in \mathcal{R}_0 \right\} \quad . \tag{5.3}$$

One has to define an orientation set $\mathcal{R}_0$ to distinguish the workspace from the translation workspace, which can be understood as the inclusion orientation workspace with $\mathcal{R}_0 = \mathbf{R}_0$.

- *Maximum Workspace*: The maximum workspace $\mathcal{W}_{max}$ is the subset of the general workspace $\mathcal{W}$ that can be reached with at least any one orientation $\mathbf{R} \in SO_3$. The maximum workspace is a special type of inclusion orientation workspace where the orientation set $\mathcal{R}_0 = SO_3$ includes all possible orientations, i.e.

$$\mathcal{W}_{max} = \left\{ \mathbf{r} \in \mathbb{R}^3 \mid \mathbf{y} = (\mathbf{r}, \mathbf{R}), \mathbf{R} \in SO_3 \right\} \quad . \tag{5.4}$$

It is a projection of the six-dimensional workspace to a three-dimensional Euclidian space for 2R3T and 3R3T robots. For planar robots of type 1R2T, it is a two-dimensional plot.

- *Total Orientation Workspace* $\mathcal{W}_{TO}$: The total orientation workspace contains all positions $\mathbf{r}$ where at each position all orientations $\mathbf{R}$ in a given set of orientations $\mathcal{R}_0$ belong to the workspace, i.e.

$$\mathcal{W}_{TO}(\mathcal{R}_0) = \left\{ \mathbf{r} \in \mathbb{R}^3 \mid \mathbf{y} = (\mathbf{r}, \mathbf{R}) \,\forall\, \mathbf{R} \in \mathcal{R}_0 \right\} \quad . \tag{5.5}$$

The dimension is two for planar robots and three for spatial robots. One has to provide the values of the orientation set $\mathcal{R}_0$ to make the workspace definition meaningful (Fig. 5.2). The total orientation workspace is a handy and intuitive description to define the workspace requirements for the generally six-dimensional workspace.

- *Dextrous Workspace* $\mathcal{W}_D$: The dextrous workspace consists of all positions $\mathbf{r}$ at which every orientation $\mathbf{R} \in SO_3$ of the platform can be generated. The dextrous workspace is a special kind of total orientation workspace where the set $\mathcal{R}_0 = SO_3$ includes all possible orientations, i.e.

$$\mathcal{W}_{TO}(\mathcal{R}_0) = \left\{ \mathbf{r} \in \mathbb{R}^3 \mid \mathbf{y} = (\mathbf{r}, \mathbf{R}) \,\forall\, \mathbf{R} \in SO_3 \right\} \quad . \tag{5.6}$$

It seems unlikely that any cable robot has a nonzero dextrous workspace without collisions between its cables.

Besides the workspace archetypes listed above, one can define other forms of projection and reduction on the generally six-dimensional workspace such as cross sections.

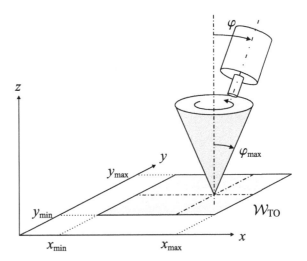

Fig. 5.2 Example for the total orientation workspace $\mathcal{W}_{TO}$. Here the possible angles φ are restricted to lie within a cone with an aperture of φ_{max}. In a practical application, this means that the tool can be tilted within a given angle by any horizontal axis

In the following, the meaning of these definitions shall be exemplified by studying a planar robot of the 1R2T type. The general workspace of these planar robots can be visualized in three dimensions. Therefore, it serves well to highlight the projections and reductions that are presented by the workspace definitions above.

In Fig. 5.3a, the general workspace of a 1R2T planar robot is visualized as a 3-D plot where the orientation angle φ of the platform is included as third axis. For a limited range of $\varphi \in [-5°; 5°]$, a parallel projection is shown in Fig. 5.3b. If we consider a slice of the general workspace by fixing the orientation of the platform to a certain value ($\varphi = 5°$), we receive the translation workspace $\mathcal{W}_{CO}$ shown in Fig. 5.3c as area in $\mathbb{R}^2$. The maximum workspace is obtained if one projects the general workspace in the direction of the orientation φ. The maximum workspace (Fig. 5.3d) can be understood as the 2-D shadow cast by the general 3-D workspace if a parallel light is cast in the direction of φ-axis. When computing the total orientation workspace (Fig. 5.3e), we define an interval $[\varphi_{min}, \varphi_{max}]$ and thereby an orientation set $\mathcal{R}_0$ for the orientation of the platform. Then, one considers a certain position $\mathbf{r} = [x, y]^T$ to belong to the total orientation workspace $\mathcal{W}_{TO}$ if every orientation in the given set $\mathcal{R}_0$ belongs to the general workspace. This is much more restrictive than for the maximum workspace where it is sufficient when at least one orientation in the interval belongs to the workspace (Fig. 5.3e). Finally, the inclusion orientation workspace $\mathcal{W}_{IO}$ for the orientation set $\mathcal{R}_0$ is depicted in Fig. 5.3f.

The notion of *workspace aspects* for parallel robots is introduced by Chablat and Wenger [95] and is useful to describe that the general workspace may consist of separated regions. Thus, Chablat and Wenger define an aspect $\mathcal{W}_A$ of the workspace such that an aspect is a connected, singularity-free region inside the general workspace $\mathcal{W}_A \subset \mathcal{W}$. The notion of aspects makes clear that the workspace consists of distinct regions and the problem of connection between regions is essential in the understanding. Aspects play an important role in classification of multiple solutions of

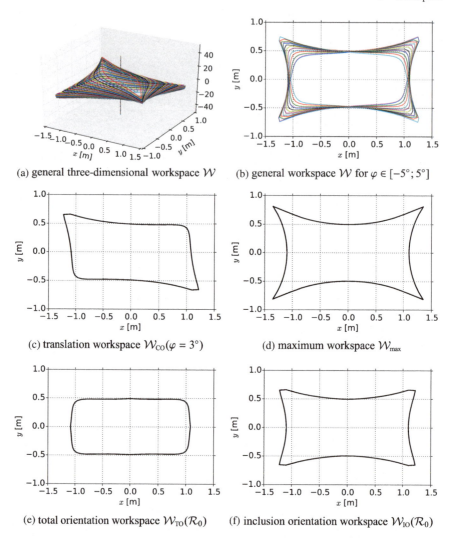

(a) general three-dimensional workspace $\mathcal{W}$ (b) general workspace $\mathcal{W}$ for $\varphi \in [-5°; 5°]$

(c) translation workspace $\mathcal{W}_{\mathrm{co}}(\varphi = 3°)$ (d) maximum workspace $\mathcal{W}_{\mathrm{max}}$

(e) total orientation workspace $\mathcal{W}_{\mathrm{TO}}(\mathcal{R}_0)$ (f) inclusion orientation workspace $\mathcal{W}_{\mathrm{IO}}(\mathcal{R}_0)$

Fig. 5.3 Examples of the different types of workspace for a simple planar robot based on testing wrench-feasibility. The orientation set is $\mathcal{R}_0 \in \{\mathbf{R} \in SO_3 \mid \mathbf{R} = \mathbf{R}_z(\varphi),\ \varphi \in [-5; 5]\}$

forward and inverse kinematics. However, the intuitive assumption that each branch of the forward kinematic relates to exactly one aspect is wrong. In contrast, for conventional parallel robots a singularity-free assembly mode change is possible [85, 519].

Octrees are found to be a suitable data model for modeling the connection between regions. Also, the workspace hull (Sect. 5.5) essentially captures only the workspace aspect around a given center. For conventional parallel robots, the aspects are separated by singularity surfaces. For cable robots, one has additionally to consider the wrench-closure or wrench-feasibility. A frequent cause for multiple aspects occurs in the orientation workspace where a 180° rotation of the mobile platform with respect to the home position may allow to control the platform although no controllable path can be found between these two aspects of the workspace. General properties and methods to compute the number and shape of aspects are open problems for cable robots.

5.1.4   Representation of the Workspace

Mathematically speaking, the workspace can be considered as a set $\mathcal{W}$ with (in most cases) infinitely many poses. In practice, we have to compute some kind of approximations of this set. To store and organize the resulting data, one has to choose a data model. In the following, we give a list of typical data models ordered by increasing complexity:

- *Discrete forms*: A very simple method just collects a finite number of poses belonging to the workspace. Typically, these poses are taken from a regular or adaptive grid. Also, random sampling can be used. The computation is simply done by sampling the space and memorizing the poses that belong to the workspace (Fig. 5.4a).
- *Mashed discrete forms*: The basic idea is the same as for discrete forms. Additionally, one stores the connection of the neighboring elements (Fig. 5.4b). A typical data model for this is an octree or higher-dimensional triangulations. The dimension of the mashed discrete forms equals the dimension of the workspace.
- *Solid geometry*: A continuous representation of space that is capable to exhaustively fill the space. This representation is more complicated to generate and analyze compared to the former ones. For computation schemes which evaluation covers full regions of space, one stores the information in solid geometry (Fig. 5.4c). Typical geometries are boxes (as taken from intervals) or a simplex that fill higher-dimensional spaces. Additionally, the connection to neighboring objects can be stored in the data model. CAD file formats are mature data models of this kind.
- *Hull representation*: The hull, border, or boundary of the workspace is the $n-1$-dimensional surface of the original space. For the types of the workspace listed in Sect. 5.1.3, the hull is either a surface or a curve that can be approximated by triangles or line strips, respectively (Fig. 5.4d). Again, CAD file formats are a good choice to store and process such workspace data.
- *Polynomial form*: The workspace is modeled by a set of multivariate polynomial equations and the data model consists of the coefficients of these polynomials. This approach is used in Sect. 5.6.

Conversion between the data model, listed above is essentially possible but may be involved depending on the desired translation. As a rule of thumb, it is relatively simple to convert a data model into the discrete form and difficult to receive the polynomial form for other data models. However, in Sect. 5.6.3 a surprising method is presented that allows to identify such a conversion from some special points.

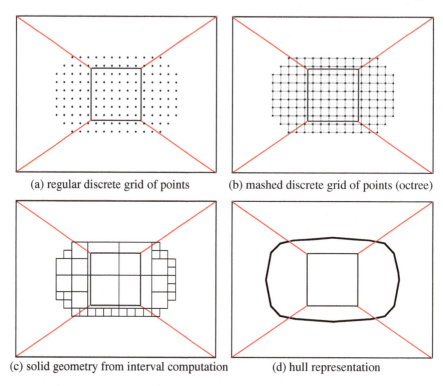

(a) regular discrete grid of points   (b) mashed discrete grid of points (octree)

(c) solid geometry from interval computation   (d) hull representation

Fig. 5.4 Data models to represent the workspace for the example of a planar 1R2T cable robot

5.2 Criteria for Workspace

There are different criteria to decide if a pose belongs to the workspace. The most important criterion that is specific to cable robots is whether the robot can be statically balanced through positive tension in the cables. To do so, there are a couple of sophisticated criteria to analyze the workspace based on the structure matrix (see e.g. [187]). From a kinematic point of view, it is also important whether the platform is in a singular configuration and if collisions with internal or external obstacles occur. A problem which is mostly ignored in the literature is that it is very difficult to construct anchor points for the cables which really allow for infinitely large angles. In practice, one has to take into account such restrictions as limited deflection angles both on the proximal and distal anchor points. Such problems have already been addressed for conventional parallel robots and we can partly apply these techniques to cable robots. Finally, the length of the cables and the spooling capacity of the winch may limit the workspace. In the following, a number of tests are presented to decide in a binary manner whether a pose belongs to the workspace or not.

5.2.1 Wrench-Closure Workspace

A pose $\mathbf{y}$ belongs to the *wrench-closure workspace* $\mathcal{W}_C$ [183] or *controllable workspace* [474] if for every wrench $\mathbf{w}_P \in \mathbb{R}^n$ a distribution of cable forces $\mathbf{f} > \mathbf{0}$, $\mathbf{f} \in \mathbb{R}^m$, so that

$$\mathbf{A}^T(\mathbf{y})\mathbf{f} + \mathbf{w}_P = \mathbf{0} \quad \text{with} \quad \mathbf{f} > \mathbf{0} \quad . \tag{5.7}$$

Thus, the following method can be used to determine an index κ to check for existence and quality of the workspace. If the matrix $\mathbf{A}^T(\mathbf{r}, \mathbf{R})$ has the maximal rank equal to the degrees-of-freedom n of the mobile platform for a given pose $(\mathbf{r}, \mathbf{R})$, let $\mathbf{k}$ be an arbitrary nonzero element of the one-dimensional kernel (or nullspace) of $\mathbf{A}^T$, i.e. $\mathbf{A}^T\mathbf{k} = \mathbf{0}$ with $\mathbf{k} \neq \mathbf{0}$. One can determine such an element $\mathbf{k}$ for example by application of a singular value decomposition of the matrix $\mathbf{A}^T$. Then, the index κ is defined as

$$\kappa = \begin{cases} \frac{\min(\mathbf{k})}{\max(\mathbf{k})} & \text{if} \quad \min \mathbf{k} > 0 \\ \frac{\max(\mathbf{k})}{\min(\mathbf{k})} & \text{if} \quad \max \mathbf{k} < 0 \\ 0 & \text{otherwise.} \end{cases} \tag{5.8}$$

For $\kappa = 0$, the pose $(\mathbf{r}, \mathbf{R})$ does not belong to the workspace and for $\kappa = 1$ the forces in all cables are equal providing an optimal transmission. Note that the index rates the distribution of forces within the cables. To check for the wrench-closure workspace, one computes κ and compares it to a small constant ε_c that serves as a threshold and defines the required quality of the workspace. If the robot is of CRPM type, it is sufficient for the wrench-closure workspace to test if all components of $\mathbf{k}$ have equal signs.

5.2.2 Wrench-Feasible Workspace

The concept of wrench-feasible poses is introduced in Sect. 3.4.3 and is extended to the workspace here. The *wrench-feasible workspace* (WFW) is defined in [134, 188, 474] as follows: The wrench-feasible workspace is the set $\mathcal{W}$ of poses $\mathbf{y}$ of the mobile platform for which for any wrench $\mathbf{w}_P \in \mathcal{Q}$ there exists a vector of cable tension $\mathbf{f}$ in $\mathcal{C}$ so that

$$\mathbf{A}^T\mathbf{f} + \mathbf{w}_P = \mathbf{0} \quad . \tag{5.9}$$

The pose $\mathbf{y}$ is called *wrench-feasible* if it allows at least one solution $\mathbf{f} \in \mathcal{C}$. However, it is atypical to find exactly one solution if $m > n$. In most cases, either no solution or infinitely many solutions exist. To test if a pose belongs to the wrench-feasible workspace, any method introduced in Sect. 3.7 to compute a force distribution can be employed. The closed-form method (Sect. 3.7.5) has been proven to possess excellent numerical performance for fully-constrained robots but leads to finding only a subset of the workspace, especially when applied to suspended robots. The Dykstra method (Sect. 3.7.4) in turn shows robust numerical performance in finding the full workspace but requires significantly more computation time. The wrench set method (Sect. 3.4.7) is also a good choice but suffers from poor numerical performance. For workspace computation, it is not important to compute continuous solutions and, thus, linear programming (Sect. 3.7.1) can also be employed.

5.2.3 Cable Length

The maximum length of cables on the winches or, to be more precise, the maximum stroke that can be controlled with the winch limits the workspace. A similar limitation applies to cable robots using pulley tackles with linear drives. When designing a new cable robot, it is simple to determine the necessary length of the cables a priori and choose appropriate winches. Still, the problem of determining the restrictions caused by limited length of the cables arises when the winches are given or the robot shall be reconfigured. In the setting, it is also necessary to consider the minimum length of the cables: Only the stroke of the winches is given by its mechanical design. Therefore, one can use a longer cable on such a winch where a minimal cable length l_0 is still outside the winches if the winch is fully coiled and, thus, the platform cannot be pulled closer to the proximal anchor point.

It is straightforward to check for a given pose for the minimum l_{min} and maximum cable length l_{max} using the inverse kinematics. This is simply achieved through testing

$$l_{min} \leq \varphi_i^{IK}(\mathbf{r}, \mathbf{R}) \leq l_{max} \tag{5.10}$$

for each cable. For a constant orientation $\mathbf{R}_0$ and the limitation induced by the maximum length of the cable, the feasible region for each cable is a sphere and thus is convex. Hence, the intersection for all m cables is also a convex set. To verify a convex polyhedron such as a box, it is enough to test the vertices of this set. However, this simplification does not apply for the minimum criterion since the respective constraints are not convex.

5.2.4 Dynamic Workspace

The concept of the *dynamic workspace* is introduced in [29] for planar cable robots and has lately also been presented for two degrees-of-free robots [176] and for spatial 3T robots [169]. The problem of dynamic workspace is also tackled implicitly in the motion planning for under-constrained robots [309, 310]. In the latter work, dynamic effects are understood as more restrictive limits of the workspace rather than a concept for enlarging the workspace.

The concept of the dynamic workspace includes dynamic inertia forces and thus, the dynamic workspace is also related to trajectory planning issues. Contrary to the wrench-feasible workspace, the dynamic workspace cannot be defined for single poses but is related to the robot's ability to move along a path with defined velocity and acceleration. Gosselin shows that there exist several parametric trajectories where the dynamic equilibrium is fulfilled in every point on the trajectory. Analyzing the dynamics for a three cable 3T robot, Gosselin shows that the inequalities

$$(\mathbf{r} \times (\mathbf{a}_2 - \mathbf{a}_3) + (\mathbf{a}_2 \times \mathbf{a}_3))^\mathrm{T} \, (\ddot{\mathbf{r}} - \mathbf{g}) > 0 \qquad (5.11)$$

$$(\mathbf{r} \times (\mathbf{a}_3 - \mathbf{a}_1) + (\mathbf{a}_3 \times \mathbf{a}_1))^\mathrm{T} \, (\ddot{\mathbf{r}} - \mathbf{g}) > 0 \qquad (5.12)$$

$$(\mathbf{r} \times (\mathbf{a}_1 - \mathbf{a}_2) + (\mathbf{a}_1 \times \mathbf{a}_2))^\mathrm{T} \, (\ddot{\mathbf{r}} - \mathbf{g}) > 0 \qquad (5.13)$$

must hold true, where $\mathbf{g} = [0, 0, -g]^\mathrm{T}$ is the gravitational acceleration. Then, different parametric curves can be substituted into the inequalities above to test if the motion can be generated by the robot without violating the force limits of the robot. For the example suspended 3T robots in [169], circular and linear motions are proposed as well as motions on the surface of a cylinder and on a sphere. From the numeric examples presented in the paper [169], it becomes clear that the workspace can be increased significantly beyond the footprint of the machine frame by exploiting the platform's inertia. This paper also addressed methods to reach such stable trajectories from standstill by blending.

5.2.5 Singularities

The definition of singularities is introduced in Sect. 4.2.4. Using these criteria, one can check whether a given pose is singular or regular. More advanced techniques target at describing the geometric form of the singularities, e.g. using Grassmann geometry [317, 322].

One can basically test for singular configuration by computing the rank of the structure matrix and compare it to the maximum rank, i.e. if rank $\mathbf{A}(\mathbf{r}, \mathbf{R})^\mathrm{T} < n$, the robot is in a singular configuration. Since singularities can be zero-dimensional points in the motion group, it might be difficult to detect singular configurations. Using interval analysis, it is basically possible to verify if any pose in a finitely large region of the motion group is in a singular configuration.

5.2.6 Cable–Cable Interference

Under the topic *cable interference*, all kind of collision issues between two cables are collected. Maeda [305] used a simple geometric pose-dependent test for interference by computing the geometric distance between two cables for the control of the WARP demonstrator. Merlet [320] is the first to do workspace studies for cable interference where cable-cable and cable-platform intersections were taken into account for the computation of the constant orientation workspace. Perreault [382] presents criteria for workspace optimization based on wrench-closure and interference between cables as well as cables and moving bodies. This work is later [381] extended to a highly efficient global method to compute the region of cable interferences for the constant orientation workspace. Ghasemi [165] addresses the determination of the collision free workspace of spatial cable robots taking into account both cable-cable and cable-platform interferences. Aref [14] performs workspace computation for suspended robot designs taking into account collision and restriction from workpieces in the workspace. A guaranteed test for interference based on interval analysis is presented by Blanchet [39]. Nguyen [361] describes a number of geometry based intersection tests.

Wischnitzer [495] analyzes kinematics under the effects of cable collisions where no friction is assumed to act on the cables in contact. Otis [374] takes up the idea of allowing cables to cross. In contrast to the approach from Wischnitzer, Otis proposes to release the forces in one of the colliding cables and to continue motion with one of the colliding cables slack and without effective force on the platform while keeping the other cable effective. Furthermore, strategies how to select the active and the released cable are described.

Lahouar [272] analyzes aspects of collision-free path-planning for cable robots. Collision is also tackled by Verhoeven [473] where especially the design aspect is considered. It is advised as a design rule to connect as many cables as possible to the same proximal or distal anchor point. Doing so clearly avoids collision between the respective cables. However, such geometry can hardly be realized from a mechanical engineering point of view.

A very interesting technique to calculate the influence of cable interference within the constant orientation workspace is presented by Perreault [381]. Through purely geometric considerations, it is possible to determine the loci of cable-cable interference from the geometry of the frame $\mathbf{a}_i$ and the relative geometry of the mobile platform $\mathbf{b}_i$. The main concept of this approach is the simple fact that two cables can interfere only if the corresponding anchor points $\mathbf{a}_i$, $\mathbf{a}_j$, $\mathbf{b}_i$, $\mathbf{b}_j$ lie in a common plane. Since the anchor points on the frame are fixed in space, the plane can be constructed as follows: Calculate the normal vector $\mathbf{n}$ that is perpendicular to the lines $\overline{A_i A_j}$ and $\overline{B_i B_j}$

$$\mathbf{n} = (\mathbf{a}_j - \mathbf{a}_i) \times (\mathbf{b}_j - \mathbf{b}_i) \quad . \tag{5.14}$$

An implicit equation for a point $\mathbf{r}$ belonging to the plane E is then

$$E : (\mathbf{r} - \mathbf{a}_i) \cdot \mathbf{n} = 0 \quad . \tag{5.15}$$

As a model of the possible interference region, one can compute the normals of the connection lines between proximal and distal anchor points from

$$\mathbf{a}_{ij} = \frac{\mathbf{a}_j - \mathbf{a}_i}{||\mathbf{a}_j - \mathbf{a}_i||_2} \quad \text{for} \quad i, j = 1, \ldots, m \quad i \neq j \tag{5.16}$$

$$\mathbf{b}_{ij} = \frac{\mathbf{b}_j - \mathbf{b}_i}{||\mathbf{b}_j - \mathbf{b}_i||_2} \quad \text{for} \quad i, j = 1, \ldots, m \quad i \neq j. \tag{5.17}$$

If $\mathbf{a}_{ij}$ and $\mathbf{b}_{ij}$ are not parallel, one can construct two triangles

$$T_{ij}^+ : \mathbf{x} = \mathbf{a}_j - \mathbf{b}_i + \lambda \mathbf{a}_{ij} + \nu \mathbf{b}_{ij} \tag{5.18}$$

$$T_{ij}^- : \mathbf{x} = \mathbf{a}_i - \mathbf{b}_j - \lambda \mathbf{a}_{ij} - \nu \mathbf{b}_{ij}, \tag{5.19}$$

with $\lambda, \nu > 0$. Exploiting the normalized length of the vectors $\mathbf{a}_{ij}$ and $\mathbf{b}_{ij}$, one practically chooses a length for λ and ν in the range of the size of the robot to receive finitely large triangles with the critical interference region. These triangles can be used for visual or automatic detection of cable-cable interference. For many robot designs, one can see from first glance if the triangles are within the workspace of interest or outside. Further information on dealing with the special cases with parallel vectors can be found in [381].

5.2.7 Cable-Platform Collisions

Beside the limitations from the distal connection point, collisions between the cable and the platform are generally possible. Given the shape of the mobile platform, e.g. as CAD data, it becomes obvious that collisions between the mobile platform and the cables might occur. An approach to detect such collisions has lately been proposed by Tempel [463]. The idea is to convert approximate the CAD model through triangles, which can be done with every modern CAD system, and to export such data, e.g. as STL. Then, one has to compute a look-up table for each distal anchor point and perform a polar decomposition of the triangles extracted from the CAD geometry. Using the sorting method proposed for the cable span (see Sect. 5.5.6), one can find a simple but efficient table to compare the actual cable direction vector with a look-up table. Since the look-up table can be stored efficiently as a simple table with critical deflection angles, it is convenient to test a given direction vector. In contrast to the cable span approach, the direction vector $\mathbf{u}_i$ is intended to lie outside of the region described by the look-up table.

5.2.8 Restrictions on the Cable Anchor Point

An assumption made in many cable models is that cables are assumed to have no bending stiffness and thus allow for an arbitrarily small bending radius. Real cables suffer from significant fatigue if certain design rules are violated. Therefore, one has to consider the deflection angle of the cables both on the platform and on the base. In a couple of robots, the distal ends of the cables are connected to universal or spherical joints. Here, the limitations of these joints have to be taken into account instead. At the proximal end of the cable, pivoting pulleys are typical to guide the cable when leaving the winch. In many cases, we can model these limitations by considering the cable direction represented by a vector $\mathbf{u}_i$ in a local coordinate system $\mathcal{K}_{A,i}$ and $\mathcal{K}_{B,i}$. Now, let the proximal frame be arranged as introduced in the section on pulley kinematics (see Sect. 7.2.1). One can compute the deflection and wrapping angles as discussed in the section on inverse pulley kinematics and compare these values against the feasible range for the angles β_i and γ_i which reflect the requirement to have a minimum and maximum wrapping angle on the pulley and a limited rotation capacity of the pulley. Geometrically speaking, the cable direction vector $\mathbf{u}_i$ has to fulfill a restriction that can be understood as being inside a cone or pyramid.

For the distal anchor point, the connection between the cable and the platform is usually realized through a spherical joint, a universal joint, or a swivel bolt. Let the platform frame $\mathcal{K}_{B,i}$ with the transformation matrix $\mathbf{R}_{B,i}$ be aligned so that the cable direction vector without deflection coincides with the z-axis of $\mathcal{K}_{B,i}$. It is straightforward to compute the pose $(\mathbf{R}, \mathbf{r})$ dependent deflection angle $\gamma_{B,i}$ with respect to the nominal angle from

$$\cos \gamma_{B,i} = {}^{0}\mathbf{e}_{ZB,i} \cdot \mathbf{u}_i = (\mathbf{R}_{B,i}\,\mathbf{e}_z) \cdot \frac{\mathbf{a}_i - \mathbf{r} - \mathbf{R}\mathbf{b}_i}{||\mathbf{a}_i - \mathbf{r} - \mathbf{R}\mathbf{b}_i||_2} \ . \tag{5.20}$$

A remarkable distinction of the different mechanisms used to connect the cable to the platform is that universal joints and spherical joints restrict the maximum feasible value for $\gamma_{B,i}$ to a specific value which depends on the technical details of the machine element and is around $\gamma_{max} = 45°$. In contrast, a swivel bolt has a kinematic singularity for $\gamma_{B,i} = 0$ and the region around the extended position must be avoided. In turn, this limits its use for cable robots. The maximum angle for cables can easily reach

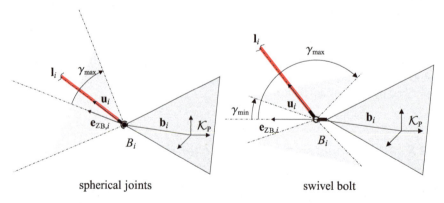

spherical joints   swivel bolt

Fig. 5.5 Feasible deflection angles for the cable on the platform for spherical joints and swivel bolts

$\gamma_{max} = 135°$ where collisions with other parts of the platform become more likely than reaching this value. The effect is visualized in Fig. 5.5 where the feasible area for the cable is shown both for a spherical joint and a swivel bolt.

5.3 Classification of Algorithms for Workspace Determination

As defined above, the workspace $\mathcal{W}$ is a nontrivial set of poses. In this section, we discuss how one can compute the workspace. There are different approaches for the determination of the workspace of cable robots with exemplary references:

- Determination of points on a discrete spatial grid [477]
- Search strategies on dynamically generated grid [389]
- Determination of the boundary of the workspace with analytic or numeric methods [24]
- Geometric methods [322]
- Validation of finite continuous sets such as intervals or higher-dimensional boxes [68, 182].
- Closed-form methods [473].

Three main concepts can be distinguished: *Discretization methods* investigate the workspace at a finite number of different poses which are basically zero-dimensional entities. These can employ any of the discretization techniques discussed earlier in Sect. 5.1.4. *Continuous methods* try to identify for a set of poses whether they belong to the workspace. The set is usually described by ranges of values. The sets can be inspired by geometric interpretation (e.g. boxes, spheres) or by formal approaches such as interval analysis. Geometric methods belong to the continuous methods and derive workspace-related properties on well-defined geometric objects. Then, geometric computations, such as intersections and unions, are used to compose the workspace. A main advantage of continuous methods is that if the patches used in the continuous methods have the same dimension as the general workspace itself, the workspace can be exhaustively covered in a finite number of computational steps. Finally, one can determine the $n - 1$-dimensional *workspace boundary*. In the following, we discuss the different approaches in more detail.

5.3.1 Discretization Methods

The discretization method for workspace determination is wide-spread and straightforward to apply on examples. Most authors used regular grids to approximate the workspace. The basic idea is to test the workspace at a number of poses that are nodes of a grid. This leads to a list of zero-dimensional objects. If the nodes are generated from a regular grid (e.g. a Cartesian, spherical, or cylindrical grid, see

Fig. 5.6a, b) there is a native connection to neighboring nodes. In regular grids, for each considered parameter of platform's pose, one defines a range from minimal to maximal values. Then, this range is tested with a given step width. Other ways to define the grid are random sampling of the workspace and adaptive methods where additional poses are defined between known poses to refine the search. While regular grids are relatively easy to analyze but numerically costly to compute, more general structures allow for better computational efficiency but present additional challenges to interpret the results.

Every property that can be calculated for a given pose can be used as criterion for the discretization method. If any one neighboring node is not part of the workspace, the current node belongs to the boundary of the workspace. Anyway, there are numerous situations where it is hard to determine the topological structure of the workspace, for example if there is a void close to the boundary. The advantages of discretization methods are

- A straightforward implementation and also relatively simple visualization of the results. Different grids can be used. More advanced methods produce data models that cannot be visualized with standard software.
- No assumptions about the topological structure of the workspace are needed a priori. Anyway, if the topological structure of the workspace is actually complicated, discretization delivers little insight into the structure.
- Little requirements for a pose property to be used for workspace calculation. Each criterion that allows a Boolean evaluation can be used. No special assumptions have to be made for the underlying mathematical or geometric structure.
- The volume of the workspace can be estimated by the number of samples belonging to the workspace. Thus, measuring coverage compared to a desired shape and size is straightforward.

Contrary the following disadvantages must be taken into account:

- The determination of the workspace boundary is complicated and inaccurate. Furthermore, the boundary is usually not smooth.
- No information about the space between the nodes of the grid is gained.
- Time consuming computation. Especially refining the resolution of the grid is expensive to compute when orientation is included.
- Poor sensitivity to changes in computation settings, such as changes in the geometry as well as algorithm and technical parameters. Grid methods are especially insensitive to infinitesimal changes in the parameters making it inappropriate to compute the derivatives of the workspace under changes in the robot or algorithm parameters.

Fig. 5.6 Constant
orientation workspace of a
example 1R2T planar robot
computed with different
kinds of discretization grids

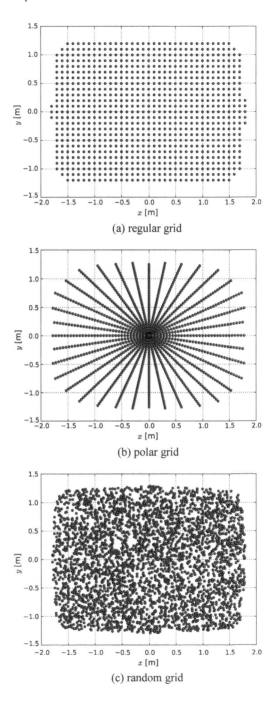

(a) regular grid

(b) polar grid

(c) random grid

5.3.2   Analytical Methods for Determination of the Workspace Boundary

To apply analytic methods for workspace computation is necessary to formulate a connection between the pose and the parameters of interest. This is in general quite difficult. When considering workspace criterions such as wrench-feasibility, singularities, or interference, the decision if a pose belongs to the workspace is derived from a lengthy algorithmic computation with iterations and distinction of cases. This is straightforward to do in a computer program but analytic mathematical tools cannot be applied in such situations.[1]

Verhoeven [473] proposes an analytical method for the determination of the boundary of the workspace. Moreover, a formula is derived to calculate the boundary of the wrench-closure workspace from a system of univariate polynomial inequations (see Sect. 5.6.1). Unfortunately, the general expressions of the equations are so complex that it seems out of reach to deal with these equations even when using advanced computer algebra systems. Beside the practical use of these equations, it provides some insight in the structure of the workspace because it shows that the wrench-closure workspace is bounded by polynomial surfaces which a degree of n. Some methods to exploit the structure of special problems are detailed in Sect. 5.6.

A subset of the analytic methods reduces the mathematical description of the workspace to lower-dimensional objects such as planes. Then, one can apply a mathematical tool to analyze the curves that are generated from the intersection of the workspace with the plane. Perreault [381] uses such an approach to describe the region of intersection with planes. The computation of the workspace hull by triangulation [389] can be understood as practical example of this approach.

5.3.3   Geometrical Methods

Geometrical methods derive a geometric interpretation of the limitations of the workspace. The limits that arise from the maximum cable length are good examples to understand the geometrical approach. We consider a cable robot of the 3T type. The main limitation through the cable length implies that the distance between the platform characteristic point and the anchor points on the frame lies in the interval given by the minimum and maximum cable length. For each cable i, this can be interpreted as a hollow sphere centered around the respective anchor point A_i (Fig. 5.7). The overall workspace of the robot is then the intersection of all m spheres. This simple example only takes the limitation from the cable length into account. A considerable collection of such tools for workspace determination is described by Merlet

---

[1]For computer codes, techniques such as automatic or symbolic differentiation can be applied. Here, such approaches are not understood to be conventional algebraic or analytic methods. Even if one succeeds in extracting the required symbolic expressions, such equations are usually so long that it is pointless to work with them.

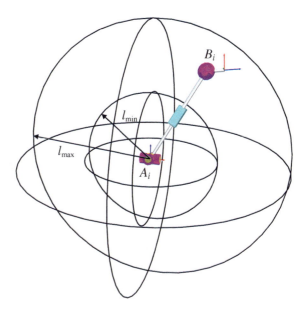

Fig. 5.7 Reachable workspace of a single cable bounded by its minimum l_{min} and maximum l_{max} length

[322] but is also used by other authors. This includes methods for considering the reachable workspace as well as limitations of the active and passive joints are proposed that can be applied to cable robots. If the geometric problem is reduced to the two- and three-dimensional space, sophisticated geometry kernels as used in modern CAD tools can efficiently deal with the geometric operations. Also, raytracer have been used to compute the workspace (Fig. 5.8). Because of the practical need to limit this procedure to the three-dimensional space, it is mostly applied to compute the translation workspace.

5.3.4 Continuous Methods

From the presentation of the workspace criteria, it becomes clear that at least some properties have a nontrivial connection between the geometry of the robot, the pose, and the respective quality. Continuous methods are based on mathematical tools that extend a point-wise investigation of the operational space to techniques that apply for regions with a finite volume that equals a set of infinitely many poses. Especially interval analysis has proven useful for the analysis and parameter synthesis of parallel robots including cable robots. A short introduction to interval analysis is attached in the appendix (see Appendix B). Such continuous methods for workspace determination of cable robots are presented by Bruckmann [68], Merlet [409], Blanchet [39], and Gouttefarde [182, 188]. Lately, Lamine [279] also applied rigorous interval computations for the workspace as parts of a design method.

Fig. 5.8 Example of the determination of the translation workspace of a parallel robot with six legs and six degrees-of-freedom through geometric computations with the raytracer Povray

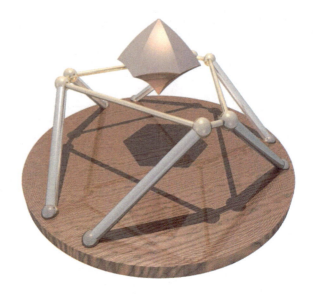

In the following sections, two main approaches are followed: workspace computation based on interval analysis is introduced which provides a rigorous but relatively slow tool to compute the workspace and its properties. Another section is dedicated to rapid computation of the workspace hull. The approach is subject to some restricting assumptions but allows for very fast and versatile usage and has proven to be very applicable to analyze and develop demonstrators. Using the pose-dependent properties which are presented in Sect. 5.2, it is trivial to compute the regular grids and we only present some results of grid-based workspace computing since the implementation of some nested for-loop need no further discussion. In the following section, the application of interval analysis to workspace computation is presented.

5.4   Continuous Workspace Analysis

Most methods discussed so far are used to analyze the behavior of cable robots at discrete poses, i.e. to give insight into the properties of the robot on a point-wise base. Such methods give precise but locally limited results. When sampling the workspace, one does not get any information of the poses lying between the sample points. We cannot cover the full workspace with such methods since we have to rely on a finite number of evaluations. Even if a long computational time is acceptable, it remains unclear if the whole desired workspace is detected, as there is no guarantee that points lying between those which are sampled fulfill the investigated criteria. Formal proofs that guarantee certain properties between two sampling points are possible from a mathematical point of view. However, such approaches are hardly used in robotics. For parallel robots, it is known that they can change their properties within the workspace. Singularities for example are defects in the transmission behavior of the robot that can be point-shaped or have the structure of lower-dimensional surfaces. The probability of finding such a configuration by random or regular grids tends to be zero since the singularity surfaces fill a lower-dimensional subspace of the robot's workspace. Thus, the absolute ratio between the points of this lower-dimensional manifold and the space where it is embedded is zero and so is the statistical probability of finding such a point by random sampling. Contrary to serial robots, the singularities of parallel robots can be located inside the workspace. Therefore, the results of discrete evaluations cannot be extended to the whole workspace.

Therefore, we introduce methods based on interval analysis that provide verified information about a domain of the workspace, i.e. about a connected open subset of the workspace. This idea is put forward for parallel robots by Hao and Merlet [197], and further carried out by the author [388, 395, 397] as well as by Bruckmann [68, 74, 78]. In the following, we extend the framework presented in earlier works to cable robots. A more detailed introduction to interval analysis is given in Appendix B. Since these domains have a finite volume, one can exhaust a given workspace in finite evaluations. This is the main benefit from using interval analysis. Performing a numerical computation for a limited number of data sets provides us with a rigorous and guaranteed information about infinitely many poses. Therefore, one can analyze the properties of the workspace in a *continuous* manner. Especially, one can verify (or refute) that certain constraints are fulfilled in every pose belonging to the workspace while taking the worst case behavior of the constraints into account. When typical implementations of interval analysis are applied, such computations can even be done in a robust way that takes the round-off errors of the computer into account. Interval methods can be applied to many calculations. Anyway, some criteria do not give meaningful answers when investigated through interval analysis. This is because interval analysis is very dependent and sensitive on the mathematical formulation of the problem. The interval computations are guaranteed to enclose the correct solution but might not lead to usable results in acceptable computation times. In this setting, we have to note that other methods suffer from such disadvantages as well but it might be more difficult to realize that the results are not accurate at all. The interval methods introduced here might fail to find the desired solution but this shortcoming is always reported rather than neglected.

In the next section, we introduce the notion of a *constraint satisfaction problem* (CSP) [196] which will be the basis of all interval algorithms that we discuss. Then, we describe the interval algorithms to solve such problems. Finally, we derive the constraints[2] for the CSP from the kinematic model of the cable robot. Each constraint represents a property of the robot that is taken into account in the CSP. Especially the constraints for workspace analysis are introduced, taking into account typical kinematic, force-closure, interference, and technological requirements. Finally, some computational examples are discussed.

---

[2]It must be noted that the constraints considered here are not identical with the definition of a constraint as it is often used in the kinematic analysis.

5.4.1   Algorithms for Solving Constraint Satisfaction Problems

The aim of the proposed approach is to provide a versatile tool for both analysis and synthesis of the cable robots. This approach allows to cover all phases of the design beginning with the analysis and preliminary estimations and ending with the optimization of cable robot. All necessary steps can be performed on the same constraints and additional constraints and effects can be added when needed without reconsideration of the model. Firstly, the problem

$$\mathbf{\Phi}(\mathbf{c}, \mathbf{v}) > \mathbf{0} \quad \forall \, \mathbf{v} \in \mathcal{X}_v \tag{5.21}$$

is considered, where the *calculation variables* are collected in the vector $\mathbf{c}$ and the *verification variables* are represented by the vector $\mathbf{v}$. The set $\mathcal{X}_v$ is referred to as *verification domain*. The problem of finding all feasible solutions of $\mathbf{\Phi}$ is called *constraint satisfaction problem* and basically determines the set

$$\mathcal{X}_c = \left\{ \mathbf{c} \in \mathbb{R}^n \mid \mathbf{\Phi}(\mathbf{c}, \mathbf{v}) > \mathbf{0} \quad \forall \, \mathbf{v} \in \mathcal{X}_v \right\} \tag{5.22}$$

for a given CSP $\mathbf{\Phi}$. Therefore, we denote algorithms to compute the solution set $\mathcal{X}_c$ of the CSP as *CSP solver*. In this section, the pose $(\mathbf{r}, \mathbf{R})$ of the robot is normally identified with the calculation variables because we consider the workspace to be the sought set. In this setting, the solution set $\mathcal{X}_c$ equals the workspace $\mathcal{W}$ of the robot and different types of the workspace, such as the translation workspace or the total orientation workspace, can be determined when appropriately choosing $\mathbf{c}$ and $\mathbf{v}$. The objective of the CSP solver is to determine all poses of the robot that fulfill given requirements where especially the criteria discussed in Sect. 5.2 are relevant for workspace quality. Therefore, the requirements for cable robots are rewritten so that they can be employed in the CSP as given by Eq. (5.21).

Interval analysis [196, 410] has proven to be a powerful tool to deal with CSP and we will use it for workspace determination of cable robots through this section [68]. The basic idea of the following algorithms is to apply an interval evaluation of the constraints $\mathbf{\Phi}$ for a box $\widehat{\mathbf{c}}$ to receive guaranteed information about all points $\mathbf{c} \in \widehat{\mathbf{c}}$ enclosed by the box. For this purpose, the interval evaluation $\widehat{\mathbf{h}} = \mathbf{\Phi}(\widehat{\mathbf{c}})$ and the result $\widehat{\mathbf{h}}$ can be classified as follows (Fig. 5.9):

Fig. 5.9 Evaluation of an inequality $\Phi > 0$ with a CSP-solver

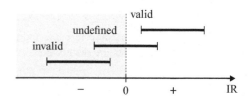

- If $\inf \widehat{\mathbf{h}} > \mathbf{0}$, then the inequality $\boldsymbol{\Phi}(\widehat{\mathbf{c}}) > \mathbf{0}$ is fulfilled in *every* point $\mathbf{c} \in \widehat{\mathbf{c}}$. Such a box $\widehat{\mathbf{c}}$ is called *valid*.
- If $\sup \widehat{\mathbf{h}}_i < \mathbf{0}$, then the inequality $\boldsymbol{\Phi}(\widehat{\mathbf{c}}) > \mathbf{0}$ is not fulfilled in *any* point $\mathbf{c} \in \widehat{\mathbf{c}}$. Thus, the box $\widehat{\mathbf{c}}$ is called *invalid*.
- In any other case, we cannot get a reliable statement. Due to the inherent overestimation of interval analysis, the box might consist of valid as well as invalid points. We do not even know if valid or invalid points are included at all. Such boxes are called *undefined*.

It shall be emphasized that the special properties of interval evaluation indeed proof the validity or invalidity with mathematical rigorousness. This fundamental property of interval analysis is used in the following algorithms to approximate the solution set.

5.4.1.1 Generic CSP Solver

In the following, a branch-and-bound algorithm based on interval analysis is presented to compute the set $\mathcal{X}_c$ of the CSP which has the type given in Eq. (5.21). The problem is solved by computing guaranteed bounds for smaller problems that are successively constructed from the original problem by subdivision (*branching*). This approach is then applied to the verification and calculation of the workspace of cable robots. The conditions for a pose to belong to the workspace are written as constraints that form a system of inequalities $\boldsymbol{\Phi} > \mathbf{0}$ which are in turn systematically evaluated. The main task is to find the solution set

$$\mathcal{X}_c = \left\{ \mathbf{c} \in \mathcal{X}_s \subset \mathbb{R}^n \mid \boldsymbol{\Phi}(\mathbf{c}) > \mathbf{0} \right\} \tag{5.23}$$

that fulfills all considered conditions. In this setting, we denote the set $\mathcal{X}_s \subset \mathbb{R}^n$ as *search space*. The generic algorithm consists of the following steps:

Algorithm 1: `Generic Interval CSP-Solver`

---

1. Save an approximation $\{\widehat{\mathbf{c}}_1, \ldots, \widehat{\mathbf{c}}_n\}$ of the search space $\mathcal{X}_s$ in the list $\mathcal{L}_T$.
2. Create empty lists $\mathcal{L}_S, \mathcal{L}_I, \mathcal{L}_F$ for the solution set ($\mathcal{L}_S$), for the invalid boxes ($\mathcal{L}_I$), and the set of undersized boxes with a diameter small than the threshold ($\mathcal{L}_F$).
3. If the list $\mathcal{L}_T$ is empty, terminate the algorithm.
4. Extract the next box $\widehat{\mathbf{c}}$ from the list $\mathcal{L}_T$.
5. If diam $\widehat{\mathbf{c}} < \varepsilon$, i.e. the width of all components of the box $\widehat{\mathbf{c}}$ is smaller than a given threshold ε, the box is undersized; store the box in the list $\mathcal{L}_F$; go to step (3).
6. If available, apply *prune* and *bound improvement* operations to the box $\widehat{\mathbf{c}}$.
7. Evaluate the constraint $\widehat{\mathbf{h}} = \boldsymbol{\Phi}(\widehat{\mathbf{c}})$.

8. If $\inf \widehat{\mathbf{h}} > \mathbf{0}$, i.e. the infimum of $\widehat{\mathbf{h}}$ is positive for all constraints which means that the constraint is valid for every $\mathbf{c} \in \widehat{\mathbf{c}}$, store $\widehat{\mathbf{c}}$ as solution in the list $\mathcal{L}_S$; go to step (3).

9. If a $\sup \widehat{h}_i < 0$ exists, i.e. at least one supremum is negative, then no solution exists $\mathbf{c} \in \widehat{\mathbf{c}}$ and the box $\widehat{\mathbf{c}}$ is stored in $\mathcal{L}_I$; go to step (3).

10. Split the box $\widehat{\mathbf{c}}$ into m sub-boxes $\{\widehat{\mathbf{c}}_1, \ldots, \widehat{\mathbf{c}}_m\}$ and store these new boxes in the list $\mathcal{L}_T$; go to step (3).

The data flow between the lists of the generic solver is illustrated in Fig. 5.10 for better reference. The initial search space is placed in the list $\mathcal{L}_T$ and sequential evaluations aim at finally placing every box $\widehat{\mathbf{c}}_i$ in one of the three lists $\mathcal{L}_S$, $\mathcal{L}_I$, and $\mathcal{L}_F$. If no guaranteed statement can be made, the box is bisected and both parts are placed in $\mathcal{L}_T$. The behavior of the algorithm is controlled by the choice of the threshold ε which gives the lower bound on the size of boxes to be processed. Therefore, ε controls the trade-off between accuracy and computation time. The *prune* operations are techniques to directly reduce the size of the box, i.e. removing invalid parts from the box. More information on this topic can be found e.g. in [196]. This algorithm is the conceptual basis of different types of solvers discussed below. The set $\mathcal{X}_s$ to be

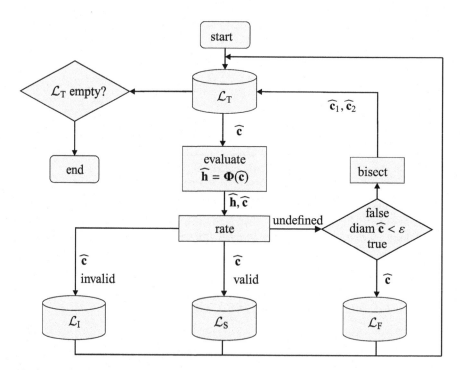

Fig. 5.10 Flowchart of the generic CSP solver

investigated is stored as a list of boxes at the beginning of the algorithm. The initial approximation $\mathcal{L}_T$ of the search space used in step 1 typically consists of exactly one box that is defined from the search range for each variable. Then, one uses interval analysis to compute guaranteed bounds for the possible values of the function. If these bounds are strictly positive, then the constraint is fulfilled for every point in the box. If the bounds are strictly negative, then the constraint is violated by every point in the box. Thus, the box is either stored in the solution set or put into the set of invalid boxes. If zero is enclosed by the interval, we cannot get reliable information in this step. This is the expected situation at the beginning of the solving procedure. Therefore, the box is subdivided into smaller boxes that are fed back into the evaluation loop. Clearly, the algorithm is well suited for recursive implementation. In any case, the presented sequential is used because it is better suited for some optimizations and for parallel execution (see [399]). A proper management of the list $\mathcal{L}_T$ is important with respect to memory consumption. Especially the operations *insert* and *extract* shall be implemented in a way such that the list behaves like a stack.

Now, we will derive two important specializations of the generic solver that are called *verify* and *calculate*.

5.4.1.2 Verify Algorithm

The verification is the simplest form of the problem statement. Given a verification set $\mathcal{X}_v$, we have to check for a given **c** if

$$\Phi(\mathbf{c}, \mathbf{v}) > 0 \quad \forall \mathbf{v} \in \mathcal{X}_v \tag{5.24}$$

holds true. In other words, one has to test if *all* boxes in the list $\mathcal{L}_T$ fulfill the inequality $\Phi > 0$. Therefore, the result of the verification procedure is expected to be just true or false. In practice, it can happen that the decision cannot be made with the given threshold ε_v because all boxes have been subdivided until all are smaller than ε_v. In this case, the result is undefined. We call the result *finite*. In any of the above-mentioned cases, we are not interested in the content of the lists $\mathcal{L}_S$ and $\mathcal{L}_I$. Once a single box is rated invalid or finite, the algorithm is terminated and returns false or finite, respectively. If all boxes from the list $\mathcal{L}_T$ are successfully processed, the algorithms returns true. As a consequence the following variant of the generic solver can be deduced:

Algorithm 2: `Verify`

---

1. Store the boxes to be verified $\{\widehat{\mathbf{v}}_1, \ldots, \widehat{\mathbf{v}}_n\}$ in the list $\mathcal{L}_T$.
2. If the list $\mathcal{L}_T$ is empty, terminate the algorithms and return *true*.
3. Extract the next box $\widehat{\mathbf{v}}$ from the list $\mathcal{L}_T$.
4. If diam $\widehat{\mathbf{v}} < \varepsilon_v$, i.e. the width of a component of the box $\widehat{\mathbf{v}}$ is smaller than the given threshold ε_v, terminate the algorithm and return *finite*.

5. If *prune* operations can successfully reduce the width of $\widehat{\mathbf{v}}$, terminate the algorithm and return *false*.
6. Evaluate the constraint $\widehat{\mathbf{h}} = \boldsymbol{\Phi}(\widehat{\mathbf{v}})$.
7. If $\inf \widehat{\mathbf{h}} > \mathbf{0}$, i.e. the infimum of $\widehat{\mathbf{h}}$ is positive for all constraints which means that the constraint is valid for every $\mathbf{v} \in \widehat{\mathbf{v}}$, discard $\widehat{\mathbf{v}}$; go to step (2).
8. If a component $\sup \widehat{h}_i < 0$ exists, i.e. at least one supremum is negative, then no solutions $\mathbf{v} \in \widehat{\mathbf{v}}$ exist; terminate the algorithm and return *false*.
9. If a $\sup \widehat{h}_i < 0$ exists, i.e. at least one supremum is negative, then no solution exists $\mathbf{v} \in \widehat{\mathbf{v}}$ and the box $\widehat{\mathbf{v}}$ is stored in $\mathcal{L}_\mathrm{I}$; go to step (2).

The version specialized in this way has the advantage over the generic form that it consumes less memory and it detects invalid boxes faster. Especially the consumption of memory is significantly reduced in the verify form. Compared to the total number N of boxes to be evaluated by the algorithm, only $\mathcal{O}(\log N)$ boxes have to be stored in the list $\mathcal{L}_\mathrm{T}$ leading to a very moderate use of memory. Contrary, if one seeks information for which values the constraints are violated, then one has to use the generic form of the algorithm.

5.4.1.3 Calculate Algorithm

The algorithm `calculate` solves a slightly modified problem compared to the algorithm `verify`. In the following, the solution set of the CSP

$$\mathcal{X}_c = \left\{ \mathbf{c} \in \mathcal{X}_s \subset \mathbb{R}^n \mid \boldsymbol{\Phi}(\mathbf{c}, \mathbf{v}) > \mathbf{0} \right\} \tag{5.25}$$

has to be found for a given $\mathbf{v}$. For the calculation, the search space $\mathcal{X}_s$ is given as a list $\mathcal{L}_\mathrm{T}$ of boxes, and the set $\mathcal{X}_c$ is sought that shall be approximated with the list $\mathcal{L}_\mathrm{S}$. Thus, the set approximated through the elements of the list $\mathcal{L}_\mathrm{S}$ converges to $\mathcal{X}_c$ if the threshold $\varepsilon_c \to 0$. The bisection process for the boxes is continued until the size of the boxes is smaller than the threshold ε_c. Otherwise, the algorithm would make infinitely many steps to approximate the solution set by infinitely small boxes. It depends on the application if the list $\mathcal{L}_\mathrm{F}$ of the finite boxes is required for further evaluation. Since the lists $\mathcal{L}_\mathrm{I}$ and $\mathcal{L}_\mathrm{F}$ may require a significant amount of memory, one can discard these boxes in many cases. If one wants to refine the calculation in later steps, one has to store the $\mathcal{L}_\mathrm{F}$ boxes. The specialized calculation algorithm then reads:

Algorithm 3: `Calculate`

---

1. Save an approximation $\{\widehat{\mathbf{c}}_1, \ldots, \widehat{\mathbf{c}}_n\}$ of the search space $\mathcal{X}_s$ in the list $\mathcal{L}_\mathrm{T}$.

2. Create empty lists $\mathcal{L}_S, \mathcal{L}_I, \mathcal{L}_F$ for the solution set ($\mathcal{L}_S$), for the invalid boxes ($\mathcal{L}_I$) and the set of undersized boxes with a diameter smaller than the threshold ($\mathcal{L}_F$).

3. If the list $\mathcal{L}_T$ is empty, terminate the algorithm.

4. Extract the next box $\widehat{\mathbf{c}}$ from the list $\mathcal{L}_T$.

5. If diam $\widehat{\mathbf{c}} < \varepsilon$, i.e. the width of all components of the box $\widehat{\mathbf{c}}$ is smaller than a given threshold ε, the box is undersized; store the box in the list $\mathcal{L}_F$; go to step (3).

6. If available, apply *prune* and *bound improvement* operations to the box $\widehat{\mathbf{c}}$.

7. Evaluate the constraint $\widehat{\mathbf{h}} = \mathbf{\Phi}(\widehat{\mathbf{c}})$.

8. If inf $\widehat{\mathbf{h}} > \mathbf{0}$, i.e. the infimum of $\widehat{\mathbf{h}}$ is positive for all constraints which means that the constraint is valid for every $\mathbf{c} \in \widehat{\mathbf{c}}$, store $\widehat{\mathbf{c}}$ as solution in the list $\mathcal{L}_S$; go to step (3).

9. If a sup $\widehat{h}_i < 0$ exists, i.e. at least one supremum is negative, then no solution exists $\mathbf{c} \in \widehat{\mathbf{c}}$ and the box $\widehat{\mathbf{c}}$ is stored in $\mathcal{L}_I$; go to step (3).

10. Split the box $\widehat{\mathbf{c}}$ into m sub-boxes $\{\widehat{\mathbf{c}}_1, \ldots, \widehat{\mathbf{c}}_m\}$ and store these new boxes in the list $\mathcal{L}_T$; go to step (3).

5.4.1.4 Hybrid Algorithm

Oneeneeds a combination of `verify` and `calculate` for a couple of practically relevant CSP. An example of such combination is the determination of the total orientation workspace $\mathcal{W}_{TO}$ or the design problem of finding all robots with a given workspace. For each position of that workspace, we have a set of orientation to check. Therefore, a position is only said to belong to the total orientation workspace if every pose from the set belongs to the workspace. If at least one infeasible orientation can be found, then the position does not belong to the workspace. The procedure can also be applied if ranges for the properties, such as available velocities, accelerations, or wrenches, must be achieved at all poses in the workspace. Mathematically speaking, we have to find all solutions of the problem of the form

$$\mathcal{X}_c = \left\{ \mathbf{c} \in \mathcal{X}_s \subset \mathbb{R}^n \mid \mathbf{\Phi}(\mathbf{c}, \mathbf{v}) > \mathbf{0} \quad \forall \mathbf{v} \in \mathcal{X}_v \subset \mathbb{R}^m \right\} \quad . \tag{5.26}$$

A solver for this problem can be constructed by using a nested version of the `calculate` and `verify` algorithm. We define both verification and calculation variable for the same problem. The structure of the algorithm is based on the `calculate` scheme. Then, we call the `verify` algorithm instead of evaluating the constraint system, where the current box $\mathbf{c}$ is used to define the search space. The algorithmic structure is as follows:

Algorithm 4: Hybrid CSP-Solver

---

1. Save an approximation $\{\widehat{\mathbf{c}}_1, \ldots, \widehat{\mathbf{c}}_n\}$ of the search space $\mathcal{X}_s$ in the list $\mathcal{L}_T$.
2. Create empty lists $\mathcal{L}_S, \mathcal{L}_I, \mathcal{L}_F$ for the solution set ($\mathcal{L}_S$), for the invalid boxes ($\mathcal{L}_I$), and the set of undersized boxes with a diameter smaller than the threshold ($\mathcal{L}_F$).
3. If the list $\mathcal{L}_T$ is empty, terminate the algorithm.
4. Extract the next box $\widehat{\mathbf{c}}$ from the list $\mathcal{L}_T$.
5. If diam $\widehat{\mathbf{c}} < \varepsilon$, i.e. the width of all components of the box $\widehat{\mathbf{c}}$ is smaller than a given threshold ε_c, the box is undersized; store the box in the list $\mathcal{L}_F$; go to step (3).
6. If available, apply *prune* and *bound improvement* operations to the box $\widehat{\mathbf{c}}$.
7. Call the algorithm Verify for the box $\widehat{\mathbf{c}}$ and the set $\mathcal{X}_v$.
8. If Verify returns *valid*, then all constraints $\boldsymbol{\Phi}(\mathbf{c}, \mathbf{v}) > 0$ are fulfilled for all $\mathbf{c} \in \widehat{\mathbf{c}}$ and all $\mathbf{v} \in \mathcal{X}_v$; store $\widehat{\mathbf{c}}$ as solution in the list $\mathcal{L}_S$; go to step (3).
9. If Verify returns *invalid*, then no $\mathbf{c} \in \widehat{\mathbf{c}}$ fulfills the constraints $\boldsymbol{\Phi}(\mathbf{c}, \mathbf{v}) > 0$ for all $\mathbf{v} \in \mathcal{X}_v$; store $\widehat{\mathbf{c}}$ in the list $\mathcal{L}_I$; go to step (3).
10. If Verify returns *finite*, then the width of the box $\widehat{\mathbf{c}}$ is too large; split the box $\widehat{\mathbf{c}}$ into m sub-boxes $\{\widehat{\mathbf{c}}_1, \ldots, \widehat{\mathbf{c}}_m\}$ and store these new boxes in $\mathcal{L}_T$; go to step (3).

Having defined the structure of the solver algorithms, some supporting steps are discussed in the following.

5.4.1.5 Bisection of Boxes

An elementary operation of the interval algorithms is the splitting of one box $\widehat{\mathbf{b}} \in \mathbb{I}^n$ into m smaller boxes $\{\widehat{\mathbf{b}}^{(1)}, \ldots, \widehat{\mathbf{b}}^{(m)}\}$. There are different criteria to achieve this. Here, we restrict ourselves to subdivision into two sub-boxes, which is called *bisection* (Fig. 5.11). In general, more complicated strategies can be used to generate sub-boxes. In a bisectional step, one maintains all components of the box $\widehat{\mathbf{b}}$ except for the jth component, which is divided at the position $b_j^* \in \widehat{b}_j$. Thus, we receive two sub-boxes

$$\widehat{\mathbf{b}} = \widehat{\mathbf{b}}^{(1)} \cup \widehat{\mathbf{b}}^{(2)} \quad \text{with} \tag{5.27}$$

$$\widehat{\mathbf{b}} = [\widehat{b}_1, \ldots, \widehat{b}_j, \ldots, \widehat{b}_n]^{\mathrm{T}}, \tag{5.28}$$

$$\widehat{\mathbf{b}}^{(1)} = [\widehat{b}_1, \ldots, [\inf \widehat{b}_j; b_j^*], \ldots, \widehat{b}_n]^{\mathrm{T}}, \tag{5.29}$$

$$\widehat{\mathbf{b}}^{(2)} = [\widehat{b}_1, \ldots, [b_j^*; \sup \widehat{b}_j], \ldots, \widehat{b}_n]^{\mathrm{T}}. \tag{5.30}$$

To determine the point where the box is split, one mostly uses the center of the interval

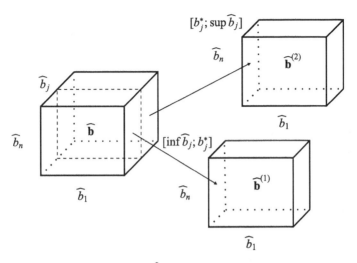

Fig. 5.11 Bisection Operation for the Box $\widehat{b}$ in the jth Component

$$b_j^* = \text{mid}\,\widehat{b}_j = \frac{1}{2}(\text{inf}\,\widehat{b}_j + \text{sup}\,\widehat{b}_j) \quad . \tag{5.31}$$

For the bisection, one has to choose the component to be split. A simple and fast method is to use the component j with the greatest diameter diam b_j. If the components in the interval box are inhomogeneous in their order of magnitude, we have to introduce weighting factors g_j for each component of the interval vector. For example, if length and angle have been collected in one interval vector, we might have to deal with length in the size of tenth of meters where the angles are in the range of π. To search all components with a comparable resolution, we have to choose the component j to be split so that diam $(g_j\widehat{b}_j)$ to be the maximum, where g_j is the respective weighting factor.

Instead of choosing the component with the largest diameter, one can use a round robin method, where the component to be split is cyclically changed.

5.4.1.6 Efficiency Improvement for Interval Algorithms

The basic CSP solver presented above is simple but leads to very long computational time and large memory consumption. Therefore, a number of heuristics have been developed to speed up the computation and to reduce the memory usage. Just some important methods are listed for the sake of brevity. Consistency tests can be used to shrink or even discard boxes in the calculation scheme. This is achieved by separating the function Φ_i into two parts where for each part an interval evaluation is computed. If the ranges computed for the separate parts are fully disjunct, the box cannot contain a solution. If one of the functions can be inverted, one can shrink the interval where solutions are possible. Another efficient tool is to compute interval evaluations on the Jacobian matrix of the function $\boldsymbol{\Phi}$. If the bounds of the interval evaluation have equal signs, the function is strictly monotonic and thus takes its extremal values at the boundaries of the interval. Thus, one can directly eliminate overestimation and compute sharper bounds.

5.4.2   Constraints for Interval Workspace Analysis

In this section, we present different criteria for workspace analysis that can be expressed by constraints of the form $\boldsymbol{\Phi} > \mathbf{0}$. For some criteria, the expression of an inequality is natural. If the criterion is expressed by an equation, it has to be rewritten to a mapping that is positive if the pose belongs to the workspace. Anyway, the only necessary condition is that the constraint can be expressed in interval form. If the condition can be written so that the expression is a closed-form or even differentiable, we can use more efficient techniques to evaluate the constraint. We will begin with the simplest conditions in the next section. Most of the conditions presented in this section can also be used for real-valued analysis of a given pose. However, in this case, the result applies only to a zero-dimensional object rather than a set of values.

5.4.2.1   Limited Cable Length

The restrictions of a pose imposed by minimum and maximum cable length are quite simple. We will call the related criterion *reachable workspace* which is a technological restriction of a cable robot design. Using the closed-form solution of the inverse kinematics from Eq. (4.2), we receive the constraint

$$l_{\min} \leq |\mathbf{a}_i - \mathbf{r} - \mathbf{R}\mathbf{b}_i| \leq l_{\max},$$

which can be rewritten into $2m$ inequalities

$$\left. \begin{array}{l} \boldsymbol{\Phi}_i^{\mathrm{IK}} : |\mathbf{a}_i - \mathbf{r} - \mathbf{R}\mathbf{b}_i| - l_{\min} > 0 \\ \boldsymbol{\Phi}_{(m+i)}^{\mathrm{IK}} : l_{\max} - |\mathbf{a}_i - \mathbf{r} - \mathbf{R}\mathbf{b}_i| > 0 \end{array} \right\} \quad \text{for} \quad i = 1, \ldots, m, \qquad (5.32)$$

where $l_{\min}$ and $l_{\max}$ are the minimal and maximal cable length, respectively. We can extend this constraint to take into account the influence of guiding pulleys by exchanging the expression coming from the inverse kinematics (see Sect. 7.2.1).

5.4.2.2   Wrench-Feasibility

The criterion for testing the wrench-feasibility of poses with intervals is clearly the most important but also involved problem in using the CSP framework. Bruckmann

[68] developed a test for wrench-feasibility based on the so-called *existence algorithm*. The existence algorithm is a variant of the `verify` algorithm presented above where the search is stopped as soon as one feasible solution is found. To apply this strategy to cable robots, one uses the equation of the structure matrix as function $\widehat{\Phi}$ and searches for existence of feasible force distribution $\mathbf{f}$ by brute force. Although this algorithm is rather ineffective in terms of computation burden, its strategy is perfectly in line with all benefits of the CSP framework.

Here, we follow a more elegant way proposed by Gouttefarde [182] who presented an interval-based wrench-feasibility test that allows to decide if a whole region described by an interval box belongs to the wrench-feasible workspace. According to Gouttefarde, one can test if the linear system formed by

$$\widehat{\mathbf{A}}^{\mathrm{T}}\mathbf{f} = -\widehat{\mathbf{w}} \qquad (5.33)$$

with the interval structure matrix $\widehat{\mathbf{A}}^{\mathrm{T}}$ is *strongly feasible*. Note that the interval form $\widehat{\mathbf{w}}$ of the wrench is a specific kind of the wrench set Q which is formed by an axis-aligned bounding box given through the ranges of the intervals. Here, strong feasibility of the equation means that for every matrix $\mathbf{A}^{\mathrm{T}} \in \widehat{\mathbf{A}}^{\mathrm{T}}$ and every wrench $\mathbf{w}_{\mathrm{P}} \in \widehat{\mathbf{w}}$ there exists a distribution of cable force $\mathbf{f} \in \mathcal{F}$ so that $\mathbf{A}^{\mathrm{T}}\mathbf{f} + \mathbf{w}_{\mathrm{P}} = \mathbf{0}$.

Hence, the strong feasibility of the interval equation (5.33) is a sufficient condition for all poses used to compute the interval structure matrix to be fully included in the wrench-feasible workspace. The criterion is said to be strong since the interval matrix $\widehat{\mathbf{A}}^{\mathrm{T}}$ is in general an overestimation, i.e. in general there exist some matrices $\mathbf{A}^{\mathrm{T}} \in \widehat{\mathbf{A}}^{\mathrm{T}}$ that do not correspond to any pose $(\mathbf{r}, \mathbf{R})$ of the robot. The property is called strong feasibility because it is a sufficient, but not necessary criterion.

Rohn [150] provides a test to check strong feasibility. To describe the test, one introduces a set $\mathcal{Y}_n$ with vectors which components are either -1 or 1. There are exactly 2^n permutations $\mathbf{y}_i \in \mathcal{Y}_n$. Given an interval matrix $\widehat{\mathbf{A}}$, one defines for each vector $\mathbf{y}_i$ a corresponding matrix $\mathbf{A}_{\mathrm{Y}}$ which is defined component-wise as

$$A_{\mathrm{Y},ij} = \inf A_{ij} + \left(\sup A_{ij} - \inf A_{ij}\right)(1 - y_i)/2 \qquad (5.34)$$

and the 2^n matrices defined in that way are called the vertices of $\widehat{\mathbf{A}}$. In a similar manner, one defines the vertices of an interval vector $\mathbf{b}_i$ by

$$b_{\mathrm{Y},i} = \inf b_i + (\sup b_i - \inf b_i)(1 - y_i)/2 \quad , \qquad (5.35)$$

which are the 2^n vertices of the box defined by the interval vector. Now, according to Rohn's theorem, the interval system is strongly feasible if, and only if, all linear systems $\mathbf{A}_{\mathrm{Y}}^{\mathrm{T}}\mathbf{f} = -\mathbf{f}_{\mathrm{Y}}$, $y \in \mathcal{Y}_n$ are all feasible. To clarify the conclusion, Gouttefarde shows that one can assess infinitely many poses contained in a box by testing only a finite number of linear systems, namely 2^n systems [182]. Since n is a fixed and relatively small number for a given robot, the computational costs for these tests is constant in the context of workspace computation. To support the computation, further tests are presented [182] for proving that a box is fully outside the workspace along with heuristics used in the interval algorithm.

By proving the feasibility of these linear systems, one has also proven that no singular configuration is inside the given box of poses. Summing up the findings above, the wrench-feasibility test can be used as a single criterion $\widehat{h}_i = \widehat{\boldsymbol{\Phi}}(\mathbf{c}, \mathbf{v})$ in the above-mentioned framework if it returns the interval $\widehat{h} = [1; 1]$ which evaluates to valid on successful testing of the pose, and $\widehat{h} = [-1; -1]$ if the box can be fully discarded (invalid), and $\widehat{h} = [-1; 1]$ in all other cases (undefined). Note that an interval implementation is only required for computing the interval structure matrix $\widehat{\mathbf{A}}^{\mathrm{T}}$. The computations afterwards can be done with conventional linear algebra.

5.4.2.3   Dexterity Requirements

The dexterity of a robot plays an important role and thus should be taken into account for workspace analysis and parameter synthesis. Most of the considerations concerning dexterity are coupled to the calculation and analysis of the pose-dependent Jacobian matrix $\mathbf{J}$ which is essentially the negative transpose of the structure matrix $\mathbf{A}^{\mathrm{T}}$. Typical indices for the dexterity are based on the eigenvalues, on the singular values, on the determinant, or on the condition number of the matrix. These dexterity measures are somehow related to each other and their computation is either time consuming, iterative, or even both. Furthermore, eigenvalues and the determinant can only be applied to robots with $m = n$ cables.

We use a slightly relaxed but efficient dexterity test here [394]. Let $\mathbf{J} = -\mathbf{A}$ be the Jacobian matrix of the inverse kinematics of the cable robot thus mapping velocities of the platform $\dot{\mathbf{y}}$ to velocities in the cables $\dot{\mathbf{l}}$. For a given platform velocity in world coordinates, it is important not to exceed the maximal feasible velocity $\dot{\mathbf{l}}_{max}$ of the cables. Therefore, we need a relation

$$\dot{\mathbf{l}}_{max} > \mathbf{J}\dot{\mathbf{y}} \quad . \tag{5.36}$$

Introducing compatible norms for the vectors yields the desired inequations

$$||\dot{\mathbf{l}}_{max}|| = ||\mathbf{J}\dot{\mathbf{y}}|| \leq ||\mathbf{J}|| \, ||\dot{\mathbf{y}}|| \quad . \tag{5.37}$$

If we choose the Euclidian norm $|| \cdot ||_2$ for vectors and the spectral norm for the matrices, we basically receive the known forms of dexterity measures discussed above. We consider the inverse kinematics and thus the maximal velocities may be generated by each actuator individually. Therefore, it is enough to limit each row of the matrix which leads to the use of the infinity norm $|| \cdot ||_\infty$. This norm is defined as

$$||v||_\infty = \max_{1 \le i \le n} |v_i| \quad \mathbf{v} \in \mathbb{R}^n \tag{5.38}$$

$$||A||_\infty = \max_{1 \le i \le n} \sum_{j=1}^{m} |a_{ij}| \quad \mathbf{A} \in \mathbb{R}^{n \times m} \quad . \tag{5.39}$$

Now, the world coordinate velocity $\dot{\mathbf{y}}$ is normalized to 1 and the norm of the Jacobian matrix $\mathbf{J}$ is considered. We deal with the $\max(\cdot)$ operator in Eq. (5.39) by checking all rows of the matrix separately to be smaller than the given maximal transmission factor j_{max} and we receive n simple conditions for the CSP as follows

$$\sum_{k=1}^{m} |\mathbf{J}_{ik}| < j_{max} \quad \forall \ 1 \le i \le n \quad , \tag{5.40}$$

where $\mathbf{J}_{ik}$ are the elements of the Jacobian matrix. Thus, we find a simple method to check for dexterity requirements over the whole workspace as inequalities enabling us to add the n additional requirements to our CSP $\widehat{\mathbf{\Phi}}$.

5.4.2.4 Cable Interference

The interference of cables can restrict the workspace of cable robots significantly. Thus, it is important to verify if the workspace is free of cable interference or at least to check whether the required (mostly rectangular) workspace may be used without collisions. The most frequent problem in this field is the calculation of the distance between two finite line segments (Fig. 5.12). This problem can easily be solved for given points. However, one has to distinguish between quite a number of special cases where lines are parallel, have identical points, etc. Therefore, we use a simple interval formula that is directly related to the original distance problem. Given line segment of one cable with

$$g_i : \mathbf{x}_i = \mathbf{a}_i + \lambda_i(\mathbf{b}_i - \mathbf{a}_i), \quad \lambda_i \in [0; 1], \quad i \in 1, 2, \ldots, m \tag{5.41}$$

we are searching for the supremum of

$$d_{ij} = ||\mathbf{x}_i - \mathbf{x}_j||_2 \quad \forall \ \lambda_i, \lambda_j \in [0; 1], \quad i \ne j. \tag{5.42}$$

By introducing λ_i, λ_j as intervals, we can replace the distinction of cases by a basic interval evaluation of a simple norm function and we receive

$$d_{min} < d_{ij} = ||\mathbf{a}_i + \lambda_i(\mathbf{b}_i - \mathbf{a}_i) - \mathbf{a}_j - \lambda_j(\mathbf{b}_j - \mathbf{a}_j)||_2 \quad \forall \ \lambda_i, \lambda_j \in [0; 1], \quad i \ne j \tag{5.43}$$

as additional inequalities that take the link interference into account and we can add them to the CSP $\widehat{\mathbf{\Phi}}$ where d_{min} is our chosen safe distance between the cables.

Fig. 5.12 Distance between
two line segments in
parameter form

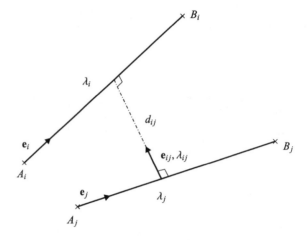

5.5   Numeric Boundary Methods

Compared to the above introduced interval methods, workspace boundary methods
rely on real valued evaluations with all their restrictions. In exchange, one gets
access to more involved numerical schemes that cannot be expressed in intervals
or the computation is too slow. A method for the determination of the boundary
of the workspace is presented by Pott [389]. The computation of the workspace
boundary aims at speed and precision rather than rigorous results or insight into the
mathematical structure of the workspace. Based on discrete investigation of single
points, a line search method is used to find the boundary of the workspace and to
iterate the boundary with arbitrary precision.

Furthermore, the workspace is represented by triangulation that allows for very
simple and accurate determination of the volume and surface. This accuracy can be
used to study the influence of the design variables (geometry of platform and machine
frame) or technical parameters such as minimum and maximum cable force.

On the one hand, workspace boundary methods are more complex than grid methods in terms of efforts for the implementation. On the other hand, they provide more
exact results. To further process the results, it is useful to convert the data to CAD
structures, where we find a rich set of tools to visualize and analyze the computed
workspace.

5.5.1   Approximation of the Workspace Boundary

Here, the translation workspace for a given orientation of the cable robot is represented by a triangulation of its hull. The idea for the determination of the workspace
is to start with a unity sphere in the estimated center $\mathbf{m}$ of the workspace and to

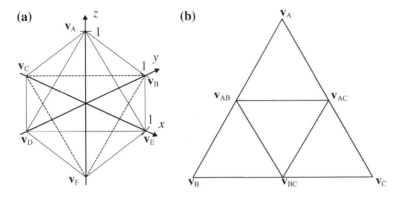

Fig. 5.13 (a) Unit octahedron (b) subdivision step for triangles

successively extend this sphere in radial direction. Clearly, this assumption may lead to an underestimation of the workspace and the estimation depends on the chosen value of **m**. On the contrary, for most technical applications, only cable robots with a compact workspace are interesting and therefore it seems reasonable to restrict a quick design procedure to such a subspace. The surface of the sphere is approximated by triangles which are created by iterative subdivision of the faces of an octahedron. Alternatively, one can also subdivide other regular polyhedrons, especially the Platonic solids with triangular facets such as a tetrahedron or an icosahedron. Especially the latter are interesting since these solids lead to a perfectly regular structure for the hull where all vertices share the same number of triangles. However, doing the initialization for a tetrahedron or an icosahedron is a bit more involved.

In the first step, the eight faces of an octahedron (Fig. 5.13a) located around the point **m** are described as triplets of vertices, e.g. $F_1 = \{\mathbf{v}_A, \mathbf{v}_B, \mathbf{v}_C\}_i$. Initially, there is a set $\mathcal{L} = \{F_1, \ldots, F_8\}$ containing eight faces. These faces of the octahedron are subdivided into four congruent triangles (Fig. 5.13b). This is done by constructing the three vertices $\mathbf{v}_{AB}$, $\mathbf{v}_{AC}$, $\mathbf{v}_{BC}$ for each triangle F_i in $\mathcal{L}$ and projecting the generated vertices onto a unit sphere

$$\mathbf{v}_{ij} = \frac{\mathbf{v}_i + \mathbf{v}_j}{|\mathbf{v}_i + \mathbf{v}_j|}, \quad i, j \in \{A, B, C\}, i \neq j \quad . \tag{5.44}$$

Then, the original triangle F_i is replaced by the four triangles $(\mathbf{v}_A, \mathbf{v}_{AB}, \mathbf{v}_{AC})$, $(\mathbf{v}_B, \mathbf{v}_{AB}, \mathbf{v}_{BC})$, $(\mathbf{v}_C, \mathbf{v}_{BC}, \mathbf{v}_{AC})$, $(\mathbf{v}_{AB}, \mathbf{v}_{AC}, \mathbf{v}_{BC})$. This process is repeated n_i times thus generating a set $\mathcal{L}$ containing $n_T = 2^{2n_i + 3}$ triangles.

In the second step, the vertices of the triangles are projected onto the hull of the workspace. Starting from the estimated center **m** of the workspace, the line

$$L : \mathbf{r} = \mathbf{m} + \lambda \mathbf{v}_i \quad \lambda \in [0; r_{max}] \tag{5.45}$$

is searched by a regula falsi method for the boundary of the workspace which is defined by a given maximum search range r_{max}. For each position $\mathbf{r}$ generated by the iterations of the line search, we can compute an arbitrary Boolean workspace criterion $\kappa(\mathbf{r}, \mathbf{R})$ such as wrench-closure, wrench-feasibility, reachability, intersection, or feasible deflection (see Sect. 5.2). The search for the boundary is done by the algorithm

---

Algorithm 5: `Line search`

---

1. Let $\lambda_{min} = 0$ be the lower bound and $\lambda_{max} = r_{max}$ be the upper bound.
2. If $\lambda_{max} - \lambda_{min} < \varepsilon_L$, stop the line search.
3. Let $\lambda = \frac{1}{2}(\lambda_{max} + \lambda_{min})$ and evaluate $\mathbf{r} = \mathbf{m} + \lambda \mathbf{v}_i$.
4. Calculate the Boolean workspace criterion $\kappa(\mathbf{r})$ of the resulting position $\mathbf{r}$.
5. If κ is valid then let $\lambda_{min} = \lambda$, else let $\lambda_{max} = \lambda$.
6. Go to step 2.

Finally, one ends up with the vertex $\mathbf{v}_i^{(h)} = \mathbf{m} + \lambda \mathbf{v}_i$ approximating the hull of the workspace with an accuracy ε_L. The corresponding triangles are rendered into a new set $\mathcal{L}^{(h)}$. Such data can be easily stored in a file such as stereo-lithography data file format (STL) or virtual reality modeling language (VRML) according to ISO 14772 [221] which can be loaded and visualized with most CAD tools. If it is also required to cope with holes within the workspace, it is trivial to add a second procedure that evaluates the line from the center to the boundary with a given step size and possibly reduces the maximum range. We can even use an interval evaluation over the range of $[0; \lambda]$ to receive guaranteed results.

5.5.2 Hull Computation for Different Types of Workspace

Having defined the data model and search strategy, one can compute the different types of the workspace. The algorithm described above is straightforward to use for computing the constant orientation workspace $\mathcal{W}_{co}$ by simply setting one specific orientation $\mathbf{R}$ for the platform. If one is interested in the inclusion orientation workspace $\mathcal{W}_{IO}$ or the maximum workspace $\mathcal{W}_{max}$, one has to slightly modify step 4 in algorithm `Line Search`. A position is said to belong to the inclusion orientation workspace $\mathcal{W}_{IO}$ if any orientation in a set $\mathcal{R}$ belongs to the workspace. Thus, in the algorithm, one prepares a list $\mathcal{L}_R$ with orientations $\mathcal{R}_i$ to be checked by the algorithm's step 4 and evaluates one orientation $\mathcal{R}_i$ after the other, until one finds an orientation that belongs to the workspace. In this case, the $\kappa(\mathbf{r})$ is valid. This can be understood as a Boolean disjunction (logical: or) between the evaluation of all $\kappa(\mathbf{r}, \mathbf{R}), \mathbf{R} \in \mathcal{L}_R$. If no such entry in $\mathcal{L}_R$ is found, then the pose and thus $\kappa(\mathbf{r})$ is invalid. To compute the maximum workspace W_{max}, the set $\mathcal{R}$ must be chosen as a discretization of the full SO_3 group.

Computing valid positions for the total orientation workspace W_{TO} is done respectively but instead of searching for at least one entry in $\mathcal{L}_R$ where the workspace test is valid, one looks for one element where the test fails. In this case, $\kappa(\mathbf{r})$ evaluates to invalid, whereas successfully completing the full list $\mathcal{L}_R$ evaluates to valid. This is equivalent to the Boolean conjunction (logical: and) of all single tests $\kappa(\mathbf{r}, \mathbf{R})$, $\mathbf{R} \in \mathcal{L}_R$. Analogously to the procedure for the maximum workspace W_{max}, one can evaluate if a pose is dextrous. However, to the best of the author's knowledge, no spatial cable robot reported in the literature possess any pose that is dextrous.

5.5.3   Boolean Set Operations with the Workspace Boundary

Practical problems usually involve the consideration for more than one criterion for workspace computation. Let W_1 and W_2 be the result from the workspace computations for the same robot but distinct criteria κ_1 and κ_2, respectively. If the criteria are restrictions that need to be considered simultaneously, one is looking for the conjunction of these criteria $\kappa = \kappa_1 \wedge \kappa_2$ or, geometrically speaking, the intersection of the geometric objects represented by the workspace $W = W_1 \cap W_2$. For the workspace computed from a discrete grid, this computation is straightforward. For the hull represented through triangulation, the general intersection is much more involved. However, if the same center is used for the projection with the line search algorithm, one can do one-by-one identifications between the vertices and use the parameter λ for comparing the respective vertices. The result of the intersection is derived when choosing $\lambda = \min(\lambda_1, \lambda_2)$ for each corresponding pair of vertices in the sets W_1 and W_2.

Some problems require that at least one of two criteria κ_1 and κ_2 needs to be fulfilled, i.e. the logical connection between the two criteria is disjunction $\kappa = \kappa_1 \vee \kappa_2$. In this case, the respective geometrical operation on the two workspaces is the union $W = W_1 \cup W_2$. Based on the precondition that the two workspace hulls share a common center $\mathbf{m}$, the union can be derived from taking $\lambda = \max(\lambda_1, \lambda_2)$.

Given that the first criterion has been evaluated and the second criterion is to be evaluated, one can slightly modify the line search algorithm to compute the intersection (union) with the second criterion. Computing the intersection (union) is simply achieved by altering the first step to initialize the search range. To compute the intersection, the pair $(\lambda_{\text{min}}, \lambda_{\text{max}})$ is initialized with $(0, \lambda)$ where λ is the result of the first workspace computation. This choice limits the line search to the already determined workspace. If the union shall be determined, the search range $(\lambda_{\text{min}}, \lambda_{\text{max}})$ is initialized instead with $(\lambda, r_{\text{max}})$ which extends the search beyond the already computed workspace.

5.5.4 Computing Properties of the Workspace from the Boundary

The triangulated hull of the workspace allows for some geometric characterizations of the workspace. It is straightforward to calculate the surface $S(\mathcal{W})$, the volume $V(\mathcal{W})$, and the center of gravity $\mathbf{c}(\mathcal{W})$ of the workspace as follows

$$S(\mathcal{W}) = \frac{1}{2} \sum^{\mathcal{L}} \|(\mathbf{v}_A - \mathbf{v}_B) \times (\mathbf{v}_A - \mathbf{v}_C)\|_2 \tag{5.46}$$

$$V(\mathcal{W}) = \frac{1}{6} \sum^{\mathcal{L}} ((\mathbf{v}_A - \mathbf{m}) \times (\mathbf{v}_B - \mathbf{m})) \cdot (\mathbf{v}_C - \mathbf{m}) \tag{5.47}$$

$$\mathbf{c}(\mathcal{W}) = \frac{1}{4V(\mathcal{W})} \sum^{\mathcal{L}} (\mathbf{v}_A + \mathbf{v}_B + \mathbf{v}_C + \mathbf{m}) \;. \tag{5.48}$$

For the volume, one can find a convenient shortcut if one substitutes $\mathbf{v}_i - \mathbf{m} = \lambda_i \mathbf{u}_i$ in the parametric form with the direction $\mathbf{u}_i$ and its length from the line search λ_i. Then, the equation for the volume becomes

$$V(\mathcal{W}) = \frac{1}{6} \sum^{\mathcal{L}} \lambda_A \lambda_B \lambda_C (\mathbf{u}_A \times \mathbf{u}_B) \cdot \mathbf{u}_C \tag{5.49}$$

where the scalar value of the product $(\mathbf{u}_A \times \mathbf{u}_B) \cdot \mathbf{u}_C$ is equal for all triangles and only depends on the number of subdivisions n_T done. Thus, one finds the simple form

$$V(\mathcal{W}) = \frac{(\mathbf{u}_A \times \mathbf{u}_B) \cdot \mathbf{u}_C}{6} \sum^{\mathcal{L}} \lambda_A \lambda_B \lambda_C \tag{5.50}$$

with the constant factor $V_i^{(n_T)} = (\mathbf{u}_A \times \mathbf{u}_B) \cdot \mathbf{u}_C$.

The accurate determination of these numbers is useful for designing robots, especially if one wants to take derivatives of these indices. For computing derivatives (see Sect. 5.5.5), one can seldom compute the expressions in closed-form. If one has to rely on numerical approximation through finite differences, the computation for neighboring values should be as accurate as possible. Therefore, one has to balance the accuracy used in the line search with the step width of the finite difference so that the results are meaningful.

5.5.5 Differential Hull

When analyzing the workspace of a cable robot, an interesting aspect is how the workspace depends on the geometrical and technical parameters or, more generally speaking, how it depends on the assumptions made and the algorithm settings. In general, the workspace will be changed if the parameters are differentially altered. Therefore, doing a sensitivity analysis on the parameters influencing the result of the workspace computation is interesting and can be done efficiently based on the workspace hull model. Mathematically speaking, one may ask for the derivations of the workspace caused by infinite changes of the describing parameters such as the positions of the winches $\mathbf{a}_i$, the geometry of the platform $\mathbf{b}_i$, or the feasible forces in the cables f_{min}, f_{max}. One may also ask for the sensitivity of the constant orientation workspace $\mathcal{W}_{co}$ for changes in the orientation $\mathbf{R}_0$. Since the workspace is a continuous set, the changes in shape and size happen on its boundary. Here, the possibility is neglected that the parameter change generates a hole in the workspace which would change the workspace's topological structures. Therefore, the change in the parameters will only influence the hull of the workspace. As we have already seen when computing the workspace, it is difficult to find a closed-form solution of the workspace, hence, for computation we cannot compute the derivations symbolically. Clearly, numerical approximation using finite differences is a possible way. If we computed the workspace using discretization or interval techniques, our solution is quite insensitive towards small changes in the parameters unless we use very small thresholds for the discretization. This problem applies both for simple discretization as well as for interval analysis. In contrast, the approximation of the workspace boundary through the hull algorithm separates the granularity of the used grid from the accuracy in the computation. Once a number of triangles is chosen, the points on the hull can be efficiently computed with high accuracy.

If we now consider small changes in the geometry of the robot, we can accurately track the change in the workspace hull with moderate computational burdens. To better understand the approach, it is important to note that the steps for determining the search directions $\mathbf{v}_i$ for the hull determination can be completed before computing the values for λ_i for each vertex. Therefore, we can change the robot model by an increment $\Delta \mathbf{a}_i$ and compute the resulting value for λ_i'. A suitable approximation is

$$\delta \approx \frac{\lambda_i' - \lambda_i}{\Delta \mathbf{a}_i} \quad . \tag{5.51}$$

The concept of the differential workspace can be applied to compute the influence of the parameters on the shape of the workspace, i.e. to compute the derivatives of the vertices of the workspace, or on the derive of the properties $S(\mathcal{W})$, $V(\mathcal{W})$, $c(\mathcal{W})$ of the workspace, i.e. through finite differences. We can compute the derivation or sensitivity

$$\frac{\partial S(\mathcal{W})}{\partial p} \quad , \tag{5.52}$$

where p is any numerical parameter of the geometry of the robot, the robot's technical parameters, or an algorithm parameter. An overview of parameters for the sensitivity analysis is given in Table 5.1.

Table 5.1 Overview of the parameters to be studied with the differential hull. The list includes a number of criteria that are introduced in Chap. 7

Geometry	Technology	Algorithm setting
Proximal anchor points $\mathbf{a}_i$	Cable force limits f_{min}, f_{max}	Settings of the force
Distal anchor points $\mathbf{b}_i$	Cable length limits l_{min}, l_{max}	distribution
Cable radius r_C	Applied wrench $\mathbf{w}_P$	Algorithm (e.g. max.
Pulley radius r_R	Maximum cable deflection on	iterations)
Center of gravity $\mathbf{r}_M$	platform and base	Platform orientation $\mathbf{R}$ (for
	Cable stiffness k'	constant orientation
	Maximum cable velocity $\dot{l}_{max}$	workspace)
	Maximum cable acceleration	Range of orientation in total
		orientation workspace

5.5.6 Cable Span

A well-known disadvantage of parallel robots and especially of cable robots is that the installation space of the robot is large compared to the workspace. One reason for this drawback is that the cables occupy a huge volume if the workspace of the robot is large. This volume, that is at least temporarily taken be the cables, is called *cable span* (Fig. 5.14). The cable span is a volumetric object for all spatial robots and a flat area for a planar robot. Based on the assumption of the standard model that the proximal anchor point is a pose-independent point in space, it is clear that the cable span for the ith cable has some kind of apex at the point A_i. Based on the considerations of the workspace boundary in these sections, one can easily compute the cable span for the constant orientation workspace. Let $(\mathbf{v}_1, \ldots, \mathbf{v}_k)$ be the k vertices of the workspace hull as it is introduced in Sect. 5.5. Then, one receives all possible points for the point B_i from

$$\mathbf{v}'_j = \mathbf{b}_i + \mathbf{v}_j \quad j = 1, \ldots, k \qquad (5.53)$$

which is nothing but translating the hull by $\mathbf{b}_i$. The cable span is approximated from connecting the point A_i with each of the vectors $\mathbf{v}'_j$. The resulting geometrical object looks a bit like a noncircular cone where most of the lines defined above are lying inside the cone. To normalize the representation and also to reduce the amount of data, a polar decomposition of the cone is proposed. From the structure of the hull, we already know the projection center $\mathbf{m}$ which is in the central region of the hull. The polar decomposition aims at sorting all the lines in the span around this central line.

The procedure to compute the polar decomposition is as follows: Firstly, a coordinate frame $\mathcal{K}_{A,i}$ is constructed at point A_i whose z-axis is exactly aligned with the vector $\mathbf{m} - \mathbf{a}_i$. The x- and y-axis can be chosen arbitrarily where it is useful to use the convention introduced along with the pulley model (see Sect. 7.2.1). Then, the k lines of the span are distributed in n_s polar segments in the frame $\mathcal{K}_{A,i}$. Firstly, each line $\mathbf{s}_j = \mathbf{v}'_j - \mathbf{a}_i$ is transformed into the local frame $\mathcal{K}_{A,i}$. Then, spherical coordinates are computed as follows:

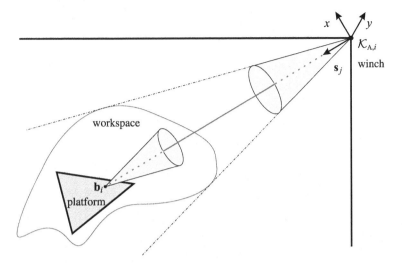

Fig. 5.14 Definition of the cable span based on the hull

Fig. 5.15 Polar
decomposition of the cable
vector to compute the cone
of the cable span for $n_\mathrm{s} = 12$
segments

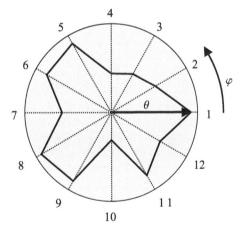

$$
\mathbf{s}_j^{(\mathrm{C})} =
\begin{bmatrix} r \\ \theta \\ \varphi \end{bmatrix}_j =
\begin{bmatrix}
\sqrt{s_x^2 + s_y^2 + s_z^2} \\
\arccos \dfrac{s_z}{\sqrt{s_x^2 + s_y^2 + s_z^2}} \\
\arctan 2(s_y, s_x)
\end{bmatrix}_j . \tag{5.54}
$$

These spherical coordinates $\mathbf{s}_j^{(\mathrm{C})}$ allow for a simple extraction of the cable span. The
k line vectors can now be sorted in ascending order of their φ-value (Fig. 5.15). Then,
n_s segments of equal size are chosen for the angle φ that represent the ranges

$$S_i = \left[\frac{i\,2\pi}{n_s}; \frac{(i+1)2\pi}{n_s} \right]_i \quad i = 0, \ldots, (n_s - 1) \quad . \tag{5.55}$$

Then, one loops through the sorted list $\mathbf{s}_j^{(c)}$ of cylinder coordinates and extracts for each range S_i the matching element with

$$\mathbf{s}_j^{(c)}\Big|_{\varphi \in S_i} \tag{5.56}$$

and stores the largest angle θ for all line vectors that belong to the respective segment. After this procedure, one has a sorted list of n_s characteristic vectors of the surface of the cable span. Connecting two neighboring vectors with the apex at $\mathcal{K}_{A,i}$ gives a triangulation of the surface of the cable span. The list of the angles over the polar coordinate is basically a look-up table to check if a vector is inside the cone.

5.5.7   Workspace Cross Sections

The workspace hull is a two-dimensional data model to characterize a three-dimensional workspace by describing its boundary. One can compute intersections between this hull and a plane to receive cross sections. Cross sections are one-dimensional curves on the surface of the workspace. We can approximate these curves with line segments that are simple to compute and also simple to process and interpret. To do so, we follow the same approach used to compute the workspace hull by triangulation. Firstly, we define a central position $\mathbf{m}$ that is expected to be inside the workspace. Secondly, we construct a number of normal vectors $\mathbf{v}_i$ that define the search direction. Contrary to the workspace hull where we need a set of normal vectors pointing to the surface of a sphere, it is straightforward to find a homogenous distribution for the normal vectors inside a circle. For the xy-plane, the vectors are given by

$$\mathbf{v}_i = [\cos(\varphi), \sin(\varphi), 0]^{\mathrm{T}} \quad \varphi \in [0; 2\pi] \tag{5.57}$$

where it is simple to generate a set of any number of spanning vertices. If the cross section shall be computed for a different normal than the z-axis, one can apply any appropriate rotation matrix $\mathbf{R}$ to the normal vectors v_i to search in other directions.

To compute the cross section, one applies a line search (see Sect. 5.5.1) to all vectors $\mathbf{v}_i$. After doing so, one finds a polygon approximating the workspace cross section simply by considering the connection line from $\mathbf{v}_i$ to $\mathbf{v}_{i+1}$ and connecting the last vector $\mathbf{v}_N$ in the list to the first $\mathbf{v}_0$. Due to their simple data model, one can easily compute the area $A(\mathcal{W})$ and the circumference $L(\mathcal{W})$ of the cross section from its vertices:

$$L(\mathcal{W}) = \sum^{\mathcal{L}} ||\mathbf{v}_A - \mathbf{v}_B||_2 \tag{5.58}$$

$$A(\mathcal{W}) = \frac{1}{2} \sum^{\mathcal{L}} |(\mathbf{v}_A \ \mathbf{v}_B \ \mathbf{e}_z)| \tag{5.59}$$

where $\mathbf{e}_z = [0, 0, 1]^T$ is the unit vector in the z-direction. To conclude, it is emphasized that the data models presented here separate the topological structure from the geometry. This facilitates the analysis of the results since explicit information on the neighbors are represented in the results data.

5.6 Analytic Boundary Determination

In the section above, we have extensively studied numerical methods to determine the boundary of the workspace. The presented numerical schemes can be applied for many workspace criteria that allow for a binary decision whether a pose belongs to workspace or not. In this section, we exploit the structure of specific criteria to compute the boundary of the workspace while considering the structure of the underlying mathematical object. As shown below, this leads to elegant solutions which are in turn restricted to the determination of the wrench-closure workspace.

5.6.1 Wrench-Closure Workspace in Closed-Form

Surprisingly, a closed-form representation of the wrench-closure workspace $\mathcal{W}$ exists [473]. The wrench-closure workspace is the union of $m - 1$ over n semi-algebraic sets given by

$$\mathcal{W} = \left\{ \mathbf{y} \in SE_3 \mid \text{rank } \widehat{\mathbf{A}}^T = n \wedge \exists_{\mathbf{h} \in \ker \widehat{\mathbf{A}}^T} \mathbf{h} > \mathbf{0} \wedge h_m > 0 \right\} \tag{5.60}$$

$$= \bigcup_{\mathcal{I} \subset \{1,...,m-1\},|\mathcal{I}|=n} \left(\begin{array}{l} \left\{ \mathbf{y} \in SE_3 \mid \det \widehat{\mathbf{A}}_{\mathcal{I}}^T > 0 \wedge (\det \widehat{\mathbf{A}}_{\mathcal{I}}^T)\widehat{\mathbf{A}}_{\mathcal{I}}^{-T}\widehat{\mathbf{a}}_m < 0 \right\} \\ \bigcup \left\{ \mathbf{y} \in SE_3 \mid \det \widehat{\mathbf{A}}_{\mathcal{I}}^T < 0 \wedge (\det \widehat{\mathbf{A}}_{\mathcal{I}}^T)\widehat{\mathbf{A}}_{\mathcal{I}}^{-T}\widehat{\mathbf{a}}_m > 0 \right\} \end{array} \right)$$

where the matrix $\widehat{\mathbf{A}}_{\mathcal{I}}^T \in \mathbb{R}^{n \times n}$ is composed of the columns with the numbers $\mathcal{I}$ from the matrix $\widehat{\mathbf{A}}^T$ and $\widehat{\mathbf{a}}_m$ is a shorthand for the mth column of $\widehat{\mathbf{A}}^T$. Each set is constituted from $n + 1$ polynomial inequalities of degree n. The set $\mathcal{I}$ represents all subsets from the columns with exactly n elements, e.g. for $m = 4$ and $n = 3$ the sets are $\{1, 2, 3\}$, $\{1, 2, 4\}$, $\{1, 3, 4\}$, and $\{2, 3, 4\}$.

Some studies on planar systems used this or similar representations to determine the wrench-closure workspace [28, 142, 143] and for spatial robots with rotational degrees-of-freedom one can hardly derive the formulas even when using computer algebra systems. However, if one restricts the consideration to the constant orientation workspace, to cross sections, or to the orientation workspace, the closed-form can be computed symbolically or numerically. Even for the general case, the closed-form presented in Eq. (5.60) reveals that the wrench-closure workspace is bounded by a large yet limited number of polynomial patches of sixth order for the spatial 3R3T case, and third order both for the constant orientation workspace of spatial robots. The translational workspace of a planar robot is bounded by quadratic functions.

5.6.2 Mathematical Structure of the Workspace Boundary

Verhoeven [473], Gouttefarde [183, 185], Stump [444], and later Sheng [436] show that the boundary of the wrench-closure workspace can be determined from algebraic expressions by evaluating the structure matrix. If the non-normalized structure matrix $\widehat{\mathbf{A}}^{\mathrm{T}}$ is used instead of $\mathbf{A}^{\mathrm{T}}$ in the expressions, the resulting terms are largely simplified. Sheng derived second or third order multivariate polynomials for the workspace boundary using computer algebra. Using this procedure, the analytic expressions N_i potentially bounding the constant orientation workspace $\mathcal{W}_{\mathrm{co}}$ are shown [436] to be for a planar robot with $m = 4$ cables

$$N_1 :\ \det(\mathbf{A}_4, \mathbf{A}_2, \mathbf{A}_3) = 0 \tag{5.61}$$
$$N_2 :\ \det(\mathbf{A}_1, \mathbf{A}_4, \mathbf{A}_3) = 0 \tag{5.62}$$
$$N_3 :\ \det(\mathbf{A}_1, \mathbf{A}_2, \mathbf{A}_4) = 0 \tag{5.63}$$
$$N_4 :\ \det(\mathbf{A}_1, \mathbf{A}_2, \mathbf{A}_3) = 0 \tag{5.64}$$

and for a spatial robot with $m = 7$ cables

$$N_1 :\ \det(\mathbf{A}_7, \mathbf{A}_2, \mathbf{A}_3, \mathbf{A}_4, \mathbf{A}_5, \mathbf{A}_6) = 0 \tag{5.65}$$
$$N_2 :\ \det(\mathbf{A}_1, \mathbf{A}_7, \mathbf{A}_3, \mathbf{A}_4, \mathbf{A}_5, \mathbf{A}_6) = 0 \tag{5.66}$$
$$\vdots \qquad\qquad \vdots$$
$$N_6 :\ \det(\mathbf{A}_1, \mathbf{A}_2, \mathbf{A}_3, \mathbf{A}_4, \mathbf{A}_5, \mathbf{A}_7) = 0 \tag{5.67}$$
$$N_7 :\ \det(\mathbf{A}_1, \mathbf{A}_2, \mathbf{A}_3, \mathbf{A}_4, \mathbf{A}_5, \mathbf{A}_6) = 0 \tag{5.68}$$

where $\mathbf{A}_i$ is the ith column of the non-normalized structure matrix $\widehat{\mathbf{A}}^{\mathrm{T}}$. Following the procedure presented in [436], a pose belongs to the workspace if a subset of the equations N_i has the same sign. This criterion is exploited later in Sect. 5.6.3.3 to quickly compute the workspace. One can essentially do the same computation for the orientation workspace by substituting a constant position into the structure matrix and receive determinants that depend on the orientation parameters rather than the position. However, the analysis for the orientation workspace is different from the translation due to the different topology of $\mathbb{R}^3$ and the rotation group SO_3 and we do not tackle this problem here.

5.6.3 Symbolic-Numeric Wrench-Closure Workspace

One can describe the boundary of the wrench-closure constant orientation workspace by second or third order polynomials for the planar or spatial case, respectively, and in the following, we derive an approach to compute such a description. The basic approach to compute the workspace boundary is as follows: Firstly, one sets up the structure matrix of the robot. Secondly, the actual geometric parameters are substituted into the formula of the robot. Thirdly, the pose parameterization is introduced to the structure matrix. Then, one can symbolically compute the determinants. Evaluating the resulting symbolic expressions yield the desired parametric curves that are the boundary of the workspace. It is straightforward to execute the above-mentioned workflow using a computer algebra system and, even for the spatial case with 6×6 matrices, one can compute the determinant for a certain robot and often also for a family of robots with a parametric representation of frame length, frame height and so on. However, if arbitrary geometry is assumed, the number of symbols in the computer algebra system becomes so large, that it can hardly be handled with current computers.

To overcome this limitation, a symbolic-numeric approach is proposed [400] which is inspired by the method of Walker and Orin [483] for the equations of motion as well as by Hiller [211] for computing the Jacobian matrix of multi-body systems. In both contributions, some kind of coefficient identification scheme is employed to extract the numerical values of an equation with known structure from numerical evaluation with carefully chosen special values. Having realized that the mathematical structure of the expressions of the workspace boundary are second or third order multivariate polynomials, we can use a pose-dependent formulation to compute values of N_i.

The surprising effect of this evaluation is that one can reconstruct the full workspace boundary from only six (planar) or 20 (spatial) local evaluations of the structure matrix and its determinants to receive a closed-form parametric representation of the constant orientation wrench-closure workspace.

The number of coefficients required to describe the workspace boundaries are shown for different number of cables and different motions pattern in Table 5.2, where 1R2T denotes the planar robot with one rotational and two translational degrees-of-freedom and 3R3T presents three translational and the rotational degrees-of-freedom. The number of coefficients results from the amount of polynomial equations, their respective degree, and the number of determinate computed from the structure matrix.

Table 5.2 Number of coefficients of the multivariate polynomials

Motion pattern	Number of cables m	Number of coefficients
1R2T	4	$4 \times 6 = 24$
1R2T	5	$4 \times 5 \times 6 = 120$
1R2T	6	$4 \times 5 \times 6 \times 6 = 720$
3R3T	7	$7 \times 20 = 140$
3R3T	8	$7 \times 8 \times 20 = 1120$
3R3T	9	$7 \times 8 \times 9 \times 20 = 10080$

5.6.3.1 The 1R2T Case

The approach for the computation of the constant orientation representation for a cable robot with four cables is as follows. For the sake of simplicity, we omit in the following an additional index for the coefficients a for each equation N_i. Each boundary equation takes the form

$$N_i(x, y) = a_{xx}x^2 + a_x x + a_{yy}y^2 + a_y y + a_{xy}xy + a_0 \qquad (5.69)$$

for a planar robot. One can numerically evaluate Eqs. (5.61)–(5.64). The identification of the coefficients $a_{xx}, \ldots, a_0$ is done by computing the determinants for six position vectors $\mathbf{r} = [x, y]^{\mathrm{T}}$ following the scheme:

- Compute the coefficient a_0 by evaluating the four determinants for the position vector $\mathbf{r} = \mathbf{0}$.
- Compute a_{xx} and a_x from the determinants received from the position vectors $\mathbf{r} = [1, 0]^{\mathrm{T}}$ and $\mathbf{r} = [-1, 0]^{\mathrm{T}}$
- Compute a_{yy} and a_y from the determinants received from the position vectors $\mathbf{r} = [0, 1]^{\mathrm{T}}$ and $\mathbf{r} = [0, -1]^{\mathrm{T}}$
- Determine a_{xy} from evaluating the position $\mathbf{r} = [1, 1]^{\mathrm{T}}$.

The numerical procedure is as follows. Compute the non-normalized structure matrix $\widehat{\mathbf{A}}^{\mathrm{T}}$ for the position $\mathbf{r} = \mathbf{0}$ and the desired orientation φ_0 and the respective numerical values of N_i for $i \in 1, \ldots, 4$ from Eqs. (5.61)–(5.64). Analyzing the polynomial expression in Eq. (5.69) reveals that substituting zero for both x and y cancels out all terms but the coefficient a_0 and thus $a_0 = N_i(0, 0)$. Secondly, one repeats the trick to identify both a_{xx} and a_x by computing $N_i(1, 0)$ and $N_i(-1, 0)$. The identification of the coefficients is slightly more complicated since we have to solve a linear 2×2 equation system whose coefficients are defined from our test poses $[1, 0]^{\mathrm{T}}$ and $[-1, 0]^{\mathrm{T}}$. Thus,

$$\begin{bmatrix} 1 & 1 \\ 1 & -1 \end{bmatrix} \begin{bmatrix} a_{xx} \\ a_x \end{bmatrix} = \begin{bmatrix} N_i(1, 0) - a_0 \\ N_i(-1, 0) - a_0 \end{bmatrix} \qquad (5.70)$$

has the simple solution

$$a_{xx} = \frac{1}{2}\left(N_i(1,0) + N_i(-1,0)\right) - a_0 \tag{5.71}$$

$$a_x = \frac{1}{2}\left(-N_i(1,0) + N_i(-1,0)\right) \quad . \tag{5.72}$$

The computation of a_{yy} and a_y with the positions $[0,1]^T$ and $[0,-1]^T$ is done respectively. In the final step, we compute a_{xy} from the position $[1,1]^T$ with the simple equation

$$a_{xy} = N_i(1,1) - a_{xx} - a_x - a_{yy} - a_y - a_0 \quad . \tag{5.73}$$

Thus, we have numerically received the exact algebraic representation of the workspace boundary by as little as computing numerically the structure matrices for six poses and determining four determinants for each structure matrix.

The identification procedure can be formalized in a matrix form as follows:

$$\mathbf{S} = \begin{bmatrix} 1 & \mathbf{0}^T & \mathbf{0}^T & 0 \\ 1 & \mathbf{A} & \mathbf{0} & 0 \\ 1 & \mathbf{0} & \mathbf{A} & 0 \\ 1 & \mathbf{1}^T & \mathbf{1}^T & 1 \end{bmatrix}, \quad \text{with} \quad \mathbf{A} = \begin{bmatrix} 1 & 1 \\ 1 & -1 \end{bmatrix} \tag{5.74}$$

and

$$\mathbf{k} = \begin{bmatrix} a_0 \\ a_{xx} \\ a_x \\ a_{yy} \\ a_y \\ a_{xy} \end{bmatrix}, \quad \mathbf{h} = \begin{bmatrix} N_i(0,0) \\ N_i(1,0) \\ N_i(-1,0) \\ N_i(0,1) \\ N_i(0,-1) \\ N_i(1,1) \end{bmatrix} \quad . \tag{5.75}$$

Thus, the coefficients of the second order multivariate polynomial are determined by the linear system

$$\mathbf{Sk} = \mathbf{h} \quad . \tag{5.76}$$

Note that the test poses are carefully chosen so that the condition number of the matrix $\mathbf{A}$ is minimized. Since numerical computation of determinants is sensitive to round-off errors, the proposed poses are expected to maximize stability. However, a prerequisite of the procedure above is that the generic test poses used in the algorithm are not singular, i.e. the determinants computed from the matrix do not vanish. If the poses are singular, a rigid body transformation can be applied to the parameters $\mathbf{a}_i, \mathbf{b}_i$ to move the reference points away from the singular loci or other test poses can be chosen where the numerical instability might be reduced.

5.6.3.2 The 3R3T Case

The procedure for the spatial robot is essentially the same as for the planar robot but, in order to avoid the tiresome algorithmic description, we only describe the linear equation formulation here. The main part in reconstructing the polynomial boundary is to solve a large system Eq. (5.76). The sought polynomial boundary $N_i(x, y, z)$ takes the form

$$
\begin{aligned}
N_i = {} & a_{xxx}x^3 + a_{yyy}y^3 + a_{zzz}z^3 + a_{xxy}x^2y + a_{xxz}x^2z + a_{xyy}xy^2 + a_{yyz}y^2z + \\
& + a_{xzz}xz^2 + a_{yzz}yz^2 + a_{xx}x^2 + a_{yy}y^2 + a_{zz}z^2 + a_{xy}xy + a_{xz}xz + a_{yz}yz + \\
& + a_x x + a_y y + a_z z + a_{xyz}xyz + a_0 \quad .
\end{aligned}
\tag{5.77}
$$

The system matrix for identifying the 20 coefficients of the polynomials reads as block matrix

$$
S =
\begin{bmatrix}
1 & 0^T & 0^T & 0^T & 0^T & 0^T & 0^T & 0 \\
1 & A & 0 & 0 & 0 & 0 & 0 & 0 \\
1 & 0 & A & 0 & 0 & 0 & 0 & 0 \\
1 & 0 & 0 & A & 0 & 0 & 0 & 0 \\
1 & C & D & 0 & B & 0 & 0 & 0 \\
1 & C & 0 & D & 0 & B & 0 & 0 \\
1 & 0 & C & D & 0 & 0 & B & 0 \\
1 & 1^T & 1^T & 1^T & 1^T & 1^T & 1^T & 1
\end{bmatrix}
\tag{5.78}
$$

where the first and last column as well as the first and last row are scalars. Furthermore, 0 is a matrix with zero elements of appropriate size and 1 is a matrix having a 1 in each element. In contrast, the other columns and rows are constructed each from 3×3 matrices from the following matrices

$$
A =
\begin{bmatrix}
1 & 1 & 1 \\
-1 & 1 & -1 \\
8 & 4 & 2
\end{bmatrix}
\tag{5.79}
$$

$$
B =
\begin{bmatrix}
1 & 1 & 1 \\
-1 & -1 & 1 \\
-1 & 1 & -1
\end{bmatrix}
\tag{5.80}
$$

$$
C =
\begin{bmatrix}
1 & 1 & 1 \\
-1 & 1 & -1 \\
1 & 1 & 1
\end{bmatrix}
\tag{5.81}
$$

$$
D =
\begin{bmatrix}
1 & 1 & 1 \\
-1 & 1 & -1 \\
-1 & 1 & -1
\end{bmatrix} \quad .
\tag{5.82}
$$

From the structure of matrix S, one can see that it is block lower triangular. Obviously, both A and B are regular, therefore S is also regular. Inverting or solving a linear system with $\mathbf{S}$ can be done efficiently. Basically, one can apply an algorithm similar to the procedure described in the section above by computing first the coefficient a_0, then the triple a_{xxx}, a_{xx}, a_x from a 3×3 system and so on.

The vector of the sought coefficients $\mathbf{k}$ of the constraint and the right-hand side $\mathbf{h}$ of the equation read

$$
\mathbf{k} = \begin{bmatrix} a_0 \\ a_{xxx} \\ a_{xx} \\ a_x \\ a_{yyy} \\ a_{yy} \\ a_y \\ a_{zzz} \\ a_{zz} \\ a_z \\ a_{xxy} \\ a_{xyy} \\ a_{xy} \\ a_{xxz} \\ a_{xzz} \\ a_{xz} \\ a_{yyz} \\ a_{yzz} \\ a_{yz} \\ a_{xyz} \end{bmatrix}, \quad \mathbf{h} = \begin{bmatrix} N_i(0,0,0) \\ N_i(1,0,0) \\ N_i(-1,0,0) \\ N_i(2,0,0) \\ N_i(0,1,0) \\ N_i(0,-1,0) \\ N_i(0,2,0) \\ N_i(0,0,1) \\ N_i(0,0,-1) \\ N_i(0,0,2) \\ N_i(1,1,0) \\ N_i(-1,-1,0) \\ N_i(1,-1,0) \\ N_i(1,0,1) \\ N_i(-1,0,-1) \\ N_i(1,0,-1) \\ N_i(0,1,1) \\ N_i(0,-1,-1) \\ N_i(0,1,-1) \\ N_i(1,1,1) \end{bmatrix} . \tag{5.83}
$$

The coefficients of the polynomial can now be determined from the simple linear system Eq. (5.76). Computing the coefficients of the wrench-closure workspace of a spatial cable robot with seven cables thus requires the following steps:

- Numerically determine the structure matrix for the 20 positions listed in Eq. (5.83).
- For each of these matrices, extract the seven 6×6 determinants as described in Eq. (5.65) to generate the vectors $\mathbf{h}$.
- Solve the system Eq. (5.76) to compute the coefficients for each of the seven polynomials.
- The seven vectors $\mathbf{k}$ contain in their 140 elements the full information on the constant orientation wrench-closure workspace of the robot.

The computational efforts of the main steps consist of setting up 20 structure matrices, computing 140 determinants, and solving seven 20×20 linear systems. Solving the linear system can be done in linear computation time due to the almost triangular structure.

The procedure can be generalized to robots with more than seven cables. In this case, one has to compute more determinants from the structure matrices.

5.6.3.3   Efficient Hull Computation

As introduced in Sect. 5.5, a triangulation of the workspace boundary is an efficient data model to further process the results from workspace determination. As shown above, the vertices of the triangles are projected onto the hull of the workspace. Starting from the estimated center $\mathbf{m}$ of the workspace, the line

$$L_i : \mathbf{r}_i = \mathbf{m} + \lambda_i \mathbf{v}_i \quad \lambda_i > 0 \tag{5.84}$$

is searched for its root. Instead of the regula falsi based line search used to fine the boundary of the workspace, one can do better with the parametric representation derived above. Since the recently used workspace criteria can only be evaluated as Boolean test of complex numerical algorithm, we used a regula falsi line search. Due to the algebraic form of the workspace boundary, we propose to substitute the line Eq. (5.84) into the surface equation Eq. (5.69) proving the following expression which reads for the planar case

$$\left(a_{xx} v_{xi}^2 + a_{xy} v_{xi} v_{yi} + a_{yy} v_{yi}^2 \right) \lambda_i{}^2$$
$$+ \left(2a_{xx} m_x v_{xi} + a_{xy} m_x v_{yi} + a_{xy} m_y v_{xi} + 2a_y m_y v_{yi} + a_x v_{xi} + a_y v_{yi} \right) \lambda_i$$
$$+ a_{xx} m_x^2 + a_{xy} m_x m_y + a_{yy} m_y^2 + a_x m_x + a_y m_y + a_0 = 0. \tag{5.85}$$

Analyzing this lengthy expression reveals the simple form of a quadratic equation in λ_i. Here, we earn again the benefits of the algebraic formulation since the boundary of the workspace is computed by just solving the polynomial with the well-known formula

$$\lambda_i^{1,2} = -\frac{p}{2} \pm \sqrt{\left(\frac{p}{2}\right)^2 - q} \ . \tag{5.86}$$

According to the assumptions made for the hull computation, we use the smallest positive value of $\lambda_i^{1,2}$ received for any one polynomial N_i. If the roots are complex or all negative, we set $\lambda_i = 0$. In the latter case, the projection center is not part of the workspace.

For the spatial case, one can do essentially the same where the final solving for λ_i requires to compute the closed-form solution to a third order polynomial. However, in both cases we have shown that all computation steps from the geometry of the robots to the triangulation of the constant orientation workspace can be executed in closed-form with simple mathematical tools.

Even more, the triangulated hull of the workspace allows for some geometric characterizations of the workspace. It is straightforward to calculate the surface $S(\mathcal{W})$ and the volume $V(\mathcal{W})$ of the workspace using Eqs. (5.46) and (5.47), respectively. Eventually, even these expressions are received in a constant number of computational steps without approximation except for the assumption that the triangulation for the quadratic and cubic surface is exact. However, the polynomial form of the workspace boundary allows to compute and bound the error for the triangulation.

Table 5.3 Geometrical parameters for a robot with seven cables: platform vectors $\mathbf{b}_i$ and base vectors $\mathbf{a}_i$

Cable i	Platform vector $\mathbf{b}_i$	Base vector $\mathbf{a}_i$
1	$[-0.125, 0.0, 0.0]^{\mathrm{T}}$	$[0.0, 0.0, 0.0]^{\mathrm{T}}$
2	$[-0.125, 0.0, 0.0]^{\mathrm{T}}$	$[4.0, 0.0, 0.0]^{\mathrm{T}}$
3	$[0.0, 0.25, 0.0]^{\mathrm{T}}$	$[4.0, 3.0, 0.0]^{\mathrm{T}}$
4	$[0.0, 0.25, 0.0]^{\mathrm{T}}$	$[0.0, 3.0, 0.0]^{\mathrm{T}}$
5	$[-0.125, 0.0, 0.0]^{\mathrm{T}}$	$[0.0, 0.0, 2.0]^{\mathrm{T}}$
6	$[-0.125, 0.0, 0.0]^{\mathrm{T}}$	$[4.0, 0.0, 2.0]^{\mathrm{T}}$
7	$[0.0, 0.25, 0.0]^{\mathrm{T}}$	$[2.0, 3.0, 2.0]^{\mathrm{T}}$

Fig. 5.16 Constant orientation wrench-closure workspace of the example design with seven cables computed from the closed-form of the workspace boundary

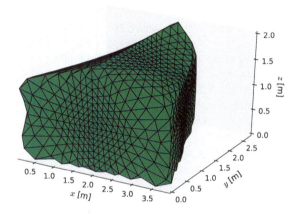

5.6.3.4 Computation Results

The workspace of the cable robot IPAnema 1 is determined for verification purpose using the algebraic expression method. The robot has seven cables and its geometrical parameters are given in Table 5.3. The determined constant orientation wrench-closure workspace $\mathcal{W}_{\mathrm{co}}$ is depicted in Fig. 5.16.

In order to determine the computational costs of the proposed method, an implementation in C++ is employed. In order to compute the matrix operations, including the evaluation of the determinants, the eigen 3 library[3] is used. The computation time is assessed on an Intel Core i5-3320M 2.6 GHz, Visual C++ 2010 using a single thread. A first test for a planar robot reveals computation times of around 0.12 ms per constant orientation evaluation with 36 points on the boundary and 0.26 ms for a resolution with 360 points. The computation time for computing the coefficients of the workspace polynomials without workspace computation is estimated to be 0.025 ms. Testing the components of the vector base of the matrix kernel to have the same sign leads to a computation time of 10 ms for 360 poses.

---

[3]Gaël Guennebaud, Benoît Jacob and others, *Eigen v3*, http://eigen.tuxfamily.org, 2010.

The evaluation of wrench-feasibility using the fast closed-form method [396] requires for 360 vertices on the hull in the linear search with regula falsi line search around 12.5 ms.

Comparing the computation times derived here, it seems that the proposed symbolic-numeric approach is very fast. To the best of the author's knowledge, no faster method has been reported in the literature.

5.6.4 Analytic Determination of the Workspace for a Planar Robot

The above-mentioned closed-form of the wrench-closure workspace [473] is applicable when considering a cable robot where some of the cable forces are fixed. From a practical point of view, this holds true if some of the cables are statically balanced by e.g. using counterweights to generate the cable forces. Then, it is possible to compute the boundary of the workspace in a symbolic form following roughly the approach given in Sect. 5.6.1. We exemplify this method with a planar robot. Consider a cable robot with $m = 4$ cables and a rectangular frame with length $2l$ and width $2b$. Then, the structure matrix can be written as follows:

$$\mathbf{A}^{\mathrm{T}}(x, y) = \begin{bmatrix} \frac{-l-x}{(l+x)^2+(b-y)^2} & \frac{l-x}{(l-x)^2+(b-y)^2} & \frac{l-x}{(l-x)^2+(b+y)^2} & \frac{-l-x}{(l+x)^2+(b+y)^2} \\ \frac{b-y}{(l+x)^2+(b-y)^2} & \frac{b-y}{(l-x)^2+(b-y)^2} & \frac{-b-y}{(l-x)^2+(b+y)^2} & \frac{-b-y}{(l+x)^2+(b+y)^2} \end{bmatrix} . \tag{5.87}$$

If two of the four cables are tensed by counterweights, the tension in two cables is constant. We apply the weights to cable 1 and 4. Thus, the following vector

$$\mathbf{f} = [m_1g \quad f_2 \quad f_3 \quad m_2g]^{\mathrm{T}} \tag{5.88}$$

results for the cable forces, where m_1, m_2 are the mass of the counterweights, respectively, and g is the gravity acceleration. Computing $\mathbf{A}^{\mathrm{T}}\mathbf{f}$ gives a system of two equations in two unknowns. In the following, we assume $m_1 = m_2$ for the sake of simplicity. Solving the resulting linear system renders a closed-form expression for the unknown cable forces

$$f_2 = f_1 \frac{A_2 B_2}{C} \tag{5.89}$$

$$f_3 = f_1 \frac{A_3 B_3}{C} \tag{5.90}$$

where

Fig. 5.17 Force-closure workspace for a simple 2T robot with $m = 4$ cables. The grey rectangle represents the robot's frame. **(a)** The two cables on the left side are tensed with constant forces of $f_1 = f_4 = 30\,\text{N}$. **(b)** The two cables on the left side are tensed with constant forces of $f_1 = f_4 = 30$ N. The workspace boundaries are shown for the minimum cable force $f_{\min} = \{0, 5, 10, 15, 20, 25, 30\}\,\text{N}$

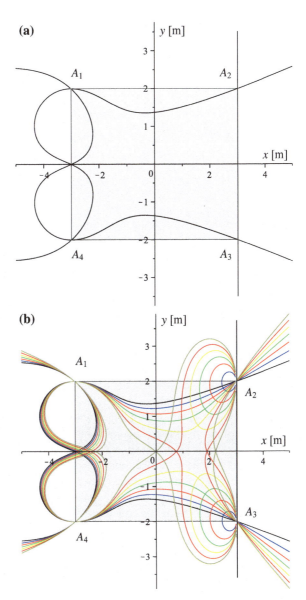

$$f_1 = mg \tag{5.91}$$

$$A_2 = (l - 2x)^2 + (b - 2y)^2 \tag{5.92}$$

$$A_3 = (l - 2x)^2 + (b + 2y)^2 \tag{5.93}$$

$$B_2 = bl^3 + 2yl^3 + 3bl^2x + 4l^2xy + bly^2 + lb^3 + 3lbx^2 \tag{5.94}$$
$$+ 2lx^2y - 2y^3l + by^2x + 2b^2yx + xb^3 + bx^3$$

$$B_3 = bl^3 - 2yl^3 + 3bl^2x - 4l^2xy + bly^2 + lb^3 + 3lbx^2 \tag{5.95}$$
$$- 2lx^2y - 2y^3l + by^2x - 2b^2yx + xb^3 + bx^3$$

$$C = b(l - x)\left((l + x)^2 + (b + y)^2\right)\left((l + x)^2 + (b - y)^2\right). \tag{5.96}$$

A position x, y belongs to the force-closure workspace if $f_2 > 0$ and $f_3 > 0$. In Fig. 5.17a, the boundary of the force-closure workspace is depicted for $l = 3$, $b = 2$, $f_1 = f_4 = 30$. The influence of the pretension in the cable 1 and 4 is studied in Fig. 5.17b where the force-feasible workspace is shown for different values of the minimal force $f_{\min}$.

5.7 Workspace Studies

In this section, we present a number of workspace studies in order to characterize the workspace of different robot architectures. Furthermore, we compare the influence of the number of cables, the applied wrench, and the force distribution method. Different types of workspace are compared.

In the following, we analyze how different criteria, parameters, and methods influence the results of the workspace computation. Therefore, we compare the workspace of different robot geometries and parameter settings, including changes in the minimum and maximum cable forces. For these studies, we analyze the accuracy for the workspace approximation (quality of the algorithms), the computational time, the volume, the surface, and the center of gravity of the workspace as well as the workspace's bounding box. Furthermore, we study the properties of the workspace algorithms and force distribution methods with respect to the used workspace criterion (wrench-closure, wrench-feasibility, wrench-feasible sets/available net wrench, collisions, stiffness, singularities) as well as the type of the workspace (constant orientation $\mathcal{W}_{CO}$ for different orientations, orientation workspace $\mathcal{W}_O$, total orientation $\mathcal{W}_{TO}$) and the applied wrench $\mathbf{w}_P$.

If not further specified in the case studies, the following parameters and assumptions are applied in order to make the different results comparable:

- For the influence of the parameters, we use some of the reference designs as described in Sect. 8.4.1.
- The geometry of the robot frame is normalized so that the volume of its axis-align bounding box equals $1\,\text{m}^3$. If the architecture of the frame is flat, the respective dimension of the platform is used for the computation of the bounding box.

- Lower and upper limit f_{min}, f_{max} of the cable forces is 1 and 10 N, respectively.
- The payload of the platform is chosen relatively to the upper cable force limit f_{max} with 0, 25, 50, 100, 150% in different directions. If not specified otherwise, no wrench $\mathbf{w}_P = \mathbf{0}$ is applied.
- If computation time is not considered, we use a force distribution method (Dykstra) that delivers a feasible distribution if one exists.
- When considering the constant orientation workspace $\mathcal{W}_{co}$, the platform is not rotated with respect to the reference design (i.e. the rotation matrix equals the identity matrix $\mathbf{R} = \mathbf{I}$).

When we address spatial robots, the geometrical analysis of the workspace becomes involved. Firstly, one can hardly make diagrams which represent the five or six-dimensional workspace in a form that is easy to understand. Furthermore, many effects, such as interferences, singularities, and limitations on the anchor points, influence the workspace in a counter-intuitive way. We use, if possible, the constant orientation workspace and total orientation workspace instead.

5.7.1 Cable Force Limits

The studies for different cable force limits are motivated from the considerations in Sect. 3.4.5 which shows that there are a couple of technical matters that restrict both the minimum and maximum cable force. In the following, the impact of the minimum f_{min} and maximum f_{max} cable forces on the shape and size of the workspace $\mathcal{W}_{co}$ is exemplified with the IPAnema 1 design. In the nominal model, cable forces between $1 \leq f \leq 10$ N are considered. We use the Dykstra method to compute force distributions in order to decide if a pose belongs to the workspace. We compute the constant orientation workspace $\mathcal{W}_{co}$ for $\mathbf{R} = \mathbf{I}$. No wrench is applied on the platform $\mathbf{w}_P = \mathbf{0}$. The workspace is approximated using the hull algorithm with $n_T = 5$ recursive subdivisions leading to 8192 triangles with an accuracy $\varepsilon = 0.0001$ m for the line search for the boundary of the workspace.

To study the influence, we kept the lower force constant, whereas the upper force was varied. The ratio between the lower and upper force is called $\kappa = \frac{f_{max}}{f_{min}}$. Firstly, we evaluate volume and surface of the workspace for some values with $\kappa \in [2; 10^5]$ as well as the computation time. In Fig. 5.18, the results of the evaluation are shown on a logarithmic scale for κ. It can be seen that around $\kappa = 1000$ the maximum for workspace and surface is reached and little gains for the volume of the workspace can be achieved using higher values. The second diagram shows the dependency of the computation time on the ratio κ. Since the Dykstra method is an iterative scheme, the computation time varies with the force settings. One can see that for κ around 50, there is a peak in the computation time where the computation time converges to an average value for high values of κ. A geometric comparison of cross sections of the workspace is shown in Fig. 5.19. All cross sections have been taken into the xy-plane with $z = 1$ m. We can see that for $\kappa = 1000$, most of the theoretical limits given by

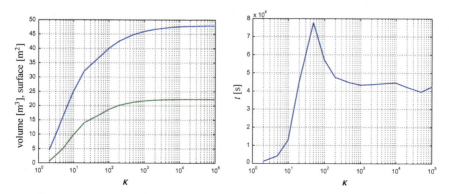

Fig. 5.18 Volume and surface (left) and computation time (right) of the workspace of the nominal IPAnema 1 design for different ratio κ between the minimum and maximum cable forces

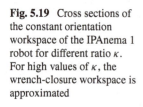

Fig. 5.19 Cross sections of the constant orientation workspace of the IPAnema 1 robot for different ratio κ. For high values of κ, the wrench-closure workspace is approximated

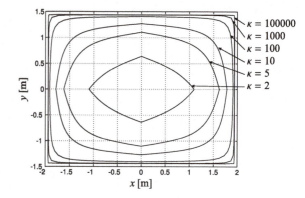

the wrench-closure workspace can be exploited. To compare shape and size of the workspace, spatial plots are given in Fig. 5.20 for different κ. Relating the volume of the wrench-closure workspace as upper limit of the IPAnema 1 to the restrictions named in Sect. 3.4.5, one finds that the achievable size of the workspace is between 60% in an optimistic setting and 10% in a very conservative setting (see Table 5.4).

5.7.2 Platform Load

When considering the wrench-feasible workspace, the load of the platform plays an important role. The load exerts a consistent wrench in the direction of the acting gravity vector, usually congruent with the negative z-axis of the inertial frame. In the following, we analyze how different wrenches influence size and shape of the workspace. In order to study the influence of the load, we apply different forces to the platform where we choose the load relative to the maximum cable force f_{max}.

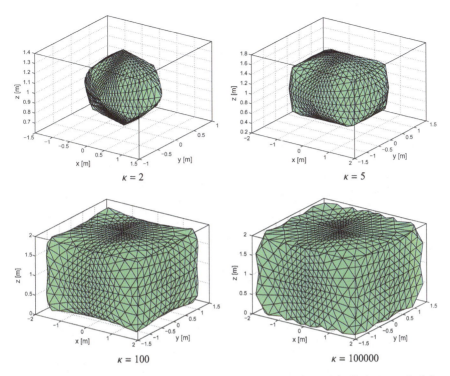

$\kappa = 2$

$\kappa = 5$

$\kappa = 100$

$\kappa = 100000$

Fig. 5.20 Comparison of the workspace hull of IPAnema 1 computed by Dykstra method for $\kappa \in \{2, 5, 100, 100000\}$. Iteration depth is reduced to 4 for the sake of clarity in the pictures

Table 5.4 Volume $V(\mathcal{W}_{\text{CO}})$ of the constant orientation workspace with $\mathbf{R} = \mathbf{I}$ of the IPAnema 1 robot computed with Dykstra method for different cable force limits

$f_{\min}$ (N)	$f_{\max}$ (N)	Volume $V(\mathcal{W})$ (m³)	Relative volume (%)	Description
0	∞	21.99	100	Wrench-closure workspace
10	180	13.28	60.4	IPAnema 1 with dynamics gearbox
100	720	7.45	33.9	IPAnema 1 with payload gearbox
40.6	180	4.25	19.3	IPAnema 1 with eigenfrequency limits
1	10	9.67	44.0	Reference force settings $\kappa = 10$

When considering the workspace of a loaded robot, care must be taken of how the workspace is defined. By applying a load, the workspace is also translated in the direction of the load, eventually leading to a suspended robot configuration. This can be easily understood when comparing the cable robot with a crane. If the applied forces are in negative z-direction right below the center of the robot frame, the cables share the load and are therefore able to withstand quite high forces. At the same time, the workspace degenerates to a long tube under the frame. Measuring the volume of the workspace creates irritating results since workspace volume is added without practical usage.

In the analysis presented in Fig. 5.21, only the part of the workspace inside the robot's frame is considered to make the changes in size and shape of the workspace evident. For 25% load, there is little effect on the size of the workspace. For 50% load, the workspace is already translated greatly in the direction of the acting force. Still, an acceptable volume remains inside the robot frame. At 100% load, the workspace hardly reaches the center of the robot frame. Therefore, only very few poses in the center of the workspace are able to withstand both plus and minus 100% of the maximum cable force in z. For even higher forces, some workspace remains on the very border of the robot frame but there is little connection to the unloaded workspace. When thinking about applications, this can cause problems because the robot might not be able to reach a working position as long as it is not loaded. However, at the same time, these effects make suspended robots effective.

When loading the platform with a torque, the deformation is quite different from the effect of forces. In Fig. 5.22, the results of different torques M_z on the cross sections of the IPAnema 1 constant orientation workspace are depicted. The maximum torque that leaves a very small region of the workspace is $M_z = 0.8$ Nm. An interesting effect occurs for smaller torques in the interval $M_z \in [0; 0.2]$ Nm: Here, we can observe an increase in the size of the workspace at the corners (top right, bottom left) of the workspace before higher torques shrink the workspace towards the center of the robot's frame. For the two other corners, we observe at the same time a continuous decrease in the size of the workspace.

After loading single forces and torques to the platform, we study the influence of available net wrenches. In this case, the platform has to provide every wrench from a given set. Only if all wrenches are feasible, the pose is said to belong to the workspace. This can be understood as an intersection of all workspaces computed for each wrench within the set. As shown by Bouchard [62], it is possible to verify if an ellipsoid with possible platform forces is fully enclosed by the available wrench set. In Fig. 5.23, the workspace for an ellipsoid with its main half-axis of length $[f_a, f_a, f_a]$ and a torque $\tau = 0$ is shown. More precisely, the wrench set $\mathcal{Q}$ is given by

$$\mathcal{Q} = \left\{ \mathbf{w}_\mathrm{P} = \begin{pmatrix} \mathbf{f}_\mathrm{P} \\ \boldsymbol{\tau}_\mathrm{P} \end{pmatrix} \in \mathbb{R}^6 \;\middle|\; \left(\frac{f_x}{f_a}\right)^2 + \left(\frac{f_y}{f_a}\right)^2 + \left(\frac{f_z}{f_a}\right)^2 \leq 1, \quad \boldsymbol{\tau} = \mathbf{0} \right\}$$

$$(5.97)$$

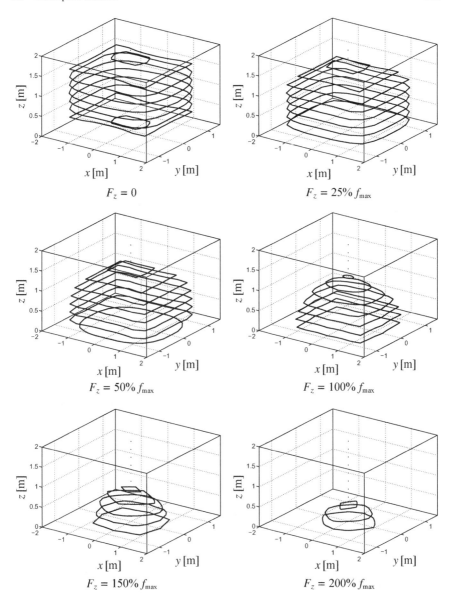

Fig. 5.21 Cross section of the constant orientation workspace of IPAnema 1 computed by Dykstra method for $f_z \in \{0, 25, 50, 100, 150, 200\}\%$ of f_{max}

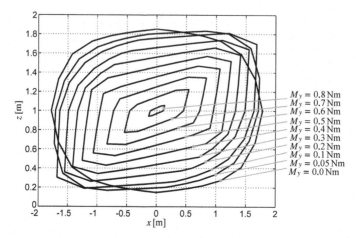

Fig. 5.22 Cross sections in y-direction of the constant orientation workspace of IPAnema 1 computed with Dykstra method for $M_y \in \{0, 0.05, 0.1, 0.2, \ldots, 0.8\}\,\mathrm{Nm}$

where f_a is the half-axis of the ellipsoid. Comparing the workspace with its counterpart of loading the robot with a single direction, a large limitation becomes clear. If the load direction is known and constant, e.g. because it is caused by the mass of the platform, the workspace has a reasonable size. In contrast, if the forces may act in arbitrary direction, the motion capabilities of the robot are significantly reduced. When loading the robot with 10% of the maximum cable force in any direction, the volume of the workspace is reduced to 67%, for 20% $f_{\max}$ only 47% volume remain, and for 50% $f_{\max}$ only 14% of the workspace is available. Loading the robot with the full cable forces is only possible at the very center of the workspace.

We find a similar situation when we consider ranges for the applied torque $\boldsymbol{\tau}_{\mathrm{P}}$. When loading the platform with an arbitrary torque in the set

$$Q = \left\{ \mathbf{w}_{\mathrm{P}} = \begin{pmatrix} \mathbf{f}_{\mathrm{P}} \\ \boldsymbol{\tau}_{\mathrm{P}} \end{pmatrix} \in \mathbb{R}^6 \;\middle|\; \mathbf{f}_{\mathrm{P}} = \mathbf{0}, \quad \tau_x^2 + \tau_y^2 + \tau_z^2 \leq \tau_a^2 \right\} , \qquad (5.98)$$

one receives the workspace shown in Fig. 5.23. Finally, we analyze the workspace for a mixture of forces and torques. For $f_a = 2\,\mathrm{N}$ and $\tau_a = 0.1\,\mathrm{Nm}$, the workspace shown in Fig. 5.24 is possible for the IPAnema 1 robot.

5.7.3 Platform Orientation

The influence of the mobile platform's orientation on the constant orientation workspace is studied for the IPAnema 1 system. Firstly, we compare the size and shape of the constant orientation workspace $\mathcal{W}_{\mathrm{co}}$ for different orientations (Figs. 5.25

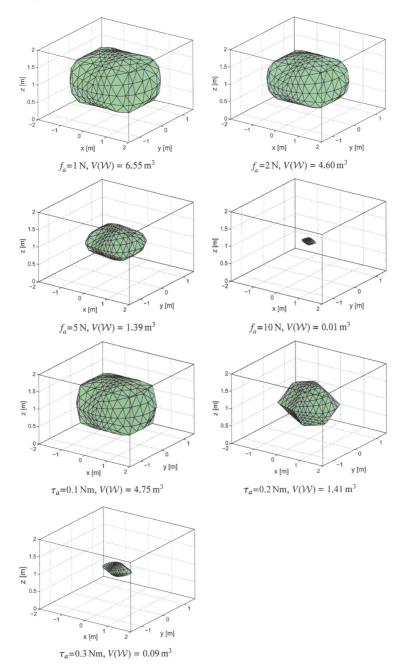

Fig. 5.23 Workspace for different available wrench sets $\mathcal{Q}$ of the IPAnema 1 robot where f_a is the half-axis of the force ellipsoid and τ_a is the half-axis of the torque ellipsoid

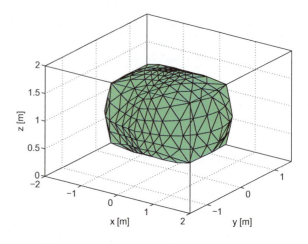

Fig. 5.24 Workspace of the IPAnema 1 robot for an available wrench sets defined by $f_a = 2\,\mathrm{N}$, $\tau_a = 0.1\,\mathrm{Nm}$. The volume of the workspace is $V(\mathcal{W}) = 3.47\,\mathrm{m}^3$

and 5.26). The Dykstra method is used for the evaluation to check feasibility. The recursive subdivision depth n_T for the surface is three iterations leading to 258 vertices. One can see that the volume V of the constant orientation workspace slightly decreases between 0° and 10°. For larger rotation angles α, β, there is a significant reduction in the size of the workspace. For angles larger than 20°, the workspace almost vanishes. At least it is so small and thin that it cannot be used in practical applications.

In Fig. 5.26, the volume and surface of the constant orientation workspace $\mathcal{W}_{\mathrm{CO}}(\mathbf{R})$ for $\mathbf{R} = \mathbf{R}_\mathrm{X}(\alpha)\mathbf{R}_\mathrm{Y}(\beta)$ is mapped over the rotation angles α and β in the range from 0° to 10°. The recursive subdivision depth is $n_\mathrm{T} = 3$ and the Dykstra method is used for the wrench-feasibility test. As expected from plots of the workspace, there is a rapid decrease in the volume of the workspace when the platform is tilted.

We consider the total orientation workspace $\mathcal{W}_{\mathrm{TO}}$, i.e. all poses that can be reached with all orientations in a given set $\mathcal{R}$. This workspace is the intersection of the constant orientation workspace for every orientation in the orientation set $\mathcal{R}$. Figure 5.27 shows the total orientation workspace for the given orientation sets.

In Fig. 5.28, a cross section plot with a rough estimate of the volume of the orientation workspace is plotted over an xy cross section. The plot is computed for closed-form wrench-feasibility with $f_i \in [f_{\min}; f_{\max}] = [1; 10]\,\mathrm{N}$ and no applied wrench $\mathbf{w}_\mathrm{P} = \mathbf{0}$. The volume is mapped from the span of the possible orientation range when performing pure rotations about x-, y-, and z-axis. The plot does not take into account poses where the orientation $\mathbf{R} = \mathbf{I}$ does not belong to the workspace.

5.7.4 Computation Method for Force Distribution

Here, different methods for computing the wrench-feasible workspace are compared. As already pointed out in Sect. 3.3, the region of convergency is a property of a force

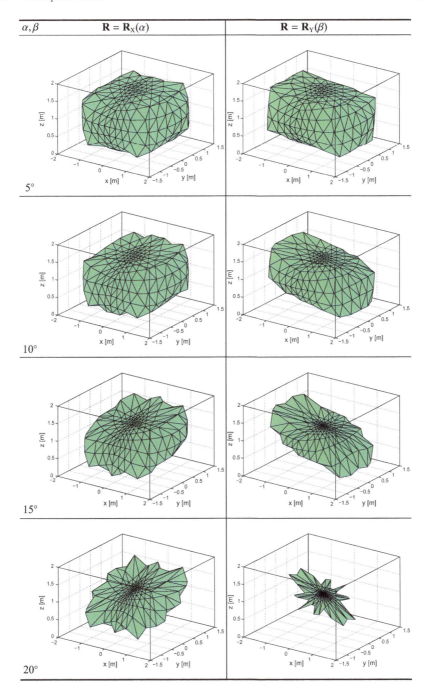

Fig. 5.25 Comparison of the constant orientation workspace $\mathcal{W}_{\mathrm{CO}}$ of IPAnema 1 for different orientations **R** of the platform

Fig. 5.26 Volume $V(\mathcal{W}_{\mathrm{CO}})$ of the constant orientation workspace for different orientations $\mathbf{R} = \mathbf{R}_{\mathrm{X}}(\alpha)\mathbf{R}_{\mathrm{Y}}(\beta)$ of the mobile platform

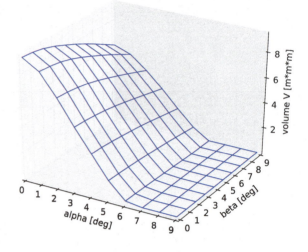

Fig. 5.27 Wrench-feasible total orientation workspace of the IPAnema 1 robot computed with the Dykstra method for $\mathcal{R} = \{\mathbf{R} \in SO_3 \mid \mathbf{R} = \mathbf{R}_{\mathrm{X}}(\alpha)\mathbf{R}_{\mathrm{Y}}(\beta),\ |\alpha|,\ |\beta| < 5°\}$

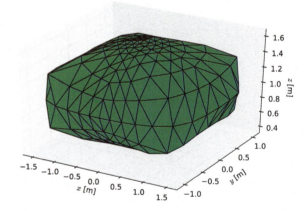

Fig. 5.28 Rough estimation of the volume of the orientation workspace. The diagram shows an xy cross section with $z = 1\,\mathrm{m}$ for the robot IPAnema 1

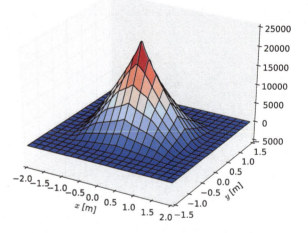

distribution method. To compare the closed-form method with the Dykstra method, the following workspace problem is considered. The geometry of the robot is taken from the nominal parameters of the IPAnema 1 robot (see Table 9.1). The wrench-feasible constant orientation workspace $\mathcal{W}_{co}$ with $\mathbf{R} = \mathbf{I}$ is computed for the cable force limits $f_{min} = 1\,\text{N}$ and $f_{max} = 10\,\text{N}$ and an applied wrench $\mathbf{w}_P = \mathbf{0}$. We obtained the results shown in Table 5.5. From the numbers in the table, we can see that the largest workspace is received with Dykstra method, wrench set method, and advanced closed-form method. The first and second method are low but reliable while the latter delivers its results quicker. The best computational times can be achieved with the closed-form methods, however, the workspace obtained is significantly smaller. For the case study in the table, only 60% of the workspace volume is found with the closed-form method. Contrary, the closed-form solution outperforms the Dykstra method in terms of computational time by a factor of more than 30.

Table 5.5 Results from the workspace computation with different methods for the IPAnema 1 robot. The used computer is an Intel Core i5-2520M, 2.50 GHz (single threaded). The workspace is computed through hull triangulation with 2048 triangles and 1026 vertices

	Dykstra closed-form		Advanced closed-form	Quadratic programming	Wrench set
Computation time [s]	3.2	0.092	0.158	3.126	10.029
Volume V [m^3]	9.671	5.795	9.658	9.590	9.671
Surface S [m^3]	24.716	17.899	24.729	24.675	24.717
Illustration	Figure 5.29a	Figure 5.29b	N/A	N/A	N/A

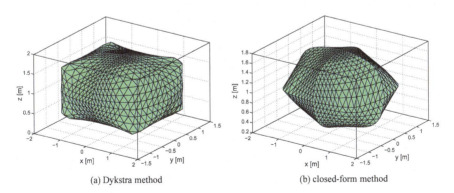

(a) Dykstra method   (b) closed-form method

Fig. 5.29 Hull of the constant orientation workspace of the IPAnema 1 robot computed with Dykstra method (left) and closed-form method (right)

5.7.5 Differential Hull Studies

In this section, a case study of the use of the differential hull approach is presented. The case study is based on the IPAnema 1 robot and the differential hull is computed for the partial derivations of the workspace hull for changes in the x-component of the first proximal anchor point $\mathbf{a}_1$. Using the differential hull, the change in shape and size of the workspace can be determined. Therefore, the algorithm computes a finite difference approximation for the differential

$$d = \frac{\partial \mathcal{W}(a_{1x})}{\partial a_{1x}} \tag{5.99}$$

where the differences are actually generated as differences in each vertex $d\mathbf{v}_i$ of the hull. The results of such a computation are visualized in Fig. 5.30. Red lines in the diagrams indicate regions with positive values of the derivatives $d\mathbf{v}_i$ and thus a growth in the workspace. In contrast, blue lines represent negative derivatives which correlate with a local decrease in workspace volume. In Fig. 5.30b, the same results are shown in order to highlight the region with negative derivatives that are occluded by the hull in the left plot since the negative derivatives are pointing inwards from the surface of the workspace. To compute the hull, the threshold for the line search is $\varepsilon = 10^{-6}$ and the finite difference in $\mathbf{a}_{1x}$ was $\Delta \mathbf{a}_{1x} = 10^{-3}$. The absolute values of the finite differences range between -0.001418 and 0.001277 which indicates at maximum a one-to-one relation between the changes in the geometry and the changes in the workspace.

The computation of the differential hull is very fast; the determination of the case study took around 30 ms on a Core i5-3320M with 2.6 GHz. Therefore, all partial

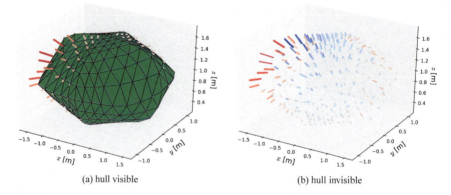

(a) hull visible   (b) hull invisible

Fig. 5.30 Differential hull of the constant orientation workspace $\mathcal{W}_{\mathrm{co}}$ of the IPAnema 1 robot computed with closed-form method for a finite difference in the x-component of the first proximal anchor point $\mathbf{a}_{ix}$. The magnitude of the differences is amplified to make the effect of the change visible. (a) The plot shows the hull with the normal lines indicating magnitude and sign of the finite difference. (b) The same analysis but with an invisible hull

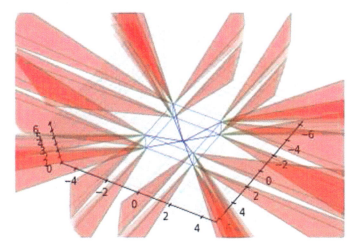

Fig. 5.31 Areas of possible cable-cable interference (red) for the constant orientation workspace of the IPAnema 1 robot. The robot geometry (blue) is sketched in the center of the plot for better reference to the geometric relations

derivatives of the workspace volume, surface, and bounding box can be determined in less than one second making the evaluation of these differences an interesting tool for the design of robots.

5.7.6 Cable–Cable Interference

The regions of possible cable-cable interference is depicted in Fig. 5.31. The region is computed with the algorithm from Perreault [381]. As one can see from the plot, the IPAnema 1 is unaffected by cable-cable interference. All triangles are pointing outward from the machine frame (marked as blue rectangle in the plot) which indicates that cable-cable interferences are not an issue for this robot. The computation time is very fast and, for the global intersection shown in the figure, the computation is by far below 1 ms.

One can easily compute the regions of intersection for a variety of orientations $\mathcal{R}$ providing families of curves with the regions of intersection. This relates to the effective restriction of the total orientation workspace. Here, the author refrains from drawing a respective plot with a family of curves because little can be seen from such a diagram. Firstly, because many overlapping faces are difficult to distinguish without rotating or animating it and, secondly, because most of the interference areas are grouped around the robot frame occluding the reference points A_i and B_i.

Fig. 5.32 Constant and total
orientation workspace of a
non-rectangular planar cable
robot

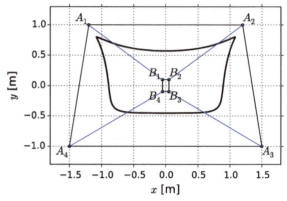

constant orientation workspace $\mathcal{W}_{\mathrm{CO}}$

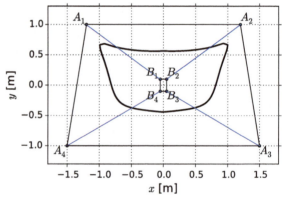

total orientation workspace $\mathcal{W}_{\mathrm{TO}}$

5.7.7   Planar Robots

In this section, we present some samples of the workspace of cable robots with
1R2T motion pattern. These robots are quite simple to study since their workspace
is a two- or three-dimensional manifold. Thus, it can be drawn without assumptions
and simplification in conventional diagrams providing an intuitive insight into size
and shape of the robot's workspace.

In Fig. 5.32, the constant orientation workspace and the total orientation workspace of a simple non-symmetric robot is depicted. One can see from the diagram
the counter-intuitive change in the shape of the workspace where the anchor points
A_1 and A_2 are moved from the corners of the rectangle towards the middle of the top
side.

5.8   Conclusion

The workspaces of cable robots are nontrivial geometric objects. In the general case, the workspace is a six-dimensional object in the Euclidian motion group that is complex both to compute and to model. Introducing projections and slices of the workspace into two and three dimensions helps to break down the complicated concept to something that can be handled for practical applications.

In this chapter, we have discussed a couple of tools to describe and compute the workspace of cable robots. As for other robotic devices, workspace computation of cable robots depends on the criteria required to accept a pose. For cable robots, the main restriction in terms of workspace size and shape is the need for positive tension. This restriction also implicitly tackles the effect of payload which is in general also pose-dependent. The limitations on the maximum motion in terms of cable length are less important and can be easily taken into account. Geometric restrictions such as collisions amongst cables and between the cables and the mobile platform are considerable limitations for the motion of the cable robot.

To describe the workspace, different models are discussed. Discretization methods such as regular grids are straightforward to compute and interpret. However, grids lack accuracy, sensitivity, and rigorous coverage. In turn, interval methods allow for guaranteed results on regions of the workspace. The main pose-dependent properties, such as wrench-feasibility, cable length, cable-cable collision, and deflection angles on the cables, are provided for usage in continuous workspace computation without holes. However, interval-based workspace computation is rigorous but slow. Thus, interval methods provide reliable results when little is known a priori about the robot design.

Some approaches for computing triangulation and algebraic curves of the boundary of the workspace are proposed. Triangulation is a highly efficient tool to quickly compute an approximation of the workspace hull and to store the data so that it can be taken up by standard engineering tools, including CAD, for planning and further processing. The algorithms proposed for hull computation allow for simple consideration of the constant orientation workspace, the maximum workspace, and the total orientation workspace. All pose-dependent criteria can be employed in the workspace hull computation. The computed hull is very accurate and sensitive to changes in the parameters if no holes are present or the evaluation criteria are known to be convex. A main drawback of the hull method is that it makes some strong assumptions about the shape and also the topology of the workspace. The hull method fails if the workspace is not simply connected or if it is not obvious to find a starting point for the search. In contrast, it is an efficient tool to tune the design parameters for well-understood architectures.

For fully-constrained cable robots, many important questions about the workspace of cable robots are understood and there are efficient algorithms to compute the workspace of cable robots. However, many of these problems remain open when the assumptions of the standard model are not fulfilled, i.e. for robots with sagging cables or for under-constrained robots with less cables than platform degrees-of-freedom.

Chapter 6
Dynamics

Abstract This chapter deals with the dynamic modeling of cable robots. Firstly, the system structure is presented. Then, models for the mechanical and electrical subsystems are introduced. Finally, results from simulation and experimental verification are presented.

6.1 Introduction

Cable-driven parallel robots are able to generate high velocities and accelerations due to their very small inertia of the moving parts. Furthermore, large workspace and high payloads are possible due to efficient force transmission through the cables. Therefore, it is proposed by many authors to use cable robots in applications that are huge, ultra-fast, or heavy-duty. Modeling and simulation of the dynamic behavior allow to study such extreme scenarios quickly, with little risk, and in a cost-efficient way. Therefore, a mechatronic model of cable robots is desirable to plan and virtually prototype large-scale or high-dynamical robots. Further applications of the dynamic simulation are structural and geometrical design of new robots. Dynamic simulation is a key element in the development and optimization of control strategies such as position control, force control as well as to test motion planning. For a complete virtual prototyping system, other pose-dependent properties of the cable robot can be found in chapters on kinematics and statics (Chaps. 3 and 7).

   A dynamic model extends the kinematic considerations by explicitly taking into account the influence of time on the behavior of the mechanical and also electrical subsystems of the cable robot. In the kinematic model, we sometimes use the concept of maximum velocities and accelerations. The origin of these limits can be found in the dynamic model, i.e. the change rates of many quantities are limited by physical laws. For the kinematic model, we have an input/output mapping that transfers an input motion at certain joints or bodies into an output motion at joints or bodies. The transfer function is based on the robot's mechanical properties and the equations are algebraic expressions. For parallel robots such as cable robots, we often receive

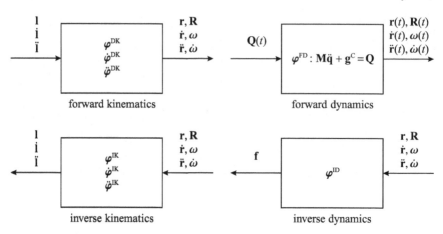

Fig. 6.1 The four main mappings in kinematics and dynamics

implicit equations that can only be solved numerically. In Fig. 6.1, an overview of the input-output relations of the different mappings is given.

6.1.1   Types of Dynamic Models

In the *forward dynamic model*, we have a mapping from input forces and torques to the output motion of the cable robot where the motion can be described as position, velocity, and acceleration of joints, points, or bodies. The sought forward dynamic model of the cable robot is a system of ordinary differential equations where the mechanical equations naturally are second order differential equations. One can use the forward dynamic model to simulate the cable robot, i.e. to predict how it moves if one connects it to virtual motors with given force signals. The simulation requires to integrate the differential equations and one usually applies numerical methods such as Runge–Kutta to solve the initial value problem.

The *inverse dynamics* characterizes the reverse mapping where the motion is given and the forces and torques that generate this motion are sought. The inverse dynamic model consists of algebraic equations and can sometimes be computed in closed-form. The equations are useful for system design as well as a model-based controller in feed-forward control strategies.

6.1.2   Review of Literature

In the literature, most researchers used Newton-Euler methods and accordingly d'Alembert's principle or Lagrange equations of the second kind to analyze the dynamics of cable robots. In many papers, the dynamics of the cable robot is investigated to design, test, and optimize the control system. Some more advanced works also try to deduce some kind of model-based controller. In contrast to kinematics where there is a huge difference in modeling under-constrained and over-constrained cable robots, the dynamic modeling only slightly differs. Most authors assume massless and linear elastic cables. Then, the motion between the winches and the platform becomes decoupled and one can easily determine the forces that are applied to the mobile platform.

As one of the first, Kawamura develops a dynamic model for his cable robot Falcon [237, 239]. It is also shown that cables with nonlinear stiffness are suitable for stable operations of the prototype. Yamamoto and Yanai [500–502] introduce inverse dynamics for a cable robot with movable proximal anchor points.

Beside from the dynamic effects of the platform and the cables, the drive-trains and winch mechanics can affect the dynamic behavior of the cable robot. Gallina [157] presents a dynamic model of a planar 2T cable robot with three and four cables including dynamics of the electric actuators. Then, the model is used in a simulation for the design of a PD controller. Cong [383] derives the dynamic modeling of a planar four cable robot including the dynamics of the winch units and Newton-Euler equations for the platform. A dynamic model and hardware-in-the-loop simulation for an over-constrained robot taking into account elastic cables, nontrivial pulley kinematics, and system dynamics of both actuators and the control system can be found in [344]. As an alternative in the modeling, the dynamic model of the robots IPAnema 3 and IPAnema 3 Mini using Lagrange equations is established by Kraus [259] who also models friction in the pulleys including nonlinear friction and a Dahl friction model. Based on this model, vibration is studied both in simulation and in practice. Recently, Tempel [462] contributed a dynamic model of the cable robot IPAnema 3.

A couple of works tackle the problem of motion planning for under-constrained cable robots. The dynamics of under-constrained cable robots for the prototype Cablev is used for simulation and control in [206, 306]. For the analysis of the dynamics, it is proposed to use flatness based methods [207, 309, 524]. A similar study presents a dynamic model and controller design [4]. As one of the first works, Barrette [29] establishes dynamic equations in order to formulate a dynamic equilibrium to exploit dynamic aspects of the workspace including trajectories through regions outside of the wrench-closure and wrench-feasible workspace. Cyclic stable trajectories are derived by Gosselin [174, 176]. Actually, no equations of motions are set up in this context. Cunningham [105] presents dynamics and motion planning for an under-constrained cable robot with only one cable. Korayem [251] uses Newton-Euler equations to set up the equations of motion where the dynamic model is used to compute the optimal trajectory to carry a payload. Fahham [137, 138, 518] derives the equation of motion using Newton-Euler equations for an over-constrained planar robot and transforms the differential equations to a one-dimensional path form in order to optimize motion planning. Wang [487] uses a dynamic model for a suspended robot with six cables in a RoboCrane-like configuration for motion planning. Lefrancois [284] provides the dynamic equations using Lagrange approach for an under-constrained robot with three degrees-of-freedom and only one cable. The model is used for motion planning and real-time control. A dynamic model using techniques from multi-body system dynamics of the cable robot Segesta is introduced [73, 139].

Rahimi [408] presents an experimental investigation on the dynamic behavior of an under-constrained suspended cable robot and determined settling times around 15 s for point-to-point motions. Beside others, Diao and Ou [116] study hardware-in-the-loop simulation and eigenfrequencies of a spatial cable robot with seven cables. Zheng [525] presents dynamics and vibration analysis of the suspension systems for wind tunnels where the eigenfrequencies are computed to be between 3.1 and 17.5 Hz for the simulation model. Bedoustani [32] uses Newton-Euler approach to derive the equations of motion taking into account linear elastic cables and damping effects in the cables. Then, Diao [120] provides a dynamic model to analyze both longitudinal and transversal vibrations of the cables. Later, the simplified dynamic equations with linearized stiffness are presented [121]. The eigenfrequencies are computed from the generalized eigenvalue problem. This paper also reported energy distributions for the vibrations and eigenfrequencies for different robots derived from simulations. For the parameters used in this study, the smallest eigenfrequency is determined to be around 20 Hz. Another experimental evaluation is presented by Kraus who shows that the dynamics of the drive-trains of the IPAnema 2 robot with cascaded motion control can be modeled as PT_1 systems with dead time [265].

Shiang [437] applies Lagrange equations of the second kind for dynamic analysis of a 3T robot with four cables. Afshari and Meghdari [3] use Lagrange equations to set up the equations of motion of a RoboCrane-type cable robot. Xianqiang [511] uses Lagrange equations of the second kind for dynamics of a hybrid cable robot which platform is constrained by some passive joints. Bedoustani [31] uses Lagrangian dynamics formulation where the change in the effective cable mass is taken into account and the author argues that the change must be taken into account for accurate simulation.

Many authors derive and implement dynamic models in order to study and design the controller of cable robots. A dynamic model and fuzzy control for a cable robot with six cables is presented as model for radio telescopes [528, 530]. Du [230] presents dynamic modeling of a cable robot for the development of a nonlinear control scheme. Khosravi also uses Lagrange equations of the second kind to set up the equations of motion taking into account cable elasticity for a planar cable robot [243–245]. Then, a controller in configuration (cable) space is designed and its stability is analyzed with Lyapunov's second method.

The connection between the time-continuous dynamics of cable robots and the time-discrete nature of digital control is lately addressed by Merlet [331, 334]. In this contribution, it is conjectured that the dynamic behavior of cable robots is heavily influenced by high frequency effects in the cables which have not yet been experimentally verified.

Zhang [520] presents a dynamic model based on Newton-Euler approach for a suspended cable robot with six cables and its linearization. Aside from Newton-Euler and Lagrange's energy method, a suspended robot is analyzed by Ya [524] using Lagrange equations of the first kind where the cable constraints are introduced as Lagrange multipliers. Another remarkable modeling approach is presented by Park [380], who proposes to use a linear complementarity problem (LCP) formulation to model the unilateral constraints imposed by the cables for dynamics. The authors analyze tension distributions and stability based on this LCP formulation. Aref [12] presents the dynamics of KNTU cable robot using Newton-Euler equations where an implicit formulation with special purpose integrator is used.

Although there are a large number of commercial and free dynamic simulation engines available, most work is presented with special purpose implementation in a high level computer language or in a general purpose numeric tool such as MATLAB, Scilab, or Octave. Wang [486, 487, 522] analyzes dynamics and stiffness of a planar 1R2T cable robot using the multi-body simulation systems MATLAB and ADAMS taking elastic cables into account. Later, Tang [457] presents a dynamic analysis of a six degrees-of-freedom robot with seven cables using ADAMS. To model the cables with a multi-body system, each cable is approximated through ten cylinder elements. Lately, Michelin [340] presents a dynamic simulation including an elastic finite element cable model and pulleys of the robot CoGiRo using XDE and MAT-LAB/Simulink. An open source framework for dynamics simulation of cable robots is lately proposed by Lau [111].

The nontrivial dynamic effects of cables such as sagging are tackled in a few works. Du [126, 234, 235] presents partial differential equations for sagging cables and derives an analogous model based on lumped mass and spring-damper elements that provide a multi-body system with ordinary differential equations for the cable dynamics. Duan [128, 129] presents the dynamic modeling based on similar partial differential equations for the cable for the model of the FAST telescope.

Dynamic models are applied to some application problems. Diao [117, 119] recalls a dynamic model for workspace consideration and force distribution analysis. Notash [364] derives the inverse dynamics equations of a cable-driven serial manipulator using Lagrange equations of the second kind. Taghirad [453] analysed the dynamics of the LAR cable robot using Newton-Euler equations where an aerostat is used and tethered to the ground by multiple cables. Liu [294, 295] uses Newton-Euler equations to generate the equations of motion of a cable robot for rehabilitation. Huang [407] presents Newton-Euler equations for an aircraft model. Oh [368, 369] proposes a dynamic model of a cable robot with two platforms inspired by a crane design that are both connected to the machine frame and to one another. Then, the dynamic model is used for controller design. Later, coupled dynamics of a helicopter and a suspended cable robot is presented [370].

Using dynamic models of cable robots is commonly spread in the literature and centered on using the mobile platform as free floating body that is constrained by cables. The equations of motion are established through Newton-Euler or Lagrange methods in almost all contributions. More in-depth results on the components are comparably rare. The cables are usually modeled as massless springs where small papers address how to gain the physical parameters of the cables such as Young's modulus. Some authors extend their considerations to model winch mechanics, pulleys including friction, and the electrodynamics of motors. Dynamic simulation of the cables using a finite element approach are proposed but yet lack experimental verification. Also, hardware-in-the-loop models are considered but little is validated to ensure the quality that is crucial to use the simulation as substitution for real experimental work.

In the following, the system architecture and the main subsystems of cable robots are introduced, mostly summarizing the findings from the literature. The considered complexity of the system is such that a realistic simulation is possible while achieving real-time efficiency.

6.2   System Structure

A cable robot is a mechatronic system consisting of a mechanical part and an electrical part (Fig. 6.2). The mechanical part includes the mobile platform that is connected by m cables guided by pulleys to the winches. The electrical part consists of m servo motors and position controllers. The governing numerical control is not further modeled here. Further details on different control algorithms can be found in the literature [259]. Its generated set-point position signal θ_i with $i = 1, \ldots, m$ is used as reference signal for the cascaded controller.

The dynamic behavior of the subsystems of a cable robot can be described by ordinary differential equations (ODEs) of first or second order. For simulation and numerical integration, the equivalent state space representation is obtained by transforming the high order differential equations into a system of first order ordinary differential equations. The overall system structure is shown in Fig. 6.2 with its forward

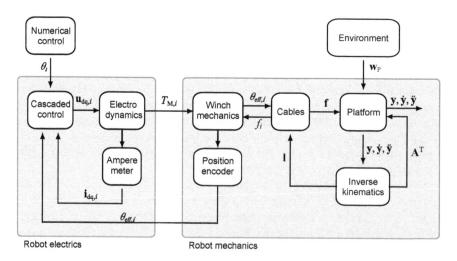

Fig. 6.2 Structure and interfaces of the mechatronic model for cable-driven parallel robots

dynamics and inverse kinematics. Furthermore, the modeled subsystems with their associated input and output quantities are depicted.

On the electrical side, the servo motors are described by their electrodynamics with the supply voltage $\mathbf{u}_{dq,i}$ as input quantities and the measured motor currents $\mathbf{i}_{dq,i}$ and torques $T_{M,i}$ as output quantities. The measured motor currents are fed back to the inner current control of the cascaded control, while the torque $T_{M,i}$ acts on the drum inside the winch together with the cable force f_i and therefore both are considered as input quantities to the winch mechanics. The drum angle $\theta_{eff,i}$ and rotary velocity $\dot{\theta}_{eff,i}$ are used as output quantities of the winch subsystem. The drum angle correlates with the rotor angle needed for the outer position and velocity control loop. Describing the platform pose by generalized coordinates $\mathbf{y}$ allows to determine the cable lengths $\mathbf{l}$ and the structure matrix $\mathbf{A}^T$ [473] by inverse kinematics. The platform motion is determined by the cable forces $\mathbf{f}$ as well as by the applied force $\mathbf{f}_P$ and torque τ_P which act on the tool center point.

6.3 Modeling of Robot Mechanics

To derive the equations of motion for the mechanical components of a cable robot, one can use recursive Newton-Euler equations or Lagrange equations of the second kind. Both mechanical principles are equivalent and result in equivalent equations. However, when actually computing the equations of motion, the methods differ in how simple they can be used to include different kinds of cable models and subsystems into account.

In order to formulate the dynamic equations of the cable robot, one has to choose the generalized coordinates $\mathbf{q}$. The generalized coordinates are the independent quantities in the dynamic system whereas other states of the system depend on the generalized coordinates. In the field of robotics, the *configuration space* and the *operational space* are widely used formulations.

The *configuration space* is given by the set of actuator variables such as the joint angles of a serial manipulator, the cable length of a cable robot, or the length of the struts of a parallel robot. In the configuration space, we can easily model the dynamic properties such as inertia and damping which are directly related to the relative motion of the actuated joints. Contrary, one has to calculate the dependent motion of the kinematic chain including the end-effector from the joint coordinates. This can be done efficiently and in closed-form for any serial kinematic chain. For parallel robots, the computational effort for solving the forward kinematics is significant and the solution is not unique. Furthermore, it might be difficult to track the numerical solution that belongs to the configuration at hand (Chap. 4).

The *operational space* is given by the Cartesian coordinates of the robot's end-effector. Therefore, it is straightforward to formulate the equations of motion for the platform. Given the motion of the end-effector, we have to deal with the dependent motion of the remainder of the robot (i.e. the cables). To compute these values, one can use inverse kinematics which can be solved efficiently in closed-form for most cable robots. In some cases, the calculation may even be omitted since the cables present relatively light elements that can be neglected. The formulation in operational space is also favorable since there is no need to distinguish between kinematically under-constrained, fully-constrained, and over-constrained robots. The platform is simply modeled as a free rigid body in space on which a given number of cable forces are applied to.

The proposed dynamic model is formulated in the operational space (Cartesian space) and it employs inverse kinematics and forward dynamics which are easy to solve and fast to compute. Based on this approach, the equations can be set up in closed-form such that computation is basically possible in real-time. The generalized coordinates of the platform pose are chosen as

$$ y = \begin{bmatrix} \mathbf{r} \\ \mathbf{Q} \end{bmatrix} , \tag{6.1} $$

where $\mathbf{r}$ is the position vector with respect to the inertial frame $\mathcal{K}_0$ and

$$ \mathbf{Q} = [q_0 \ \underbrace{q_x \ q_y \ q_z}_{q_1}]^{\mathrm{T}} = [q_0 \ \mathbf{q}_1]^{\mathrm{T}} \tag{6.2} $$

is the platform rotation described by a quaternion $\mathbf{Q}$ in order to avoid singularities and to provide numerical stability. The remaining derivation uses formulas that omit the imaginary units of the quaternion $\mathbf{Q}$ and handle it as ordinary four-dimensional vector. It should be noted that quaternions have to fulfill the normalization constraint

$$ \mathbf{Q}^2 - 1 = 0 \tag{6.3} $$

if used to parameterize the rotation group SO_3. The associated rotation matrix $\mathbf{R}$ is derived from the quaternion $\mathbf{Q}$ as shown in [430]. Using the skew-symmetric matrix

$$ \tilde{\mathbf{q}} = \begin{bmatrix} 0 & -q_z & q_y \\ q_z & 0 & -q_x \\ -q_y & q_x & 0 \end{bmatrix} \tag{6.4} $$

that is associated with the cross-product. The rotation matrix $\mathbf{R}$ can be computed from the quaternion $\mathbf{Q}$ with

$$ \mathbf{R} = \mathbf{I} + 2q_0\tilde{\mathbf{q}} + 2\tilde{\mathbf{q}}^{\mathrm{T}}\tilde{\mathbf{q}} . \tag{6.5} $$

Furthermore, the relation between the angular velocity ω and the time derivative $\dot{\mathbf{Q}}$ of the quaternion becomes

$$ \omega = \mathbf{P}\dot{\mathbf{Q}} , \tag{6.6} $$

where the transformation matrix $\mathbf{P}$ is

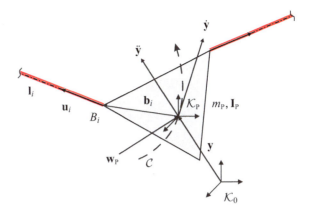

Fig. 6.3 The dynamic properties of the mobile platform when moving along a path $\mathcal{C}$

$$\mathbf{P} = 2\left[-\mathbf{q}_1 \quad q_0\mathbf{I} + \tilde{\mathbf{q}}\right] \in \mathbb{R}^{3\times4} \quad. \tag{6.7}$$

Using a parameterization of SO_3 with four parameters such as quaternions introduce an additional constraint which may be avoided using a three-parameter representation such as Euler angles or roll-pitch-yaw angles. All constraint-free parameterizations have singularities and their numerical stability may vary. Unfortunately, this is a general dilemma when dealing with rotations: One has to choose between singularity-free and constraint-free formulations which either have advantages and disadvantages over the others. Since most of the following considerations can easily be reformulated using any parameterization, we stick to quaternion equations.

6.3.1 Mobile Platform

The mobile platform is described as a free floating rigid body without kinematic constraints (Fig. 6.3). As shown in Sect. 3.3, the structure equations holds true for the stationary state

$$\mathbf{0} = \mathbf{A}^\mathrm{T}\mathbf{f} + \mathbf{w}_\mathrm{P} \tag{6.8}$$

with the applied wrench $\mathbf{w}_\mathrm{P} = [\mathbf{f}_\mathrm{P}^\mathrm{T} \quad \boldsymbol{\tau}_\mathrm{P}^\mathrm{T}]^\mathrm{T}$. In the dynamic state, the left-hand-side depends on the motion of the mobile platform. Using Newton-Euler formulation for the platform dynamics yields a differential algebraic equation system (DAE)[1] with six second-order differential equations, one algebraic equation, and seven unknowns which can be written in a compact form as follows

---

[1] If we use a three-parameter representation of the rotation group, we would receive a system of ordinary differential equations of second order instead at the cost of artificial singular configurations.

$$\left.\begin{array}{r} \mathbf{M}\,\mathbf{T}\ddot{\mathbf{y}} - \mathbf{D}\,\mathbf{T}\dot{\mathbf{y}} + \mathbf{g}^c = \mathbf{A}^T\mathbf{f} + \mathbf{w}_P \\ \mathbf{Q}^2 - 1 = 0 \end{array}\right\} . \qquad (6.9)$$

The matrix $\mathbf{T}$ is a transformation matrix with

$$\mathbf{T} = \begin{bmatrix} \mathbf{I} & \mathbf{0} \\ \mathbf{0} & \mathbf{P} \end{bmatrix} , \qquad (6.10)$$

where $\mathbf{P}$ is the transformation as defined in Eq. (6.7). The mass matrix $\mathbf{M}$ is given by

$$\mathbf{M} = \begin{bmatrix} m_P\mathbf{I} & \mathbf{0} \\ \mathbf{0} & \mathbf{I}_P \end{bmatrix} , \qquad (6.11)$$

where m_P and $\mathbf{I}_P$ are the mass of the mobile platform and its inertia tensor, respectively. In this setting, the inertia tensor is given in the coordinates of the base frame $\mathcal{K}_0$. The damping is characterized by the matrix

$$\mathbf{D} = \begin{bmatrix} \mathbf{D}_{\mathrm{lin}} & \mathbf{0} \\ \mathbf{0} & \mathbf{D}_{\mathrm{rot}} \end{bmatrix} , \qquad (6.12)$$

which coefficients $\mathbf{D}_{\mathrm{lin}}$ and $\mathbf{D}_{\mathrm{rot}}$ are linear and rotational damping, respectively. The damping reflects friction of the platform in its enclosing medium as well as friction in the cable attachment points. For cable robots moving at relatively low speed, the friction in air is most probably negligible. But for ultra-high speed as well as operation in thicker medium such as water, the influence of friction is significant. Finally, the generalized centripetal and Coriolis forces are collected in the vector

$$\mathbf{g}^c = \begin{bmatrix} \mathbf{0} \\ \mathbf{I}_P\dot{\mathbf{P}}\dot{\mathbf{q}}_1 + \tilde{\omega}\mathbf{I}_P\omega \end{bmatrix} . \qquad (6.13)$$

Deriving Eq. (6.3) twice with respect to time yields

$$\mathbf{Q}^T\dot{\mathbf{Q}} = 0 \qquad (6.14)$$

$$\mathbf{Q}^T\ddot{\mathbf{Q}} + \dot{\mathbf{Q}}^2 = 0 \qquad (6.15)$$

and one can transform the differential-algebraic equations into a system of ordinary differential equations of the form

$$\underbrace{\begin{bmatrix} m_P\mathbf{I} & \mathbf{0} \\ \mathbf{0} & \mathbf{I}_P\mathbf{P} \\ \mathbf{0} & \mathbf{Q}^T \end{bmatrix}}_{\mathbf{M}} \ddot{\mathbf{y}} = \underbrace{\begin{bmatrix} \mathbf{D}_{\mathrm{lin}} & \mathbf{0} \\ \mathbf{0} & \mathbf{D}_{\mathrm{rot}}\mathbf{P} \\ \mathbf{0} & \dot{\mathbf{Q}}^T \end{bmatrix} \dot{\mathbf{y}} + \begin{bmatrix} \mathbf{w}_P + \mathbf{A}^T\mathbf{f} - \mathbf{g}^c \\ \mathbf{0} \end{bmatrix}}_{\mathbf{k}} . \qquad (6.16)$$

Fig. 6.4 Concept of
normalization for a
quaternion **Q** by orthogonal
projection onto the constraint
manifold C. The projection
in the figure is only a
concept in three-dimensional
space whereas quaternions
are four-dimensional objects

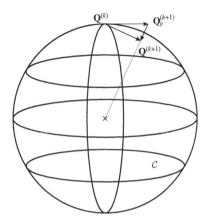

Solving the linear second-order equation $\mathbf{M\ddot{y}} = \mathbf{k}$ and subsequent integration of $\mathbf{\ddot{y}}$ involves drift effects caused by differentiation of Eq. (6.3) which can be dealt with by projecting $\mathbf{Q}$ and $\mathbf{\dot{Q}}$ onto the constraint manifold, respectively (Fig. 6.4). For the quaternion $\mathbf{Q}$, this orthogonal projection can be simply done by normalization

$$\mathbf{Q}^{(k+1)} = \frac{\mathbf{Q}_{\mathrm{P}}^{(k+1)}}{\left\| \mathbf{Q}_{\mathrm{P}}^{(k+1)} \right\|_{2}} \tag{6.17}$$

and for the derivative $\mathbf{\dot{Q}}$

$$\mathbf{\dot{Q}}^{(k+1)} = \mathbf{\dot{Q}}_{\mathrm{P}}^{(k+1)} - \frac{\mathbf{Q}^{(k+1)} \mathbf{\dot{Q}}_{\mathrm{P}}^{(k+1)}}{\left(\mathbf{Q}^{(k+1)} \right)^{2}} \mathbf{Q}^{(k+1)} \quad . \tag{6.18}$$

6.3.2 Cables

The cables show different dynamic effects which depend on their length, the dynamics, the material, and pretension. The following effects influence the dynamic behavior of a cable:

- *Elastic deformation*: Due to finite stiffness of the cable material, the effective cable length $l_{\mathrm{eff},i}$ depends on the cable tension f_i. The stiffness of the cable can be linear or nonlinear [239]. Damping is also caused due to friction inside the cables. Elastic deformation occurs especially for robots with very long cables [235] and robots with a high ratio of $f_{\min}$ and $f_{\max}$ admissible cable forces.
- *Thermal elongation*: The cable length changes depending on the temperature of the cable, moreover magnitude of elongation depends on the cable's material as

well as on the temperature change. Metallic cables stretch with higher temperature where this does not apply in general to all synthetic fibers. Some fibers, such as carbon fiber, may also contract under increasing temperature. Thermal elongation appears for example in outdoor applications. Aside from environmental effects, changes in cable temperature are also induced from high friction from the winch's drum and its pulleys as well as thermal sources in the winches or from the process or environment.

- *Creeping*: is a special form of deformation behavior of the cable material. For some material such as polyethylene, creeping is a hysteresis effect causing reversible elongation of the cable for some hours or days. Other materials show permanent deformations. For some materials, this effect can be mitigated by racking, i.e. pretension with a certain fraction of the breaking load for a longer time. Creeping effects depend mostly on the material.
- *Sagging*: static deformation of the cable caused by the weight of the cable. This effect is important for very large robots, for compliant cables, and for robots with very low minimum cable forces $f_{\min}$ [530].
- *Transversal vibration*: cables oscillate like the strings of a guitar causing high-frequency changes in the effective length $l_{\text{eff},i}$ and in the tension f_i of the cable. This effect is important for robots moving with high accelerations and especially if the cable's weight m_C cannot be neglected with respect to the platform mass m_P.
- *Longitudinal vibration*: cables oscillate along their axes. For large-scale robots, the cable material's sonic speed can present a noticeable delay and thus reduce the dynamic bandwidth of the cable robot. The same happens if robots shall be actuated with ultra-high accelerations.
- *Imperfect flexibility*: Cables are usually modeled to have negligible bending and torsional stiffness. Especially, cables made from steel show a finite bending stiffness on small pulleys. Furthermore, hysteresis effects occur in practice when the cable leaves the drum and the pretension is too small.

In the simplest model, cables can be understood as ideal unilateral constraint that transmit the pulling force from the winch to the platform without delay and loss. For this ideal model, the transmission function of the cable is given by the signum function such that a force is transmitted if the force is positive and no force applies to the platform if the computed force is negative. Although this approach can be implemented straightforward, it makes simulation difficult because the transmission function is discontinuous and shows a highly nonlinear behavior. Little additional efforts have to be undertaken to model a linear elastic deformation of the cable. Beside explicit modeling of the above-mentioned effects, some effects can be approximated by means of characteristic lines and look-up tables.

In the following, the cable is modeled as parallel spring-damper-system with a variable spring rate c_i and damping rate d_i, due to the changing effective cable length $l_{\text{eff},i}$. Eigendynamics of the cable is neglected at this point, i.e. it is assumed that the tensile force of the cable only depends on its elongation. Thus, neither longitudinal nor transversal vibration within the cable are taken into account. In any case, elastic modeling of cables allows to describe the vibrations of the mobile platform.

The standard model for the cables describes a unilateral transmission of forces, i.e. the cables can only transmit pulling but not pushing forces. Therefore, the cable force f_i is modeled as a piecewise function that reads

$$f_i(\mathbf{y}, \theta) = \begin{cases} c_i(l_{\text{eff},i}(\theta))\Delta l_i(\mathbf{y}, \theta) + d_i(l_{\text{eff},i}(\theta))\Delta \dot{l}_i(\mathbf{y}, \theta) & \text{for} \quad \Delta l_i > 0 \\ 0 & \text{for} \quad \Delta l_i \leq 0 \end{cases} \quad . \quad (6.19)$$

The effective initial cable length $l_{\text{eff},0,i} = l_{\text{N},i}(\mathbf{y}_0)$ is calculated by the nominal cable length

$$l_{\text{N},i}(\mathbf{y}) = \|\mathbf{l}_i(\mathbf{y})\|_2 + l_{\text{R},i}(\mathbf{y}) + l_{\text{G},i} \quad , \quad (6.20)$$

whereas $l_{\text{R},i}$ describes the cable length resting on the pulley and $l_{\text{G},i}$ describes the cable length inside the winch as shown in Fig. 6.5. With the unwound cable length $l_{\text{D},i} = r_i\theta_i$, the effective cable length reads $l_{\text{eff},i} = l_{\text{eff},0,i} + l_{\text{D},i}$ and the difference between the nominal cable lengths and the unwound cable lengths can be written as

$$\Delta l_i(\mathbf{y}, \theta_i) = l_{\text{N},i}(\mathbf{y}) - l_{\text{N},i}(\mathbf{y}_0) - l_{\text{D},i} \quad . \quad (6.21)$$

Deriving Eq. (6.21) with respect to time for $i = 1, \ldots, m$ yields

$$\Delta \dot{\mathbf{l}} = \mathbf{A}\mathbf{T}\dot{\mathbf{y}} - \dot{\mathbf{l}}_{\text{D}} \quad . \quad (6.22)$$

As mentioned above, the spring coefficient of the cable is not constant, if the cable is wound on a drum. Therefore, the effective free cable length is the basis for the determination of the current spring constant c_i. As it can be seen from Fig. 6.5, the cable length effective for forward and inverse kinematics code can differ from the

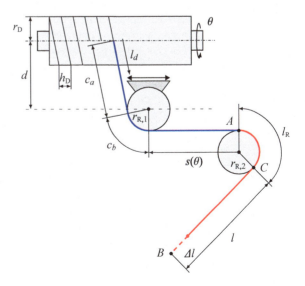

Fig. 6.5 Kinematics of the IPAnema winch with drum, spooling units, and two guiding pulleys

free cable length which has to be used to determine the stiffness of the cable. For the kinematic code, only this part of the cable that is attached to the second pulley is important and in practice, one calibrates the cable length such that the position sensors refer to the length between the point A_i and B_i. Although the length of the cable inside the winch is subject to implementation details, it is still necessary for determination of the spring coefficient c_i. Therefore, one has to take into account the additional values given in Fig. 6.5 where most of these lengths are constant for the IPAnema winch.

For practical computations, it is important to determine the numerical value of the spring constant of a cable material from measurement. In spite of decades of research on steel cables, it is still difficult to predict the elastic deformations of a yet unknown cable. Instead, one has to rely on tables, such as given by Feyrer [149]. For materials other than steel such as polyethylene, polyamide, or carbon fibers, literature values are even more difficult to find. As an estimate, one can determine the spring constant from the effective cross section A_c and Young's modulus E_c by

$$c_i = \frac{E_c A_c}{l_{i,0}} \quad , \tag{6.23}$$

where $l_{i,0}$ is the original length of the cable. For cables with high spring constants, one can use the measured length instead.

6.3.3 Winch Mechanics

The dynamics of the winch unit depends on the mechanical implementation. A typical winch with a servo drive, gearbox, and drum clearly differs from a pulley mechanism with a linear-direct drive (Fig. 6.6). In the following, we focus on winches of the first type and including motor, gearbox, drum, and supplementary moving parts such as guiding mechanism. In the following, we assume the cable robot has m identical winches with identical nominal parameters such as radii and moments of inertia. It is straightforward to use different values for each individual winch but the additional index i is discarded in the following equations for the sake of clarity.

The mechanics of the winch is modeled as a rigid multi-body system with one degree-of-freedom. Therefore, the equations of motion become quite simple where the determination of exact parameters for the inertia and friction may render more complex in practice.

The kinematic transmission in the winches depends on the gearbox ratio ν_{PG} and the effective radius of the drum r_D. When determining the effective radius of the drum r_D, one has to take into account that the effective radius depends on both, the diameter of the drum itself and the diameter of the current cable. If cables of different radii are used in the winch, then the distance of the neutral fiber from the axis of the drum changes. This effects in turn the force generated in the cable as well as the

Fig. 6.6 Dynamic model of
the IPAnema winch with
inertia parameters and
transmission factors

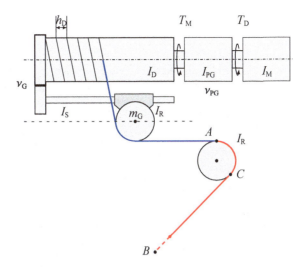

length and velocity of the cable. The relation between the rotation of the motor shaft
θ_M, and thus often also the position sensor, and the rotation of the drum θ_D is given
by

$$\theta_D = \nu_{PG}\theta_M \quad .\tag{6.24}$$

For the transmission between the drum and the cable, one has to take into account
the helical coiling of the cable onto the drum. This correction factor is important
since the position sensor is often integrated in the drum or into the motor. Therefore,
deviations between the perimeter and the helical sum up with each turn of the drum
and lead to a significant error in the cable length. Using the effective radius of the
drum r_D and the pitch h_D of the drum, the length of the cable becomes

$$l = \sqrt{1 + \left(\frac{h_D}{2\pi}\right)^2}\, r_D\theta_D\tag{6.25}$$

and thus the overall linear transmission ratio ν_w is given by

$$l = \underbrace{\sqrt{1 + \left(\frac{h_D}{2\pi}\right)^2}\, r_D\nu_{PG}}_{\nu_w}\,\theta_M \quad .\tag{6.26}$$

The winch dynamics is primarily determined by the winch mechanic's moment of
inertia I_w which can be determined by summing up the respective values of all parts
of the winch (Fig. 6.6). The moment of inertia of the winch renders

$$I_{\mathrm{W}} = v_{\mathrm{PG}}(I_{\mathrm{M}} + I_{\mathrm{PG}}) + I_{\mathrm{D}} + v_{\mathrm{G}}\left(2I_{\mathrm{S}} + m_{\mathrm{G}}\left(\frac{h_{\mathrm{D}}}{2\pi}\right)^{2}\right) + 2I_{\mathrm{R}} \quad, \tag{6.27}$$

where I_{M} is the moment of the motor's rotor, I_{PG} is the moment of inertia of the gearbox with respect to the motor, I_{D} is the moment of inertia of the drum, I_{S} is the moment of inertia for the displacement spindle, m_{G} is the mass of the linear traveling carriage, and I_{R} is the moment of inertia for the pulleys. The gear ratio v_{G} is the transmission between the drum and linearly displaceable carriage for each pulley.

Furthermore, the dynamic behavior of the winches depends on the frictional torque $T_{\mathrm{F}}(f)$, which itself depends on the cable force f. With the cable force on the one side and the motor torque T_{M} on the other side, the drum acceleration follows with

$$\ddot{\theta}_{\mathrm{eff}} = \frac{r_{\mathrm{D}}f + T_{\mathrm{M}} + T_{\mathrm{F}}(f)}{I_{\mathrm{W}}} \quad. \tag{6.28}$$

6.3.4 Lagrange Function for Platform and Cables

We have established the equations of motion using Newton-Euler methods in the previous section. However, it is worthwhile to compute the Lagrange function consisting of the kinetic and potential energy of the system as an alternative approach for deriving the dynamic equations of the cable robot. Using the Lagrange function, one can derive the equations of motion for the system but the Lagrange function is also related to kinematics and statics, as we will see. Thereby, the consideration of energy reveals some interesting links between kinematics, statics, and dynamics for cable robots.

The Lagrangian $L = T - U$ is the difference between the system's potential energy $U(\mathbf{q})$ and the kinetic energy $T(\mathbf{q}, \dot{\mathbf{q}})$, where the potential energy only depends on the system's configuration in terms of the current position of the generalized coordinates $\mathbf{q}$. In general, the kinetic energy is a function of both the generalized position $\mathbf{q}$ and the generalized velocities $\dot{\mathbf{q}}$.

Since the cables are modeled to be elastic elements, there are two subsets of coordinates in the system. The motion of the platform $\mathbf{y}$ has n degrees-of-freedom (six in the spatial case) and for each winch one independent coordinate is contributed to the vector $\mathbf{\Theta}_{\mathrm{D}}$. Thus, the generalized coordinates of the dynamic system are given by $\mathbf{q} = [\mathbf{y}^{\mathrm{T}}, \mathbf{\Theta}_{\mathrm{D}}^{\mathrm{T}}]^{\mathrm{T}} \in \mathbb{R}^{m+n}$.

The potential energy for a cable robot consists of the potential energy in the cables which is modeled in the following as linear springs. The effect of the mass of the cables is usually neglected in dynamics. To compute the potential energy in the cables, one has to distinguish between the geometrical distance $\overline{A_i B_i}$ between the proximal anchor points A_i and the distal anchor points B_i and the real cable length. The ideal distance is computed from standard inverse kinematics $\mathbf{l}_{\mathrm{G}} = \mathbf{\Phi}^{\mathrm{IK}}(\mathbf{y})$ as a function of the platform pose $\mathbf{y}$ and it is denoted by geometrical cable length. Furthermore, the unstretched cable connected to the motor in the winch has a length that depends on the independent generalized coordinate of the motor $\boldsymbol{\Theta}_D$ and is denoted by the vector $\mathbf{l}_w \in \mathbb{R}^m$ where $\mathbf{l}_w$ is a function of the winch kinematics. For well-designed winches, this relation is linear and depends on the drum radius, the cable radius, the drum pitch, and the spooling unit. Then, one computes $\mathbf{l}_w = \nu_w \boldsymbol{\Theta}_D$ from the effective winch transmission ratio ν_w. In general, this relation is nonlinear, e.g. if diagonal pull is accepted in order to compensate for the drum's pitch or if the cable is winded in multiple layers. For winches without cable guidance, the transmission may even be nondeterministic adding random uncertainties to the model.

Now, let $\Delta\mathbf{l} = \mathbf{l}_G(\mathbf{y}) - \mathbf{l}_w(\boldsymbol{\Theta}_D)$ be the difference between the real cable length $\mathbf{l}_w$ and the geometric distance $\mathbf{l}_G$. According to [245, 259], the potential energy of the cables reads

$$U = U_P + U_c = U_P + \frac{1}{2}\Delta\mathbf{l}^T K_c \Delta\mathbf{l} \tag{6.29}$$

$$= U_P + \frac{1}{2}(\boldsymbol{\Phi}^{IK}(\mathbf{y}) - \mathbf{l}_w(\boldsymbol{\Theta}_D))^T K_c (\boldsymbol{\Phi}^{IK}(\mathbf{y}) - \mathbf{l}_w(\boldsymbol{\Theta}_D)) \quad , \tag{6.30}$$

where U_P is the potential energy of the mobile platform and U_c is the potential energy stored in the cables. The potential energy of the platform is computed from

$$U_P = m_P g \, \mathbf{r} \cdot \mathbf{e}_z \quad , \tag{6.31}$$

where for the sake of simplicity the reference point of the platform frame $\mathcal{K}_P$ is assumed to be at the platform's center of gravity. The kinetic energy T of the cable robot is

$$T = \frac{1}{2}\dot{\mathbf{y}}^T \mathbf{M}(\mathbf{y})\dot{\mathbf{y}} + \frac{1}{2}\dot{\boldsymbol{\Theta}}_D^T \mathbf{M}_w \dot{\boldsymbol{\Theta}}_D \quad , \tag{6.32}$$

where $\mathbf{M}(\mathbf{y})$ is the position-dependent generalized mass matrix of the mobile platform and $\mathbf{M}_w$ is a diagonal matrix with the generalized mass of the moving parts of the drive-train. This generalized mass collects the inertia of the motor, gearbox, drum, and respective fraction of the spooling unit as well as the mass of the inner pulleys. All these inertia properties are lumped together in one condensed mass.

Thus, one can derive the equations of motion using the Lagrange equations of the second kind as follows

$$\frac{d}{dt}\left(\frac{\partial L}{\partial \dot{q}_i}\right) - \frac{\partial L}{\partial q_i} = Q_i \quad i = 1, \ldots, n \tag{6.33}$$

and receives the equations of motion. The eigendynamics of the winch may simply be neglected in this formulation by computing only the operational space part of the equation and considering the motion of the winch $\boldsymbol{\Theta}_D$ as rheonomic constraint.

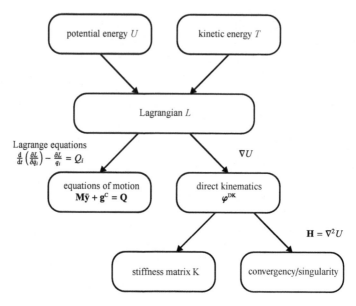

Fig. 6.7 Kinematics and dynamics is coupled through energy

6.3.5 Forward Kinematics and Dynamics

The Lagrangian function L reveals a connection between kinematics and dynamics (Fig. 6.7). Let the motion of the motors be the rheonomic input of the dynamic system. For a given position of the motor shafts, the free cable length is determined and we do the following thought experiment: An idealized controller keeps the current position of the motors without error for a sufficiently long time. If the wrench externally applied to the platform is constant e.g. only gravity is applied, then all kinetic energy in the system is eventually dissipated over time due to friction and the platform finally rests in an equilibrium pose. This pose $\mathbf{y}$ is only defined by locally minimizing the potential energy U in the system. Note that the energy functional may have multiple local minima and the one approached by the platform depends on the initial configuration of the platform and the trajectory on which the kinetic energy is dissipated. This implication is well in line with results from analysis of forward kinematics that postulated multiple solutions as well the experimental result that the platform is displaced to a neighbouring pose. As carried out in Sect. 4.3, this minimum is exactly the sought solution of the forward kinematics. Thus, the energy based solution of forward kinematics can be understood as the solution of the steady state dynamic equation and we have already discussed efficient methods to compute this equilibrium pose even under real-time requirements. Also, the Hessian matrix $\mathbf{H}$ according to Eq. (4.48) can be interpreted in view of that connection. For forward kinematics, one wants the Hessian matrix to be positive definite because then the objective function is locally convex and the minimization problem can be solved numerically stable. From dynamics point of view, positive definiteness of the Hessian matrix **H** is related to the stability of the pose with reaction forces pushing the platform into the equilibrium pose. In turn, an indefinite Hessian matrix **H** indicates an instable robot pose.

In a similar way, the stiffness of the cable robot is coupled to the shape of the potential energy vector field. One can compute the linearized stiffness from the second order derivative of the potential energy as follows

$$\frac{\partial^2 U}{\partial \mathbf{y}^2} = \mathbf{K} \quad , \tag{6.34}$$

where **K** is the linearized stiffness matrix. Clearly, if nonlinear springs are considered, the resulting computation becomes more involved but allows to approximate more advanced models with a linear stiffness matrix.

6.4 Modeling of Robot Electro-Mechanics

The robot's electrical system includes the drive-trains with embedded position controllers for the servo motors, the amplifiers, and the servo motors. The amplifiers use pulse-width modulation to provide the motors with the necessary rotating three-phase voltage. As the winches are mechatronic systems, one has to take into account control issues for the dynamic modeling. Here, we assume a decoupled cascaded control architecture for position control for each winch. This kind of control scheme is used in industrial servo drives although it does not present the most recent state of the art in drive control. For the simulation model, the amplifiers and motors are assumed as ideal devices without dynamic behavior. An analogous model for the drive-trains with a PT_1 behavior with decay is proposed by Kraus [259, 265] for the whole drive-train.

The permanent magnet synchronous motors are modeled as simplified motors without damping windings, iron loss, and with symmetrical star-connected motor windings (see Fig. 6.8). To realize field-orientated control, the electrodynamics differential equations are transformed from the stator's three phase system into a two phase rotor fix frame using Clarke and Park transformation [152]. Introducing the

Fig. 6.8 Equivalent circuit diagram of a servo motor

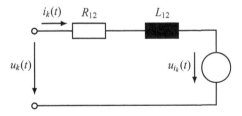

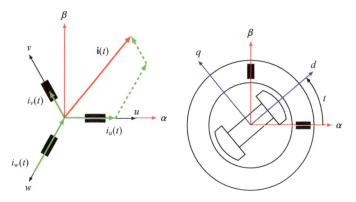

Fig. 6.9 Model of the electric servo motor

winding resistance R_{12} and the flux linkage $\boldsymbol{\psi}_{dq}$, the voltage differential equation respective the dq-frame (Fig. 6.9) reads

$$\dot{\boldsymbol{\psi}}_{dq} = \mathbf{u}_{dq,i} - R_{12}\mathbf{i}_{dq,i} - \dot{\theta}_{\text{eff},i}\mathbf{T}_{1}\boldsymbol{\psi}_{dq} \quad , \tag{6.35}$$

where $\mathbf{u}_{dq,i}$ describes the input voltage controlled by the upstream cascaded control and the matrix $\mathbf{T}_1$ is defined as

$$\mathbf{T}_1 = \begin{bmatrix} 0 & -1 \\ 1 & 0 \end{bmatrix} \quad . \tag{6.36}$$

The motor current

$$\mathbf{i}_{dq,i} = L_{12}^{-1}\left(\boldsymbol{\psi}_{dq} - \boldsymbol{\psi}_{R,dq}\right) \tag{6.37}$$

is used for the inner current control loop, where L_{12} is the winding inductance and $\boldsymbol{\psi}_{R,dq}$ is the rotor flux linkage caused by the permanent magnets. Considering the pole pair number Z_P, the motor torque is obtained by

$$T_{M,i} = \frac{3}{2}Z_P\left(\mathbf{T}_1\boldsymbol{\psi}_{dq}\right)\mathbf{i}_{dq,i} \quad . \tag{6.38}$$

The position control calculates the reference value for the downstream velocity control by the set-point θ_i and the effective angle $\theta_{\text{eff},i}$ using the controller amplification k_θ

$$\dot{\theta}_{\text{ref},i} = k_\theta\left(\theta_i - \theta_{\text{eff},i}\right) \quad . \tag{6.39}$$

For the velocity control loop, a proportional-integral controller with an amplification of $k_{\dot{\theta}}$ and a time constant $k_{T\dot{\theta}}$ is used. Calculation of the control deviation

$$\Delta\dot{\theta}_i = \dot{\theta}_{\text{ref},i} - \dot{\theta}_{\text{eff},i} \tag{6.40}$$

leads to the desired motor current

$$\mathbf{i}_{\text{dq,ref},i} = \begin{bmatrix} 0 \\ k_{\dot{\theta}} \left(\Delta \dot{\theta}_i + k_{T\dot{\theta}}^{-1} \Delta \theta_i \right) \end{bmatrix} \quad . \tag{6.41}$$

The reference value for the d-axis is set to zero, since a motor current along the d-axis has no influence on the motor torque. As with the velocity, the current is controlled by a proportional-integral controller with an amplification of $\mathbf{k}_{\text{dq}}$, a time constant $\mathbf{k}_{\text{Tdq}}$, and the control deviation

$$\Delta \mathbf{i}_{\text{dq},i} = \mathbf{i}_{\text{dq,ref},i} - \mathbf{i}_{\text{dq, eff},i} \tag{6.42}$$

yielding the supply voltage

$$\mathbf{u}_{\text{dq},i} = \begin{bmatrix} k_{\text{D}} \left(\Delta i_{\text{D},i} \right) + k_{\text{Td}}^{-1} \int \Delta i_{\text{D},i} \, dt \\ k_{\text{Q}} \left(\Delta i_{\text{Q},i} \right) + k_{\text{Tq}}^{-1} \int \Delta i_{\text{Q},i} \, dt \end{bmatrix} \quad . \tag{6.43}$$

6.5 Implementation and Validation

For the validation experiments, the cable-driven parallel robot IPAnema 1.5 is used which provides a six degrees-of-freedom end-effector with seven or eight cables and focuses on industrial applications in the field of material handling (Fig. 6.10).

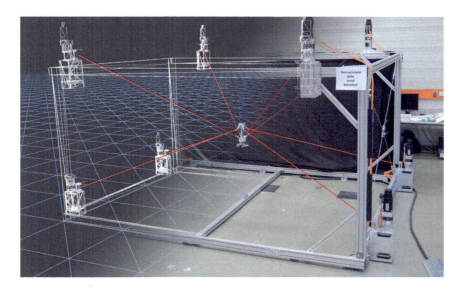

Fig. 6.10 Experimental setup of the cable robot IPAnema 1.5 with eight cables

Table 6.1 Measured IPAnema 1.5 geometrical parameters used for the validation of the dynamic model: base vectors $\mathbf{a}_i$ and mobile platform vectors $\mathbf{b}_i$

Cable i	Base vector $\mathbf{a}_i$ [m]	Platform vector $\mathbf{b}_i$ [m]
1	$[-1.537, 1.664, 1.172]^{\mathrm{T}}$	$[-0.05, 0.05, -0.05]^{\mathrm{T}}$
2	$[1.424, 1.745, 1.163]^{\mathrm{T}}$	$[0.05, 0.05, -0.05]^{\mathrm{T}}$
3	$[1.493, -1.573, 1.153]^{\mathrm{T}}$	$[0.05, -0.05, -0.05]^{\mathrm{T}}$
4	$[-1.474, -1.660, 1.159]^{\mathrm{T}}$	$[-0.05, -0.05, -0.05]^{\mathrm{T}}$
5	$[-2.054, 1.197, -0.601]^{\mathrm{T}}$	$[-0.05, 0.05, 0.05]^{\mathrm{T}}$
6	$[1.977, 1.343, -0.613]^{\mathrm{T}}$	$[0.05, 0.05, 0.05]^{\mathrm{T}}$
7	$[2.049, -1.166, -0.618]^{\mathrm{T}}$	$[0.05, -0.05, 0.05]^{\mathrm{T}}$
8	$[-1.975, -1.225, -0.609]^{\mathrm{T}}$	$[-0.05, -0.05, 0.05]^{\mathrm{T}}$

Table 6.2 Parameters of the dynamic model for the robot IPAnema 1.5

Parameter	Symbol	Value	Unit
Rotor moment of inertia	I_{M}	$2.8 \cdot 10^{-4}$	$\mathrm{kg\,m^2}$
Planetary gear moment of inertia	I_{PG}	$2.63 \cdot 10^{-4}$	$\mathrm{kg\,m^2}$
Threaded spindle moment of inertia	I_{S}	$2.29 \cdot 10^{-4}$	$\mathrm{kg\,m^2}$
Drum moment of inertia	I_{D}	$67.42 \cdot 10^{-4}$	$\mathrm{kg\,m^2}$
Pulley moment of inertia	I_{R}	$0.06 \cdot 10^{-4}$	$\mathrm{kg\,m^2}$
Threaded spindle pitch	h_{S}	0.004	m
Mass of pulley carriage	m_{C}	0.8	kg
Planetary gear transmission ratio	ν_{PG}	3	-
Gear transmission ratio	ν_{G}	2	-
Effective drum radius for cable with $r = 1.5$ mm	r_{D}	0.0479	m
Winch moment of inertia	I_{W}	$88.54 \cdot 10^{-4}$	$\mathrm{kg\,m^2}$
Platform mass	m_{P}	2.676	kg
Platform tensor of inertia about x-axis	$\mathbf{I}_{\mathrm{P}}^{(xx)}$	0.0261	$\mathrm{kg\,m^2}$
Platform tensor of inertia about y-axis	$\mathbf{I}_{\mathrm{P}}^{(yy)}$	0.0268	$\mathrm{kg\,m^2}$
Platform tensor of inertia about z-axis	$\mathbf{I}_{\mathrm{P}}^{(zz)}$	0.0073	$\mathrm{kg\,m^2}$
Spring rate of the cable at length 5.037 m	c_{ref}	9644	N/m
Damping coefficient	d_{ref}	8000	Ns/m

The geometrical parameters of this robot are given in Table 6.1 and the parameter values for the dynamic modeling are given in Table 6.2. The values for the electrical subsystem can be found in Table 6.3. The winches are equipped with permanent magnet synchronous motors IndraDyn S by Bosch-Rexroth. The control system is based on the PC-based real-time extension RTX and an adopted NC-controller by ISG (Stuttgart, Germany). For the hardware-in-the-loop simulation, the interpolation cycle time is 2 ms. A more detailed description of this cable robot is given in Sect. 9.3.1.

Table 6.3 Parameters of the electrical model for the robot IPAnema 1.5 using servo drives by Bosch-Rexroth of type MSK050B-0600 and appropriate inverters

Parameter	Symbol	Value	Unit
Torsional constant	k_M	0.90	Nm/A
Voltage constant	$k_{U,1000}$	55.0	V/min
Winding resistance	R_{12}	3.3	Ω
Winding inductance	L_{12}	19.90	mH
Pole pairs	Z_P	4	-

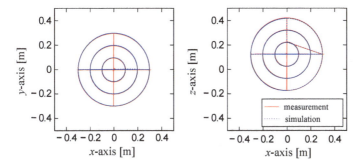

Fig. 6.11 Two-dimensional view of platform trajectories obtained by simulation (dotted line) and measurement (solid line)

The model is implemented by the use of Simulink and MATLAB, whereas each subsystem is modeled individually and connected as indicated in Fig. 6.2. For real-time simulation, the real-time coder of Simulink is used to generate source and header files, which provide public functions to access the simulation model. The programming languages C and C++ can be chosen. Embedding the simulation model in the RTX real-time environment is straightforward and can be done by including the header files in a main program which calls the simulation function. To integrate the system at runtime, a fourth order Runge-Kutta method is employed with fixed time step size.

Validation of the robot model is done by comparison of the actual platform trajectory with a simulated trajectory using a Leica Absolute Tracker AT901-MR for position determination with a sample rate of 1 ms and an accuracy of 0.025 mm. Figure 6.11 shows the measured and simulated circular trajectory respective the xy- and xz-plane with a maximum diameter of 600 mm and a velocity of 0.9 m/s while Fig. 6.12 shows the platform movement in direction of the x-axis with respect to time. The maximum positional deviation of the simulated trajectory respective the measured trajectory amounts to 5 mm for the movement in the xy-plane and 9 mm for the xz-plane. Thus, the accuracy of the simulation is the order of magnitude of

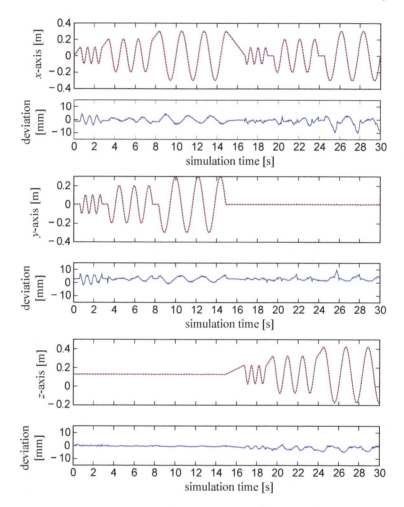

Fig. 6.12 Comparison of the simulated (dotted line) and measured (solid line) platform motion in direction of the x-, y-, z-axis with respect to time

the absolute accuracy of the robot (see also Sect. 9.3.1.4). The motor's internal current sensor is used to obtain the motor torque for comparison against the simulated torque with a sample rate of 1 ms (Fig. 6.13). The difference between simulation and measurement may result from the uncertain initial cable tension, the neglected compliance of the robot frame, and subsidence of cable and winch mechanics.

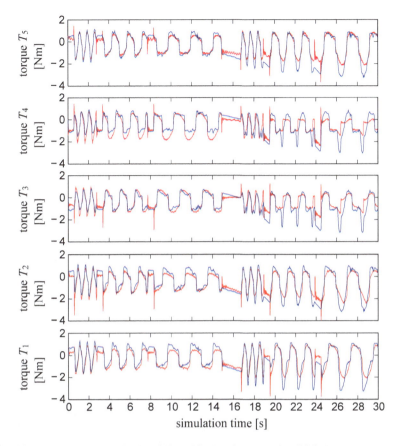

Fig. 6.13 Comparison of the simulated (dotted line) and measured (solid line) servo motor torque for axis 1–5 with respect to time

6.6   Conclusions

In this chapter, the dynamic model for cable robots is introduced. The model consists of the mechanical subsystems taking into account platform dynamics, cable elasticity, and mechanical winch dynamics. The equations of motion are set up using Newton-Euler and Lagrangian method. Furthermore, the characteristics of the motors and controllers are introduced and modeled in a simple but efficient way. The comparison between simulation results and measurements with the IPAnema 1.5 prototype shows good agreement. Thus, such model can be used in similar settings to plan and validate new geometries for cable robots as well as to design force-control.

The model structure for motors and control seems appropriate to be used for models of different scale. When using the electric model for robots with a very high system dynamics, the simplification in the motors and the controller may lead to increasing errors in the simulation.

Contrarily, for the mechanical model of the robot, it is expected that it is also valid for high velocities and accelerations. When scaling up the size of the robot, it is expected that the cable model is the weakness because it does not take into account large-scale effects such as sagging and geometrically large vibrations. Here, the cable is introduced as stateless system without dynamic behavior.

An open problem in dynamic simulation is to account for a realistic cable model under dynamic loads. As shown in the next chapter, the steady-state cable models such as the well-known Irvine model can be employed in kinematics and statics. However, one can observe a swaying motion of the sagging cable when the platform is laterally accelerated. Currently, no efficient models are available to simulate this behavior. Also, wave-formed motion of the cables are not captured by current models and certainly require the introduction of multiple elastic degrees-of-freedom for the cable. Thus, new formulations are sought to cope with these effects.

Chapter 7
Kinematics with Nonstandard Cable Models

Abstract In this chapter, we deal with the extension of the standard kinematic model by taking realistic assumptions for the cables into account. The modeling of nontrivial winch kinematics with guiding pulleys is addressed in Sect. 7.2. The consideration of cable mass leads to sagging (Sect. 7.3) and the finite stiffness of the cables causes elastic effects in the cables (Sect. 7.4).

7.1 Introduction

The standard kinematic model for cable robots is discussed in detail in the previous chapters. Applying the respective kinematics code to the controller enables operation of the cable robots. However, different maneuvers of the robots such as fast motion, poses at the boundary of the workspace, or very large robots show significant deviations from the expected behavior of the standard model. Typical discrepancies include positioning errors, insufficient stiffness, vibration, and slack cables. Clearly, the foundation of the standard model comprises a couple of assumptions that are violated in practice. The assumptions made for the standard model are stated explicitly in Sect. 3.1.2. Therefore, a natural line of action is to include some of the neglected effects in the model in order to receive advanced models.

The scope of this chapter is to partly extend the standard model towards more realistic but also more involved effects. We have to clearly state that some of the effects are still subject to recent research and are either not fully understood or no tools are available to deal with the mathematical problems unveiled so far. Furthermore, the state of the art beyond the standard model is less structured. Many separate approaches are presented e.g. to deal with elasticity in inverse kinematics where no consolidated model is considered that is applicable to other problems such as singularities, workspace, etc. Also, the cross relation between many of the effects has not yet been unveiled. Approaches to deal with elastic cables, sagging cables, pulley kinematics, and under-constrained robots have been presented in the literature.

© Springer International Publishing AG, part of Springer Nature 2018

A. Pott, *Cable-Driven Parallel Robots*, Springer Tracts in Advanced
Robotics 120, https://doi.org/10.1007/978-3-319-76138-1_7

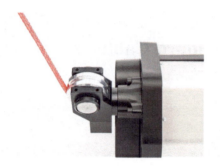

Fig. 7.1 External guiding pulley used with IPAnema 3 robot (side and isometric)

However, the approaches cannot be combined and it is an open issue how to deal with under-constrained robots that use pulley mechanisms.

This chapter is dedicated to three effects beyond the standard model: elastic cables, hefty cables, and pulley kinematics. The former two effects belong to the cable model, i.e. one tries to refine the understanding of the behavior of the cable where the latter effect is a global aspect in the kinematics of cable robots where geometric behavior of the robot is coupled to the static behavior. *Elastic* effects in the cable include linear and nonlinear change of the cable length depending on the cable force as well as creeping effects of the cable that cause length changes over longer time periods and also caused through aging of the cable.

A model of a *hefty* cable takes the effect of the cable mass into account. As a consequence of this mass, the cable is subject to sagging and vibration. Sagging describes the static deformation of the cable under gravity while vibration is a dynamic effect causing reference elements in the cable to perform oscillations in the direction of the cable (longitudinal vibration) and orthogonal to the direction of the cable (transversal vibration).

Finally, real cables have a finite bending stiffness and thus cannot be bent sharply. Therefore, cables are guided over curved surfaces typically realized by *pulleys* (Fig. 7.1). These additional geometric constraints are mostly effective at the proximal end of the cable and disturb the kinematic behavior of the robot. Thus, kinematic transformations are sought to take into account the curved form of the cable at the anchor points.

An under-constrained cable robot is a structure where the number of cables is less than the degrees-of-freedom of the end-effector. When suspending a load with less than six cables, the load finally rests at a certain pose. The underlying problem is no longer purely geometric since the stability of the static equilibrium now has a direct impact on the motion of the platform.

It is out of the scope of this work and hardly addressed in the literature what is the thermal behavior of cable robots, how elasticity of the machine frame and the mobile platform affect the motion of the robot.

Comparing the standard model with the advanced cable models leads to the problem of considering what really happens with a cable robot. One has to recall that every model is a simplification of the real physical behavior. From an engineering point of view, a model should be as simple as possible and as complex as required – and no more. A model of the cable or of the entire cable robot always has some limitations. A good model makes clear what effects have been taken into account and what simplifications have been made. Care must be taken when a model predicts yet unobserved effects. In this case, it is crucial to perform an experimental validation to distinguish modeling artifacts from relevant system properties.

The development of advanced models targets at a more accurate description of the behavior of the cable robot where the final goal is often related to increasing the accuracy of the robot. However, one will favor a simple model over a complex one if the gain in accuracy is not useful in the context of an application. There is no intrinsic property of a model that can tell us, how much the robot differs from reality. Therefore, experiments are required to validate the quality and the applicability of a model. In certain situations, one has to rely on simple models even if it becomes clear that the model has some shortcomings. One may have to stick to simple models because of limitations in the computation capacity, to fulfill real-time constraints, or because of technical limitations of the computer system where the model should be implemented. The latter restrictions often arise from a lack of openness in industrial equipment. Balancing such limitations against the potential of new ideas is the core of implementing innovative concepts.

7.2 Kinematics for Pulley Mechanisms

Cables are flexible and versatile construction elements that are used in civil engineering applications such as bridges, buildings, cablecars, and elevators. Nevertheless, there are important design rules when using cables that have to be taken into account such as minimum feasible bending radius and applicable dynamic loads. Therefore, one has to integrate elements such as pulleys to allow for acceptable durability as well as safety. Although mostly neglected in theoretical studies, almost every cable robot uses some kind of cylindric surfaces or pulleys to guide the cables. A perfect imitation of a spherical joint would require a sharp edge which cuts like a knife into the cable and would be subject to wear itself. Thus, curved guiding surfaces are necessary to achieve a reasonable lifetime of the cable and also for the guiding element. Some researchers use almost point-shaped guidance systems such as ceramic eyes [210] but very short operating life of only some dozen hours are reported for such installations. Guidance systems for cables are common machine elements in the field of transport as well as intralogistics, and mandatory design rules such as ISO 4308-1:3003(E) [222] and F.E.M. 1.001 [145] apply for cranes. As a rule of thumb, the radius of the pulley must be in the range between ten to twenty times the radius of the cable. Therefore, it becomes clear that in many applications the radius of the pulleys is relatively small but not negligible.

Only few authors have addressed the influence of guiding pulleys on the kinematics of cable-driven parallel robots. Jeong [226] already mentions the effect of pulleys

Fig. 7.2 Serial chain as analogous model for the guiding pulley

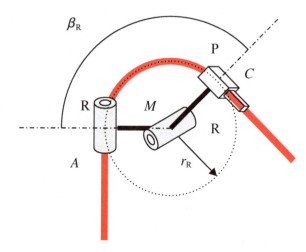

on the kinematic performance and accuracy but it seems that pulleys are not taken into account in the proposed forward kinematics code.

Bruckmann [72] derives an inverse kinematics transformation to cope with pulleys. This approach is enhanced by the author [391] to provide different formulations of the inverse pulley kinematics which are later also used in workspace computation and for statics consideration. Furthermore, Schmidt [434] develops also forward kinematics code considering pulleys. The influence of pulleys is also taken into account for the dynamic simulation of cable robots [344, 462]. Zoso presents a suspended planar robot where the kinematic and dynamic modeling includes pulleys [534]. Von Zitzewitz [532] proposes a kinematic model and a forward kinematics algorithm for an over-constrained cable robot with a pulley which is slightly different from the approach from Pott [391] and Bruckmann [72].

A notable result is presented by Gouttefarde [181, 189] who proposes a kinematic code that allows to take into account both the effect of pulleys and sagging of the cables in inverse kinematics.

A remarkable paper related to steel cables guided on pulleys is published by Lu [499] who analyzes the bending stiffness of cables and their capability to drive a drum or pulley when wrapped around it. Kraus [260] considered, aside from the kinematic influence, also the friction effects in the pulleys using Coulomb and Dahl friction models and showed that pulleys considerably contribute to the energetic efficiency of cable robots. Some considerations are already discussed in Sect. 3.4.5 related to pretensioning and wrapping the cable around the pulley and to drive the pulley without slip.

Lately, Gonzalez-Rodriguez [167] reconsidered the effect of pulleys for planar robots and proposes a robot design using pulleys both on the proximal and distal anchor points. When considering the resulting kinematic chain, the deviation introduced through a single pulley is eventually canceled.

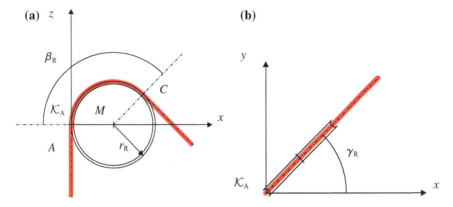

Fig. 7.3 Definition of coordinate frame $\mathcal{K}_A$ and variables for pulley kinematics

Guiding pulleys compromise the need for a feasible bending radius to prevent damaging and to achieve a kinematically well-defined behavior that can be modeled with acceptable efforts. In the standard model, simple constraints arise from the assumption of a fixed point for the proximal cable end in space. In the pulley model, the point is constrained to a *horn torus*, i.e. a torus with equal minor and major radii r_R. This also allows to define an equivalent multi-body kinematic chain with an RRP (Fig. 7.2) where R represents a revolute joint and P stands for a prismatic joint. The modeling with standard kinematic software is involved since the length of the cable is distributed amongst the rotation of the second revolute joint that mimics the rotation of the pulley and the prismatic joint that represents the cable length after leaving the pulley.

The parameters and coordinate frame used to exactly define the kinematics and geometry of a guiding pulley on the winch are depicted in Fig. 7.3. In the remainder of this section, we omit the index i for the reference points, frames, angles, and lengths for the sake of clarity. The equations hold true for all legs of the robot. The pulley kinematics realize a two degrees-of-freedom motion of the virtual point C where the cable leaves the pulley. The first revolute joint is aligned with the z-axis of frame $\mathcal{K}_A$ to pan the pulley. The second revolute joint is the pulley itself and its joint axis is initially aligned with the y-axis of frame $\mathcal{K}_A$. The center of the second rotation is located at point M. The distance between the two joint axes is the effective radius r_R and it is assumed that the two joint axes are perpendicular to each other.

The fixed point A in the origin of the coordinate frame $\mathcal{K}_A$ is the characteristic point of the pulley kinematics and both its position $\mathbf{a}$ as well as its orientation $\mathbf{R}_A$ are considered to be given as design parameters. The cable coming from the winch arrives at the pulley at point A and wraps around the pulley with an effective radius r_R, i.e. the radius that applies to the neutral fiber in the center of the cable. In the following, we assume that r_R is the effective radius, i.e. the radius resulting from both the actual radius of the pulley and the radius of the cable.[1] In general, the actual radius of the pulley groove is smaller than r_R. Note that this holds true only if the geometric profile of the pulley and the radius of the cable match perfectly. The cable leaves the pulley at point C and the angle between point A and C is denoted by β_R. The coiling of the cable is only stable if $0 < \beta_R < \pi$. In its initial position, the pulley is located in the xz-plane of frame $\mathcal{K}_A$. The initial position is defined as the orientation of the pulley mechanisms so that its mechanical design works perfectly and the feasible rotation about z-axis is equal in positive and negative direction. In the initial position, β_R is measured in positive direction about the y-axis of $\mathcal{K}_A$.

In Fig. 7.3b, one can see the rotated pulley. The panning angle of the rotation is denoted by γ_R and is taken in positive direction about the z-axis of frame $\mathcal{K}_A$. The exact definition of the angles β_R and γ_R with respect to $\mathcal{K}_A$ is crucial for both the formulation of the kinematic codes and the consideration of collisions between the cable and the pulley mechanism. The orientation of $\mathcal{K}_A$ is expressed by the rotation matrix $\mathbf{R}_A$ and it is a design parameter of the whole robot since it is not trivial to build pulley mechanisms that allow for large deflection of the cables. Furthermore, a pulley mechanism cannot be exited in arbitrary direction making the orientation of the installation relevant. If the angle β_R is smaller than a certain threshold, the cable is at risk of uncontrollably leaving the pulley and will be severely damaged from friction with the housing. Selecting a proper pulley geometry is subject to the design procedure of the robot while the initial orientation of it needs to be carefully chosen to avoid restrictions from collisions in the pulley kinematics.

Similar to the proximal pulley, we define a coordinate frame on the platform (Fig. 7.4). Here, we have a frame $\mathcal{K}_B$ that is located at the distal end B of the cable. The z-axis of frame $\mathcal{K}_B$ defines the initial orientation of the cable guidance mechanisms on the platform. Therefore, one can easily define the deflection angle of the platform. Once again, this angle is important for the consideration of collisions between the cable and the platform. Thus, the orientation $\mathcal{K}_B$ becomes a design parameter as well. The design aspects to determine good orientations for the proximal and distal frames is addressed in Chap. 8. If an universal joint is used at the distal end of the cable, the x- and y-axis of the frame $\mathcal{K}_B$ correspond to the first and second axis of the universal joint. If a swivel bolt is used, the first axis of the swivel bolt is aligned with the z-axis of $\mathcal{K}_B$.

On the platform, one can use guiding surfaces instead of pulleys since the cable undergoes no motion in its longitudinal direction. Therefore, a horn can be used to fix the cable (Fig. 7.5). The reference point $\mathbf{b}_i$ is located at the bottleneck between the circular guiding surfaces. This kind of fixture on the platform is favorable especially for planar robots. The geometry is generalized to the spatial case by using a toroidal surface similar to the eyelets used on the proximal anchor points. Although the cable

---

[1] To be precise, one has to also take into account that the cross section of the cable is deformed under tension. This effect is especially significant for synthetic fiber cables and if the profile of the pulley groove does not exactly match the diameter of the cable. The kinematic impact of ovalization of the cable on the drum is analyzed by Schmidt [435] when the cable is coiled onto the drum under changing tension. We neglect this effect in the following.

Fig. 7.4 Definition of the coordinate frame $\mathcal{K}_B$ on the mobile platform

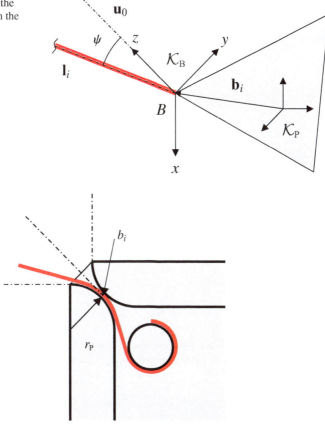

Fig. 7.5 Cable guidance on curved surfaces on the mobile platform

end is fixed inside the platform, there still occurs some slip between the guiding surface and the cable leading to wear on the cable.

7.2.1 Inverse Kinematics

Using the definitions of the local frame $\mathcal{K}_A$, one can derive the kinematic equations in closed-form as follows [391]: Firstly, one has to compute the vector to the point B with respect to frame $\mathcal{K}_A$, where frame $\mathcal{K}_A$ shall be both the origin and the reference frame for the vector. One finds

$$^A\mathbf{b} = \mathbf{R}_A^T(\mathbf{R}^P\mathbf{b} + \mathbf{r} - \mathbf{a}) \quad, \tag{7.1}$$

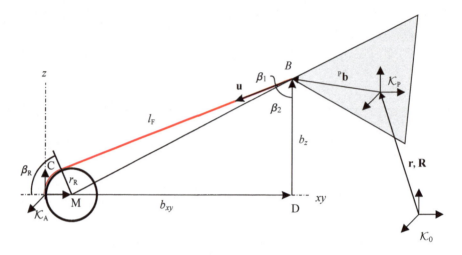

Fig. 7.6 Kinematics of the pulley mechanism in the plane defined by z-axis of $\mathcal{K}_A$ and the distal anchor point B

where $^A\mathbf{b} = [b_x, b_y, b_z]^T$ is the vector from point A to point B in frame $\mathcal{K}_A$ and $^P\mathbf{b}$ is the vector to point B with respect to the end-effector frame $\mathcal{K}_P$. Then, one has to consider the plane defined by the z-axis of frame $\mathcal{K}_A$ and the point $\mathbf{b}$ which is shown in Fig. 7.6. The corrected cable length taking into account the cable wrapped around the pulley becomes

$$l = \beta_R r_R + l_F \; , \tag{7.2}$$

where β_R is the wrapping angle around the pulley, r_R is the effective pulley radius, and l_F is the free cable length from point C to point B. Considering the two right-angled triangles in Fig. 7.6, one receives

$$(b_{xy} - r_R)^2 + b_z^2 = \overline{MB}^2 = l_F^2 + r_R^2 \; , \tag{7.3}$$

where $b_{xy} = \sqrt{b_x^2 + b_y^2}$ and b_z are the coordinates of the point B with respect to frame $\mathcal{K}_A$ in a cylindrical coordinate system. Thus, the free cable length is given by

$$l_F = \sqrt{(b_{xy} - r_R)^2 + b_z^2 - r_R^2} = \sqrt{b_{xy}^2 - 2b_{xy}r_R + b_z^2} \; . \tag{7.4}$$

To solve the inverse kinematics, one needs the angle β_R which can be computed as follows: Considering the tetragon CMDB, we find two angles to be right-angled. Therefore, one concludes that the enclosed angle $\beta_1 + \beta_2$ at point B equals the sought complementary angle β_R at point M. Using trigonometric functions yields

$$\beta_{\mathrm{R}} = \beta_1 + \beta_2 = \arccos \frac{l_{\mathrm{F}}}{\sqrt{(b_{xy} - r_{\mathrm{R}})^2 + b_z^2}} + \arccos \frac{b_z}{\sqrt{(b_{xy} - r_{\mathrm{R}})^2 + b_z^2}} \ . \quad (7.5)$$

Thus, we receive the following closed-form solution for the cable length l from combining the equations above to

$$l = \left(\arccos \frac{\sqrt{b_{xy}^2 - 2b_{xy}r_{\mathrm{R}} + b_z^2}}{\sqrt{(b_{xy} - r_{\mathrm{R}})^2 + b_z^2}} + \arccos \frac{b_z}{\sqrt{(b_{xy} - r_{\mathrm{R}})^2 + b_z^2}} \right) r_{\mathrm{R}} + \sqrt{b_{xy}^2 - 2b_{xy}r_{\mathrm{R}} + b_z^2} \ . $$

$$(7.6)$$

Further reduction in the computational costs can be achieved by using the addition theorem for $\arccos(\cdot)$. It is worthwhile to mention that one can set up similar formulas using either $\arctan(\cdot)$ or $\arcsin(\cdot)$ in Eq. (7.5) as follows

$$\beta_{\mathrm{R}} = \arcsin \frac{r_{\mathrm{R}}}{\sqrt{(b_{xy} - r_{\mathrm{R}})^2 + b_z^2}} + \arcsin \frac{b_{xy} - r_{\mathrm{R}}}{\sqrt{(b_{xy} - r_{\mathrm{R}})^2 + b_z^2}} \qquad (7.7)$$

$$= \arctan \frac{r_{\mathrm{R}}}{l_{\mathrm{F}}} + \arctan \frac{b_{xy} - r_{\mathrm{R}}}{b_z} \qquad (7.8)$$

where both formulations require a distinction of cases when b_z changes its sign. Using $\arctan(\cdot)$ and the respective addition theorem gives a very compact expression which is only valid for positive b_z. The advantage of the presented formula (7.5) is that one gets the symbolic derivative for the first-order kinematics without additional efforts.

The first rotation angle γ_{R} of the pulley about the z-axis of frame $\mathcal{K}_{\mathrm{A}}$ is simply computed from

$$\gamma_{\mathrm{R}} = \arctan2 \, (b_y, b_x) \quad . \qquad (7.9)$$

To calculate the normal vector $\mathbf{u}_{\mathrm{R}}$ along the cable, one rotates a negative unit vector $-\mathbf{e}_z$ about the z-axis with the following transformation matrices

$$\mathbf{u}_{\mathrm{R}} = -\mathbf{R}_{\mathrm{A}} \, \mathbf{R}_{\mathrm{Z}}(\gamma_{\mathrm{R}}) \, \mathbf{R}_{\mathrm{Y}}(\beta_{\mathrm{R}})\mathbf{e}_z \quad , \qquad (7.10)$$

where $\mathbf{R}_{\mathrm{Y}}(\beta_{\mathrm{R}})$ and $\mathbf{R}_{\mathrm{Z}}(\gamma_{\mathrm{R}})$ are the elementary rotation matrix about the y- and z-axis, respectively. One receives the simple form

$$\mathbf{u}_{\mathrm{R}} = -\mathbf{R}_{\mathrm{A}} \begin{bmatrix} \cos(\gamma_{\mathrm{R}}) \sin(\beta_{\mathrm{R}}) \\ \sin(\gamma_{\mathrm{R}}) \sin(\beta_{\mathrm{R}}) \\ \cos(\beta_{\mathrm{R}}) \end{bmatrix} \qquad (7.11)$$

for the direction of the cable.

7.2.2   Structure Equation and Pulley Kinematics

Considering the force and torque equilibrium for the platform driven by a pulley mechanism leads to the well-known definition of the structure equations of the standard model (see Sect. 3.3)

$$\mathbf{A}^{\mathrm{T}}\,\mathbf{f} + \mathbf{w}_{\mathrm{p}} = \mathbf{0} \quad, \tag{7.12}$$

where $\mathbf{A}^{\mathrm{T}}$ is the pose-dependent structure matrix, $\mathbf{f}$ is the vector of the positive cable forces, and $\mathbf{w}_{\mathrm{p}}$ is the applied wrench at the platform. When considering a pulley model for the robot, the basic linear structure of the equation is maintained where one has to use a modified unit vector $\mathbf{u}_{\mathrm{R}}$ for the direction of the cables as given by Eq. (7.10). Thus, the structure matrix $\mathbf{A}_{\mathrm{R}}^{\mathrm{T}}$ for pulley kinematics becomes

$$\mathbf{A}_{\mathrm{R}}^{\mathrm{T}} = - \begin{bmatrix} \mathbf{R}_{\mathrm{A},1} \begin{bmatrix} \cos(\gamma_{\mathrm{R},1})\sin(\beta_{\mathrm{R},1}) \\ \sin(\gamma_{\mathrm{R},1})\sin(\beta_{\mathrm{R},1}) \\ \cos(\beta_{\mathrm{R},1}) \end{bmatrix} & \cdots & \mathbf{R}_{\mathrm{A},m} \begin{bmatrix} \cos(\gamma_{\mathrm{R},m})\sin(\beta_{\mathrm{R},m}) \\ \sin(\gamma_{\mathrm{R},m})\sin(\beta_{\mathrm{R},m}) \\ \cos(\beta_{\mathrm{R},m}) \end{bmatrix} \\ \mathbf{b}_1 \times \mathbf{R}_{\mathrm{A},1} \begin{bmatrix} \cos(\gamma_{\mathrm{R},1})\sin(\beta_{\mathrm{R},1}) \\ \sin(\gamma_{\mathrm{R},1})\sin(\beta_{\mathrm{R},1}) \\ \cos(\beta_{\mathrm{R},1}) \end{bmatrix} & \cdots\; \mathbf{b}_m \times \mathbf{R}_{\mathrm{A},m} \begin{bmatrix} \cos(\gamma_{\mathrm{R},m})\sin(\beta_{\mathrm{R},m}) \\ \sin(\gamma_{\mathrm{R},m})\sin(\beta_{\mathrm{R},m}) \\ \cos(\beta_{\mathrm{R},m}) \end{bmatrix} \end{bmatrix}$$

$$\tag{7.13}$$

where the definition of $\mathbf{u}_{\mathrm{R},i}$ given by Eq. (7.10) is used and the angles $\beta_{\mathrm{R},i}$, $\gamma_{\mathrm{R},i}$ are computed from Eqs. (7.5) and (7.9), respectively.

7.2.3   Forward Kinematics Code

The kinematic model for pulleys has an influence on the forward kinematics according to the changed geometric relations as indicated by Eq. (7.6). In contrast to the constraint equations received for the standard model, the model equations for pulley kinematics are non-algebraic and all known results for the number of solutions of the forward kinematics problem do not apply for pulley kinematics. Therefore, it is an open problem if any and how many solutions exist. Geometric considerations show that the constraint surface is the rotation of an Archimedean spiral (Fig. 7.7a, b). If the radius of the pulley is small compared to the length of the cable, this spiral (Fig. 7.7) is quite close to the spherical constraint surfaces of the standard model. Taking into account that the cable undergoes small elastic deformations, solutions to the geometric problem are expected and one can use similar assumptions for the forward kinematics as described for the standard model (see Sect. 4.3).

A numerical method to compute the forward kinematics taking pulleys into account is presented by Schmidt [434]. The following algorithm goes along this work. This iterative scheme for forward kinematics is derived from the inverse kinematics with pulleys as given in Sect. 7.2.1. Since these extended equations are still solved

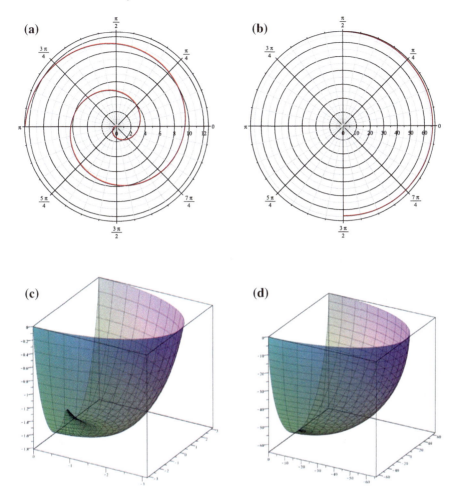

Fig. 7.7 The constraints imposed by the pulley models are based on an Archimedean spiral. The upper line shows polar plots of the Archimedean spiral for small (**a**) and large (**b**) cable length and the resulting constraint surface for small and large cable length. The lower line shows the respective constraint surface for forward kinematics for short (**c**) and long (**d**) cables

in closed-form, one can apply the approach for forward kinematics as given for the standard model. Firstly, the inverse kinematics equations are rearranged to form a system of implicit equations

$$\nu_{R,i}(\mathbf{l}, \mathbf{r}, \mathbf{R}) = l_{R,u}^2 - l_i^2 = 0 \quad \text{for} \quad i = 1, \dots, m \quad , \tag{7.14}$$

where the inverse kinematics solutions and the given cable length are subtracted. Secondly, we argue that minimizing the potential energy in the cables allows to search for a pose $(\mathbf{r}, \mathbf{R})$ so that the sum of all squared differences becomes minimal. Then, the objective function is

$$\varphi_R^{DK}(\mathbf{l}) = \min_{\mathbf{r},\mathbf{R}} \sum_i^m v_{R,i}^2(\mathbf{l},\mathbf{r},\mathbf{R}) \quad . \tag{7.15}$$

The determination of such a minimum is computed using a Levenberg-Marquardt method. In this setting, the formulation from Sect. 7.2.1 of a closed-form equation of the inverse kinematics without a distinction of cases is highly favorable since it largely simplifies the determination of the Jacobian matrix and the generation of its kinematic code. One can use the initial guess from the interval bounding method (Sect. 4.41) for the Levenberg-Marquardt optimizer. In contrast to the kinematic code of the standard model, the generated C program is significantly longer. The evaluation of the objective functions has more than 50 lines of code and the analytic Jacobian matrix is longer than 150 lines. However, since the code is composed from simple mathematical operations, the executing time is hardly influenced.

7.2.4 Results

The geometrical parameters of the IPAnema 1 robot used for this study are given in Table 9.1. For this robot, the radius of the cable is $r_c = 0.002$ m and the effective radius of the pulleys of the real robot is $r_R = 0.05$ m. All local frames $\mathcal{K}_{A,i}$ of the base anchor points A_i at the winches run parallel to the world frame $\mathcal{K}_0$ (see Fig. 7.8), thus all rotation matrices are $\mathbf{R}_{A,i} = \mathbf{I}$.

Table 7.1 Comparison of the workspace volume and surface for different radii of the pulley r_R of the IPAnema 1 robot

Pulley radius r_R [m]	Volume V [m^3]	Relative volume [%]	Surface S [m^2]	Relative surface [%]
0.000	5.81488	100.00	17.92656	100.00
0.001	5.81287	99.96	17.87905	99.73
0.010	5.80754	99.87	17.89538	99.83
0.025	5.79722	99.70	17.87482	99.71
0.050	5.77644	99.34	17.75667	99.05
0.150	5.68850	97.83	17.50093	97.63
0.250	5.57410	95.86	17.17330	95.80
0.350	5.42254	93.25	16.76406	93.52
0.400	5.32749	91.62	16.49955	92.04

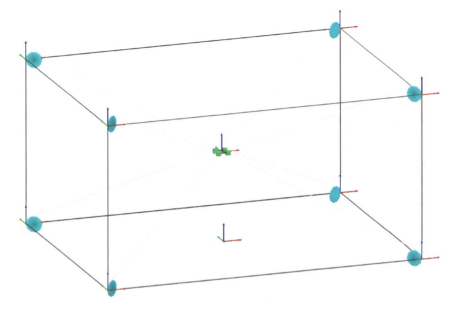

Fig. 7.8 Kinematics of the IPAnema 1 robot with guiding pulleys at the proximal anchor points

7.2.4.1   Cable Length Error

The deviation between the standard model and the pulley model for inverse kinematics is depicted in Fig. 7.9, where the diagrams show the difference between both
inverse kinematics codes along a trajectory for different radii r_R of the pulley. One
can see that the cable length computed from the extended formula is always longer
than the standard model. This is clear because the way around the pulley is longer
than the direct connection between the points $\overline{A_i B_i}$. The relation for the radius r_R of
the pulley as well as the deviation $\Delta l = l - l_R$ for the kinematic models is shown in
the lower diagram (Fig. 7.9). The ratio is almost constant for the considered interval
of pulley radii $r_R \in [0.1; 0.01]$ m. Thus, the dependency between the additional
length of the cable and the radius of the pulley is approximately linear in this range.

7.2.4.2   Influence on Force Distribution

In this section, the difference in the force distribution that arises from the static model
is analyzed taking the guiding pulleys into account. To calculate the force distribution,
the following closed-form formula is used (see Sect. 3.7.5). For the example below,
the bounds $f_{min} = 1\,\mathrm{N}$, $f_{max} = 10\,\mathrm{N}$, are used and an external wrench $\mathbf{w}_p = \mathbf{0}$ is
applied. Figure 7.10 shows the comparison for the forces f_1 in cable 1 when moving
along a trajectory for different radii r_R of the pulley. The differences are again in the

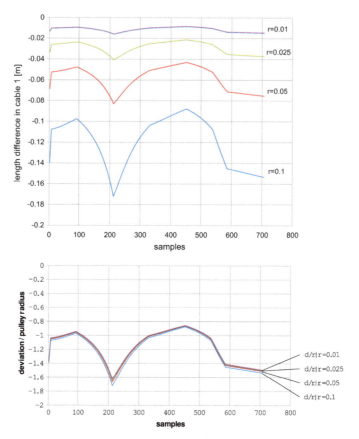

Fig. 7.9 Deviation between the standard kinematic model and the pulley model for different radii r_R of the pulley in cable 1. The upper diagram shows the absolute difference where the low diagram shows the ratio between difference and radius of the pulley

range of some percentage and the magnitude of the difference is linear for typical sizes of the pulleys. For the practical use in force control, the influence is assumed to be less important since the error in the cable forces is of the same magnitude as the measurement error caused from typical force sensors.

7.2.4.3 Deviations of the Workspace

In this section, the influence of the pulleys on the size and shape of the workspace is studied. To compare the results from the workspace calculation with and without pulleys, we use the triangulation of the workspace's hull (see Sect. 5.5.1). Although the triangulation lacks the verified nature of interval computations [182], it can be computed with high accuracy at moderate computational times of some seconds.

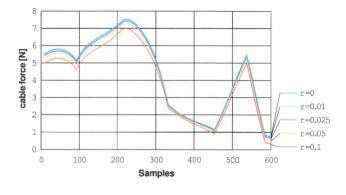

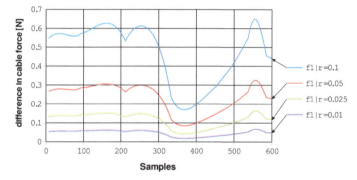

Fig. 7.10 Deviation between the standard kinematic model and the pulley model for the forces f_1. The upper diagram shows the absolute cable forces in cable 1 for a pulley radius of $r_R = \{0, 0.01, 0.025, 0.05, 0.1\}$ m where the lower diagram shows the difference between the standard model and different radii of the pulley

This improved sensitivity to small changes in the geometry makes it feasible for the comparison here. This accuracy allows to measure in detail the influence of the design variables (geometry of platform and machine frame) or technical parameters such as minimum and maximum cable force.

To study the influence of pulleys and especially the influence of radius r_R on the workspace, the hull of the workspace is computed and the performance criteria surface $S(\mathcal{W})$ from Eq. (5.46) and volume $V(\mathcal{W})$ from Eq. (5.47) are employed to compare the results for different radii r_R of the pulleys. To check for wrench-feasibility of a pose **y**, Eq. (3.5) is evaluated using the Dykstra method (see Sect. 3.7.4) and the determined force is compared to the force limits as given in the previous section. The parameters of the workspace algorithms are set as follows: The iterations depth for the recursive refinement of the hull is chosen to be six leading to 16386 vertices and 32768 triangles. The accuracy for the line search is $\varepsilon = 10^{-4}$ m so that the first four digits of the performance indices are meaningful. The computational results from the example are given in Table 7.1. In this evaluation, we used even larger radii

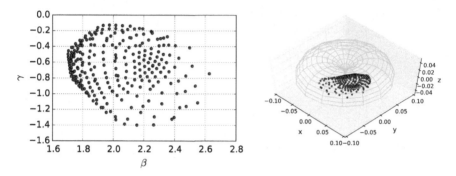

Fig. 7.11 Left: Deflection angles β_R and γ_R for the first winches of the IPAnema 1 robot throughout the workspace. Right: Actual proximal anchor points C_1 where the cable leaves the pulley in the local frame $\mathcal{K}_{A,1}$

for the pulleys than before. The relative error of the workspace volume is less than 2% for realistic values of the pulley's radius r_R.

Figure 7.11 shows the actually occurring deflection angles $\beta_{R,1}$ and $\gamma_{R,1}$ in the pulley mechanism for the IPAnema 1 robot. The sample poses are chosen from the hull of the workspace thus covering the extremal positions of the pulley. One can see that the panning angle γ_R of the pulley is in the range $[-\frac{\pi}{2}; 0]$ thus pointing to the inside of the machine frame. The considered winch $i = 1$ is an upper winch located at the top of the robot frame. Thus, the wrapping angle is $\beta_{R,1} \in [\frac{\pi}{2}; \pi]$ where the cable always wraps at least a quarter of the pulley. Only a small part of the toroidal surface is actually used. Thus, the region where the point C_i may be located is notably smaller than the torus.

7.2.4.4 Forward Kinematics

Simulation and experimental tests [434] with the extended kinematics code show that it is applicable to the controller. To test convergency, 5000 random poses in the workspace of the IPAnema 2 robot are used. Compared to the standard model, the extended pulley kinematics is slightly slower but runs stable (Fig. 7.12). Experimental results on the improved accuracy show that using a pulley kinematic code in the controller improves the accuracy of the robot by 21% leading to an absolute accuracy for the IPAnema 2 robot of 17.5 mm and a repeatability of 0.51 mm. The computation time for the forward code is determined to be in the range between 80 and 130 μs which allows for operation in a real-time system.

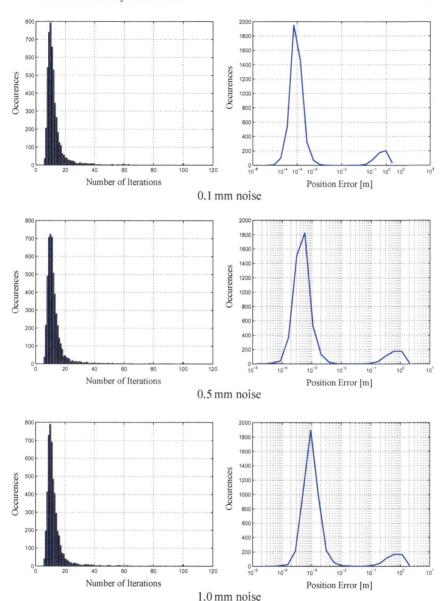

Fig. 7.12   Numerical results for the forward kinematics with pulleys using Levenberg-Marquardt algorithm. Left column: Histogram of number of iterations for noise 0.1 mm, 0.5 mm, 1 mm on the cable length. Right column: Histogram of position error for noise 0.1 mm, 0.5 mm, 1 mm on the cable length

7.2.5 Summary

Using an extended modeling for pulleys, the differences of the kinematics, statics, and workspace between the simplified and extended model are studied. It turns out that the difference of the volume and surface of the workspace is less than 10% even for huge pulley radii and in the range of 1% for typical pulleys. Considering other unconsidered uncertainties, the influence on the workspace may be neglected in many applications. The comparison of the inverse kinematics codes and thus the expected accuracy of the robot are influenced in the same magnitude. The deviations between standard and pulley model are almost linear in the considered range for the pulley radius. However, while deviations in the predicted workspace size in range of millimeters are negligible, the same magnitude is significant for accuracy. Thus, using the proposed pulley model is an efficient tool for improving the accuracy of the IPAnema 1 robot by around 25% compared to the standard model. The formulation developed for inverse kinematics is successfully used for the forward kinematics code. Solving the respective optimization problem with the Levenberg-Marquardt algorithm allows for application in the real-time controller. The shortening of the cables caused by guiding the cable around the pulley significantly increases the inner tension in the robot and, thus, disturbs the force equilibrium of the mobile platform.

Open issues in the field for considering pulleys are their influence on the location of singularities and the theoretical foundation of the forward kinematics.

7.3 Kinematics with Sagging Cables

A common observation for long cables in civil engineering applications such as bridges and high voltage lines is that the cables undergo significant deformation caused by the weight of the cables. This so-called *sagging* of the cables results from the mass m_c of the cable itself and the influence of gravity on that mass as well as the negligible bending stiffness of the cable. If the ratio between the length and the mass of the cable is such that sagging becomes an issue, the cable is called *hefty*. Especially for large-scale cable robots, one cannot assume that the cables form perfect lines in space. In practice, sagging of cables leads to a coupling between the geometrical constraints imposed by the cable length and pulley kinematics on the one hand side as well as the static and elastic effects of the cables on the other hand side. Thus, to describe sagging of cables in cable robots, one has to consider kinematics and statics simultaneously since there is a strong dependency between both. From this point of view, the effect of sagging shares some properties with the modeling of elastic cables as well as the modeling of under-constrained robots where kinematics and statics also are interconnected.

A couple of contributions are dedicated to study sagging for cable robots. The compensation for sagging cables in a measurement device is proposed by Jeong

[226], however, no method for implementation is given. The research on sagging is strongly influenced and initialized by the planning and development of the FAST cable robot. This huge robot is the archetype of large-scale cable robots. In Kozak [256, 257], the kinematic modeling with cable sagging and the resulting changes in stiffness are addressed and the model used for cable robots is basically adopted from the reference book on civil engineering by Irvine [218].

Korayam [252] presents the modeling for elastic and hefty cables for suspended robots. Du [234] presents the partial differential equations of the cables for the FAST telescope and derives ordinary differential equations for the equations of motion. Further studies of the authors [126, 231–233] aim at forward kinematics, statics, stiffness, and control of cable-driven parallel robots with sagging cables for application with large-scale cable robots. Recently, Nguyen presented the modeling of sagging cables for the robot CoGiRo [360, 361]. An extension for the effect of pulleys in the sagging model is added to the model [189]. The relation between sagging and stiffness is also considered by Yuan [513–515], who derives a stiffness model for suspended cable robots with six cables taking sagging of the cables into account. Additionally, the dynamic effects and the eigenfrequencies are determined.

A simplified model for sagging is proposed by Gouttefarde [179, 181] where a parabolic model is used to approximate the sagging cable where the elastic reactions of the cables are neglected. Based on that model, a simplified static analysis of the robots is performed. Dallej [107] presents a vision-based control approach for the simplified sagging model and a mean translational error of 8.2 mm is reported for the CoGiRo setup.

Inverse kinematics with $m \leq 6$ sagging cables is addressed by Merlet [332] who points out that by introducing the catenary equations of the sagging cable, the inverse kinematics problem is composed from non-algebraic equations. In general, it is not clear if the solution space is dense. Also, the forward kinematics problem is addressed by Merlet [333, 335] where interval analysis is used to compute solutions for this problem. The numeric examples presented in the work show that multiple solutions can exist.

Yao [506] proposed an approach to compensate for sagging and presents experimental results for the scaled model of the FAST telescope with design size of some 45 m. A repeatability of 1 mm to 1.7 mm is reported determined for the scaled model. Later, Hui [213] proposed a kinematic code for inverse kinematics of the full-size FAST telescope taking sagging for a suspended robot with six cables into account. A stiffness model for robots with sagging cables is proposed by Arsenault and applied to a suspended 2T robot [18]. A similar approach is pursued by Ottaviano [377] who analyzed the kinematics of a planar cable robot with two cables taking sagging of hefty elastic cables into account.

A numeric evaluation using the software package XDE is presented by Michelin [340]. This dynamic model also takes into account vibrations in the cables where very long computation times roughly 3600 times slower than real-time are reported. However, this approach expresses a connection to simulate cable robots with conventional simulation software.

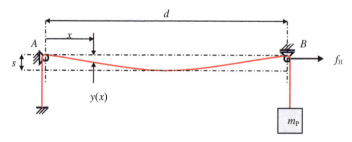

Fig. 7.13 Schematic drawing of the cable sagging when considering cable mass and a constant cable force

Duan [128] mentions the modeling with partial differential equations, however, the formulation can hardly be extracted from the paper. Interestingly, no study analyzes the partial differential equations directly but tries to convert the partial differential equations to a system of ordinary nonlinear differential equations.

7.3.1 Modeling of Sagging Cables

In the following, the modeling of the cable under gravity force is introduced. If cable mass is taken into account, the shape of the cable, its end-point position, and the cable forces become coupled. In the static equilibrium, the gravity forces in the cable cause a displacement of all cable elements leading to a cable shape that is entirely below the straight line between the cable's end-points A_i and B_i. This effect is called *sagging* and is modeled in this section. In the literature [218, 374], it is proposed to neglect sagging if the cable tension f_i is notably larger than the weight of the cable, i.e. the criterion

$$f_i \gg g \varrho'_c l_i \tag{7.16}$$

has to be fulfilled to ignore sagging, where ϱ'_c is the linear density or weight per length of the cable. The gravity acceleration is g and l_i is the free cable length.

A simplified situation for cable sagging is depicted in Fig. 7.13 where a horizontal cable is fixed on the left side at A and guided around a pulley on the right side at B. The radius of the pulley is neglected. Thus, the cable sags over the horizontal distance d between the proximal and distal anchor point A and B. To tense the cable, a mass m_p is subject to gravity and puts a constant force on the cable without constraining the cable length. The cable is assumed to have mass m_c, which may be given by its density ϱ_c or as specific weight per length ϱ'_c, sometimes also called linear density ϱ'_c. The mass of the vertical part of the cable behind the pulley B shall be lumped in the mass element m_p. The figure shows the cable in equilibrium. A well-known result from classical mechanical engineering is that the form of the cable is a catenary. The

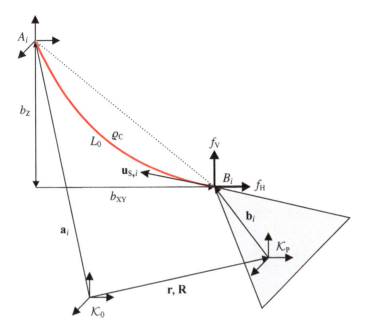

Fig. 7.14 Kinematics of a cable robot with sagging cables

largest deviation from the straight line is denoted with s and occurs in the middle of the cable if the ideal line is horizontal.

Sagging affects different properties of the cable robot. In order to reduce sagging to a reasonable level, one has to choose a compromise between pretension in the cables and the feasible slack span between the real cable shape and the ideal straight line between the anchor points (Fig. 7.14). This difference causes an error in the effective cable length $l_{s,i}$ of the sagging cable and the distance between the anchor points represented by the cable length l_i of the standard model. Furthermore, the direction $\mathbf{u}_{s,i}$ of the cable force applied to the platform is altered by sagging of the cable. At the same time, the cable mass m_c changes the effective force acting on the platform. Thus, some disturbance in the equilibrium of the platform is expected.

Here, we discuss the phenomenological relations between tension, cable mass, and sagging. Therefore, the cables are considered to be inextensible, i.e. the cable is perfectly stiff and $E_c A_c \rightarrow \infty$ is assumed for the spring constant of the cable where E_c is the Young's modulus of the cable and A_c is the effective cross section of the cable. At the same time, the cable is assumed to be perfectly flexible and the bending stiffness $E_c I_c = 0$ of the cable vanishes where I_c is the second moment of area of the cable. For moderate sagging with a small curvature of the cable, the bending stiffness of the cable is ignored but the bending stiffness becomes an issue when modeling the wrapping of the cable around a pulley or around the winch drum.

7.3.2   Inelastic Horizontal Cable Model

Firstly, the catenary equation for a cable under gravity without elastic elongation is discussed. The derivation goes along with the considerations that can be found e.g. in Irvine [218] and Feyrer [149]. The shape of the sagging cable (Fig. 7.13) is determined from the ordinary differential equation

$$\iint g_C \sqrt{1 + \left(\frac{dy}{dx}\right)^2} \, dx^2 = f_H y \tag{7.17}$$

where $g_C = g\varrho_C'$ is the gravity force of the cable per length, $y(x)$ is the function of the shape of the cable over the length coordinate x, and f_H is the horizontal cable force. Integration of this equation provides the general solution for the inelastic cable

$$y(x) = \frac{f_H}{g_C}\left(\cosh\left((x + C_1)\frac{g_C}{f_H}\right) - C_2\right) \tag{7.18}$$

with two integration constants C_1 and C_2 that must be chosen to fulfill the geometric boundary conditions so that the end-point B_i of the cable has the specified coordinates. With an appropriate choice of the coordinate system, the solution for Eq. (7.17) in the horizontal case can be found to be

$$y(x) = \frac{f_H}{g_C}\left(\cosh\left(\frac{g_C x}{f_H}\right) - 1\right) \quad . \tag{7.19}$$

The curve of the horizontal cable $y(x)$ is exemplified in Fig. 7.15. The choice of the coordinate system is such that the curve touches the origin at its minimum. Thus, the magnitude of the maximum sag is computed from evaluating $y(\frac{d}{2})$ and one receives

$$s = y\left(\frac{d}{2}\right) = \frac{f_H}{g_C}\left(\cosh\left(\frac{gl}{2 f_H}\right) - 1\right) \quad . \tag{7.20}$$

The specific gravity force per length g_C of the cable is computed from

$$g_C = g\varrho_C' = g\varrho_C A_C = g\varrho_C \pi r_C^2 \tag{7.21}$$

where g is the inertial acceleration, ϱ_C and ϱ_C' are the density and the linear density of the cable, respectively, and A_C is the effective cross section of the cable. Cables are not homogeneous due manufacturing from fibers or wires. Therefore, one uses the effective linear density g_C that also includes the braiding of the cable. In practice, cable manufacturers provide the material constant g_C for their products. Sample data for Dyneema are given in Table 7.2 and for a typical stainless steel cable 18×7 in Table 7.3.

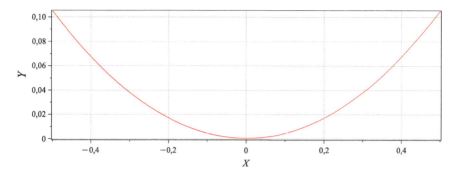

Fig. 7.15 Sagging in [mm] of a Dyneema cable with a diameter of 2.5 mm and a cable force of $F = 50\,\mathrm{N}$

Diameter d_C [mm]	Breaking load [daN]	Specific weight g_C [kg/100 m]
1.0	195	0.09
1.5	230	0.13
2.0	410	0.18
2.5	580	0.35
3.0	950	0.46
4.0	1300	0.70
5.0	2600	1.30
6.0	4300	2.30
8.0	5300	3.50
9.0	7500	4.90
10.0	9000	6.00
12.0	11900	7.20
14.0	14500	9.50
16.0	19200	13.10

Table 7.2 Breaking load and specific weight of Dyneema cables with a diameter between 1–16 mm for LIROS D-PRO Dyneema cables. All parameters according to manufacturer's data. LIROS GmbH, Berg, Germany

Computation of the derivative of y with respect to the length coordinate x is straightforward and yields

$$y' = \frac{dy}{dx} = \sinh \frac{g_C x}{f_{\mathrm{H}}} \quad . \tag{7.22}$$

Then, one computes the cable length relative to the distance d between the anchor points. This is achieved by computing the integral

$$l = \int_{-\frac{d}{2}}^{\frac{d}{2}} \sqrt{1 + \left(\frac{dy}{dx}\right)^2}\, dx \tag{7.23}$$

Table 7.3 Minimum breaking load and specific weight of stainless steel cables 18 × 7 for winches according to DIN 3069 [124]

Diameter d_C [mm]	Breaking load [daN]	Specific weight g_C [kg/100 m]
3.0	523	3.62
4.0	906	6.43
5.0	1328	10.00
6.0	1921	14.50
6.5	2227	16.40
7.0	2700	19.70
8.0	3417	25.70
10.0	5329	40.20
12.0	7763	57.90
14.0	10750	78.80
16.0	13531	103.00

which basically means that the following integral must be solved

$$l = \int \sqrt{1 + \sinh^2 z}\, dz \quad .$$ (7.24)

The integral is solved in closed-form yielding the solution

$$l = \frac{f_H}{g_C} \frac{\sqrt{1 + \sinh^2 \frac{g_C x}{f_H}}\, \sinh \frac{g_C x}{f_H}}{\cosh \frac{g_C x}{f_H}} = \frac{f_H}{g_C} \sqrt{1 + \sinh^2 \frac{g_C x}{f_H}}\, \tanh \frac{g_C x}{f_H}$$ (7.25)

and by substituting the borders $-\frac{d}{2}$ and $\frac{d}{2}$ into the function, one finds the cable length l_C in closed-form as a surprisingly simple expression

$$l_C = 2\frac{f_H}{g_C} \sinh \frac{g_C d}{2 f_H} \quad .$$ (7.26)

In order to emphasize the constant parameters of the cable sagging, one substitutes $\mu = \frac{g_C}{f_H}$ and receives

$$l_C = \frac{2}{\mu} \sinh \frac{\mu d}{2}$$ (7.27)

$$y = \frac{1}{\mu} (\cosh(\mu x) - 1) \quad .$$ (7.28)

Since sagging is especially important for large-scale robots, it is worthwhile to also consider the effect of sagging and elastic elongation of the cable.

7.3.3 Irvine's Elastic Cable Model

The effect of sagging and elasticity can also be modeled for the cable by solving an ordinary differential equation where additionally the linear elastic effects of the cables are taken into account. A hefty and elastic cable is connected to a proximal anchor point A with the local coordinates $\mathbf{a} = \mathbf{0}$. Then, the following relations hold for the distal end B of the cable $\mathbf{b} = [x_B, z_B]^T$ and the forces $f_C = [f_H, f_V]^T$ at point B (see e.g. [218, 332, 361, 431])

$$x_B = f_H \left(\frac{L_0}{E_c A_c} + \frac{\sinh^{-1}\left(\frac{f_V}{f_H}\right) - \sinh^{-1}\left(f_V - \frac{g_c L_0}{f_H}\right)}{g_c} \right) \tag{7.29}$$

$$z_B = \frac{\sqrt{f_H^2 + f_V^2} - \sqrt{f_H^2 + (f_V - g_c L_0)^2}}{g_c} + \frac{f_V L_0}{E_c A_c} - \frac{g_c L_0^2}{2 E_c A_c} \tag{7.30}$$

where L_0 is the unstrained cable length, E_c is the Young's modulus of the cable, and A_c is the cable cross section. If one assumes the physical cable constants L_0, E_c, A_c, g_c to be known, one eventually receives a cable model that describes the position of the point B in the xz-plane of the frame $\mathcal{K}_{A,i}$ as a function of the horizontal and vertical force applied at the platform as

$$\begin{bmatrix} x_B \\ z_B \end{bmatrix} = \mathbf{\Phi}^c(f_H, f_V) \quad, \tag{7.31}$$

where the function of the cable model $\mathbf{\Phi}^c : \mathbb{R}^2 \to \mathbb{R}^2$ is not algebraic but can be computed in closed-form.

As for the pulley kinematics, one has to compute the plane of the sagging cable. If one assumes that gravity acts in negative z-direction of the world frame $\mathcal{K}_0$, the orientation $\gamma_{s,i}$ of the xz plane of $\mathcal{K}_{A,i}$ is computed following Eq. (7.9) from

$$\gamma_{s,i} = \arctan2 \left(\mathbf{l}_i \cdot \mathbf{e}_Y, \mathbf{l}_i \cdot \mathbf{e}_X \right) \quad, \tag{7.32}$$

where $\mathbf{l}_i = \mathbf{a}_i - \mathbf{R}\mathbf{b}_i - \mathbf{r}$ is the cable vector as used in the standard cable model. The rotation matrix $\mathbf{R}_i$ for transforming the cable plane to world coordinates reads

$$\mathbf{R}_i = \begin{bmatrix} \cos \gamma_{s,i} & -\sin \gamma_{s,i} & 0 \\ \sin \gamma_{s,i} & \cos \gamma_{s,i} & 0 \\ 0 & 0 & 1 \end{bmatrix} \tag{7.33}$$

and can also be expressed using only the vector $\mathbf{l}_i$ without trigonometric functions

$$\mathbf{R}_i = \frac{1}{\sqrt{(\mathbf{l}_i \cdot \mathbf{e}_\text{X})^2 + (\mathbf{l}_i \cdot \mathbf{e}_\text{Y})^2}} \begin{bmatrix} \mathbf{l}_i \cdot \mathbf{e}_\text{X} & -\mathbf{l}_i \cdot \mathbf{e}_\text{Y} & 0 \\ \mathbf{l}_i \cdot \mathbf{e}_\text{Y} & \mathbf{l}_i \cdot \mathbf{e}_\text{X} & 0 \\ 0 & 0 & \sqrt{(\mathbf{l}_i \cdot \mathbf{e}_\text{X})^2 + (\mathbf{l}_i \cdot \mathbf{e}_\text{Y})^2} \end{bmatrix} . \quad (7.34)$$

In the local frame $\mathcal{K}_{\text{A},i}$, the cable force vector is composed from $\mathbf{f}_i = [f_{\text{H},i}, 0, f_{\text{V},i}]^\text{T}$. Next, one rewrites the equilibrium condition of the platform to include the cable forces derived from the sagging cable model as given in Eq. (7.31). The forces applied to the mobile platform through the sagging cable are given by

$$-\begin{bmatrix} \mathbf{R}_i \mathbf{f}_i \\ \mathbf{Rb}_i \times \mathbf{R}_i \mathbf{f}_i \end{bmatrix} = \begin{bmatrix} \mathbf{R}_i \mathbf{e}_\text{X} & \mathbf{R}_i \mathbf{e}_\text{Y} \\ \mathbf{Rb}_i \times \mathbf{R}_i \mathbf{e}_\text{X} & \mathbf{Rb}_i \times \mathbf{R}_i \mathbf{e}_\text{Y} \end{bmatrix} \begin{bmatrix} f_{\text{H},i} \\ f_{\text{V},i} \end{bmatrix} \quad i = 1, \dots, m \quad (7.35)$$

and one receives an equation that is roughly similar to the structure equations for the standard model as follows

$$\mathbf{A}_\text{s} \mathbf{f}_\text{s} + \mathbf{w}_\text{P} = \mathbf{0} \quad , \quad (7.36)$$

where the matrix $\mathbf{A}_\text{s} \in \mathbb{R}^{6 \times 2m}$ takes the form

$$\mathbf{A}_\text{s} = \begin{bmatrix} \mathbf{R}_1 \mathbf{e}_\text{X} & \mathbf{R}_1 \mathbf{e}_\text{Y} & \mathbf{R}_2 \mathbf{e}_\text{X} & \dots & \mathbf{R}_m \mathbf{e}_\text{X} & \mathbf{R}_m \mathbf{e}_\text{Y} \\ \mathbf{Rb}_1 \times \mathbf{R}_1 \mathbf{e}_\text{X} & \mathbf{Rb}_1 \times \mathbf{R}_1 \mathbf{e}_\text{Y} & \mathbf{Rb}_2 \times \mathbf{R}_2 \mathbf{e}_\text{X} & \dots & \mathbf{Rb}_m \times \mathbf{R}_m \mathbf{e}_\text{X} & \mathbf{Rb}_m \times \mathbf{R}_m \mathbf{e}_\text{Y} \end{bmatrix}$$
$$(7.37)$$

with the cable force vector

$$\mathbf{f}_\text{s} = [f_{\text{H},1}, f_{\text{V},1}, f_{\text{H},2}, \dots, f_{\text{H},m}, f_{\text{V},m}]^\text{T} \in \mathbb{R}^{2m} \quad (7.38)$$

as well as the applied wrench $\mathbf{w}_\text{P} = [0, 0, -m_\text{P}\, g, 0, 0, 0]^\text{T}$.

For inverse kinematics of spatial cable robots, one eventually receives three unknown variables for each cable: firstly, the sought unstrained cable length l_i and secondly, the horizontal $F_{\text{H},i} \geq 0$, and vertical cable tension $F_{\text{V},i} \geq 0$. In total, the entire nonlinear system contains $3m$ unknown variables. In terms of equations, one has six equilibrium conditions for the mobile platform and $2m$ equations from the cable model, thus a total of $n + 2m$ equations. Therefore, the system can only be solved directly for $m = 6$ cables where for $m < 6$ cables, one has an over-constrained system with probably no solution.

Solving the nonlinear system is numerically involved since the equations are quite complex while the contained terms with $\sinh^{-1}(\cdot)$ along with the physical constants in largely different orders of magnitude are sources of numerical instability. Some verified numerical results computed with interval analysis are presented by Merlet [332] where the computation time is around 10 s which is rather time consuming compared to the standard model where we have computation times around one million times faster.

7.3.4 Summary

A deformation model of elastic cables with notable weight is currently the most advanced but also most challenging cable model. However, the basic assumption is that the cable is at static equilibrium and sags in a single vertical plane. Experiments with lateral motion of the cable robots show additional swaying of the cable and even helical shapes of the cables in the presence of abrupt acceleration. More involved cable models as those used with finite element methods or with multi-body chain models are essentially able to capture such effects. The computation time for dynamic simulation increases rapidly and today their application is rather limited. Since such models do not assume a steady-state behavior of the robots, they cannot be integrated with conventional controller architectures that are based on a static kinematic transformation.

7.4 Kinematics with Elastic Cables

Especially if cables are long, the effect of elongation under load cannot be neglected. Although the structural stiffness of the cable robot is high, the absolute stiffness values for the cables are comparably small due to the small cross section of a cable. If the cables are long or the robot is operated with significant changes in the payload or the process forces, notable displacements can be observed on the platform. Besides elongation in the cables, also the winches, the machine frame, and the mobile platform are sources of elastic deformations disturbing the motion of the robot. For high precision applications as well as for large-scale manipulators, the elongations of cables are addressed in the following in the context of kinematic transformation. As shown in Chap. 6 on dynamics, most authors addressed elastic cables in the dynamic models while the effects of cable elongation are ignored in most kinematic studies. One study taking the effect of elastic cables into account is undertaken by Merlet [323, 324] where an interval-based algorithm is presented to compute inverse kinematics for elastic cables. Implicitly, the elastic effects in the cables are also tackled for the forward kinematics where an elastic behavior of the cables is assumed to compensate for uncertainties [390].

7.4.1 Inverse Kinematics

The inverse kinematics of fully-constrained robots with elastic cables becomes more involved than the solution of the standard model introduced in Sect. 4.2.1. However, the problem can be addressed based on the concepts introduced in the context of statics and workspace as described in this section. Considering an elastic cable model maintains the linear shape of the cable while relating the cable tension to its elongation. Thus, the direction vector of the cable $\mathbf{u}_i$ remains independent from the tension and therefore the structure matrix $\mathbf{A}^{\mathrm{T}}$ remains unaffected from the tension. In order to compute the cable length, one can follow this procedure.

Firstly, the desired pose $(\mathbf{r}, \mathbf{R})$ is chosen and direction vectors $\mathbf{u}_i$ are computed from Eq. (3.3). One easily takes into account the effects of pulleys by using Eq. (7.11) instead. Having determined the direction of the cable, the pose-dependent structure matrix $\mathbf{A}^{\mathrm{T}}$ is set up in order to evaluate the statics. Using one of the methods introduced for force distribution, one generates a feasible force distribution $\mathbf{f}$. This can be done efficiently in closed-form (see Sect. 3.7.5). In contrast to inverse kinematics of the standard model, this step can only be executed if the desired pose $(\mathbf{r}, \mathbf{R})$ belongs to the wrench-closure workspace $\mathcal{W}$ where one can evaluate the inverse kinematics of the standard model for any pose. Using the elastic model of the cable, one can now compensate for the elongation in the cables. The linear elongation Δl_i of the cable is computed from

$$\Delta l_i = k_i f_i \quad , \tag{7.39}$$

where k_i is the stiffness of the cables that results from

$$k_i = \frac{E_{\mathrm{c}} A_{\mathrm{c}}}{l_i + l_{0,i}} \quad . \tag{7.40}$$

The parameters of the cable stiffness are introduced in Sect. 3.8. Having computed the elongation of the cable Δl_i, one can easily make a first order estimate for the corrected cable length $\mathbf{l}_{\mathrm{E}}$ by reducing the elongation from the cable length computed with the standard model

$$\mathbf{l}_{\mathrm{E}} = \boldsymbol{\Phi}^{\mathrm{IK}}(\mathbf{r}, \mathbf{R}) - \Delta \mathbf{l}(E_{\mathrm{c}}, A_{\mathrm{c}}, \mathbf{l}_0) \quad . \tag{7.41}$$

Given the uncertainties of the elastic parameters, the errors of the first order estimation can be neglected. As shown by Kraus [259], one can also compensate for the elastic reaction by means of control where the tension needs to be measured for this approach. This is achieved by controlling the tension to set-point values computed from the desired tension distribution.

Dealing with elastic cables as described above is done in closed-form. Yet, the computational efforts are notably increased since one has to additionally set up the structure matrix and evaluate the force distribution. For a linear elastic cable force model, the determined cable length is unique for a given force. However, as shown in Sect. 3.6, there exist infinite force distributions and thus it depends on the approaches and objectives how to choose a unique solution. Additionally, one has to take into account the wrench $\mathbf{w}_{\mathrm{p}}$ applied to the mobile platform. This wrench can be caused by the payload, d'Alembert inertia forces, and process forces. It is straightforward to take these effects into account if the value is known. From the perspective of the kinematics code within the controller, the problem arises how to determine it in practice.

In this setting, the behavior of the force distribution algorithms is interesting close to the boundary of the workspace. Some methods such as the closed-form methods provide continuous and smooth (but of course infeasible) force distributions if poses outside the workspace are evaluated, whereas approaches such as the barycenter method or the advanced closed-form method provide no solutions or solutions being far away from the last feasible solution making the behavior of the kinematic code very sensitive to errors and noise.

7.4.2 Forward Kinematics

For over-constrained cable robots, the computation of the forward kinematics with elastic cables is only slightly altered because the basic procedure presented in Sect. 4.3 implicitly takes the concept of force distribution into account. The linear elastic model of the cables is considered for the forward kinematics by using a physical measure of the potential energy instead of a simple root mean square.

As discussed in Sect. 4.3, the forward kinematics can be solved by minimizing the potential energy U of the cable robot. Now using a physical model of the cables, the potential energy of the robot is $U = U_\mathrm{P} + \sum_i U_i$ where the potential energy of the cables is

$$U_i = \frac{1}{2} k_i v_i^2 \quad , \tag{7.42}$$

k_i being the stiffness of the i-th cable. The potential energy of the platform is

$$U_\mathrm{P} = m_\mathrm{P} g (\mathbf{r} \cdot \mathbf{e}_z) \tag{7.43}$$

where m_P is the platform mass, g is the gravity acceleration, and $\mathbf{e}_z$ is the unit vector in the direction of gravity. When the mass of the platform is taken into account as proposed through the potential energy, the solution estimated through the forward kinematics depends on the ratio between the stiffness of the cables and the platform mass. To consider the stiffness of the cables, one has to compute k_i according to Eq. (7.40) for each cable taking into account both the given cable length and additionally the constant length between the winches and the last pulley. Thus, one has to optimize the function

$$\varphi^{\mathrm{DKE}}(\mathbf{l}) = \min_{\mathbf{r}, \mathbf{R}} m_\mathrm{P} g (\mathbf{r} \cdot \mathbf{e}_z) + \sum_i^m k_i (l_i, l_{0,i}) \, v_i^2 (\mathbf{l}, \mathbf{r}, \mathbf{R}) \tag{7.44}$$

which can be done by using the Levenberg-Marquardt algorithm. Again, the solution found by this optimization is in general a local one, whereas the energy functional U has multiple local minima. In contrast to the standard model, no results are known how many minima exist depending on the number of cables m and motion pattern. It also remains an open problem to analyze under which conditions the solution branch is changed. If the kinematic code is used in the controller to estimate the current pose of the platform after small changes in the cable length, one can employ the last pose $(\mathbf{r}, \mathbf{R})$ of the platform as initial value for the next iterative optimization. In this case, the minimization of the potential energy U is meaningful as it physically describes in terms of energy how the platform is displaced from the last pose to the new one.

As for the elastic inverse kinematics, one needs an estimate for the current weight of the platform including a possibly changing payload as well as the stiffness of the cables. Compared to the forward kinematics of the standard model, the elastic effects of the cables do not introduce new theoretical problems. However, practical usage of the elastic models is complicated because of the additional parameters required to compute the model. Interestingly, small uncertainties in the stiffness parameter of the cables have a moderate impact on the estimated pose. However, the usually high stiffness of the cables makes the computation sensitive to measurement errors in the cable length as such deviations induce high elastic forces and thus have significant impact on the cable force.

7.5   Conclusions

Kinematic models that take the effect of panning pulleys into account have been proposed. Although the inverse kinematics equations are more involved than the equations for the standard model, one can still solve them efficiently in closed-form. If required, the structure matrix and differential kinematics can be symbolically computed. Thus, a numerical code for forward kinematics is derived and implemented with real-time efficiency. However, since the inverse kinematics equations are no longer algebraic, no theoretical results are received yet that predict existence and number of solutions for forward kinematics. Application of the pulley model in dynamics is straightforward and one can also take pulleys into account in workspace computation and force distribution. However, the differences to the standard model are mostly marginal.

Linear elastic cable models have been investigated for inverse kinematics. In contrast to the pulley models, elastic models connect some aspects of statics to the kinematic equations, making the problem involved. The kinematic codes used for forward kinematics can be understood as elastic models of the cables since the potential energy stored in the elastic cables is minimized in order to find a solution for the forward kinematics. The linear model of the cables is somewhat understood, however, it remains an open issue to deal with nonlinear elastic deformations where steel cables show some nonlinear elongations and synthetic fiber cables are subject to distinct nonlinear elastic effects. In dynamics, one usually employs linear or nonlinear elastic cable models since the incorporation of these effects is relatively simple in the dynamic formulation if an operational space formulation of the dynamic equations is used.

The effect of the cable mass on the kinematic behavior of the cable robot closely links static considerations to the kinematic problems. This effect is of notable importance when large-scale robots are considered, where the cables are long and comparatively heavy. Two cable models are predominantly applied, the inelastic and the elastic catenary curve. The underlying partial differential equations can be integrated in closed-form to reveal the well-known formulas based on hyperbolic cosine. Thus, numerical evaluation of the model is done efficiently. However, the formulation assumes a quasi-static state for the cable without transversal vibrations, neither in the direction of gravity where sagging occurs nor in the horizontal direction. Considering the static equilibrium constraints and the cable model as a system of equations, again, a non-algebraic structure is revealed and no theorems are known on the existence and number of solutions. The stiffness of cable robots can be evaluated taking the effect of sagging into account. After all, the computation is considerably more expensive than for the standard model.

The main challenge in the field of advanced cable models remains in achieving a more accurate model of the cable robot while considering a set of nontrivial assumptions on the cable. Although a number of models exist to consider sagging, elastics, and pulleys, one can hardly employ the models to predict the behavior of a new robot without experimental validation and identification of the parameters in time-consuming measurements.

A couple of concepts that are widely understood for the standard model are entirely open issues for nonstandard cable models. The theoretical analysis of kinematics with advanced cable models is in its infancy. No propositions are made on the number of solutions for the forward kinematics. Other open issues include the structure of the workspace, the location and also classification of singularity, an applicable formulation for the dynamics of nonstandard cables, and calibration as well as parameter identification methods. Especially for the problem of singularties, it is conjectured that the mathematical structure of advanced cable models introduce a new kind of singularity for cable robots which has no equivalent in conventional serial or parallel robots.

Chapter 8
Design

Abstract In this chapter, the design procedure of cable robots is addressed through a methodology presented in Sect. 8.2. Firstly, the application requirements that need to be satisfied by a cable robot are considered in Sect. 8.3. Secondly, different reference models for cable robots are reviewed in Sect. 8.4 that serve as parametric templates. These models are facilitated by different algorithms for geometry synthesis. Approaches based on optimization and interval analysis are presented in Sect. 8.5. The mechanical design of cable robots is discussed in Sect. 8.6.

8.1 Introduction

When working with cable-driven parallel robots, it seems that their design is one of the most challenging tasks. As we have seen in the first chapters, the analysis of kinematics, statics, and workspace involves a lot of advanced mathematical tools. Throughout the previous chapters, the problem of determining the robot properties is tackled if the physical parameters of the robot are given. For robot design, one has to solve the inverse problem and determine the robot geometry from given robot properties. For example, given a certain shape and size of the workspace, the task is to find a geometry for a robot so that this robot's workspace is generated. Machine design is a core problem of engineering, since it answers the question of how to build a machine for a specific purpose.

The design procedure of a cable robot is by far more time-consuming than applying any of the previously mentioned methods for analyzing a cable robot. This is due to the fact that robot design involves a couple of analysis methods which are required to efficiently work in cooperation. Just using (not researching) an approach or algorithm presented in previous chapters is expected to be done in hours or days. In contrast, applying a design procedure is a matter of at least weeks or even months for a research prototype and clearly beyond a person-year for a product.

The mechanical design of a new generation of winches took around 18 person-months for the IPAnema 3 robot. An iteration of the design of a research demonstrator

Table 8.1 Estimated and real efforts for design and project execution related to creating a cable robot and its components collected over around a decade. Care must be taken when comparing the projects since they also reflect different stages of experience in designing cable robots

Phase/component	Description	Duration
Winch	Design of the IPAnema 1 winch	6 person-months
Winch	IPAnema 1 winch: mechanical optimization, evaluation, ready for setup in IPAnema 1 demonstrator	6 person-months
Winch	Design of the IPAnema 3 winch	18 person-months
Platform	Simple platform for general purpose, parameter generation, mechanical design, integration of the cable-ends	1 person-month
Demonstrator	Shelf robot: geometry design, winch and controller adaption, winch manufacturing, initial operation	12 person-months
Demonstrator	IPAnema 2 planar: mechanical design of machine frame and linear actuation scheme, controller configuration, platform design, initial operation	15 person-months
Demonstrator	IPAnema 2 design for Automatica 2010. Retrofitting drive-trains and robot geometry of the IPAnema 1 system for the trade fair	6 person-months
Demonstrator	IPAnema 3 demonstrator setup	12 person-months
Demonstrator	IPAnema 3 Mini	9 person-months
Demonstrator	Copacabana robots for research purpose (geometry design, winch manufacturing, CE procedure, initial operation)	6 person-months
Special purpose machinery	MPI motion simulator including safety design for person transportation	18–24 person-months
Special purpose machinery	EXPO robots (parameter design, trajectory verification, programming system, support in initial operation and calibration)	18 person-months
Product	Large-scale crane (project outline)	15 person-years (estimate)

(without the aforementioned winch design) is estimated in the scale of one to two person-years. The full design of a large-scale robot for outdoor operation in product quality is estimated to be around 15 person-years. An overview of measured and estimated research and development time is given in Table 8.1. From these considerations, it becomes clear that a design method is about establishing efficient and applicable simulation tools as well as procedures. From today's perspective, an automated design procedure with a black-box behavior, where one can simply put in the requirements and receive a ready-to-use robot, is out of scope. Even for simpler serial robots, the design procedure is not automated yet. For example, consider the design of houses by architects which has been performed millions of times since the ancient times. Being an architect is a respected profession only caring for the design and customization towards the application's need. Thus, the focus for robot design is on efficient tools and applicable methods rather than fully automated programs.

There is a rich literature on the design of cable robots. However, many papers touch the problem of design but only present their experimental work without providing a rationale behind the design decisions or the methodology.

8.1.1 Literature on Parameter Synthesis and Optimal Design

The central challenge in the geometrical design of cable robots is the determination of the position of proximal and distal anchor points so that the robot fulfills given requirements including, but not limited to, workspace and stiffness, while avoiding auto-collisions and singularities. Such approaches are called synthesis procedures and have been studied by a number of authors. Merlet [336] sketches a design procedure based on interval analysis originating from conventional parallel robots [197] for dimensioning cable robots. This basic design procedure is extended for a conventional parallel robot to employ multiple technical requirements in a modular way in [394], and to multiple design criteria and combined with global optimization [399]. A transfer of the method to cable robots is proposed by Gouttefarde [186] where also an interval-based branch-and-prune method is employed to compute variants of the robot that have a prescribed wrench-feasible workspace. Although the underlying workspace test is different from the one proposed by Bruckmann, the basic design procedure is essentially the same. A design procedure based on formulating the design problem as constraint satisfaction problem (CSP) is presented by Bruckmann [74] where a combination of CSP and optimization is used to overcome the long computation times of interval CSP solving. Bruckmann [75] uses interval analysis and constraint programming to compute geometrical designs for a given workspace or given task. Lately, Lamine [279] employed such an interval analysis-based design procedure for planar and spatial cable robots for a given wrench-feasible workspace. The author performs optimal design for two geometric parameters in the planar case and four parameters in the spatial case, leading to computation times of 1:33 h and 14:26 h, respectively.

Azizian [19–22] proposed to design the geometry of planar and spatial cable robots so that a given box is inside the wrench-closure workspace. The design approach is based on a direct test if a given region of the desired workspace can be generated by the sought robot. Linear relaxation techniques are applied for the test. Using this test as performance criterion, the dimensional synthesis is achieved. For the planar case, computation times of some 10 s are reported. Gouttefarde [180] proposes a design methodology in multiple steps with a pre-selection of discrete candidates and different performance criteria including cable-cable collisions, a given available wrench set, and a given workspace where a combinational approach is used. A similar approach is followed by Gagliardini [155] who approached the problem of geometrical design related to reconfiguration of the robot by considering a combinational problem of discrete geometric variants in order to optimize stiffness for a task. Lafourcade [269, 270] employs a geometrical procedure to design a robot with seven or nine cables for a given workspace where the procedure is largely based on the assumption that a spatial robot has only three distinct distal anchor points and up to three cables share a common distal anchor point.

In contrast to parameter synthesis, the basic idea of optimal design is driven by the assumption that one can assess the quality of the robot through a performance metric that makes two competing design variants strictly comparable. Consequently, different approaches are elaborated in the literature to find the best robot through the mathematical tool of optimization. Williams [490, 493] discusses the design of planar robots with four cables where wrench generation and cable-cable interference are taken into account. The design procedure is based on an exhaustive search with constant steps in the three design parameter and checking if a grid of discrete points belongs to the wrench-closure workspace. Fattah [141, 142] presents an approach to the design of planar cable robots where optimization with respect to workspace size and the global condition index is compared. A performance index to compare the workspace of different cable robots applicable for optimal design is proposed by Verhoeven [473]. According to the presented results, this index serves well for assessing the full six-dimensional workspace with position and orientation. Pusey [405, 406] analyzes the design of a 6-6 cable robot and compares different robots with respect to the workspace and global conditioning index. The simplex algorithm is employed by Pham [384] to optimize the geometry of a planar cable robot with respect to the size of the workspace where force distribution and stiffness are considered. Hassan [198] proposes to optimize the geometry of a cable robot by minimizing the difference between the highest and lowest cable force amongst the cables. Thus, a measure similar to the standard deviation for the cable forces is used as performance index. Aref [15] discusses the optimal design of cable robots and compares different strategies where multi-objective optimization is proposed for cable robots since different criteria need to be considered. Therefore, a cost function based on dexterity, collision free workspace, and wrench-feasible workspace is employed. In order to solve the optimization problem, genetic algorithms and pattern search are proposed. Two performance indices are proposed by Tang [458] to be used for robot design called *all cable tension distribution index* (ACTDI), which is essentially the standard variation of the cable forces, and the *global tension distribution index* (GTDI). Fang [139] employed an optimal design procedure to maximize the wrench-closure workspace for the robot Segesta and combines research strategies based on the Powell method and simulated annealing algorithm to optimize a five parameter model of Segesta.

Different authors employed *genetic algorithms* for the optimization where little results are generated as to how the results from genetic optimization differ from gradient-based optimization. Duan [127] performs an optimal design procedure using genetic algorithms for a model of the FAST where different criteria are taken into account. In this approach, requirements are included to also limit cable sag. Yangmin [505] proposes to use genetic algorithms for the optimization of planar cable robots with four cables with respect to the global dexterity index and to the overall stiffness index. A linearly weighed objective function composed from these indices is applied.

A similar idea is pursued by Fahham [137] where a genetic algorithm is applied to design a planar cable robot for optimal trajectory time taking into account constraints on the cable forces as well as on the cable velocities. An optimization strategy from the field of artificial intelligence is used by Xiaoling [498] who proposes to optimize the geometry of a spatial cable robot with four cables with least square-support vector regression method which is commonly used in machine learning. Since the author does not provide the computation time, the effectiveness can hardly be compared with conventional optimization methods.

Bahrami [26] discusses optimal design of a 3T robot with four cables taking into account the volume of the workspace, the global dexterity index, and energy consumption. Genetic algorithms are used to solve the optimal design problem. Ouyang [378] optimizes the geometry of a simple length-width-height parametric model of a cable robot with eight cables so that the robot has a given workspace.

The problem of reconfiguration is related to the problem of optimal design. However, in reconfiguration only parameters of the robot are varied which are expected to be easily changeable in an existing physical robot. Typical parameters are the location of the proximal anchor points. Nguyen [363] addresses the reconfiguration of a cable robot by moving the proximal attachment points in order to optimize the kinetostatic performance. Nguyen targets to customize the translational workspace as well as the orientation workspace, the available velocities and accelerations at the end-effector, as well as the available wrench set. The idea of providing specific motion capabilities is taken a step further by Gagliardini [153] who introduces the concept of the twist-feasible workspace, i.e. the workspace where all twists from a given twist-set can be generated by the robot. Both wrench-feasible and twist-feasible workspace are subject to the genetic algorithm. The discrete reconfiguration of cable robots [154, 156] is addressed by choosing from sets of geometrical design parameters. The papers discuss a procedure with multiple steps to find the feasible design for reconfiguration taking different task constraints into account. Although both are theoretically and practically possible, no results are published on adaption of the drive-trains by e.g. changing the gearboxes to adjust the kinetostatic performance.

A rather specific application is investigated by Gao [158]. The optimal geometrical design of a cable-driven mechanism with four cables is presented where the platform is constrained by a central spine with springs for application as an artificial neck. In the paper, the positioning of the anchor points for minimizing the actuator forces is presented where a simple geometric model with only two parameters is employed.

For automatic as well as user-driven geometrical design, one has to use a parametric model that serves as template of the robot. Such geometric archetypes provide the relations of the proximal anchor points on the robot frame and the distal anchor points on the mobile platform. By choosing a typical number of two and twelve parameters, one receives a specimen of a robot from the family described by the archetype. Kawamura [238] develops an ultra-fast cable robot with seven cables for six degrees-of-freedom called Falcon whose cable arrangement has some similarities with the Delta robot [98]. The geometry of the Falcon is characterized by a planar rectangular robot frame and a slender bar for the mobile platform. Choe [97] proposes a cable robot with four cables and three degrees-of-freedom which can be seen as an evolution of the Falcon system. Tadokoro [451] discusses different design variants for a cable robot focusing on the form of the mobile platform. A portable cable robot with winches on trucks is proposed by Bosscher [50]. In order to constrain the orientation of the mobile platform, pairs of cables are used which are arranged similar to the parallel struts of Delta robots.

Tang [458] proposes three parametric archetypes for cable robots with seven cables where the proximal anchor points are located on two circles over and under the mobile platform. Later, Alikhani [10] proposed a robot archetype with nine cables and six actuators where three pairs of cables are guided in parallel lines. By doing so, the robot is constrained to be operated with purely translational motion. Lamaury [277] discusses the geometrical design of the ReelAx8 robot underlining the symmetries used for this robot. The paper discusses the anchor point arrangement on the mobile platform that is later used for CoGiRo. A suspended 3T system similar to the Skycam is proposed by Filipovic [151] where only three out of four cables are actuated. Two cables are conventionally fixed to the mobile platform and actuated by winches. The remaining two constraints are realized through one cable that is redirected by a pulley on the mobile platform so that one end is fixed at a proximal anchor point and the other end is connected to a winch. The idea is promising since it saves one actuator. However, no results on the size of the wrench-feasible workspace are provided.

A remarkable geometry is introduced by Miermeister [346, 402] who proposed a special geometry for cable robots that allows for unlimited rotation of the mobile platform about one axis both without cable-cable interference and with wrench-feasibility throughout the full rotation of the mobile platform. Different variants with nine to twelve cables are possible and, especially if more cables are used, the robot can perform unlimited rotation of the mobile platform within a reasonably large workspace.

A rather unconventional approach is pursued by Liu [291] who proposes to resign from coiling the cable and considers to change the efficient cable length by deforming a cable four-bar mechanism through a slider in order to improve the accuracy in the mechanism. Using three of such kinematic loops, a planar 2T robot with six cables is received which is actuated by three sliders.

8.1.2 Dimensioning of Components and Hardware Design

Surprisingly, the problem of properly dimensioning the main components of cable robots is hardly addressed in the literature. It seems that only Kraus [263] proposes a method to dimension the drive-trains of cable robots based on velocity-force charts for optimal design of the actuators. Some remarks on cable requirements and selection for cable robots are presented by Weis [489]. Further references on designing winches and cables guidance elements as well as guidelines for the selection of cables are not elaborated for the special operating conditions of cable robots. Instead, one has to derive such information from text books [149, 204] and applicable norms [222].

More material is published on the design of laboratory test-beds and demonstrator systems. The scope of such papers is usually limited to listing the installed components. When it comes to the development of cable robots, one can find a number of conceptual drawings, CAD illustrations, and pictures for specific parts such as winches, cable guidance systems, mobile platforms, and machines frames. In contrast, a design methodology is hardly tackled. Contributions focusing on components can hardly be separated from those that target at specific applications: The first contribution from Landsberger [280] presents the primal concept of a cable-driven parallel robot as well as the physical prototype of the first robot which consists of a central constraining linkage and six antagonistic cables. Cong [383] proposes a winch design where the whole drum is translated instead of a spooling unit in order to keep the point of cable attack fixed in space. Pott [404] presents the conceptual design of the IPAnema winches. The winch design employs a spooling unit traveling parallel to the drum to allow for accurate and repeatable coiling of the cable. Compared to Cong's design, the winch can be built more compact for the same cable stroke but requires an additional guiding pulley. The problem of two cables sharing one proximal anchor points is addressed by Fassi [140] who presents the outline of a double winch unit where two cables share almost the same proximal anchor point. Since eyelets are used to guide the cable, the approach is limited to soft and thin cables. According to other experimental results, such a design is expected to generate excessive wear on the cables. Billette [38] proposes a design for a winch with two actuators where one of the actuators produces large impetus for simulating impacts in a haptic interface. An unconventional idea is put forward by Yeo [509], presenting a design for mechanical components to vary the stiffness of cable-driven parallel robots. Also, the integration of this device into a planar cable robot is described. Baoyan [27] investigates additional counter-weights to the mobile platform of the FAST robot in order to enlarge the orientation workspace. A similar approach is taken up by Zitzewitz [144, 533] who proposes to employ passive springs or counter-weights in the electric actuators to optimize the performance of the drive-trains of cable robots. By presenting of the drive-train, the operation point of the servo-drives is shifted so that both the positive as well as the negative quadrant of the velocity-torque space are exploited. Thereby, the required motor size is reduced and energy efficiency is improved at the cost of higher inertia and reduced dynamics.

A cable robot design with movable proximal anchor points on either guideways or mobile vehicles is analyzed by Zhou [526]. Korayem [253] presents the cable robot ICaSbot which is a suspended cable robot with six cables. A full robot system, including kinematics, control, and user interface [254], is presented and performance measurements according to ISO 8283 [220] are performed.

8.1.3 Case Studies and Applications

A couple of practical results and conceptual case studies are presented in the context of design. Lindemann [290] proposes the haptic device called Texas 9-String, consisting of a serial kinematic chain used as a joystick. It is constrained by nine cables to measure the motion and to generate force feedback. Some technical details on the hardware design of the NIST roboCrane are presented by Bostelman [54]. Then, Albus [6] describes the redesign of the RoboCrane system towards the so-called SPIDER configuration and provides technical details on the implementation. Geometrical design and development of real-time controller system is discussed for the cable robot Segesta [139, 210]. Technical details on the Marionet robot family are presented by Merlet [324, 336] who employs pulley tackles for the actuation of the cable robot and compared the pulley tackles to winches with respect to different criteria [327].

Some applications in the field of medical devices and rehabilitation are addressed in the literature. Ottaviano [376] discusses the design requirements of a 4-4 cable robot for applications in hospitals and a small-scaled model of the handling application is presented. Rosati [423] proposes to use a suspended cable robot with four cables for arm rehabilitation and presents results from application [421, 422]. Then, it is proposed to use movable proximal anchor points in order to enlarge and shape the workspace of 2T cable robots [424]. The design of the Sophia-4 cable robot for neurorehabilitation is described [425, 517]. Morris [351] develops a planar cable robot with three cables for generating forces for physical therapy assistance.

The design of large-scale systems imposes additional requirements. Buterbaugh [81] presents a prototype for positioning the radio target for antenna testing using a suspended robot with six cables similar to the RoboCrane design. Rui [508] performs design pre-studies for a 3T suspended cable robot with four cables as case study for the FAST telescope. This research is later generalized to six cables [459, 460]. Sensitivity studies are performed for the geometric parameters of the FAST robot where sagging is taken into account for the modeling of the long cables. Bruckmann presents the development of a high-bay storage retrieval machine [70] as well as a suspension system for wind tunnels [69]. The optimization of the cable length for a robot with moving proximal anchor points is addressed by Sturm [445]. The achievable motion range of the actuators is crucial for the effectiveness of the cable robot. Thus, the resulting robot can be understood as an over-constrained cable robot counterpart of the conventional parallel robots known as Hexaglide [205]. Tempel [463] presents planning approaches and a safety system for a cable robot for an entertainment application that needs to fulfill strict safety regulations.

Bostelman [56] describes some possible applications of the RoboCrane system as well as integration concepts with movable winches on the ceiling. Later, Maeda [305] presented the design of the WARP robot consisting of seven cables and a triangular mobile platform. Another handling application is sketched by Dekker [113] where a cable robot called DeltaBot for fast pick-and-place operations is presented. A peak performance of 150 cycles per minute and a position repeatability of 0.1 mm is reported.

Bosscher [49] proposes the C^4 robot for 3D-printing of buildings. The proposed design has twelve cables and the proximal anchor points of the lower eight cables can be moved on linear guideways in order to avoid collision between the cables and the printed structure. To guide the mobile platform, four pairs of cables are geometrically parallel so that the orientation of the mobile platform is locked. Capua [83] proposes a four cable suspended robot called SpiderBot where the winches are located on the mobile platform. The proximal anchor points can be fixed and released with suction cups to the ceiling. By proper fixing and releasing of the cables, the robot can climb under the ceiling.

The reconfiguration of a suspended cable robot with six cables is discussed by Zi [529] where the proximal anchor points can be moved on the circumference of a circle. The contribution is remarkable in being amongst the few works that also experimentally investigated reconfiguration.

Some authors apply cables to actuate serial kinematic chains. Kossowski [255] addresses the design of a hybrid cable-actuated structure. Mroz [352] proposes a robot design where a serial chain is actuated by parallel attached cables. The work is remarkable as it addresses the design methodology from requirements through kinematic and static analysis to controller design. Lim [287] discusses the design of the actuation of a universal joint by four cables and different arrangements of the cables are compared.

Liu [292, 293] proposes a design concept for a cable robot with four cables and two actuators where additional springs are integrated into the pulley mechanisms in order to distribute the forces amongst the cables. A similar approach is pursued by Khakpour [240] where differentially driven cable robots are discussed instead of spring-loaded pulley tackles. Therefore, multiple winches are connected to a single drive-train, thus, constraining the motion and reducing the costs for the system at the same time. Springs are considered to additionally constrain the mobile platform in order to further reduce the number of actuators. Lately, this idea of directly adding springs to the mobile platform to shape the workspace is evolved by Duan [131].

Summarizing the findings from the review of literature, one can conclude that many aspects in the design and realization of cable robots are considered in the literature. However, the approaches are at best loosely connected. In the following, a holistic approach is presented putting a couple of the methods reviewed above into a design procedure.

8.2 Product Development for Cable Robots

The procedure presented here is an adaption of known development methods for application to cable robots. It is basically derived from the V-model used, amongst others, in mechatronic component design (Fig. 8.1). The elements of the procedure have their origin in accepted design strategies like *systems engineering* [471], *software engineering*, and *design of mechatronic systems* [472]. Therefore, these methods are the foundation for developing a suitable method for the design of cable robots. Today, the design of cable robots is hardly a systematic procedure due to lack of knowledge and experiences with existing systems and past developments. One has to consider that cable robots are highly modular robotic systems which have to be adopted to special fields of application. Taking into account the variety of applications

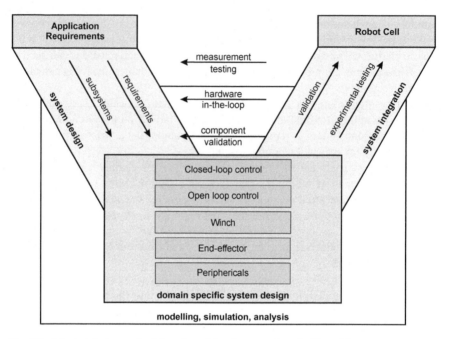

Fig. 8.1 Adopted design methodology for cable robots based on the V-model

presented in Chap. 2, the procedure must be customized to the field of application. Furthermore, the design and development procedure for a robotic system is a process that lasts several months or even years. Therefore, a list of key tasks in the development is presented and the tasks are related to the algorithms and methods presented in previous chapters.

The development procedure is based on the V-model described by VDI 2206 (Fig. 8.1, [472]). Beginning from the application requirements, the technical domain specific subsystems are defined. For cable robots, the domains are mechanical engineering for the winches, the machine frame, and the end-effector. The instrumentation of the end-effector is application specific and it is out of the scope of this work to include the process specific aspects. Control engineering is the basis for open and closed-loop control for the robot controller. The development of a proper user interface and high levels of control require also aspects of software engineering. Additionally, design and dimensioning of the cables require some knowledge of civil engineering.

In the first phase, the application requirements have to be translated to the component specifications. This requires mapping the required workspace or motion path to a suitable geometrical design. Payload, dynamic forces, and process forces need to be considered for the winch, platform, cable, and frame. The integration of the subsystem into the robotic system starts when all subsystems are properly defined. Once the domain specific design for the subsystems is finalized, the performance of the subsystems is tested against the component requirements. This can be done by using hardware-in-the-loop simulation as no physical robot is available at this stage. In particular, for cable robots with extreme requirements such as huge workspace, large payloads, or high dynamic requirements, simulations of the system are the only way to validate the subsystem design. Additionally, experiments are prepared and accompanied by simulations to validate the subsystem design. Shortcomings in the design of the subsystem must be settled through iterations before further integration in the robotic system.

An overview of the six development phases with their respective tasks, results, and tools is shown in Table 8.2. The phases of this development are executed within the framework of the V-model. Therefore, at the end of the domain specific design phase, a validation procedure starts in order to find shortcomings in the design by comparing the components' performance to the design specification. If shortcomings become evident, relaxations have to be accepted in the design specifications or revisions have to be made for the components.

Here, the main phases in designing a cable robot system are proposed. The list below describes the cable robot specific tasks and relates methods introduced throughout this work where applicable. The six phases follow the design methodology for mechatronic systems as outlined in Fig. 8.1:

1. Application analysis: Process and requirement definition. Details on the definitions are given in Sect. 8.3.

   - Technical requirements and environment: limitation of the installation space, required size of the workspace $\mathcal{W}^R$, relevant obstacles for collision testing, temperature (Sect. 8.3).
   - Process description including velocities, accelerations, process forces, payload (Sect. 8.3) as well as intensity of usage for actuator design.
   - Process sequence and operating grade
   - Required accuracy and repeatability

   Results: draft of application and catalog of criteria (design specification).

2. System level design: cable robot system architecture (Sect. 8.4)

   - Design of the geometry of the mobile platform and the machine frame using workspace computation (Chap. 5)
   - Design layout of the robotic system (Sect. 8.5)
   - Concept for controller and selection of controller architecture
   - Dynamic simulation model of the robot (Chap. 6)

   Results: layout plan, geometry data for the robot, simulation model, archetype of the robot.

3. Domain specific design of hardware for winches, cables, drive-trains, pulleys, machine frame, mobile platform, end-effector instrumentation

   - Design and dimensioning of the fixed machine frame in order to implement the chosen values for $\mathbf{a}_i$ and $\mathbf{R}_{A,i}$

Table 8.2 Design procedure for cable-driven parallel robots

Phase	Tasks	Document	Tool, description, used methods
Application analysis	Process description, technical requirements and application demands robot process description process sequence	Design specification workspace, velocity, dynamics, payload	Workshops catalog of criteria
System level design	Application layout geometry synthesis for mobile platform and machine frame control system architecture simulation	CAD layouts, $\mathbf{a}_i$, $\mathbf{b}_i$, l_{min}, l_{max}, f_{min}, f_{max}	Templates of existing robots, WireCenter, CAD, computer aided factory planning MATLAB/Simulink
Hardware design	Design of mobile platform cable selection winch design dimensioning of drive-trains fieldbus, control platform sensor selection	CAD of platform cable type, pulleys	CAD building blocks list of suppliers bill of material
Control software development	Control system architecture development of controller models (kinematic code, position and force, sensor signal filtering and preprocessing) control system configuration design of (graphical) user interface, selection of electric components	Controller software, user interface implementation, circuit diagrams	MATLAB/Simulink TwinCAT 2/3, B&R Automation Studio MS Visual Studio, dynamic simulation
System integration	Hardware-in-the-loop simulation virtual initial operation initial operation calibration and referencing application programming and teaching risk assessment, documentation	Simulation user's manual	HIL for open-loop control and initial operation MATLAB/Simulink WireCenter CAM tools for off-line programming
Evaluation and validation	Requirement validation through measurements test usage site acceptance test	Measurement protocols, CE-certificate	Measurement devices

- Refine geometry design for mobile platform, i.e. fix how to mechanically realize the values for $\mathbf{b}_i$ as well as the static loads
- Using the required workspace hull $\mathcal{W}^R$, one computes the required cable stroke Δl_i (Sects. 5.5 and 4.2.1)
- Select the cable material and type, thus define the parameters for cable length Δl_i and thus l_{min}, l_{max}, radius r_c, material E_c, and fatigue related properties (Sect. 8.6.1)

- Design or selection of actuation system such as winch and pulley mechanism (Sect. 8.6.2). Define the pulley radius r_R.
- Select drive-trains, gearboxes, inverters, and motors as well as sensors for position and/or force measurement (Sect. 8.6.3). This defines effective f_{max} through the actuators.
- By defining the mechanical components of the robot, the minimum and maximum cable forces f_{min}, f_{max} can be determined (Sect. 3.4.5).

Results: bill of material, engineering drawings of the robot system, the robot frame, and the mobile platform.

4. Domain specific design of controller software: Design and implementation of open and closed-loop control including electrical system development

- Define controller architecture e.g. position control, velocity control, or hybrid force control as well as the framework for controller integration (proprietary or open source)
- Development, implementation, and configuration of controller modules: kinematic codes for forward and inverse kinematics (Chap. 4), closed-loop control, sensor signal processing, force controller if applicable
- Motion controller configuration and parameter tuning
- Controller logic for interfacing with peripherals; fault recovery strategies, definition of error states
- Development of user interface (UI)
- Design and implementation of the control cabinet, selection of communication technologies such as fieldbus, selection and integration of sensors, electrical integration of external peripheral
- Preliminary risk assessment and selection of safety-related sensors
- Application specific control development

Results: connection diagram for the control cabinet, software and configuration data for the controller.

5. System integration: Connection of components, assembly, testing

- Component assembly, installation, wiring, mechanical connection
- Performance validation through workspace studies and kinematic computations. The validation focuses on the feasibility of technical aspects such as deflection angles of the anchors points (Sect. 5.2.8), effective cable force limits (Sect. 5.2.2), cable-cable interference (Sect. 5.2.6), as well as cable-application interference (Sect. 5.5.6). In case of shortcomings with respect to the design specifications, the phases 2–4 have to be repeated.
- Hardware-in-the-loop simulation (Chap. 6)
- Virtual initial operation
- Real initial operation of the controller system with the robot hardware
- Accuracy improvement through referencing and calibration (Sect. 9.2)
- Application programming
- Risk assessment and risk mitigation
- Factory acceptance test (FAT)

Results: run capable cable robot, measurement protocols, performance data.

6. Evaluation and validation: Test, evaluation, initial operation, documentation and reporting

- Test measurement and validation of desired properties on the real machine
- Test operation on the desired application task
- Documentation and manuals
- CE-certification
- Site acceptance test (SAT)

Results: finalized cable robot system.

Usually, the six phases are executed subsequently where different levels of validation of the achieved system performance against the initially defined requirements are done. If the performance metric defined for the target robot system cannot be reached at the end of the development stage, one has to go back to the previous stage and refine the development. The objective of the development procedure above is to limit the needful number of iterative cycles where from today's perspective it is out of reach to avoid iterative cycles at all.

8.3 Application Requirements

There is a large number of connections between the requirements of the robotic system level and the requirements for technical parameters of the component level. This leads to a heavily coupled system which is typical for parallel robots. In Table 8.3, an overview is given how application requirements affect component requirements. These dependencies become meaningful for the design of variants as well as for reconfiguration. Particularly the connection has to be considered when changes in the requirements occur and the consequences for the robot system must be determined. Most requirements are strongly related to each other. Some are correlated but in most cases they are antagonistic and it makes sense to order the requirements by importance. In the following, the main technical requirements for cable robots are listed and their respective impact on the robot design is briefly discussed.

8.3.1 Workspace

The required workspace $\mathcal{W}^R$ is expressed in terms of geometric primitives such as a box (length × width × height), a cylinder (radius, height), a sphere (with radius), or an ellipsoid (by three semi-principal axes). For the design of planar robots, one uses rectangles, circles, and ellipses, respectively. From the perspective of the application, one has to define if the required workspace needs to be reached with one defined orientation (constant orientation workspace $\mathcal{W}_{co}$), with all orientations from a given

Table 8.3 Dependencies between application requirements and component requirements

| | Requirements for components | | | | | | |
| | Winch | | | | | Cable | |
Application requirement	Cable velocity	Cable force	Cable acceleration	Cable stroke	Deflection angle	Accuracy	Stiffness
Workspace	None	Medium	None	Strong	Strong	None	None
Payload	None	Strong	None	None	None	None	Strong
Applied forces	None	Strong	None	None	None	None	Strong
Acceleration	Weak	Strong	Strong	None	None	Medium	Strong
Velocity	Strong	Medium	Medium	None	None	Medium	None
Installation space	None	Medium	None	Weak	Strong	None	Medium
Accuracy	None	Medium	None	None	None	Strong	Strong

set $\mathcal{R}^R$ (total orientation workspace $\mathcal{W}_{TO}$), or with at least one arbitrary orientation (maximum workspace $\mathcal{W}_{max}$). The data model of the workspace as well as the type of the desired workspace are then input conditions and configuration settings in the design and verification procedure (see also Sects. 5.1.3 and 5.1.4). The application requirements for the workspace influence different specifications for the components.

The prescribed workspace has medium impact on the cable forces and the maximum strength of the cables since a larger workspace requires higher forces in the cables in order to prevent them from sagging. Another reason for higher forces is an unfavorable angle of attack of the cables on the mobile platform at the workspace boundary. An increase in the maximum feasible cable force mitigates this limitation.

A larger required workspace $\mathcal{W}^R$ heavily influences the required cable stroke Δl of the winches. Usually there is a proportional scaling in the size of the workspace and the required length of the cable. Contrary, the available cable stroke of a given winch is a strict limit for the size of the workspace that cannot be bypassed with other methods.

Changes in the required workspace size have a strong effect on the maximum deflection angles for the winches and distal anchor points on the mobile platform. These two components have to be designed carefully since restricted deflection capacities largely limit the usable workspace. Summing up, changes in the required workspace $\mathcal{W}^R$ heavily affect the robot design.

8.3.2 Payload

The payload of the robot is described by the possible weight m^R that can be supported by the mobile platform. From an application point of view, the payload is expressed as a mass in kilogram. Together with the desired acceleration, this is translated to the required wrench set $\mathcal{Q}^R$ of the robot that is analyzed in statics (Sect. 3.3), workspace considerations (Sect. 5.2.2), drive-train dimensioning (Sect. 8.6.3), and as parameter in the dynamic simulation (Chap. 6). The payload also affects the mechanical design of the mobile platform in order to carry the payload. Depending on size and distribution of the load as well as on the motion requirements of the robot, one has to also take into account the moment of inertia on the mobile platform Θ^R for computing the wrench set $\mathcal{Q}^R$.

The design of the robot depends on the required payload of the application. The available payload depends in general on the pose of the robot but for the sake of simplicity, it is usually assumed that only the smallest feasible payload is considered that can be carried in the whole desired workspace and one does not take advantage from higher payload at special positions.

The payload of the robot has a strong impact on the selection of winches, drive-trains, and cables. As a rule of thumb, there is a proportional scaling of the required payload to the required capacities of the winches and the cables. Since the rated power of the motors is a main cost-driver, payload needs careful attention.

Additionally, one has to consider the elasticity of the cables along with the payload. Higher payloads and thus higher tension in the cables lead to larger elongation of the cables. This can disturb the control system and compromise accuracy. After changing the requirements for payload, one has to reconsider the static deviation of the robot and elastic effects caused by dynamic behavior of the robot. The experimental results from Kraus [259] show that loading and unloading the feasible payload already change the performance of the control system. Clearly, geometric reconfiguration of the robot has an even larger influence.

8.3.3 Applied Forces and Torques

Applied wrenches on the mobile platform arise from the process that is executed such as mating forces in assembly or cutting forces in machining. The applied forces and torques are described by enclosing the required wrench $\mathbf{w}_P^R$ by a box or by a hyper-ellipsoid. Again, one receives a desired wrench set $\mathcal{Q}^R$ that serves as input parameters for statics (Sect. 3.3), workspace considerations (Sect. 5.2.2), drive-train dimensioning (Sect. 8.6.3), and for dynamic simulations (Chap. 6). Thus, the applied wrench on the mobile platform has a similar effect on the robot as changes in the payload. In contrast to the payload, the wrench set generated by applied wrenches does not depend on the acceleration of the mobile platform. Examples of such definitions are given in the workspace studies (see Sect. 5.7.2).

Since the applied wrench has to be counteracted by the winches, the drive-trains, and the cables, one must address changes in the applied wrench $\mathbf{w}_P^R$. For estimating new feasible cable force limits, this coupling effect can be considered to be linear.

Also for the applied wrench, one has to take the elastic effects of the cables into account. Even if higher forces can be generated with a higher breaking load of the cables, the elastic reactions at the mobile platform and thus the positioning errors are larger. Therefore, one has to check if the larger elastic effects are feasible for the application.

8.3.4 Acceleration

The feasible dynamics of the mobile platform are governed in terms of the linear a_{max} and angular accelerations α_{max} of the mobile platform. Similar to the requirements for the applied wrench, the specification of the acceleration a_{max}, α_{max} can be described by a box or ellipsoid. For that purpose, one has to use both platform weight m_P and moment of inertia $\mathbf{I}_P$ as well as the payload to compute the equivalent wrench set. Using d'Alembert's formula, one computes the inertia wrench $\mathbf{w}_I$ for the mobile platform by

$$\mathbf{w}_I = \begin{bmatrix} m_P \mathbf{a}_P \\ \mathbf{I}_P \alpha_P + \omega_P \times \mathbf{I}_P \omega_P \end{bmatrix} \tag{8.1}$$

where ω_P, $\mathbf{a}_P$, α_P are the vectors with the required angular velocity, the linear acceleration, and the angular acceleration of the mobile platform, respectively. Thus, using above-mentioned d'Alembert's formula, one determines the required wrench set $\mathcal{Q}^R$ for a given scenario. This wrench set $\mathcal{Q}^R$ is then taken into account in statics and workspace computation as well as for dynamics evaluation. The required accelerations of the end-effector affect the following subsystems.

The cable forces are mainly influenced by the cable acceleration. The required acceleration is directly governed by the maximum force f_{max} that can be generated by the winch since the generalized inertia seen by the winch is the connecting factor between acceleration and cable force. The same holds true for limitations in the acceleration capacities of the winch. As a rule of thumb, there is a proportional relation between the linear accelerations of the end-effector and the maximum acceleration provided by the winch.

When considering higher accelerations, there are additional requirements for the mechanical parts of the winches. In such cases, one has to consider the dynamics of the cable and auxiliary parts of the winch such as the guiding pulleys and the panning joint of the pulley. Higher platform accelerations require additional actions for robust and reliable operation of the cable when it is guided around the pulleys. For ultra-fast motion, the centrifugal forces of the cable need to be taken into account when the cable is guided over pulleys or coiled onto the drum.

Higher accelerations indirectly influence the inertia of the cable since higher forces lead to thicker cables and thereby to high cable inertia forces. The guiding pulleys and winch drums also grow with the cable diameter and their respective impact on the inertia is increased. If very high accelerations are required, one has to additionally consider the cable stiffness. For ultra-high accelerations, the connection between stiffness and bandwidth of the cables needs to be considered. Otherwise, the commanded motion of the actuators leads only to elastic reactions in the cables rather than generating the desired motion of the mobile platform.

8.3.5 Velocity

The velocities of the mobile platform or, more precisely, the twist are described either by a box in terms of $||\mathbf{v}||_\infty < v_{max}$ or by an ellipsoid with its main axes. This is equivalent to the definition of a twist-set and a twist-feasible workspace [153]. The specified velocity of the mobile platform affects the following subsystems of the cable robot:

The required velocity directly determines the cable velocity of the winch which needs to be at least as high as the end-effector velocity to realize the required end-effector velocities in all poses.

In order to realize the required acceleration along a path, the end-effector velocities constrain the required accelerations of the winches. Even if there is no acceleration of the mobile platform along a linear path, the nonlinear transmission of the inverse kinematics generates requirements on the acceleration of the winches. If the path is curved, additional force requirements according to Eq. (8.1) arise for the acceleration and for the forces of the cables.

High velocities of the cable put notable prerequisites on the control system in order to allow accurate and smooth generation and tracking performance of the robot. Considering typical controller cycle times of 1 ms, the set-point samples have already a distance of around 1 mm for robots with medium velocity of $v = 1$ m/s. For ultra-fast cable robots that reach 10 m/s and more, control tracking becomes an issue. This holds especially true for fully-constrained cable robots since small control errors lead to high parasitic forces in the cables due to high internal tension.

For cable robots with very high velocities, extra demands show up for the robust guidance of the cable. In experimental tests, it is observed that cable velocities of 10 m/s and more lead to nonnegligible centrifugal forces in the cables both on the pulleys and on the drum. Thereby, additional cable forces must be generated that counteract the centrifugal forces. In such cases, the cables may leave the pulleys and take off from the drum. Such phenomena severely undermine the reliable operation of the robot.

8.3.6 Installation Space

Cable robots in general have a complex connection between cable forces, shape and size of the workspace, and the installation space required for the robot. The installation space specifies the envelope around the winches. Mathematically speaking, one can define the installation space as the convex hull of the proximal anchor points A_i plus the size of the winch units. In practice, it is common to consider the size of a rectangular box enclosing all winches. For some robot designs, the workspace is larger than the convex hull of the proximal anchor points A_i. This holds true for many suspended robots. However, also fully-constrained robot designs such as the Falcon design show this property (Sect. 8.4.4). If the maximum installation space is given, it is usually assumed that the winches have to be placed on the surface of the installation space. This restriction is tackled by the use of parametric robot models as introduced in Sect. 8.4. For these robots, the installation space needs special treatment. If the installation space is given or changed, the following properties of the robot are influenced.

The installation space mainly specifies the required length of the cables. If the installation space is increased, longer cables are required even if the desired size of the workspace is maintained. Since the length of the cables is coupled to the cable's stiffness, one has a medium influence of installation space on robot stiffness.

A strong connection must be taken into account between the installation space and the workspace. Roughly speaking, there is a linear scaling between the size of the workspace and the size of the installation space. For large-scale robots, additional factors arise. To maintain the stiffness, the additional length of the cables needs to be compensated by thicker cables which in turn increase weight and pretension to prevent sagging. Therefore, there is a strong connection between the installation space and the workspace.

The cable stroke of the winches is moderately influenced by the installation space compared to the size of the workspace. Increasing the installation space while maintaining the size of the workspace puts little additional restriction on the cable stroke of the winches since the cable can have an arbitrarily long remaining length when fully coiled on the winch.

The deflection angles of the cable guidance system strongly depend on the ratio between installation space and workspace. Especially if the workspace is large compared to the installation space, large deflection angles are required both on the proximal and the distal anchor points which might be difficult to realize.

8.3.7 Accuracy

The precision of a robotic system is given by its *accuracy* and *repeatability* (see Sect. 9.2 for the definition) and these performance measures directly determine the robot's ability to properly perform its task. Designing a robot for a given accuracy is difficult since accuracy can usually be determined experimentally once the robot is fully assembled. Thus, proper assessment of this important performance indicator is made at a late state of the development, making it a critical factor in the development process. The following subsystems relate to the required accuracy. Experimentally determined values for the accuracy and repeatability are given in Sect. 9.3.1.4.

The proper positioning of the end-effector through cables depends on the feasible tension in the cables. Therefore, the minimum and maximum cable forces have a medium impact on the accuracy of the robot.

Inaccurate cable length caused by imperfect mechanics, control errors, and elastic deformation directly affects the accuracy of the end-effector. Therefore, one has to design the position measurement and control system for the cable length mainly depending on the required accuracy. Additionally, all mechanical parts in the winch system need to be designed so that the actuator motion is accurately rendered to cable length.

Elastic cables allow for larger elongation. Therefore, the position accuracy is reduced if the stiffness of the cables decreases. Further reductions in the stiffness of the drive-train are caused by the winches, the cable guidance system, and the mobile platform. For large-scale robots, one has to also consider the elastic deformation of the robot frame. As a consequence, the accuracy degenerates with an increasing size of the robot.

8.4 System Design and Structural Synthesis

Once all application requirements for a new robot are qualitatively and quantitatively described, the designer has to make the decision to transform the requirements into specifications for the robot. The first step in this procedure is the selection of required motion patterns and the degrees-of-freedom of the robot. Secondly, the number of cables needs to be selected. The task in the design of the robot can be identified with the *structural synthesis* or *topological synthesis* which has widely been discussed for conventional parallel robots, see the five reference books by Gogu [166] as well as Merlet [322, Chap. 2]. In topological synthesis of cable-driven parallel robots, one has by far less options and combinations available than for the design of conventional parallel robots. This is due to the fact that the kinematic chains of cable robots are governed by the kinematic properties of the cables. The dominant structure are cable actuators that cause a change in the effective length of the cables. Only few systems use cables of constant efficient length with moving anchor points. Although different cable connection mechanisms are possible, such as universal joints, swivel bolts, and spherical joints, their kinematic properties differ marginally and it is disregarded in the structural synthesis. Instead, its selection is subject to mechanical design. Thus, structural synthesis for cable robots requires to define

- Motion pattern and degrees-of-freedom n,
- Selection of *over-constrained* (CRPM, RRPM) or *under-constrained* (IRPM) cable robots
- Choosing amongst *suspended* or *fully-constrained* cable robots
- Definition of the number of cables m
- Selection of the actuation scheme: controlling *cable length* by winches, pulley tackles, twisting or *controlling the position of the proximal anchor points* by linear guideways or levers.

The choice of motion pattern is straightforward derived from the mobility requirement of the application. The decision between IRPM and CRPM/RRPM is more involved. IRPM robots come at lower costs for the robot hardware since less winches and less motors are involved. Due to the smaller number of cables, restrictions, from collisions amongst the cables are reduced. In turn, IRPM impose significant drawbacks in terms of accuracy, stiffness, and dynamic capabilities. Furthermore, kinematics, control, and error recovery of under-constrained robots are challenging adding costs to the development of the controller system which may consume the savings achieved by the reduced actuators.

Under-constrained robots (IRPM) are always suspended whereas CRPM or RRPM may be operated in suspended configuration. Suspended configurations can be used when the load is high and the accelerations are low, e.g. in crane-like handling and assembly applications or for relatively slow positioning of sensors such as cable-suspended cameras. Suspended robots suffer from limited stiffness in direction of gravity. Furthermore, their ability of distributing forces amongst the cables is reduced. In return, suspended robots hardly interfere with obstacles below and lateral to the mobile platform.

The number of cables m shall be chosen as small as possible in order to reduce costs. However, there are reasons to employ more than the minimum number of cables. Robots of the 2T motion pattern can be realized with three cables. However, the workspace and installation space has a triangular shape which often does not fit the needs of the application. In this case, a redundantly constrained robot with four cables allows to notably enlarge the workspace and gives additional degrees-of-freedom in the design to shape the robot's properties. The same situation occurs for spatial cable robots with 3R3T motion pattern. Using seven cables for a fully-constrained robot is possible but the workspace is small compared to the installation space. Again, increasing the number of cables to eight or more allows for more suitable workspace shapes. Using more than the minimum number of cables can be useful to achieve additional advantages. The shape and size of the workspace can be customized, additional cables allow for system inherent robustness against failure in single drive-trains for applications with high demands for safety. Such applications are transportation of passengers, handling of hazardous goods, or operation in changing environments. More cables are necessary to actuate additional degrees-of-freedom on the mobile platform or to perform unlimited rotation of the platform (see Sect. 8.4.13). Finally, redundant cables can, simplified put, increase payload.

The choice of the actuation system influences the workspace, stiffness, and energy efficiency. Cable robots with winches allow for a huge workspace where linear guideways are considerably restricted. An advantage of linear actuation of the proximal anchor points is pointed out by Bruckmann [68]: Since a share of the effective cable force results from the passive reaction forces of the linear guideways, one can realize high cable tension and high stiffness with a good energy efficiency.

The choices to be made for the criteria above have to be made by an engineer's careful considerations and no quantitative measures are provided here. After fixing these properties, one can approach the problem of finding a proper geometry of the robot, which is tackled in the following sections.

8.4.1 Common Architectures and Reference Designs

To describe the geometry of a cable robot, one has to define the positions of all proximal and all distal anchor points to uniquely fix the geometry of the robot. Each cable has a proximal and distal anchor point, the coordinates of which can be chosen freely. Thus, using $6m$ geometric parameters in the spatial case and $4m$ parameters in the planar case defines a cable robot. The properties, such as workspace, kinematics, statics, singularities, and stiffness, of the fully-constrained robot must be invariant under rigid body transformations of the machine frame and also under rigid body transformations of the mobile platform. Only the effect of gravity is altered through some rotation. We neglect this fact in the following. If one changes all proximal anchor points $\mathbf{a}_i$ of the robot by translation or rotation, the robot is essentially maintained and its properties remain unchained. Thus, within the $6m$-dimensional design space of the robot, six dimensions can be removed for such transformations of the machine frame and another six dimensions can be removed for the mobile platform leaving $6m - 12$ independent design parameters in the spatial case and $4m - 6$ parameters in the planar case.

A simple convention to reflect this dimensional reduction can be achieved by choosing $\mathbf{a}_1 = \mathbf{b}_1 = \mathbf{0}$ for fixing six parameters in the spatial case or four parameters in the planar case. One can proceed with defining the second anchor point to be located on the x-axis, and thus, fixing the respective y- and z-coordinate to zero, i.e. $\mathbf{a}_2 = [a_{2x}, 0, 0]^\mathsf{T}$ and $\mathbf{b}_2 = [b_{2x}, 0, 0]^\mathsf{T}$. Thus, another four parameters are eliminated in the spatial case and two parameters vanish in the planar case. Finally, for the spatial case, the third point of the frame and the mobile platform is fixed to the xy-plane and the respective anchor points become $\mathbf{a}_3 = [a_{3x}, a_{3y}, 0]^\mathsf{T}$ and $\mathbf{b}_3 = [b_{3x}, b_{3y}, 0]^\mathsf{T}$. Thereby, we have eliminated two geometric parameters and the total number of geometric parameters is reduced by twelve in the spatial case and six in the planar case. The convention described here is widespread, however, there are also others to achieve this reduction.

For the purely translational robots implementing the motion pattern 2T and 3T, the reduction can be continued. Here, all cables need to be connected to the same distal points since this is the condition for elimination of the rotation capability of the robot. Thus, only $3m$ ($2m$) parameters can be chosen in the spatial (planar) case and removing the congruent robots through rigid body transformations leaves $3m - 6$ parameter in the spatial case and $2m - 3$ parameters in the planar case.

For the 2R3T motion pattern, one can reduce the parameter space since the design requires all distal anchor points b_i to lie on a line. Thus, the first anchor point is chosen to be $\mathbf{b}_1 = \mathbf{0}$ and each of the remaining anchor points is characterized only by its translation along the common axis. Thus, the number of free parameters is $3m - 6$ on the machine frame and $(m - 1)$ for the mobile platform which results in a total number of $4m - 7$ independent design parameters for the 2R3T class.

Although the reduction approach described above reduces the number of design parameters, a large design space is still left for robots of motion pattern 1R2T, and especially for the 3R3T class. Therefore, subclasses described by fewer parameters

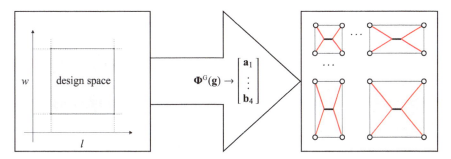

Fig. 8.2 Concept of parametric models transforming parameters $\mathbf{g} = [l, w]^T$ in the design space into robot geometries $(\mathbf{a}_1, \ldots, \mathbf{b}_4)$

are useful to describe robot families. Such parameterizations are vector functions $\mathbf{\Phi}^G : \mathbb{R}^{n_D} \to \mathbb{R}^{6m}$ of the form

$$\begin{bmatrix} \mathbf{a}_1 & \ldots & \mathbf{a}_m & \mathbf{b}_1 & \ldots & \mathbf{b}_m \end{bmatrix} = \mathbf{\Phi}^G(\mathbf{g}) \qquad (8.2)$$

that map the design parameters $\mathbf{g}_D = [g_1, \ldots, g_{n_D}]^T$ to the geometric parameters of the robot $\mathbf{a}_1, \ldots, \mathbf{a}_m, \mathbf{b}_1, \ldots, \mathbf{b}_m$. The basic procedure of generating robot designs from design parameters is depicted in Fig. 8.2. In this work, the function $\mathbf{\Phi}^G$ for different robot archetypes is represented by a table such as Table 8.5 that lists the mathematical expression for the position vectors of the proximal $\mathbf{a}_i$ and distal $\mathbf{b}_i$ anchor points. This notation is for the sake of readability, favored, over the notation as equation in the text.

A well-known parameterization of conventional parallel robots is the simplified symmetric mechanism (SSM, Fig. 8.7) [322]. This robot design is originally used for some of the first cable robots such as the RoboCrane [6]. By pre-selecting certain relations amongst the geometric parameters, useful properties of the robot are produced, e.g. a large translational or orientation workspace, a high stiffness, or little influence from cable interference. In Table 8.4, an overview of such archetypical designs is given. A number of archetypes of cable robots is illustrated in Figs. 8.3, 8.4, and 8.5. The shown designs represent typical configurations to arrange the cables. The generic type has eight cables and six degrees-of-freedom with a rectangular mobile platform and machine frame. The translational motion is quite reasonable where rotations are limited. Furthermore, the cables constrained in all directions make collisions with the environment a considerable problem. The suspended robot is a variant with six cables and six degrees-of-freedom.

The presented generic designs are understood to be templates for starting the development of a new robot or to customize the design for specific application requirements. The parametric description allows for model reduction of the design space and thus decreases the number of free design parameters to a set that is expected to generate useful robots. The parameterizations is used either for automatic search methods or for manual design where one searches for a suitable solution within

Table 8.4 Overview of archetypical robot designs

Name	m	n	Description
RoboCrane	6	6	The original NIST design is essentially an inverted simplified symmetric manipulator (SSM) with six cables
Falcon	7	6	Design by Kawamura with a flat robot frame and an orthogonal strut platform for picker applications
Segesta 7	7	6	The original Segesta robot design; a six degrees-of-freedom robot with seven cables and a planar mobile platform
IPAnema 1	7	6	The original IPAnema design with a box frame and a planar platform
IPAnema 1	8	6	Box frame with planar mobile platform where all anchor points are on the corners of the box
IPAnema 1.5	8	6	Trapezoidal frame with crossed cables and a strut-shaped platform
IPAnema 2	8	6	Trapezoidal box frame with crossed cables and trapezoidal mobile platform for handling
IPAnema 3	8	6	Box-shaped platform and trapezoidal platform for general purpose; used both for handling and haptic interaction
CoGiRo	8	6	Suspended but redundant platform for handling
CableSimulator	8	6	Box frame with icosahedra platform for improved orientation capabilities
IPAnema-Falcon	8	6	Long strut platform with crossed cables for high stiffness and orientation capacities for tilting
Endless Z9/Z12	9–12	6	Superior orientation capacities with 9 to 12 cables where one axis can be operated with unlimited rotation
French-German	12	6	A fully antagonistic design where two cables are mainly used to actuate one degree-of-freedom of end-effector motion with large orientation capacities
Segesta 9	9	6	Box frame with planar platform
Segesta 12	12	6	Experimental setup with counterweights and movable proximal anchor points
CabLev	3	6	Suspended robot with three cables and movable proximal anchor points

a meaningful class of robots. In the following section, some design templates are described in detail and typical properties are discussed.

8.4.2   Generic Redundantly-Constrained Robot Design with Eight Cables

In the following, a number of common parameterizations of the robot frames are presented. A widely used structure for the fixed machine frame is a rectangular box where the winches are located at or close to the eight corners. As a rule of thumb,

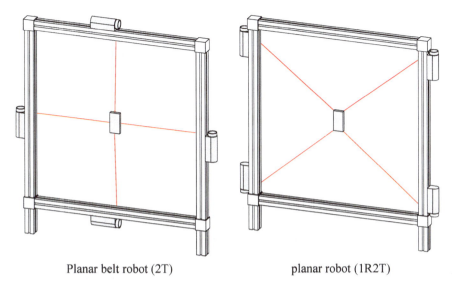

Planar belt robot (2T)   planar robot (1R2T)

Fig. 8.3 Archetypical designs of planar cable robots

the numbering scheme for the winches is clockwise and from top to bottom. Since this definition fixes the layout of the machine frame, robots with somewhat crossing cables result from choosing an appropriate geometry of the mobile platform. The box is parameterized by its length l_B, width w_B, and height h_B (Fig. 8.6). The coordinates of the anchor points are given in Table 8.5. The reference coordinate frame is in the center of the lower side of the box. The same parameterization is used for the mobile platform with the parameters length l_P, width w_P, and height h_P.

The generic design realizes a 3R3T motion pattern which can be operated both in fully-constrained and suspended configuration. If all parameters are nonzero, the robot is in 8-8 configuration. However, if one or more parameters are zero, other configurations are possible such as 8-4 (3R3T), 8-2 (2R3T), and 8-1 (3T). Setting the frame parameters to zero allows for 4-8 (3R3T), 2-8 (2R3T) and 1-8 (rather exotic). Mixing zero parameters amongst platform and base allows for unusual configurations where 4-4 (3R3T) is reasonable and some of the remaining settings are architecturally singular. In terms of wrench-feasibility, the generic robot design provides reasonable translational workspace while the size of the orientation workspace is small. Due to the small orientation workspace, cable-cable interference is mostly not the relevant limitation. Constraining the mobile platform from all sides gives good stiffness properties where collisions with other objects in the machine frame are considerable limitations. The design suffers from singularities in the center of the workspace if the geometry of the mobile platform is congruent with the robot frame.

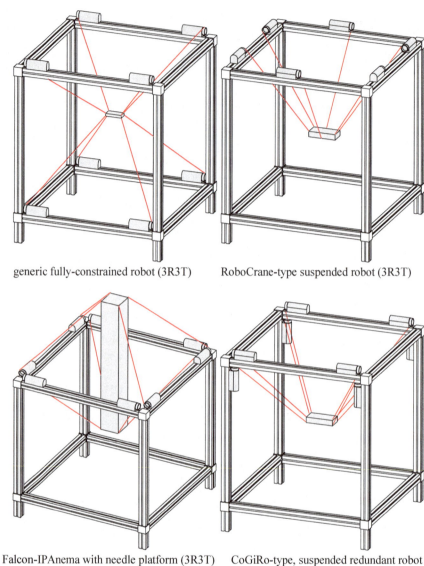

generic fully-constrained robot (3R3T)   RoboCrane-type suspended robot (3R3T)

Falcon-IPAnema with needle platform (3R3T)   CoGiRo-type, suspended redundant robot
                                                (3R3T)

Fig. 8.4 Archetypical designs of spatial cable robots

Amongst other robots, the IPAnema 1 [389] robot implements this generic design
where parameters are set to $\mathbf{g} = [4.0, 3.0, 2.0, 0.12, 0.12, 0.0]^{\mathrm{T}}$. As the height of the
mobile platform is $h_{\mathrm{P}} = 0$, one receives an 8-4 configuration.

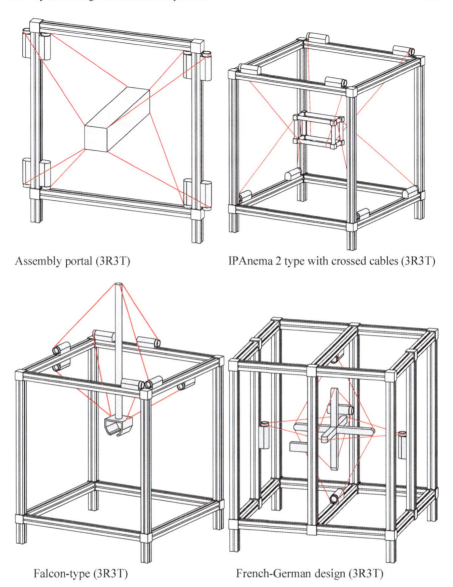

Assembly portal (3R3T)   IPAnema 2 type with crossed cables (3R3T)

Falcon-type (3R3T)   French-German design (3R3T)

Fig. 8.5 Archetypical designs of spatial cable robots

8.4.3 RoboCrane

The first cable robot demonstrator was the NIST RoboCrane design [6]. This robot system combines aspects of a multi cable crane as it can be found in many applications, such as container cranes, where each of its six cables has an independent actuation. Conventional container cranes have a special geometry with congruent

Fig. 8.6 Geometry of the
generic box frame
parameterized with length,
width, and height

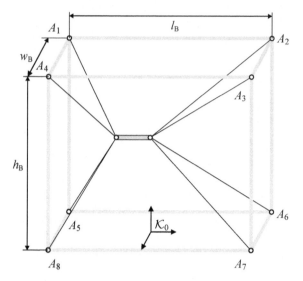

Table 8.5 Coordinates of the proximal and distal anchor points $\mathbf{a}_i$ and $\mathbf{b}_i$ with eight cables and a box-shaped geometry defined by the six parameters $\mathbf{g} = [l_B, w_B, h_B, l_P, w_P, h_P]^T$

Cable i	Base vector $\mathbf{a}_i$			Platform vector $\mathbf{b}_i$		
	x	y	z	x	y	z
1	$-l_B/2$	$w_B/2$	h_B	$-l_P/2$	$w_P/2$	h_P
2	$l_B/2$	$w_B/2$	h_B	$l_P/2$	$w_P/2$	h_P
3	$l_B/2$	$-w_B/2$	h_B	$l_P/2$	$-w_P/2$	h_P
4	$-l_B/2$	$-w_B/2$	h_B	$-l_P/2$	$-w_P/2$	h_P
5	$-l_B/2$	$w_B/2$	0	$-l_P/2$	$w_P/2$	0
6	$l_B/2$	$w_B/2$	0	$l_P/2$	$w_P/2$	0
7	$l_B/2$	$-w_B/2$	0	$l_P/2$	$-w_P/2$	0
8	$-l_B/2$	$-w_B/2$	0	$-l_P/2$	$-w_P/2$	0

machine frame and mobile platform and thus a number of their cables is geometrically parallel. From a kinematic point of view, such cranes are operated in a singular configuration and the swaying motion of the load can be identified with the self-motion or over-mobility of the kinematic chains. To overcome this problem, the RoboCrane design uses six actuated cables to move the mobile platform and adopts the geometry of the simplified symmetric manipulator (SSM) [322], see Fig. 8.7. In order to keep all cables under tension, the robot is operated in a *suspended* configuration which inverts the original SSM configuration. The SSM design is defined by four geometric parameters $\mathbf{g} = [r_B, \alpha_B, r_P, \alpha_P]^T$. The geometry of the proximal anchor points is chosen so that all points A_i are distributed on a circle with radius r_B in the xy-plane of the base frame $\mathcal{K}_0$. Then, the circle is divided into three and the anchor

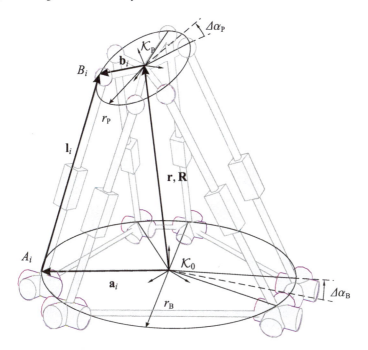

Fig. 8.7 Parameterization of the simplified symmetric mechanism (SSM)

Table 8.6 Parameterization of the SSM robot for proximal and distal anchor points and the four geometric parameters $\mathbf{g} = [r_B, \alpha_B, r_P, \alpha_P]^T$	Cable i	Base vector $\mathbf{a}_i$	Platform vector $\mathbf{b}_i$
	1	$r_B \mathbf{R}_Z(\Delta\alpha_B)\mathbf{e}_x$	$r_P \mathbf{R}_Z(\Delta\alpha_P)\mathbf{e}_x$
	2	$r_B \mathbf{R}_Z(120° - \Delta\alpha_B)\mathbf{e}_x$	$r_P \mathbf{R}_Z(120° - \Delta\alpha_P)\mathbf{e}_x$
	3	$r_B \mathbf{R}_Z(120° + \Delta\alpha_B)\mathbf{e}_x$	$r_P \mathbf{R}_Z(120° + \Delta\alpha_P)\mathbf{e}_x$
	4	$r_B \mathbf{R}_Z(240° - \Delta\alpha_B)\mathbf{e}_x$	$r_P \mathbf{R}_Z(240° - \Delta\alpha_P)\mathbf{e}_x$
	5	$r_B \mathbf{R}_Z(240° + \Delta\alpha_B)\mathbf{e}_x$	$r_P \mathbf{R}_Z(240° + \Delta\alpha_P)\mathbf{e}_x$
	6	$r_B \mathbf{R}_Z(-\Delta\alpha_B)\mathbf{e}_x$	$r_P \mathbf{R}_Z(-\Delta\alpha_P)\mathbf{e}_x$

points are displaced pairwise by angle $\pm\Delta\alpha_B$ (see Fig. 8.7). The same pattern is used for the mobile platform where the radius r_P and the angle $\Delta\alpha_P$ are used, respectively. This arrangement leads to the parameterization given in Table 8.6.

The robot design generates the motion pattern 3R3T for typical parameter sets. However, setting the platform radius r_P to zero turns the robot into pattern 3T. The RoboCrane design has a simple and well understood geometry for parallel robots and many properties such as the number of solutions of the forward kinematics and the singular surfaces have been widely studied (see [322] for an overview). The workspace of such suspended robots is relatively small and the available wrench set is limited due to the suspended configuration. The robot design is hardly limited by cable interference. Typical applications are handling and lifting tasks.

Table 8.7 Parameterization of the Falcon robot from Kawamura based on six geometric parameters $\mathbf{g} = [l_B, w_B, h_B, l_P, w_P, h_P]^T$

Cable i	Base vector $\mathbf{a}_i$	Platform vector $\mathbf{b}_i$
1	$[-l_B/2, w_B/2, h_B]^T$	$[0, 0, h_P]^T$
2	$[l_B/2, 0, h_B]^T$	$[0, 0, h_P]^T$
3	$[-l_B/2, -w_B/2, h_B]^T$	$[0, 0, h_P]^T$
4	$[-l_B/2, w_B/2, 0]^T$	$[-l_P/2, w_P/2, 0]^T$
5	$[l_B/2, w_B/2, 0]^T$	$[l_P/2, w_P/2, 0]^T$
6	$[l_B/2, -w_B/2, 0]^T$	$[l_P/2, -w_P/2, 0]^T$
7	$[-l_B/2, -w_B/2, 0]^T$	$[-l_P/2, -w_P/2, 0]^T$

8.4.4 Falcon

The Falcon design is proposed by Kawamura [237] where the robot frame is a planar rectangle and the mobile platform consists of a slender strut orthogonal to the plane of the robot frame. Thus, the cables are connected to the ends of the strut which largely reduces the effect of cable-cable and cable-platform collisions. Furthermore, the cables are directed away from the operation space which is usually located below the frame. The robot is proposed for fast pick-and-place applications. The geometry of the robot has some similarities with the well-known Delta robot design [98] although the Falcon design allows for six degrees-of-freedom. Especially the direction of the cables leaving the lower tip of the strut-shaped platform allow for little collisions between the robot and the handled goods if a gripper is mounted at the lower end.

The parameterization shown in Table 8.7 is deduced from the Falcon prototype presented in the reference paper [237]. The robot is operated in a fully-constrained configuration and the cable connection is in 7-5 configuration generating a 3R3T motion pattern. If the lateral platform parameters l_P and w_P are set to zero, the robot degenerates to a 2R3T design in 7-2 configuration. It is worthwhile to mention that the nominal parameters of the Falcon prototype generate a *flat* robot frame with zero height in z-axis and all proximal anchor points A_i lie within a common plane.

The volume of the workspace of the Falcon is notably large, however, it has triangular cross sections as it is typical for robots with seven cables. For long and slender platforms, large changes in the orientation are possible for rotations about the x- and y-axis. In contrast, rotation about the z-axis of a slender platform is hardly possible. Also, the stiffness is poor against torques about the z-axis.

8.4.5 Segesta 7

The original Segesta design (here called Segesta 7 because of its seven cables) follows the rule to connect multiple cables to one anchor point on the mobile platform in order to reduce the restriction caused by cable-cable interference [210]. The robot frame is defined by a box where the winches are located at the corners or edges

Table 8.8 Parameterization of the Segesta 7 robot based on five geometric parameters $\mathbf{g} = [l_B, w_B, h_B, l_P, w_P]^T$

Cable i	Base vector $\mathbf{a}_i$	Platform vector $\mathbf{b}_i$
1	$[-l_B/2, w_B/2, h_B]^T$	$[-l_P/2, w_P/2, 0]^T$
2	$[l_B/2, w_B/2, h_B]^T$	$[l_P/2, w_P/2, 0]^T$
3	$[0, -w_B/2, h_B]^T$	$[0, -w_P/2, 0]^T$
4	$[-l_B/2, w_B/2, 0]^T$	$[-l_P/2, w_P/2, 0]^T$
5	$[l_B/2, w_B/2, 0]^T$	$[l_P/2, w_P/2, 0]^T$
6	$[l_B/2, -w_B/2, 0]^T$	$[0, -w_P/2, 0]^T$
7	$[-l_B/2, -w_B/2, 0]^T$	$[0, -w_P/2, 0]^T$

which simplifies mechanical construction of the robot frame. The distal anchor points form a planar triangle. The robot has six degrees-of-freedom and allows for some rotations of the mobile platform. Since the seven winches cannot fully surround the entire volume of the box-shaped frame, the workspace of the robot is relatively small compared to the volume of the box and workspace cross sections in the xy-plane have a form similar to triangles. In order to reduce collisions amongst the cables, multiple cables are connected in a 7-3 configuration to common distal anchor points on the mobile platform. The robot is a CRPM design and is operated preferably in a fully-constrained configuration. The parameterization of the mobile platform and machine frame is given in Table 8.8 where the five parameter description is taken from Fang [139].

8.4.6 IPAnema 1.5

A simple modification of the box-shaped machine frame is the trapezoidal frame layout. This structure is used, amongst others, for the IPAnema 1.5 robot. The robot has a box-shaped frame where the proximal anchor points A_i are pairwise shifted along the horizontal edges of the frame. This kind of modification results in a somewhat regular shape while breaking congruency between mobile platform and machine frame. As a rule of thumb, this seems to be a good approach to avoid singular configurations in the center of the machine frame. The primary dimensions are length l_B, width w_B, and height h_B. The actual locations of the anchor points are then modified by offsets Δl_B, Δw_B, and Δh_B that shift the anchor points along the respective edges where positive values shift the winches towards the center of the box. These Δ-parameters are normally chosen to be small related to the respective length, width, and height of the robot frame. There are in total six different permutations how to distribute these offsets (see Table 8.9). In the top-bottom case, all anchor points are distributed on two parallel planes with a distance of h_B. In the right-left case, the parallel planes have a distance of l_B along the x-coordinate, and in the front-back case, the distance is w_B. Each of the configurations allows for larger rotations in some directions but restricts the orientation workspace in a perpendicular direction.

Table 8.9 Six variations of the parameterization of coordinates of the proximal anchor points $\mathbf{a}_i$ of an eight-cable modified box-shaped machine frame defined by the parameters $\mathbf{g} = [l_B, w_B, h_B, \Delta l_B, \Delta w_B]^T$, $\mathbf{g} = [l_B, w_B, h_B, \Delta w_B, \Delta h_B]^T$, and $\mathbf{g} = [l_B, w_B, h_B, \Delta l_B, \Delta w_B]^T$, respectively

Cable	Top-bottom Base vector $\mathbf{a}_i$			Bottom-top Base vector $\mathbf{a}_i$		
i	x	y	z	x	y	z
1	$-l_B/2 + \Delta l_B$	$w_B/2$	h_B	$-l_B/2$	$w_B/2 - \Delta w_B$	h_B
2	$l_B/2 - \Delta l_B$	$w_B/2$	h_B	$l_B/2$	$w_B/2 - \Delta w_B$	h_B
3	$l_B/2 - \Delta l_B$	$-w_B/2$	h_B	$l_B/2$	$-w_B/2 + \Delta w_B$	h_B
4	$-l_B/2 + \Delta l_B$	$-w_B/2$	h_B	$-l_B/2$	$-w_B/2 + \Delta w_B$	h_B
5	$-l_B/2$	$w_B/2 - \Delta w_B$	0	$-l_B/2 + \Delta l_B$	$w_B/2$	0
6	$l_B/2$	$w_B/2 - \Delta w_B$	0	$l_B/2 - \Delta l_B$	$w_B/2$	0
7	$l_B/2$	$-w_B/2 + \Delta w_B$	0	$l_B/2 - \Delta l_B$	$-w_B/2$	0
8	$-l_B/2$	$-w_B/2 + \Delta w_B$	0	$-l_B/2 + \Delta l_B$	$-w_B/2$	0

i	Right-left Base vector $\mathbf{a}_i$			Left-right Base vector $\mathbf{a}_i$		
	x	y	z	x	y	z
1	$-l_B/2$	$w_B/2 - \Delta w_B$	h_B	$-l_B/2$	$w_B/2$	$h_B - \Delta h_B$
2	$l_B/2$	$w_B/2$	$h_B - \Delta h_B$	$l_B/2$	$w_B/2 - \Delta w_B$	h_B
3	$l_B/2$	$-w_B/2$	$h_B - \Delta h_B$	$l_B/2$	$-w_B/2 + \Delta w_B$	h_B
4	$-l_B/2$	$-w_B/2 + \Delta w_B$	h_B	$-l_B/2$	$-w_B/2$	$h_B - \Delta h_B$
5	$-l_B/2$	$w_B/2 - \Delta w_B$	0	$-l_B/2$	$w_B/2$	Δh_B
6	$l_B/2$	$w_B/2$	Δh_B	$l_B/2$	$w_B/2 - \Delta w_B$	0
7	$l_B/2$	$-w_B/2$	Δh_B	$l_B/2$	$-w_B/2 + \Delta w_B$	0
8	$-l_B/2$	$-w_B/2 + \Delta w_B$	0	$-l_B/2$	$-w_B/2$	Δh_B

i	Back-front Base vector $\mathbf{a}_i$			Front-back Base vector $\mathbf{a}_i$		
	x	y	z	x	y	z
1	$-l_B/2 + \Delta l_B$	$w_B/2$	h_B	$-l_B/2$	$w_B/2 - \Delta w_B$	h_B
2	$l_B/2 - \Delta l_B$	$w_B/2$	h_B	$l_B/2$	$w_B/2 - \Delta w_B$	h_B
3	$l_B/2 - \Delta l_B$	$-w_B/2$	h_B	$l_B/2$	$-w_B/2 + \Delta w_B$	h_B
4	$-l_B/2 + \Delta l_B$	$-w_B/2$	h_B	$-l_B/2$	$-w_B/2 + \Delta w_B$	h_B
5	$-l_B/2$	$w_B/2 - \Delta w_B$	0	$-l_B/2 + \Delta l_B$	$w_B/2$	0
6	$l_B/2$	$w_B/2 - \Delta w_B$	0	$l_B/2 - \Delta l_B$	$w_B/2$	0
7	$l_B/2$	$-w_B/2 + \Delta w_B$	0	$l_B/2 - \Delta l_B$	$-w_B/2$	0
8	$-l_B/2$	$-w_B/2 + \Delta w_B$	0	$-l_B/2 + \Delta l_B$	$-w_B/2$	0

When connected to the box-shaped platform, the cables can be crossed to enlarge the workspace and to improve the stiffness of the robot. Cable-cable collisions in the translational workspace must be avoided by appropriate choice of parameters.

Table 8.10 Parameterization of the IPAnema 2 robot based on nine geometric parameters $\mathbf{g} = [l_B, w_B, h_B, h_{B0},$ $l_P, w_P, h_P, \Delta l_P, \Delta w_P]^T$

Cable i	Base vector $\mathbf{a}_i$	Platform vector $\mathbf{b}_i$
1	$[-l_B/2, w_B/2, h_B]^T$	$[-l_P/2 + \Delta l_P, w_P/2, 0]^T$
2	$[l_B/2, w_B/2, h_B]^T$	$[l_P/2 - \Delta l_P, w_P/2, 0]^T$
3	$[l_B/2, -w_B/2, h_B]^T$	$[l_P/2 - \Delta l_P, -w_P/2, 0]^T$
4	$[-l_B/2, -w_B/2, h_B]^T$	$[-l_P/2 + \Delta l_P, -w_P/2, 0]^T$
5	$[-l_B/2, w_B/2, h_{B0}]^T$	$[-l_P/2, w_P/2 - \Delta w_P, h_P]^T$
6	$[l_B/2, w_B/2, h_{B0}]^T$	$[l_P/2, w_P/2 - \Delta w_P, h_P]^T$
7	$[l_B/2, -w_B/2, h_{B0}]^T$	$[l_P/2, -w_P/2 + \Delta w_P, h_P]^T$
8	$[-l_B/2, -w_B/2, h_{B0}]^T$	$[-l_P/2, -w_P/2 + \Delta w_P, h_P]^T$

8.4.7 IPAnema 2

The trapezoidal shape of the robot frame used for the IPAnema 1.5 frame allows for crossed cable configuration that increases the size of the orientation workspace and improves stiffness. For the IPAnema 2 parameterization, the effect is maintained but generated by appropriate geometry on the mobile platform. The IPAnema 2 parameterization (Table 8.10) has eight cables for fully-constrained operation. Due to the usage of a trapezoidal platform, the robot is in 8-8 configuration exact for canceling all Δ-parameter and at least one of the three platform parameters l_P, w_P, or h_P. The 8-8 configuration requires a general purpose kinematic transformation as described in Sect. 4.3.4 and cannot exploit geometrical assumptions for the kinematic transformation. However, having only distinct points on the base A_i and on the mobile platform B_i allows to use standard mechanical construction elements for the guiding pulleys and for the connection on the platform. The frame of the robot is a simple box with the proximal anchor points at its corners. As a matter of convenience, an additional parameter for the heights of the four lower anchor points is introduced which can be neglected when appropriate requirements for the workspace are given.

The robot has a large wrench-feasible translational workspace which is free of cable-cable collision for appropriate geometric parameters. The parameterization allows for considerable rotation about the x- or y-axis depending on the length l_P and width w_P of the mobile platform. The rotation about the z-axis is compromised by the increase in the translational workspace and the improved rotational stiffness.

The crossed configuration of the cables reduces the collision between the cables and objects under the robot. Thus, handling and assembly are typical applications for the IPAnema 2 design.

Fig. 8.8 Layout and workspace of the IPAnema 3 robot

8.4.8 IPAnema 3

The IPAnema 3 parameterization (Fig. 8.8) introduces additional degrees-of-freedom in the robot design compared to the IPAnema 2 design described above by introducing two additional parameters (Table 8.11). Thus, the robot is a RRPM design with eight cables allowing for 3R3T motion. A trapezoidal shape is used for both the machine frame and the mobile platform allowing for more sophisticated compromises between translational and orientation workspace. Thus, one has a set of five parameters to form the mobile platform and the machine frame. Except for rather specific parameter settings, the robot is in 8-8 configuration with crossed cables. The geometry is well suitable for using simple machine elements in the implementation and proper dimensioning allows for a reasonably large workspace without cable-cable interference.

For moderately large values of the Δ-parameters Δl_B, Δw_B, Δl_P, and Δw_P, one can use a rectangular framework to build the mobile platform and machine frame, respectively. Different parameter variants of the IPAnema 3 geometry are used for experimental investigation of applications such as handling, assembly, and haptic interaction.

8.4.9 CoGiRo

The CoGiRo design is a suspended but redundantly restrained robot design with eight cables [276]. The mobile platform and base are both box-shaped (Fig. 8.4). CoGiRo introduces cross cables but with a scheme that differs from the one described for the IPAnema robots. The first four proximal anchor points A_1–A_4 are connected to the upper four distal anchor points but the assignment is permuted as if the robot was rotated by 90° clockwise. In turn, the proximal anchor points A_5–A_8 are connected to the lower distal anchor points where, again, the assignment is permuted counter-clockwise. Thus, the moment about the z-axis induced by the first group is

Table 8.11 Parameterization of the IPAnema 3 robot based on eleven geometric parameters $\mathbf{g} = [l_B, w_B, h_B, \Delta l_B,$ $\Delta w_B, h_{B0}, l_P, w_P, h_P,$ $\Delta l_P, \Delta w_P]^T$

Cable i	Base vector $\mathbf{a}_i$	Platform vector $\mathbf{b}_i$
1	$[-l_B/2 + \Delta l_B, w_B/2, h_B]^T$	$[-l_P/2 + \Delta l_P, w_P/2, 0]^T$
2	$[l_B/2 - \Delta l_B, w_B/2, h_B]^T$	$[l_P/2 - \Delta l_P, w_P/2, 0]^T$
3	$[l_B/2 - \Delta l_B, -w_B/2, h_B]^T$	$[l_P/2 - \Delta l_P, -w_P/2, 0]^T$
4	$[-l_B/2 + \Delta l_B, -w_B/2, h_B]^T$	$[-l_P/2 + \Delta l_P, -w_P/2, 0]^T$
5	$[-l_B/2, w_B/2 - \Delta w_B, h_{B0}]^T$	$[-l_P/2, w_P/2 - \Delta w_P, h_P]^T$
6	$[l_B/2, w_B/2 - \Delta w_B, h_{B0}]^T$	$[l_P/2, w_P/2 - \Delta w_P, h_P]^T$
7	$[l_B/2, -w_B/2 + \Delta w_B, h_{B0}]^T$	$[l_P/2, -w_P/2 + \Delta w_P, h_P]^T$
8	$[-l_B/2, -w_B/2 + \Delta w_B, h_{B0}]^T$	$[-l_P/2, -w_P/2 + \Delta w_P, h_P]^T$

counterbalanced by the second group. A similar archetype is also proposed by Aref [15]. This layout allows at certain positions for a large orientation workspace and, in experimental tests, objects are rotated by more than 90° about the z-axis. The special arrangement of the cables allows the mobile platform to travel rather close to the boundary of the supporting machine frame giving the robot a very good ratio between installation space and wrench-feasible workspace. Although the robot is suspended, a considerable structural stiffness can be achieved. The parameterization for the CoGiRo robot is given in Table 8.12.

It must be noted that some algorithms for force distribution and forward kinematics that are efficient for fully-constrained cable robots require modifications to be applied to redundantly-constrained but suspended cable robots such as CoGiRo. Thus, the choice of the geometry has some impact on the tools which are applicable for analysis and design.

8.4.10 Cable Simulator

For the design of a driving and flight simulator, a new mechanical construction for the mobile platform is proposed in order to create a stiff but light-weight framework for the mobile platform [343]. While many other cable robots are designed to interact with their environment, the motion simulator focuses on creating the desired motion and a large installation space is desired inside of the mobile platform for the passenger and instrumentation. The proposed icosahedral geometry of the mobile platform

Table 8.12 Parameterization of the CoGiRo suspended robot based on seven geometric parameters $\mathbf{g} = [l_B, w_B, h_B, h_{B0}, l_P, w_P, h_P]^T$

Cable i	Base vector $\mathbf{a}_i$	Platform vector $\mathbf{b}_i$
1	$[-l_B/2, w_B/2, h_{B0} + h_B]^T$	$[l_P/2, w_P/2, h_P]^T$
2	$[l_B/2, w_B/2, h_{B0} + h_B]^T$	$[l_P/2, -w_P/2, h_P]^T$
3	$[l_B/2, -w_B/2, h_{B0} + h_B]^T$	$[-l_P/2, -w_P/2, h_P]^T$
4	$[-l_B/2, -w_B/2, h_{B0} + h_B]^T$	$[-l_P/2, w_P/2, h_P]^T$
5	$[-l_B/2, w_B/2, h_{B0}]^T$	$[-l_P/2, -w_P/2, 0]^T$
6	$[l_B/2, w_B/2, h_{B0}]^T$	$[-l_P/2, w_P/2, 0]^T$
7	$[l_B/2, -w_B/2, h_{B0}]^T$	$[l_P/2, w_P/2, 0]^T$
8	$[-l_B/2, -w_B/2, h_{B0}]^T$	$[l_P/2, -w_P/2, 0]^T$

Fig. 8.9 Top view of the parameterization of the mobile platform for the Cable Simulator design with an icosahedron platform in the xy-plane

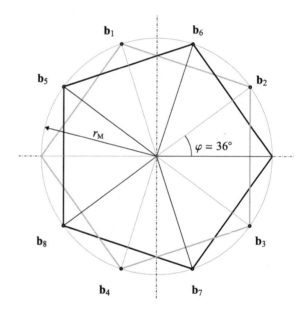

allows for a stiff exo-structure without obstacles inside while positioning the distal anchor points with little collisions. The cables are connected in an 8-8 configuration (Fig. 8.9).

In the nominal design of the robot, the upper winches are connected to the lower vertices of the mobile platform and the lower winches to the upper vertices of the mobile platform. Again, this crossed cable configuration increases the stiffness of the robot as well as its orientation capabilities. The robot has an asymmetric workspace due to the arrangement of the distal anchor points on the robot frame. However, the orientation workspace is comparably large.

The parametric model of the robot is given in Table 8.13 where the two parameters

Cable i	Base vector $\mathbf{a}_i$	Platform vector $\mathbf{b}_i$
1	$[-l_{\mathrm{B}}/2, w_{\mathrm{B}}/2, h_{\mathrm{B}}]^{\mathrm{T}}$	$\mathbf{R}_{\mathrm{Z}}(108°)\,[r_{\mathrm{M}}, 0, -h]^{\mathrm{T}}$
2	$[l_{\mathrm{B}}/2, w_{\mathrm{B}}/2, h_{\mathrm{B}}]^{\mathrm{T}}$	$\mathbf{R}_{\mathrm{Z}}(36°)\,[r_{\mathrm{M}}, 0, -h]^{\mathrm{T}}$
3	$[l_{\mathrm{B}}/2, -w_{\mathrm{B}}/2, h_{\mathrm{B}}]^{\mathrm{T}}$	$\mathbf{R}_{\mathrm{Z}}(-36°)\,[r_{\mathrm{M}}, 0, -h]^{\mathrm{T}}$
4	$[-l_{\mathrm{B}}/2, -w_{\mathrm{B}}/2, h_{\mathrm{B}}]^{\mathrm{T}}$	$\mathbf{R}_{\mathrm{Z}}(-108°)\,[r_{\mathrm{M}}, 0, -h]^{\mathrm{T}}$
5	$[-l_{\mathrm{B}}/2, w_{\mathrm{B}}/2, 0]^{\mathrm{T}}$	$\mathbf{R}_{\mathrm{Z}}(144°)\,[r_{\mathrm{M}}, 0, h]^{\mathrm{T}}$
6	$[l_{\mathrm{B}}/2, w_{\mathrm{B}}/2, 0]^{\mathrm{T}}$	$\mathbf{R}_{\mathrm{Z}}(72°)\,[r_{\mathrm{M}}, 0, h]^{\mathrm{T}}$
7	$[l_{\mathrm{B}}/2, -w_{\mathrm{B}}/2, 0]^{\mathrm{T}}$	$\mathbf{R}_{\mathrm{Z}}(-72°)\,[r_{\mathrm{M}}, 0, h]^{\mathrm{T}}$
8	$[-l_{\mathrm{B}}/2, -w_{\mathrm{B}}/2, 0]^{\mathrm{T}}$	$\mathbf{R}_{\mathrm{Z}}(-144°)\,[r_{\mathrm{M}}, 0, h]^{\mathrm{T}}$

Table 8.13 Parameterization of the Cable Simulator design with eight cables and an icosahedron platform based on four geometric parameters $\mathbf{g} = [l_{\mathrm{B}}, w_{\mathrm{B}}, h_{\mathrm{B}}, r_{\mathrm{P}}]^{\mathrm{T}}$

$$h = \frac{\cos\frac{\pi}{5}}{1 + \cos\frac{\pi}{5}} r_{\mathrm{P}} \quad \text{and} \tag{8.3}$$

$$r_{\mathrm{M}} = \frac{2\sqrt{5}}{5} r_{\mathrm{P}} \tag{8.4}$$

are used to parameterize the vertices of the icosahedron. r_{P} is the radius of the circumscribed sphere. The frame of the robot is simply described by a rectangular box by length, width, and height, respectively. Thus, the design is fully defined by only four geometric parameters.

8.4.11 IPAnema-Falcon

The IPAnema-Falcon design is a synthesis of the ideas of crossing cables, a flat frame with zero height, and a layout taking the benefits from eight cables. A slender strut is employed as mobile platform and the cables are arranged in a crossed configuration. Compared to the Falcon system, the robot is designed to operate in an 8-8 fully-constrained configuration which simplified the mechanical design with distinct anchor points both on the frame and on the mobile platform while maintaining the advantages of the Falcon robot with its compact frame. Using eight cables, the workspace of the robot can be shaped to have almost rectangular horizontal cross sections that cover a notably larger ratio of the footprint of the robot frame when compared with the conventional Falcon.

The geometric parameters of the IPAnema-Falcon design are given in Table 8.14. All proximal anchor points are fixed in a common xy-plane where two Δ-parameter are used to displace the proximal anchor points A_i along the edges. On the mobile platform, the cables are connected to the corners of a box.

The design provides high stiffness and a large orientation capacity for tilting thanks to its crossed cables. At the same time, a good translational workspace is maintained. The concept of the robot is depicted in Fig. 8.10. Note that the conceptual drawing

Table 8.14 Parameterization of the IPAnema-Falcon design with eight cables and a strut platform based on seven geometric parameters $\mathbf{g} = [l_B, w_B, \Delta l_B, \Delta w_B, l_P, w_P, h_P]^T$

Cable i	Base vector $\mathbf{a}_i$	Platform vector $\mathbf{b}_i$
1	$[-l_B/2 + \Delta l_B, w_B/2, 0]^T$	$[-l_P/2, w_P/2, 0]^T$
2	$[l_B/2 - \Delta l_B, w_B/2, 0]^T$	$[l_P/2, w_P/2, 0]^T$
3	$[l_B/2 - \Delta l_B, -w_B/2, 0]^T$	$[l_P/2, -w_P/2, 0]^T$
4	$[-l_B/2 + \Delta l_B, -w_B/2, 0]^T$	$[-l_P/2, -w_P/2, 0]^T$
5	$[-l_B/2, w_B/2 - \Delta w_B, 0]^T$	$[-l_P/2, w_P/2, h_P]^T$
6	$[l_B/2, w_B/2 - \Delta w_B, 0]^T$	$[l_P/2, w_P/2, h_P]^T$
7	$[l_B/2, -w_B/2 + \Delta w_B, 0]^T$	$[l_P/2, -w_P/2, h_P]^T$
8	$[-l_B/2, -w_B/2 + \Delta w_B, 0]^T$	$[-l_P/2, -w_P/2, h_P]^T$

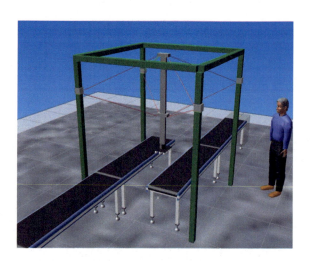

Fig. 8.10 Concept for a pick-and-place installation with a Falcon like robot

in the figure shows a cable robot with a 2×2 m footprint. Due to the efficient force transmission of cable robots, this robot can be used for picking larger and also heavier objects than a conventional picker based on the Delta robot. In the draft, a closed robot frame is mounted over two conveyers for pick-and-place operations. The design combines a high payload, large workspace, and high velocities with collision-free motion for conveyer-to-conveyer handling.

8.4.12 French-German

A design with twelve cables to generate a 3R3T robot is proposed where pairs of antagonistic cables are used to actuate one degree-of-freedom of the end-effector. The idea is developed by Verhoeven and Lafourcade [473] and is called *French-German* design due to the inventors' nationalities. The parameterization of the robot based on six geometric parameters is given in Table 8.15. The geometry of the mobile platform is based on a star-shaped structure with a bar in three orthogonal directions. The distal cable ends are located at the ends of the arms with the length l_P, width w_P, height h_P, respectively. The robot is enclosed by a rectangular frame with length l_B, width w_B, and height h_B, where the proximal anchor points are each located on the center of the surfaces of the box (Fig. 8.5). Variants of the robot can be derived by permuting the assignment between the surfaces of the proximal box with the bars on the mobile platform.

The robot is a fully-constrained highly redundant RRPM in 6-6 configuration. The design follows the idea of having an antagonistic pair with two cables for each degree-of-freedom of the platform motion. However, the actuation is not decoupled amongst the pairs of cables. With the high number of cables and a mobile platform of reasonable size, high stiffness can be achieved. The design aims at maximizing the orientation workspace. In turn, the translational workspace is limited. Using coinciding anchor points both on the platform and the base, the problem of cable-cable interference is reduced at the cost of the additional efforts to realize this special property in the mechanical design. In turn, the mobile platform is widely surrounded by the cables and actuation units make collisions with external obstacles likely. Therefore, the robot is suitable for applications where no direct contact needs to be made with the surrounding such as motion simulation or sensor testing. To the best of the author's knowledge, no physical prototype of this robot is built yet.

8.4.13 Endless Z9 and Z12

Comparing the properties of cable robots with the requirements of handling and pick-and-place tasks, it becomes apparent that large or even unlimited rotation is required in many applications. Considering the connection of many cables to the mobile platform in a spatial robot, it seems clear from intuition that unlimited rotation is impossible for a cable constrained system. However, one way to achieve this effect is to employ a platform that is similar to a crank shaft. Fixing three cables to each end of the shaft is a generic 2R3T design (see Fig. 2.3). Then, one uses an eccentric connection point on the shaft to control the rotation of the shaft (see Figs. 8.11 and 8.12). In this simple example, three cables share the same distal anchor point at each end of the shaft as well as on the crank. From a kinematic point of view, this yields exactly the desired mobility of the platform. In practice, it is rather difficult to implement a mechanical design with two or more cables sharing a common point on

Table 8.15 Parameterization of the French-German design with twelve cables based on six geometric parameters $\mathbf{g} = [l_B, w_B, h_B, l_P, w_P, h_P]^T$

Cable i	Base vector $\mathbf{a}_i$			Platform vector $\mathbf{b}_i$		
	x	y	z	x	y	z
1	0	0	$h_B/2$	$l_P/2$	0	0
2	0	0	$h_B/2$	$-l_P/2$	0	0
3	0	$w_B/2$	0	0	0	$h_P/2$
4	0	$w_B/2$	0	0	0	$-h_P/2$
5	$l_B/2$	0	0	0	$w_P/2$	0
6	$l_B/2$	0	0	0	$-w_P/2$	0
7	0	$-w_B/2$	0	0	0	$h_P/2$
8	0	$-w_B/2$	0	0	0	$-h_P/2$
9	$-l_B/2$	0	0	0	$w_P/2$	0
10	$-l_B/2$	0	0	0	$-w_P/2$	0
11	0	0	$-h_B/2$	$l_P/2$	0	0
12	0	0	$-h_B/2$	$-l_P/2$	0	0

the platform. There are two possibilities to overcome this practical problem. Firstly, one distributes the anchor points along the axis of the shaft without losing the property of infinite rotation. Secondly, one connects the cables that should share the anchor point to a common ring, e.g. the outer ring of a ball bearing. The latter solution allows for an elaborated mechanical design. Following this idea, the concept is extended as follows: A standard design for a 3R3T robot is used and a shaft is inserted into the mobile platform. By fixing three cables at an eccentric point on that shaft, one can infinitely rotate the shaft relative to the platform only by pulling on the upper three cables (Fig. 8.13). The parameterization for nine and twelve cables is given Table 8.16.

The idea is proposed by Pott and Miermeister [346, 402, 403] and can be even more generalized by allowing for additional serial degrees-of-freedom on the platform [346]. As shown in the paper, each additional degree-of-freedom requires at least one extra cable. However, one might need to add more cables for a degree-of-freedom for a given geometry of the platform for a cable robot.

Strictly speaking, the robot implementing the latter approach is no longer fully parallel but hybrid since we have introduced a serial chain on the platform. This has no practical implications and the advantage of the parallel robot such as being light-weight with actuators on the fixed base are maintained.

A sample of this concept with unlimited rotation about its z-axis in exemplified in the following. The design of the robot is based on a simple 9-3 design with a triangular frame structure and a planar mobile platform. The geometry data of the sample robot is given in Table 8.17.

In order to show the surprising possibility of unlimited rotation, a proof-of-concept example is presented below. We consider a trajectory that includes a full rotation

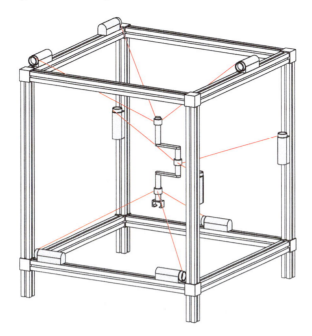

Fig. 8.11 Concept of a cable robot with unlimited rotation about its z-axis in the form of crank shaft

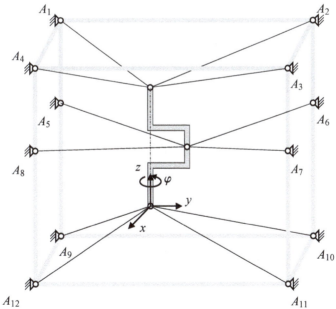

Fig. 8.12 The endless Z12 robot architecture with $m = 12$ cables and a platform with a crank

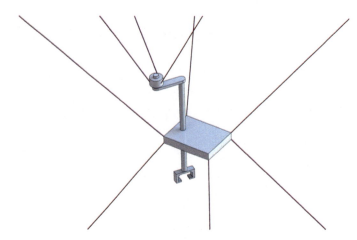

Fig. 8.13 Extension of a conventional 3R3T robot design with an eccentric shaft to allow for an unlimited rotation

Table 8.16 Geometry data for the endless Z9 and Z12 robot design for the base $\mathbf{a}_i$ and platform $\mathbf{b}_i$ anchor points

Robot	Cable i	Base vector $\mathbf{a}_i$			Platform vector $\mathbf{b}_i$		
		x	y	z	x	y	z
Endless Z9	1	$-\frac{r_B}{2}$	$\frac{\sqrt{3}}{2}r_B$	H_B	r_P	0	H_P
	2	r_B	0	H_B	r_P	0	H_P
	3	$-\frac{r_B}{2}$	$-\frac{\sqrt{3}}{2}r_B$	H_B	r_P	0	H_P
	4	$-\frac{r_B}{2}$	$\frac{\sqrt{3}}{2}r_B$	h_B	0	0	h_P
	5	r_B	0	h_B	0	0	h_P
	6	$-\frac{r_B}{2}$	$-\frac{\sqrt{3}}{2}r_B$	h_B	0	0	h_P
	7	$-\frac{r_B}{2}$	$\frac{\sqrt{3}}{2}r_B$	0	0	0	0
	8	r_B	0	0	0	0	0
	9	$-\frac{r_B}{2}$	$-\frac{\sqrt{3}}{2}r_B$	0	0	0	0
Endless Z12	1	$-r_B$	r_B	H_B	r_P	0	H_P
	2	r_B	r_B	H_B	r_P	0	H_P
	3	r_B	$-r_B$	H_B	r_P	0	H_P
	4	$-r_B$	$-r_B$	H_B	r_P	0	H_P
	5	$-r_B$	r_B	h_B	0	0	h_P
	6	r_B	r_B	h_B	0	0	h_P
	7	r_B	$-r_B$	h_B	0	0	h_P
	8	$-r_B$	$-r_B$	h_B	0	0	h_P
	9	$-r_B$	r_B	0	0	0	0
	10	r_B	r_B	0	0	0	0
	11	r_B	$-r_B$	0	0	0	0
	12	$-r_B$	$-r_B$	0	0	0	0

Table 8.17 Geometry data for the base $\mathbf{a}_i$ and platform $\mathbf{b}_i$ anchor points for the Endless Z9 robot

Cable i	Base vector $\mathbf{a}_i$ (m)			Platform vector $\mathbf{b}_i$ (m)		
	x	y	z	x	y	z
1	-1	$\sqrt{3}$	3	0.3	0	0.5
2	2	0	3	0.3	0	0.5
3	-1	$-\sqrt{3}$	3	0.3	0	0.5
4	-1	$\sqrt{3}$	2	0	0	0.4
5	2	0	2	0	0	0.4
6	-1	$-\sqrt{3}$	2	0	0	0.4
7	-1	$\sqrt{3}$	0	0	0	0
8	2	0	0	0	0	0
9	-1	$-\sqrt{3}$	0	0	0	0

about the z-axis of the mobile platform. Let $\mathbf{r} = [0, 0.5, 1.5]^{\mathrm{T}}$ be the position of the platform. Furthermore, the orientation of the platform $\mathbf{R}$ is chosen to be the elementary rotation matrix $\mathbf{R}_z(\varphi)$. The force distributions are computed using the closed-form method (Sect. 3.7.5). In Fig. 8.14, the computation results for the cable forces for all nine cables are shown for $\varphi \in [0; 2\pi]$. It can be seen that the values of all nine forces are continuous and between the force bounds $f_{\min} = 1$ and $f_{\max} = 10$. Therefore, the platform is capable of performing a full rotation about its z-axis under wrench-feasibility with optional infinite repetitions. The example proves that it is possible to design a cable robot with unlimited rotation capacities for at least one of its axes. Although it cannot be seen from the diagram, the cables do not intersect at any time.

According to [473], two cables can only intersect at one point. Since three sets of each three cables share a common anchor point, these cables cannot intersect during the motion. For the motion example in Fig. 8.14, the three sets of cables are moving in three separate layers and the cables maintain sufficient distance.

The infinite rotation capability is possible at different positions within the robot frame. To show this property of the robot design, the total orientation workspace $\mathcal{W}_{\text{TO}}$ of the robot is computed where the orientation set

$$\mathcal{R} = \{\mathbf{R} \in SO_3 \mid \mathbf{R} = \mathbf{R}_z(\varphi), \quad \varphi \in [0; 2\pi]\} \tag{8.5}$$

is used. The Dykstra method (see Sect. 3.7.4) is used for workspace testing and the workspace shown in Fig. 8.15 is determined. The volume of the workspace is $1.32\,\mathrm{m}^3$ and the shape is compact. Note that the ability for unlimited rotation about the platform's axis is maintained even if this axis is slightly tilted.

Using the cable-cable interference method described in Sect. 5.2.6, the cable-cable interference is studied. For this analysis, the following geometric parameters are used for the endless Z12 design: $r_{\mathrm{B}} = 2$, $r_{\mathrm{P}} = 0.3$, $H_{\mathrm{B}} = 3$, $h_{\mathrm{B}} = 2$, $H_{\mathrm{P}} = 0.5$, and $h_{\mathrm{P}} = 0.2$. Note that the z-coordinates of the points $\mathbf{b}_5$–$\mathbf{b}_8$ are smaller in order to

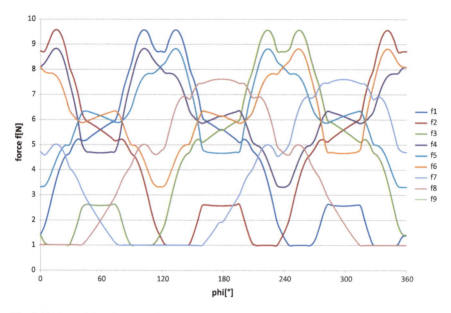

Fig. 8.14 Run of the nine cable forces f_i along the z rotation with angle φ

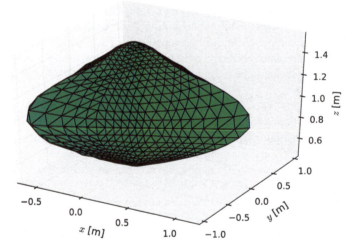

Fig. 8.15 Total orientation workspace $\mathcal{W}_{\text{TO}}$ for $\mathcal{R} \in \mathbf{R}_{\text{z}}(\varphi)$ with $\varphi \in [0; 2\pi]$ for the sample with unlimited z-rotation capability

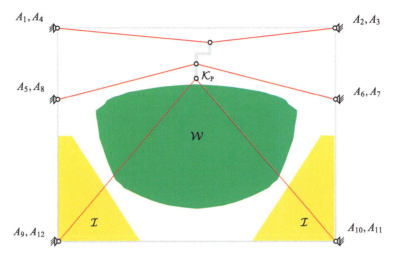

Fig. 8.16 Lateral view in the xz-plane of the region of cable-cable interference and total orientation workspace $\mathcal{W}$ of the endless Z12 robot and the regions of cable-cable interference $\mathcal{I}$

avoid collisions between the cable groups 1–4 and 5–8. It can be seen from Fig. 8.16 that the total orientation workspace and the region of interference are separated and cable-cable interference are avoided throughout the workspace.

8.5 Parameter Synthesis

After selecting an appropriate motion pattern and a respective cable robot parameterization, one needs to define the actual geometry for the robot according to the application requirements. *Parameter synthesis* is understood as the procedure to determine the geometric parameters of the cable robot from given application requirements. The overall design procedure consists of a number of phases. Firstly, all requirements and performance wishes for the desired robot system are determined (Sect. 8.3). Using the methods described throughout this book, one needs a procedure to assess whether a candidate design qualitatively possesses the desired property and, even better, to quantify the property. The desired motion pattern is chosen in *structural synthesis* and thereby the degrees-of-freedom n are defined for the robot. For the desired motion pattern and the required number of cables m, one chooses a parameterization that introduces a mapping depending on the design parameters $\mathbf{g}$ as function $(\mathbf{a}_i, \mathbf{b}_i) = \mathbf{\Phi}^{G}(\mathbf{g})$, which are introduced in Sect. 8.4.1. The main aspects in parameter synthesis are:

- Select performance criteria for the robot. The tools introduced in Chaps. 3–5 provide a variety of methods to measure all kinds of performances of the robot and to assess these performance measures against the application requirements.

Fig. 8.17 The main steps in
the parameter design
procedure

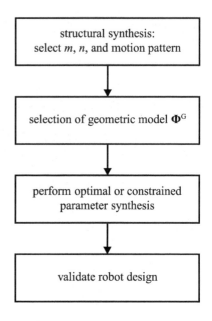

- Use a search strategy, e.g. engineer's expert knowledge, optimization, constraint satisfaction, or constrained global optimization, to find one or more parameter vectors **g** with a sufficient geometry for the cable robot.
- Validate the design **g** found by the step above. Since the analysis criteria used in parameter synthesis are compromised for computation time, a thorough analysis should be executed for the identified robot design. Furthermore, the validation procedure may include dynamic simulation (Chap. 6) and preliminary controller design which is neglected in the geometrical parameter synthesis.

After determining and validating a proper robot geometry, the design procedure continues to choose and select technical parameters as prerequisites for the mechanical design procedure. This includes especially drive-train design. For a well-designed drive-train, it is usually required to consider application requirements in depth [263]. When all technical parameters are defined, the mechanical design is approached which typically uses CAD software to prepare the parts of the robot for manufacturing (Fig. 8.17).

In the following, two main approaches are considered for the parameter synthesis: Optimal design and constraint satisfaction.

8.5.1 Parameter Synthesis as Optimal Design Problem

Kinematic properties of robots largely depend on the geometry of the robot and additionally vary notably throughout the workspace. This is especially true for both conventional and cable-driven parallel robots. Appropriate choice of the geometry is an important task [318] that is addressed by geometrical parameter synthesis.

For parameter synthesis, two different kinds of requirements occur. Firstly, some criteria have to be fulfilled unconditionally. Such criteria are called *compulsory* or *imperative* and most of these criteria arise from technical limitations. Limitations such as avoidance of singularities and interference are understood to be compulsory. Secondly, some criteria are not mandatory but *desirable* to minimize.[1] Such criteria can be quantitatively represented through an objective function or cost function. Frequently used examples are robot properties such as size of the workspace, stiffness, accuracy, dynamic capabilities, cycle time, weight, installation size, and sensitivity to errors. Nonmechanical examples for the desirable criteria are the minimization of energy consumption or simply the minimization of monetary costs.

A wide-spread approach for the design of parallel robots is the so-called *optimal design* (see e.g. [375, 387]). In this procedure, one parameterizes a class of robots through design variables $\mathbf{g}$ so that different values for $\mathbf{g}$ correspond to *variants* of the robot. The parameter models of the archetypes described in Sect. 8.4 are examples of such models and the focus in the following lies on using such models. In order to compare different variants of the same class, performance criteria k_i are defined that characterize the quality of the robot with respect to certain properties. Typical examples of such criteria are amongst others:

- Volume of the workspace,
- Stiffness of the robot,
- Dexterity indices,
- Quality of possible force distributions,
- Available wrench set,
- Accuracy,
- Energy efficiency,
- Closeness to cable-cable interference and cable-platform interference,
- Closeness to singularities.

Each property is represented by a performance index.[2] Before using a specific performance index, one must carefully consider if higher values of the index actually correspond to a robot that is better in the context of the design task. If, and only if, this assumption holds true in general, optimal design leads the way to the best robot.

One can optimize a single criterion or a combination of some criteria. Since the performance indices have different metrics by nature, one has to apply weighting factors or normalization in order to balance the influence of the single factors where usually an application-specific balancing is needed. In practice, the use of linear combination is wide-spread

---

[1]Note that in optimization minimizing or maximization of f is complementary since simply evaluating $-f$ instead of f achieves the complementary effect.

[2]There might be more that one index for measuring a property. The question what is the best index to measure dexterity of a robot is rather complicated and an intensively discussed issue in robotics. We refrain from this discussion and focus on the design procedure instead.

$$f(\mathbf{g}) = \sum_i \eta_i k_i(\mathbf{g}) \qquad (8.6)$$

where k_i are the robot design-specific performance criteria and η_i are the respective weighting factors. Some important criteria such as the existence of singularities within the workspace are characterized by Boolean values *true* or *false* which are mapped to numerical values of 1 (true) and 0 (false). Using sufficiently high weighting factors η_i, one can try to enforce the compliance with such requirements. In general, such definitions do not guarantee that compulsory requirements are met.

   This approach has some significant drawbacks. Firstly, the binary criteria make the function discontinuous. Therefore, gradient-based optimization algorithms lose efficiency in finding the optimum. Secondly, it is possible to find optimal solutions that violate compulsory requirements. Thus, a design being optimal in the sense of the objective function may be not understood to be optimal by the engineer since it has obvious drawbacks, e.g. collision in the center of the workspace. Especially when many performance criteria are involved, it is rather difficult (or impossible) to choose meaningful weighting factors η_i so that the optimum of the objective function Eq. (8.6) relates to the sought robot design. Disregarding any problem related to successfully solving the optimization problem, one has to note that every choice of the weights η_i implicates another optimal design. Hence, one implicitly predefines the optimal robot by setting the values for η_i where a priori it is not clear how the values η_i affect the result of the optimal design procedure. Weighting the objective functions is basically an act where one prioritizes the antagonistic criteria. Thus, multi-criteria optimal design is about making a compromise while concealing the rationale behind the compromise.

8.5.1.1   Local and Global Performance Indices

Many performance indices k_i depend on the robot pose $\mathbf{y} = (\mathbf{r}, \mathbf{R})$ within the workspace and can vary largely for different poses. For cable robots, typical indices are coupled to the pose-dependent structure matrix $\mathbf{A}^\mathrm{T}$ such as size of the available wrench set, manipulability, singular values of the structure matrix, and stiffness. In order to receive a global index that measures the quality of the robot design through-out the workspace $\mathcal{W}$, the integral over the workspace is computed from

$$\bar{k} = \frac{1}{V(\mathcal{W})} \int_{\mathcal{W}} k(\mathbf{y}) \, \mathrm{d}\mathbf{y} \qquad (8.7)$$

where $V(\mathcal{W})$ is the volume of the workspace, as proposed in e.g. [15]. If a dexterity measure is used in the equations above, it is called *global dexterity index* (GDI) [172]. Thus, one receives the mean value $\bar{k}$ of the performance index k over the workspace. This mean value is well-defined but even for simple indices k_i it is virtually impossible to symbolically compute the integral from Eq. (8.7). A practicable way to estimate the mean value $\bar{k}$ as global index is to use a discrete set of N sample poses $\mathbf{y}_i$ by e.g. using a regular or random grid for the poses to evaluate. Thus, the integral is approximated by the sum

$$\tilde{k} = \frac{1}{N} \sum_{i=1}^{N} k(\mathbf{y}_i) \quad . \tag{8.8}$$

Having an acceptable performance on average may not be sufficient in the design. Depending on the used index and the application, it might be acceptable to improve manipulability or stiffness on average. It is up to the designer to assess whether unbounded local deficits in a certain index are acceptable for the target application or not. Considering the standard deviation as additional statistical indicator of the homogeneous distribution of the property can mitigate the local loss in performances evaluation. However, this complicates the comparison of the performances since there is no unique way to compare two robot designs that are both characterized by mean and standard deviation, respectively.

A conservative approach to overcome the problem is to take the smallest occurring value of the performance index k within the workspace as global index leading to the global index

$$\check{k} = \min_{\mathbf{y} \in \mathcal{W}} k(\mathbf{y}) \quad . \tag{8.9}$$

This kind of approach must be used if a deficit in the performance index is unfeasible. Typical examples are dexterity indices dropping to zero in a kinematic singularity or cable-cable distances dropping to zero showing a collision.[3] Although such deficits are local, the robot design featuring such defects is unacceptable. A rigorous determination of the lower bound $\check{k}$ is possible using e.g. interval analysis or minimization. However, scanning through a regular grid of poses $\mathbf{y}_i$ gives a quick estimate for the index $\check{k}$ as follows

$$\check{k} \approx \min_{1 \leq i \leq N} k(\mathbf{y}_i) \quad . \tag{8.10}$$

Although the application of such indices based on the minimum value is mathematically feasible, some problems arise for numerical algorithms employed for optimal design since performance indices based on the minimum function are not differentiable and may even have regions of constant value spanning a certain area in the parameter space. Such properties render optimization algorithms based on gradients inefficient. As discussed in Sect. 8.5.2, constrained global optimization using interval analysis can be employed to deal with design problems having rigorous requirements for the lower bound of performance parameters.

---

[3]Therefore, it is advised to check for collisions and singularities with global methods or through rigorous tests as discussed in Sect. 5.4.

Table 8.18 Case study for optimal design of a planar robot with the geometric parameters $\mathbf{g} = [l_{\mathrm{B}}, w_{\mathrm{B}}, l_{\mathrm{P}}, w_{\mathrm{P}}]^{\mathrm{T}}$

Cable i	Base vector $\mathbf{a}_i$	Platform vector $\mathbf{b}_i$
1	$[-l_{\mathrm{B}}/2, w_{\mathrm{B}}/2]^{\mathrm{T}}$	$[-l_{\mathrm{P}}/2, w_{\mathrm{P}}/2]^{\mathrm{T}}$
2	$[l_{\mathrm{B}}/2, w_{\mathrm{B}}/2]^{\mathrm{T}}$	$[l_{\mathrm{P}}/2, w_{\mathrm{P}}/2]^{\mathrm{T}}$
3	$[l_{\mathrm{B}}/2, -w_{\mathrm{B}}/2]^{\mathrm{T}}$	$[l_{\mathrm{P}}/2, -w_{\mathrm{P}}/2]^{\mathrm{T}}$
4	$[-l_{\mathrm{B}}/2, -w_{\mathrm{B}}/2]^{\mathrm{T}}$	$[-l_{\mathrm{P}}/2, -w_{\mathrm{P}}/2]^{\mathrm{T}}$

8.5.1.2 Case Study for a Planar Robot

A simple case study of optimal design is presented in the following. The design of a planar cable robot with rectangular base and platform is exemplified. The generating function $\mathbf{\Phi}^G$ is given by Table 8.18. Defined by the task to carry instrumentation, the size of the mobile platform is predefined to be of length $l_{\mathrm{P}} = 0.4\,\mathrm{m}$ and width $w_{\mathrm{P}} = 0.2\,\mathrm{m}$. The desired workspace $\mathcal{W}^{\mathrm{R}}$ is square with edge length of 1×1 m centered around the origin. The remaining design parameters are the length and the width of the robot frame yielding the sought design parameter vector $\mathbf{g} = [l_{\mathrm{B}}, w_{\mathrm{B}}]^{\mathrm{T}}$. The cable forces are predefined to be in the range of $f \in [1; 10]\,\mathrm{N}$ and we are using a simple wrench-feasibility test (closed-form solution) for workspace assessment.

The aim of this example is to minimize the installation space of the robot, thus, the simple objective function

$$f(\mathbf{g}) = (l_{\mathrm{B}} \, w_{\mathrm{B}})p + p \qquad (8.11)$$

is employed which is the area of the machine frame where p is a penalty factor. In order to enforce the desired workspace, a binary penalty p is multiplied with

$$p = \begin{cases} 1 & \text{if the desired workspace } \mathcal{W}^{\mathrm{R}} \text{ is wrench-feasible} \\ 10 & \text{otherwise} \end{cases} \qquad (8.12)$$

Having defined this, one can evaluate a function $f(\mathbf{g})$ for different robot geometries and receive a value to compare the variants of the robot. The required workspace evaluation is done by simply checking a regular grid of 121 poses within the desired workspace $\mathcal{W}$.

Employing a simple Nelder-Mead (downhill simplex method) optimization procedure on the objective function with the initial value $\mathbf{g} = [3.0, 2.0]^{\mathrm{T}}\,\mathrm{m}$ yields the optimal robot design shown in Fig. 8.18. In this figure, the final robot design with the required workspace is depicted on the left. Aside, the parameter run over 120 iterations is plotted where one can obverse a smooth convergency towards the final values after a turbulent initial phase. The computation time of this simple optimization procedure is around 80 ms. The result of the procedure is quite reasonable.

In such a simple example, an exhaustive search provides the exact environment of the objective function in the design space around the optimum value (see Fig. 8.19). On the left lower side, one can see the *feasible* designs without penalty and on the

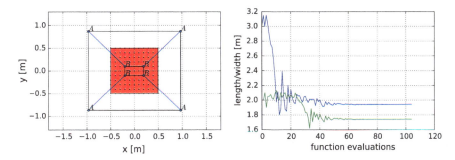

Fig. 8.18 Simple example of optimal design optimizing the installation space of the robot for a given workspace

Fig. 8.19 Map of the objective functions used for the optimal design example. The discontinuity generated by the penalty factor p can be clearly identified

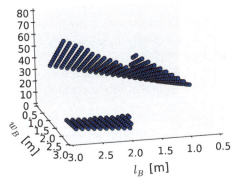

right upper side of the diagram the designs that fail in the workspace computation are visible. Therefore, the optimal design is located exactly at the transition between active and inactive penalty. Although mathematically sound, the discontinuity in the objective function makes the numerical optimization rather involved if larger parameter vectors are used in complicated cases. In the next section, a slightly varied design problem is discussed that leads to a counter-intuitive result.

8.5.1.3 Unexpected Results from Optimal Design of Planar Robots

As discussed above, the volume of the robot frame or the robot platform seem to be reasonable candidates for the objective function, i.e. in order to find the largest workspace for the smallest frame. In contrast to the design example above, it is now allowed for the function Φ^G (see Table 8.18) to change both the shape of the platform and the shape of the frame. Thus all four parameters $\mathbf{g} = [l_B, w_B, l_P, w_P]^T$ may be altered. Instead of minimizing the installation space, one asks to maximize the workspace for a given volume of the installation space. Setting up a similar procedure as above leads to a surprising result. One can generate an infinitely large

Fig. 8.20 Zero volume
robot with arbitrarily large
workspace

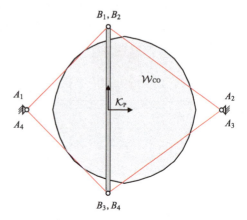

workspace with an installation space of zero and rendering the idea of optimal design pointless.

For planar and also spatial robots, one can find special designs where the volume of the convex hull of all $\mathbf{a}_i$ as well as the volume of all $\mathbf{b}_i$ is exactly zero and the robot has an arbitrarily large workspace (Fig. 8.20). Since both the y-coordinate of all $\mathbf{a}_i$ and x-coordinate of all $\mathbf{b}_i$ is zero, neither platform nor frame enclose a finite volume. However, the crossed configuration allows the platform to generate a significant workspace. Moreover, the size of the robot still depends on the nonzero length of the frame if the width w_B is zero while the objective function does not. The same result can also be received in the spatial case, where a Falcon-like design with a planar frame and a vertical but flat platform can be employed to have a finite workspace volume for a cable robot of which the volume of the convex hull of the platform and the frame vanishes. This shows that the definition of a meaningful objective function is difficult to achieve in general. Interestingly, optimization algorithms sometimes unveil such configurations to the designer.

Reconsidering the zero volume robot leads to the idea of putting constraints on the convex hull of platform and base. This kind of considerations are analyzed in Sect. 8.5.2 where the concepts *constrained optimization* and *constraint satisfaction problem* for design are discussed.

8.5.1.4   Conclusion from Optimal Design

A proper selection of the geometrical design parameters is crucial for dimensioning the cable robot. Optimal parameter design is an ambivalent tool. Carefully used, it is suitable for tuning the parameters of the robot towards a well-defined design goal. However, global search through optimal design is difficult for two reasons. On the one hand, one has a rather challenging mathematical problem which is time-consuming to set up and also time-consuming to solve globally. On the other hand, the definition of a proper optimization function is a challenge of its own and, to the best of the author's knowledge, there are no formal criteria to a priori proof that the optimum value of the objective function corresponds to a cable robot that has the desired properties. The key to optimal design is a meaningful objective function. Having constructed this function, the optimum robot is exactly defined although this robot is not yet revealed to the designer before the time consuming optimization procedure is executed. Therefore, optimal design approaches shall only be applied if one has unconditional trust in the objective function. The example above with a zero-volume robot illustrates that quite reasonable assumptions can lead to surprising and unfortunately useless results. On the other side, applying optimization to a well-defined aspect for finding the optimal parameters is a powerful tool. At the current stage, the engineer must be kept in the loop to assess the proposals coming for the optimal design before accepting them.

8.5.2 Parameter Synthesis with Interval Analysis

This section deals with algorithms and methods to derive a cable robot geometry for prescribed properties where the workspace is assumed to be given. Prescribing the main performance properties is a change in paradigm compared to the idea of optimal design. The basic approach is based on interval parameter synthesis and is firstly proposed by Hao and Merlet [197].

A procedure for geometrical parameter synthesis based on constraint programming is introduced in the following. The application requirements for the robot to be designed are considered as constraints rather than objective functions. The rationale behind this approach is that many requirements are compulsory and must be fulfilled with a certain performance level where more performance may be useful but shall be compromised in order to fulfill other mandatory requirements. Understanding application requirements as constraints makes it easier to, at the same time, take different requirements into account without the need to define an objective that has to weight the different performance criteria. Thus, the mathematical formulation reflects the underlying design problem in a more natural way. In contrast to the multi-criteria optimization problem, defining the design problem with constraints is straightforward and adding or removing additional requirements has no influence on other requirements. In the multi-criteria approach, one has to balance the objective function whenever a qualitative or quantitative change is made to the requirements. Such a constrained optimization problem for robot design takes the form

$$\text{minimize} \quad f(\mathbf{g}) \tag{8.13}$$
$$\text{subject to} \quad \mathbf{\Phi}(\mathbf{g}, \mathbf{y}) > \mathbf{0} \quad \forall \, \mathbf{y} \in \mathcal{W}^{\text{R}} \ . \tag{8.14}$$

An approach to solve this problem is discussed in the following. Here, the vector $\mathbf{g}$ collects all design variables which typically reflect the geometry of the robot. The vector $\mathbf{y}$ is the pose of the robot and $\mathcal{W}^{\text{R}}$ is a set describing the size and shape of the desired workspace. The function $f : \mathbb{R}^{N_g} \to \mathbb{R}$ is the objective function that shall be minimized. Thus, one searches for the global optimum of the function f under the constraints Φ [40, 195]. This kind of optimization problem is seldom considered since the constraints of the optimization problem are given by a *constraint satisfaction problem* (CSP), i.e. the constraints need to be fulfilled in every point of the workspace. Understanding the compulsive requirements as constraints in the optimization problem realizes the concept of taking the minimum as the performance criteria as argued in Sect. 8.5.1.1.

If the algorithms introduced in Sect. 5.4 are used to compute the solution of the CSP, one can guarantee that the requirements are fulfilled in every point of the workspace. However, one can also compromise thoroughness of the results for a reduction in computational time. In this case, hull computation or grid discretization are used to speed up the verification of the workspace (see Sect. 5.5).

The posed global optimization problem defines precisely which robot designs represented through the vector $\mathbf{g}$ are sought. In practice, it is rather complicated to test if a found vector $\mathbf{g}^*$, which locally optimizes the objective function, is also the desired global optimum of the design problem.

Different methods are applied in global optimization including *heuristic methods*, *approximation methods*, and *systematic methods* [40]. Heuristic methods are, amongst others, *genetic algorithms*, *simulated annealing*, and *Monte-Carlo methods*. These approaches include stochastic elements to support the broad search in a huge parameter space. However, such approaches cannot guarantee to find the global optimum. Furthermore, such methods are designed to deal with conventional constraints rather than having to fulfill a CSP that represents the workspace of the robot.

In the following, a method for global optimization with CSP is discussed that employs interval analysis to deal with the continuous nature of the CSP [197, 394, 399]. This algorithm combines some methods that have been already used in Sect. 5.4 with an interval algorithm for optimization. These methods are successfully used for parameter synthesis of parallel robots [398] as well as for cable robots [68]. Since one employs the CSP approach introduced for workspace analysis, one takes benefit from using the same modeling and implementation both for analysis of the robot and for geometry design. For parameter synthesis, the verification set $\mathcal{X}_v$ is identified with the desired workspace $\mathcal{W}^R$.

8.5.2.1 Parameter Synthesis as Constraint Satisfaction Problem

In the following, the interval algorithm used for workspace computation with the hybrid solver (Sect. 5.4.1.4) is applied to parameter design. This leads to a CSP of the type

$$\Phi(\mathbf{c}, \mathbf{v}) > \mathbf{0} \quad \forall \, \mathbf{v} \in \mathcal{X}_v \qquad (8.15)$$

with the constraints Φ, the calculation variables $\mathbf{c}$, the verification variables $\mathbf{v}$, and the verification set $\mathcal{X}_v$. In order to perform parameter synthesis, the geometric parameters $\mathbf{g}$ of the robot are associated with the calculation variables $\mathbf{c}$ and the world coordinates

Fig. 8.21 Example of a simple parameter relation for the length of the machine frame l_B and the length of the mobile platform l_P. One can easily include many of such *hints* from the human designer on the level of the geometry of the cable robot to eliminate roughly 50% of the parameter space without a time consuming workspace evaluation

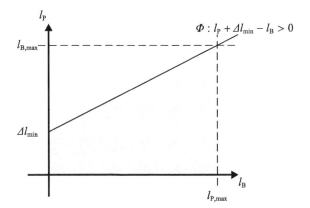

y are identified with the verification variable $\mathbf{v}$. Thus, the desired workspace $\mathcal{W}^R$ is given through the verification set $\mathcal{X}_v$. Hence, this CSP represents the task to find the set $\mathcal{X}_s$ where each $\mathbf{g} \in \mathcal{X}_s$ represents a cable robot with the desired properties, including the size and shape of the workspace $\mathcal{W}^R$, and with all performance criteria $\mathbf{\Phi}$ fulfilled in every point $\mathbf{y}$ of the workspace. One uses the same criteria as constraints that are employed in workspace computation introduced in Sect. 5.4. It is a remarkable advantage of the CSP approach that one uses a unique model for both analysis and synthesis. This holistic approach can even be used to compute valid ranges for the technical parameters of the robot as shown for parallel robots [398]. Using the interval solver introduced there, one can guarantee that the desired performance criteria are fulfilled in every point of the workspace $\mathcal{W}^R$ and that every robot in the solution set $\mathcal{X}_s$ has the desired properties (Fig. 8.21).

Constraints for Parameter Synthesis

The discussion so far is focused on employing local performance criteria in the design procedure in order to enforce the quality. These criteria originate from the analysis of the robot and connect the quality of the robot to the robot pose $\mathbf{y}$ and the robot geometry $\mathbf{g}$. Based on the parameterizations $\mathbf{\Phi}^G$ introduced in Sect. 8.4, one adds relations amongst the geometric parameters in order to speed up the computation or to exclude certain configurations, e.g. because of considerations from mechanical engineering, of the components of the robot. The simplest type of constraints are restrictions for a single parameter such as putting lower or upper limits on the parameter value. Such restrictions are practically implemented by setting an appropriate initial search space $\mathcal{X}_c$ for the parameter. Depending on the application, one is interested in considering only mobile platforms where the platform is smaller than the machine frame. Using the simple box parameterization from Sect. 8.4.2, one receives the trivial additional constraints

$$\Phi_{\text{len}} : l_{\text{B}} - l_{\text{P}} > 0 \tag{8.16}$$

$$\Phi_{\text{width}} : w_{\text{B}} - w_{\text{P}} > 0 \tag{8.17}$$

$$\Phi_{\text{high}} : h_{\text{B}} - h_{\text{P}} > 0 \tag{8.18}$$

where the parameters l_{B}, l_{P}, w_{B}, w_{P}, h_{B}, and h_{P} are the length, width, and height of the platform and base, respectively. Since the CSP solver can efficiently evaluate arbitrarily many of such constraints Φ_i, one discards a huge number of design candidates through the constraints above without performing the time consuming consideration of the workspace. Technically speaking, the hybrid CSP solver does not need to evaluate the underlying algorithm Verify that in effect consumes the largest amount of the computational time. One can consider such design parameter constraints as handy yet efficient tool for the engineer to give *hints* to the automatic solver. Typical design constraints include relations in the size of the mobile platform and the machine frame, minimal distance between anchor points if one does not want the anchor points to coincide or minimum/maximum volume constraints for the mobile platform and machine frame. Also, heuristic knowledge on parameter relations can be used to restrict the search space of the solver. However, care must be taken when restricting the parameter space so possible solutions to the design problem are not excluded. As it can be seen from the Falcon design (Sect. 8.4.4), having a platform with a larger height than the height of the machine frame allows for interesting robot properties.

Such design hints are trivial to construct but can reduce the efficient search space significantly. Each of the three hints given above reduce the search space by around 50%. Furthermore, the hints can be efficiently evaluated by the solver since typical hints have a simple mathematical structure which can be quickly exploited in consistency tests and evaluation of the constraints' derivatives.

8.5.2.2 Global Optimization

Extending the procedure above, acceptable designs are found by solving the respective CSP (see Sect. 5.4.1.4). Thus, one receives a set $\mathcal{X}_s$ of candidates that fulfill all requirements for the application. Now, the question arises how to choose the optimal robot from the solution set. For this, an objective function f is defined that is based on a measure to be unconditionally optimized. Examples of such criteria to be optimized are the overall installation space or, even more meaningful, the monetary costs of the robot. In the following, the issue is addressed how the optimal value of the objective function can be found using interval analysis.

Free Global Optimization

Before discussing the constrained global optimization, we start with the free global optimization without constraints to show the principle. Searching for a minimum of a function aims at locating a position $\mathbf{g}^*$ where the evaluations of the functions

$f(\mathbf{g}^*)$ is smaller than for any other value of $\mathbf{g}$ close by. Using conventional function evaluation, one can numerically test the function value for one specific value $\mathbf{g}$. The main problem in finding the global optimum of a function is that there exist infinite points in the search space $\mathcal{X}_c$ where one can only perform a finite number of evaluations. Interval analysis prooved to be an effective tool in global optimization since the interval evaluation of a function computes guaranteed bounds for a set with infinite points. This property is the key feature that allows to construct algorithms for global optimization using interval analysis. The basic idea is as follows: Let $\widehat{f}_1 = f(\widehat{c}_1)$ and $\widehat{f}_2 = f(\widehat{c}_2)$ be two evaluations of the function f over disjunct interval domains $\widehat{c}_1, \widehat{c}_2$. By comparison of the two images $\widehat{f}_1$ and $\widehat{f}_2$, the following conclusions for global minimization are drawn:

- $\sup f_1 < \inf f_2$: The interval $\widehat{c}_2$ cannot contain the global minimum.
- $\sup f_2 < \inf f_1$: The interval $\widehat{c}_1$ cannot contain the global minimum.

Based on this simple comparison, a *branch-and-bound* algorithm is constructed that guarantees to enclose the global minimum if it is in the search domain [196]. The algorithm is as follows:

Algorithm 6: `Generic global optimizer`

---

1. Evaluate the objective function $\widehat{h}_i = f(\widehat{c}_i)$ for each interval boxes $\{\widehat{c}_1, \ldots, \widehat{c}_n\}$ of the search space and save the pairs $(\widehat{c}_i, \widehat{h}_i)$ in a list $\mathcal{L}_T$.
2. Create a list $\mathcal{L}_S$ for the solution candidates. Set the guaranteed upper bound to $h^* = \infty$.
3. If the list $\mathcal{L}_T$ is empty, terminate the algorithm.
4. Update h^*, if $h^* > \min_{\mathcal{L}_T}(\sup \widehat{h})$, i.e. there exists a box in $\mathcal{L}_T$ with a smaller supremum.
5. Discard all pairs with $\inf \widehat{h} > h^*$ from the lists $\mathcal{L}_T$ and $\mathcal{L}_S$.
6. Extract the pair $(\widehat{c}, \widehat{h})$ from $\mathcal{L}_T$ which has the smallest infimum $\inf \widehat{h}$.
7. If available, apply *prune* and *bound improvement* operations to the box $\widehat{c}$.
8. If $\operatorname{diam} \widehat{c} < \varepsilon$ and $\operatorname{diam} \widehat{h} < \mu$, i.e. the diameter of all components of the box is larger than the threshold ε and the diameter value of the objective function is smaller than μ, the box is sufficiently small. Save $(\widehat{c}, \widehat{h})$ in the list $\mathcal{L}_S$; go to step (3).
9. Split the box $\widehat{c}$ into m sub-boxes $\{\widehat{c}_1, \ldots, \widehat{c}_m\}$, evaluate the objective function for each sub-box $\widehat{h}_i = f(\widehat{c}_i)$ and store the pairs in the list $\mathcal{L}_T$; go to step (3).

The behavior of the algorithm is governed by two important parameters which have to be chosen a priori. The threshold ε is the lower bound of the size of the boxes in the parameter space and μ is the maximum error in the objective function. When the algorithm is terminated, every pair $(\widehat{c}_i, \widehat{h}_i)$ in the list $\mathcal{L}_S$ fulfills:

$$\text{diam}\ \widehat{\mathbf{c}}_i < \varepsilon \ , \tag{8.19}$$

$$\text{diam}\ \widehat{h}_i < \mu \ , \tag{8.20}$$

$$h_* \ \leq \ \widehat{h}_i \leq h^* \ . \tag{8.21}$$

After termination of the algorithm, a guaranteed lower bound for the value of the objective function is also found. The objective function cannot take values smaller than $h_* = \min_{\mathcal{L}_S}(\inf \widehat{h})$. Summing up, the algorithm has the following remarkable properties:

- The algorithm guarantees to find the global minimum in the search space. If more than one global minimum with equal values exists, all are found.
- While searching for the global minimum, one also finds guaranteed upper and lower bounds for the value of the objective function f.
- One finds guaranteed bounds for the design parameters $\mathbf{c}$ that contain the global minimum.
- The objective function needs not to be continuous differentiable as long as an interval evaluation can be computed.
- The algorithm can be well executed on a parallel computer due to numerous independent evaluations.
- Using interval analysis, the algorithm is robust under numerical round-off errors.

A straightforward combination of parameter synthesis and global optimization is achieved by using just the resulting set of feasible designs as input for the global optimization as described above. Technically speaking, one simply assigns the list $\mathcal{L}_S$ computed with the hybrid CSP-solver as input list $\mathcal{L}_T$ of the Generic global optimizer. However, comparing the two algorithms reveals considerable similarities in their structure. When used for parameter optimization of robots, a specific difference is by far faster evaluation of the objective function f than the tiresome verification of the underlying constrained satisfaction problem. Usually, the difference of the computation time is in the range of three to five orders of magnitude. Thus, for the constrained global minimization, one can intensively test for potential improvements of the objective functions before taking the burden of computing the respective CSP. If h^* cannot be improved, the respective box $\mathbf{c}_i$ can be discarded without paying attention to the tiresome computation of the constraints.

Constrained Global Optimization

Combining the CSP solver with the global optimization leads to the following constrained global minimization problem:

$$\text{minimize} \quad f(\mathbf{c}) \tag{8.22}$$

$$\text{subject to} \quad \mathbf{\Phi}(\mathbf{c}, \mathbf{v}) > \mathbf{0} \quad \forall \mathbf{v} \in \mathcal{X}_v \ . \tag{8.23}$$

The basic structure of the hybrid CSP-solver and the *generic global optimizer* differ in the tests being applied before discarding a box. The CSP solver considers the constraint system $\Phi(\mathbf{c}, \mathbf{v}) > \mathbf{0}$ whereas the optimizer compares the objective function to the best known value h^*. Both criteria must be fulfilled to make the current box a candidate for the global optimum. For the optimization of cable robots, the evaluation of the objective function is much faster than the execution of the algorithm Verify. Therefore, the objective function is prioritized. If the box has no potential to improve the current value of h^*, the box is discarded. Once the improvement of the best value h^* is started, this rejection rule is successively improved as the value h^* gets smaller and allows for discarding more boxes.

Consistency Tests for the Objective Function

So-called filtering techniques are used as heuristics to notably speed-up the computation time in global optimization with interval analysis. Here, only the outline is mentioned where a detailed description can be found in the literature on optimization [40, 196].

When the first feasible solution $\widehat{\mathbf{c}}$ is found for the CSP, one can compute an upper bound for the objective function. This is done by simply evaluating the objective function. The supremum $h^* = \sup f(\widehat{\mathbf{c}})$ is clearly an upper bound for the global optimum. However, this bound can usually be improved simply by evaluating any discrete point $\mathbf{c} \in \widehat{\mathbf{c}}$. Clearly, it holds true that $f(\mathbf{c}) \in f(\widehat{\mathbf{c}})$ and $h'^* = f(\mathbf{c}) \leq h^*$. Since the interval evaluation is usually subject to overestimation, it is likely that picking a single point from the box provides a better bound. An efficient choice for the point $\mathbf{c}$ is the center of the box $\mathbf{c} = \mathrm{mid}\,\widehat{\mathbf{c}}$. Thus, one receives a sharp criterion to discard other boxes in the $\mathcal{L}_{\mathrm{T}}$ list.

Gradient of the Objective Function

Although the objective function does not need to be continuous or differentiable to be applied in the interval optimization algorithm, one can speed-up the computation if it is differentiable. The interval evaluation of the gradient $\widehat{\mathbf{z}} = \nabla f(\widehat{\mathbf{c}})$ contains useful information about monotonicity. If one component of the gradient $\widehat{z}_i$ is strictly positive, i.e. if $\widehat{z}_i > 0$, then the function is monotonically increasing. Therefore, the sought minimum of the objective function over that interval $\widehat{\mathbf{c}}$ must occur on the boundary of that interval. Thus, the box can be contracted to the partially degenerated interval $\widehat{\mathbf{c}}' = [\widehat{c}_1, \ldots, \inf \widehat{c}_i, \ldots, \widehat{c}_n]$ and $h'^* = f(\widehat{\mathbf{c}}')$ is a candidate for improving the global upper bound. Likewise, if one component of the gradient $\widehat{z}_i$ is strictly negative, i.e. if $\widehat{z}_i < 0$, then the function is monotonically decreasing. Again, the minimum must be on the boundary and one contracts the box to $\widehat{\mathbf{c}}' = [\widehat{c}_1, \ldots, \sup \widehat{c}_i, \ldots, \widehat{c}_n]^{\mathrm{T}}$. This reduction is made for every component in the interval evaluation of the gradient. If at least one contraction is made, one recursively evaluates the gradient for the improved box $\widehat{\mathbf{c}}'$.

Objective Function

As discussed above, the requirements for the robot are characterized to be compulsory. The constraint programming approach is employed to reduce the search space with feasible variants of the robots to those designs **g** that fulfill all given compulsory requirements. Within this set, one may now search unconditionally for a robot that fits best to the desired criteria. Now, one employs criteria like minimization of installation space or minimization of economic cost of the robot. While the minimization of the size of the robot is an intuitive assumption that can be elegantly expressed in mathematical formula, the monetary costs are a main driving factor in the product development that is derived from experience and heuristics. Thus, the former objective function can be used easily to employ tests of the global optimization procedure where the latter is more suitable for practical usage.

A pure geometric criterion for the optimization of the robot is the installation space which can be computed from the geometric parameters. As we have seen in Sect. 8.5.1.3, one has to take into account the size of the platform, machine frame, and the winches to determine a meaningful installation space. However, for suspended robots other criteria must be used to measure the size of the robot.

8.5.2.3 Synthesis for Depending Technical Parameters

Using geometrical parameter synthesis and optimal design, one can find a cable robot that fulfills the given application requirements represented by the requirements Φ. However, in this design phase assumptions on technical parameters such as the maximum cable length, available cable forces, and cable deflection angles are used. The problem of technical parameter synthesis tackles the problem how to appropriately choose these parameters. Using the CSP approach, one employs the model constraints in order to execute a synthesis procedure where the workspace $\mathcal{W}^R$ is introduced as verification set $\mathcal{X}_v$ and the parameterization of the pose is connected to the verification variables **v**. The geometrical parameters **g** are understood to be constant and the values determined through the design procedure above are assigned. In the technical parameter synthesis, the technical parameters are identified with the calculation variables **c** and one can determine the range for these parameters that are sufficient to fulfill the application requirements. Since the primary geometry synthesis already proofed that such robots exist, the technical parameter synthesis only addressed the issue of choosing optimal values for the technical parameters. Following this idea, one customizes e.g. the orientation of the coordinate for the guiding pulleys $\mathcal{K}_{A,i}$ which is described by the matrix $\mathbf{R}_{A,i}$ as well as orientation of the distal cable-end connectors given through $\mathbf{R}_{B,i}$. This parameter study yields the valid range for the orientations where one usually picks the one with the smallest deflection angles. The influence of many technical parameters is decoupled and the technical parameter synthesis aims at checking if a machine element can be chosen as cost-efficient as possible.

8.6 Hardware Design

After choosing an appropriate geometry for the cable robot, the mechanical design of
the robot hardware has to be done subsequently. Using the determined dimensions of
the robot as input parameters, notable efforts in mechanical engineering are required
to break down the geometrical specifications to machine parts such as dimensioning
the machine frame, choosing fixture elements, and achieving proper arrangements
of the electric wiring. Most of the work is not specific for cable robots and we refer
to the respective construction methods. In the following, we deal with aspects of
mechanical design that involve components that are specific to cable robots. This
includes the selection of cables including guidance and connection to the platform.
Following, design guidelines for the winches are discussed along with selection
of drive-trains. Finally, selection and integration of sensors for the cable robot is
discussed.

8.6.1 Cables

8.6.1.1 Selection of Cables

When designing the hardware for a cable robot from scratch, the cable is likely to be
the first mechanical element to be selected. A couple of other components depend on
the chosen cable's length, radius, and material. Therefore, the selection of the cable
is crucial for further mechanical design of the cable robot. For the selection of the
cable, one needs to know some key performance parameters such as

- Maximum nominal force f_{max},
- Minimum nominal force f_{min},
- Safety factor of breaking load f_{break},
- Allowed elongation of the cable,
- And number of bending cycles n_B the cable has to resist.

Some figures on the fatigue bending cycles for synthetic fiber cables can be found
in Hearle [204]. The figures are based on bending of the cable around a drum or
pulley[4] with a ratio of $r_D/r_C = 15$ and $r_D/r_C = 25$ between the cable radius r_C and the
drum radius r_D. This ratio is commonly abbreviated by D/d in the literature. High-
performance polyethylene (e.g. Dyneema) has roughly $n_B = 10^6$ bending cycles to
failure at 20% of the breaking load which drops down to around $n_B = 10^4$ cycles at
50% of the breaking load. For aramid, the situation is worse where aramid allows
for some hundred thousand bending cycles at 10% of its breaking loads which drops
below $n_B = 10^4$ cycles at 30% breaking load. For steel, one receives around $n_B = 10^6$
cycles at 10% breaking load which goes down to some twenty thousand cycles at

---

[4]In the literature on cable technology, pulleys are frequently called *sheaves* and instead of cable,
usually the technical term *(wire) rope* is used. The respective test is referred to as *cycling bending
over sheaves* (BOS) test.

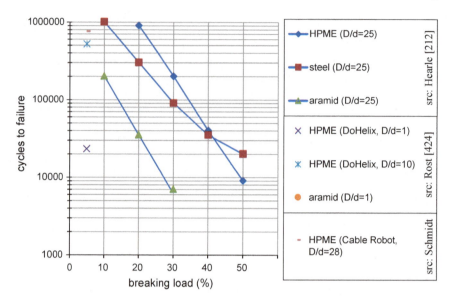

Fig. 8.22 Comparison of the cycles to failure for cable of HPME (Dyneema), aramid, and steel measured for different ratios of $r_D/r_C = D/d$. The figures are compiled from different data sets [204, 426]

50% breaking load. Preliminary tests for Dyneema cables with a diameter of 2.5 mm show a perspective for some million cycles when loading the cable with around 10% breaking load. Thus, safety factors against failure by fatigue of around 10 are typical for long-term installations of cable robots. For Dyneema cables, some $n_B = 5 \cdot 10^5$ cycles are determined at 5.5% breaking load [426] for $r_D/r_C = 10$. Experimental validations based on load cycles within a cable robot are undertaken by Schmidt where Dyneema cables with $r_C = 1.25$ mm and $r_D/r_C = 28$ are used. At a tension level of 4.7% of the cable breaking load, more than $n_B > 7.5 \cdot 10^5$ cycles are observed without failure (Fig. 8.22).

Wehr investigated recently the fatigue of synthetic fiber cables under dynamic loads [488]. Selected results from this experimental study are presented in Fig. 8.23. Note that the latter parameter set is determined for $d/D = 15$ leading to higher wear and shorter lifetime. The experimental tests were performed for relative loads of some 8% and more. Longer lifetimes are conjectured but not experimentally validated. Furthermore, the experiments from Wehr show that synthetic fiber cables are more resistant to dynamic stress. This has to be taken into account for fast cable robots that employ cable velocities of several meter per second and accelerations beyond 10 m/s^2.

The ratio r_D/r_C is generally a very important parameter for assessing the fatigue of cables. For steel cables, one finds reference values [149] which are experimentally generated for applications such as elevators and cranes. In this setting, a ratio of $r_D/r_C = 25$ is considered to be small for steel which leads to considerably large

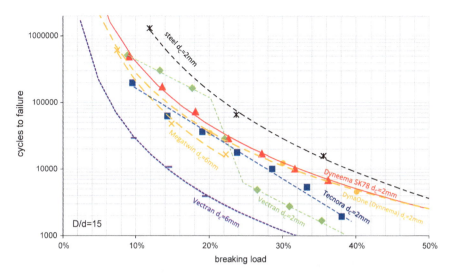

Fig. 8.23 Comparison of the cycles to failure for cables made of HPME (Dyneema), Tecnora, Vectran, and steel with diameter $d_C = \{2, 6\}$ mm. The bending ratio is $r_D/r_C = D/d = 15$. Own figure based on data from [488] (Courtesy of M. Wehr)

diameters for drums and pulleys in civil engineering constructions. The limited data sets are summarized in Fig. 8.22.

The number of feasible bending cycles for the cable may not be confused with the number of motion cycles of the robot. Even for the simplest winch design, one has two bending cycles for a linear motion of the robot. Firstly, when uncoiling the cable from the drum and secondly when redirecting the cable by a pulley.[5] In practice, the number of pulleys can be larger: Winches with spooling guides have an additional pulley. The same holds true if pulleys are used to measure the cable force or if a pulley tackle is used to actuate the cable. Also, additional guidance elements such as double-pulleys (Sect. 8.6.1.2) which improve flexibility and reconfigurability have to be taken into account for the cable selection.

A method for proper selection of cables remains an open issue if a large number of bending cycles shall be achieved without replacing cables. The figures above are preliminary reference values. For large-scale but low dynamic applications, the cables and cable guidance systems can be designed for a reasonable life-time. There is a perspective for the use of synthetic fiber cables for highly dynamic cable robots in long-term usage. To account for fatigue based on the static breaking load, factors of 10–20 are realistic and a ratio of $r_D/r_C \geq 25$ is advised. However, application specific tests for fatigue with the chosen values are mandatory to proof feasibility.

---

[5]The wear of the cable in an eyelet is by far worse since the ratio of diameter is very low for an eyelet generating excessive wear.

8.6.1.2 Proximal Cable Guidance System

Compared to other machines using running cables such as cranes or elevators, cable robots require deflecting a moving cable within a large range of directions. Thus, a machine element at the proximal anchor point A_i is needed that has a kinematically well-defined behavior in fixing the point A_i in space while allowing the cable to move unhindered in its current direction. In the following, the cable guidance system is considered independently from the actuation system since these two components can be almost unconditionally combined. A very simple guidance system consists of an *eyelet* where the cable is guided through [139]. An ideal eyelet realizes an accurate point-shaped outlet for the cable (Fig. 8.24). However, the kinematically ideal eyelet has a sharp edge, thus, it cuts like a knife into the cable causing a lot of friction and abrasion. When using eyelets, the simplified standard cable model can be justified where cables form perfect lines. In practice, one has to make a trade-off between a kinematically perfect eyelet that allows for accurate guidance of the cable and a rounding eyelet that reduces abrasive wear of the cables. Using ceramic eyelets with synthetic fiber cables, the lifetime of the cables is reported in the range of some ten hours of operations. Increasing the radius of the eyelet introduces a more involved kinematic transformation. Using a toroidal shape, the kinematic behavior of a panning pulley is mimicked. In Fig. 8.24, a conceptual CAD of a cable guided through an eyelet is shown. For moderate deflection angles of up to 45°, Fang [139] reported acceptable force loss due to friction in the eyelet where the friction increases notably when the deflection angle increases. Since the eyelet has similar contact conditions as expressed through the Euler-Eytelwein formula, an exponential connection between deflection angle and friction force is expected causing significant losses in the cable force. Depending on the friction on the eyelet, efficiency factors $\eta_E = 0.83 \ldots 0.96$ can be expected for 45° deflection and $\eta_E = 0.77 \ldots 0.95$ for 60° deflection. Almost all energy dissipated this way is fed into the cable causing heavy wear inside the cable.

Some cable guidance systems consist of one or more pulleys to redirect the cable from the actuation system into the workspace. In a typical design, the cable is guided to a panning pulley and its rotational axis is coaxial to the direction of the cable coming from the actuation system. The axis of the pulley is perpendicular to the panning axis and allows guiding the cable into a large variety of directions (see Fig. 8.25). Usage of such guiding pulleys largely reduces the wear of the cable compared to an eyelet but requires an advanced kinematic transformation (see Sect. 7.2.1).

The idea of panning pulleys can be extended to double panning pulley units where a pair of panning pulleys is mounted in one housing to provide omnidirectional feeding angles on the actuation side as well as on the distal side of the pulley unit (Fig. 8.26). In such designs, the panning axis of both pulleys must be aligned. Such units are applied for cable routing if the winches are installed on the floor. Furthermore, double pulley units are efficient machine elements that facilitate reconfiguration of cable robots. Relocating the double pulley unit is relatively easy to do by clamping the unit to different positions on the machine frame. In contrast, the winches are more bulky and reconfiguration of the winches requires also changes in the electric wiring

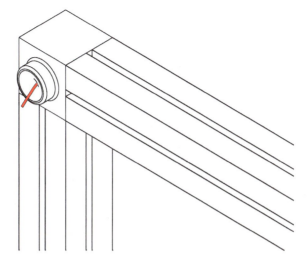

Fig. 8.24 Guiding the cable through an eyelet allows for an approximate ideal point-shaped redirection of the cable

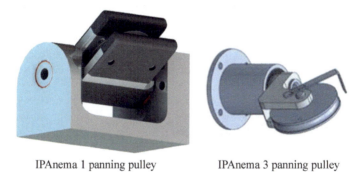

IPAnema 1 panning pulley   IPAnema 3 panning pulley

Fig. 8.25 Guiding the cable over a panning pulley reduces the wear of the cable but makes kinematics more complicated

of the robot. Beside the ease of configuration, double panning pulleys introduce an additional pulley in the drive-train, increasing cable force loss through friction, increasing fatigue in the cables, and reduce the accuracy of force measurements at the proximal end of the cable.

8.6.1.3 Platform Anchor Point Design

In order to exert forces on the mobile platform, the distal end of the cable needs to be fixed. As discussed above for the proximal anchor point, the key challenge in attaching the cable to the platform is to maintain a kinematically well-defined

Fig. 8.26 Double panning
pulley unit for
omnidirectional guidance of
the cable allows to redirect
the cable at the upper side of
the frame (picture without
cable)

behavior within a wide range of possible attack angles. On the platform, the end of
the cable needs to be fixed which is different from the proximal anchor points where
the running cables is led through a defined point. Fixing the cable to the platform
basically consists of two main aspects. Firstly, one has to make a durable connection
with the cable that allows transmitting the force from the cable into another machine
element. Secondly, one has to provide the ability to withstand the forces in different
directions. Additionally, a light-weight design on the cable side is desirable to reduce
parasitic inertia in the cable.

For this purpose, different concepts are used on the demonstrators. Some robots
simply clamp the cables between flat jaws. This is very simple to build, however
it causes excessive wear on the cable in the region where the cable is clamped and
sharply bent. The clamping design allows to approximately connect two or more
cables to the same position on the platform. It is rather difficult to set up accurate
cable length when clamping one or even more cables to the platform. The assembly procedure usually leaves significant uncertainties in the cable length requiring
subsequent calibration procedures.

The ends of both steel and synthetic fiber cables can be formed to a soft eye. Then,
a snap hook is a simple and re-usable way to connect the cable end to a ring bolt on
the platform. This link to the platform is a good solution also for reconfiguration of
the platform and for using cables in a rough environment. However, the connection
is subject to uncertainties and clearance leading to decrease in both, accuracy and
durability. Instead of metallic snap hook, one can use soft loops made from synthetic
fibers. Additionally, snap hook can cause sudden settling effects if large deflection
angles are reached.

A more elaborated but also more expensive connection consists of end sleeves for
the cables. Such end connectors add a well-defined mechanical interface such as a
plain or screw bolt to the end of the cable that can be linked by a joint to the mobile
platform. The construction of a compact and accurate end-connector is challenging.
The best connection for both steel and synthetic fiber cables requires to split the

cable into its fibers or wires and to mold the single fibers or wires into a conic housing. However, molding the cable into the end-connector requires some expertise to achieve high tensile strength. The result is a very robust and efficient connection that can be pre-assembled with some accuracy. In turn, shortening of the cables and repair of broken cable connections are more involved and require specific tools. A simpler solution is to coil the cable around a smaller bolt that is inserted transversal in the end-connector (see Fig. 8.37). For synthetic fiber cables, one can transmit forces almost as high as the molded connection can bear with good resistance to fatigue.

End-connectors can be linked to the platform through a spherical joint, a universal joint, or a swivel bolt. The two latter connectors basically consist of two revolute joints with orthogonal axes where the alignment of the axes differs with respect to the cable direction. Universal joints provide symmetric deflection capabilities with comparably small deflection angles up to 60° with small deflection angles being more preferable. In contrast, in a swivel bolt, one axis is aligned with the straight cable. This configuration causes a kinematic singularity when the swivel bolt is stressed in purely longitudinal direction. Instead, swivel bolts allow for very large deflection angles beyond 90° and provide good force transmission capacities for large deflection angles. The choice between universal joints and swivel bolts is a matter of proper design.

For planar robots, the connection to the mobile platform can be implemented with an end-connector as described above with a conventional revolute joint. Since one needs only one rotational axis, there are many simple possibilities for the mechanical integration into the mobile platform.

Instead of joints, one can guide the cable through circular profiles on the platform in order to bend the cables in a well-defined way. Connecting the cable between two cylindrical surfaces can be easily manufactured and provide a good resistance to fatigue in the cables (see Fig. 7.5). The transmission of the cable force onto the platform can then be done by winding the cable around a bollard. This idea was implemented in a planar robot in 2014 as well as proposed in [167] for use with a spatial robot (Fig. 8.27).

8.6.2 Cable Actuation Systems

8.6.2.1 Winches

The design of a winch (or hoist) is a well-understood task in mechanical engineering since winches have been used in crane applications for centuries. One distinguishes between drums with and without cable guidance system and also between coiling in single or multiple layers. Winches without cable guiding system are more cost efficient to build and the cable is coiled in multiple layers. Clearly, one can store much longer cables on such winches. These advantages come at the cost of reduced accuracy in the estimation of the wound cable length as well as high wear of the cable. At least for steel cables, the loss in lifetime of cables is found to be excessive

Fig. 8.27 Distal anchor
point on the mobile platform
based on a cable rolling up
on cam that is mounted onto
a revolute joint

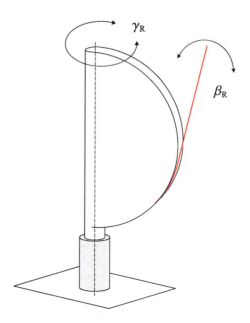

and can bring down the lifetime to 2–9% of the nominal value [149]. Thus, the cable
is destroyed some 10–50 times faster. Multi-layered spooling is generally not advised
for all kinds of longer operations (Fig. 8.28).

Winch designs for cable robots are presented for example in Rostock [206, 306]
and Duisburg [70], by Tecnalia [224], and Fraunhofer IPA [404]. Linear drives are
proposed by Surdilovic [448]. A comparison between winches and pulley tackles as
actuation scheme is made by Merlet [327]. An alternative design for a winch-based
actuation system is proposed by Cong [383] where the drum is mounted on a linear
guide on the shaft in order to keep the contact point where the cable leaves the drum
fixed in space. When rotating the drum, a spindle drive moves the drum on the shaft
to compensate for the pitch of the drum. A similar idea is depicted in Fig. 8.29 where
the motor and the drum are connected by a 3R mechanism for transmission of the
torque where the drum is mounted on a screw thread to keep the contact point fixed
in space.

To allow for accurate coiling of the cable, this section focuses on single layer
winches. In the following, a review of the essential dimensions of a winch is presented. The length Δl_{D} of a cable coiled onto the drum with diameter d_{D} is

$$\Delta l_{\mathrm{D}} = n_{\mathrm{w}}\sqrt{d_{\mathrm{D}}^2\pi^2 + h_{\mathrm{D}}^2} \ , \tag{8.24}$$

where n_{w} is the number of windings on the drum and h_{D} is the pitch of the drum
(Fig. 8.30). Clearly, the pitch h_{D} must be greater than the diameter d_{c} of the cable.
For example for a 6 mm synthetic fiber cable, a pitch $h_{\mathrm{D}} = 7$ mm is used in the
IPAnema 3 winch and for a 2.5 mm synthetic fiber cable a pitch of $h_{\mathrm{D}} = 3$ mm works
well. For steel cables, the norm for crane hoists [222] can be used to find sufficient

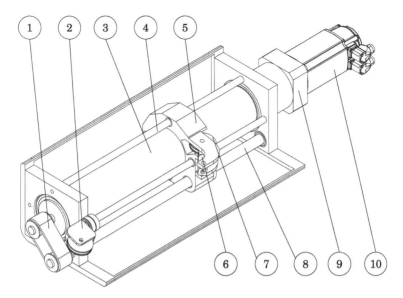

Fig. 8.28 Concept for a servo-controlled winch with integrated force sensing: (1) transmission belt, (2) guiding pulley, (3) drum, (4) linear guidance, (5) spooling unit, (6) force sensor, (7) guiding pulley, (8) spindle, (9) planetary gearbox, (10) servo motor

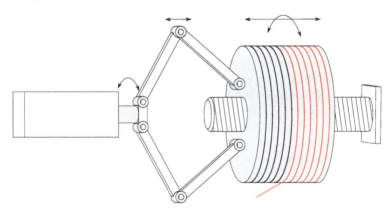

Fig. 8.29 Movable winches based on a four-bar linkage and a movable drum on a thread screw

values. The minimum length of the drum l_D is

$$l_D = n_w h_D + l_{D0} \quad , \tag{8.25}$$

where l_{D0} is a positive length that collects the unused length of the drum. This additional length is required to fix the cable on the drum, as safety margin for the end-of-travel sensors, and for regions that cannot be accessed on the drum e.g. because of installation space of the cable guidance system. If the cable is clamped onto the

Fig. 8.30 Development
drawing of the cable on the
drum

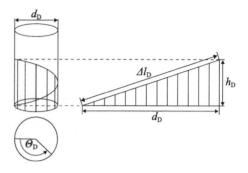

drum, some $n_E = 3\ldots5$ windings should be used to reduce the effective forces
acting on the clamping. The exact relation of the force at the end of the cable f_E and
cable force f is determined from the formula by Euler-Eytelwein

$$f \le f_E e^{\mu_D \alpha} \quad , \tag{8.26}$$

where μ_D is the coefficient of friction for the cable on the drum and the angle $\alpha = 2\pi n_E$ is determined from the number of windings n_E around the drum. With a medium
coefficient of friction of $\mu_D = 0.1$ for Dyneema on steel, we find a reduction of the
force f_E by around 6.5 for three windings and by around 23 for five windings.
See Table 3.7 for the friction coefficients of different cable materials. Due to the
significant reduction in the cable force at the end, the cable can be clamped to the
drum without damage and only moderate forces are applied to the cable.

The diameter of the drum must be chosen according to different effects. Firstly, the
diameter is a linear factor for the computation of the overall transmission index of the
winch. Since a typical servo motor has a high nominal angular velocity, gearboxes,
or belt drives are used to reduce the velocity of the cable. Therefore, small diameter
of the drums are favorable to support the transmission to smaller velocities of cable.
The motor motion is transformed to cable velocity by

$$i = \frac{\pi d_D}{2 v_{PG}} \omega_M \quad , \tag{8.27}$$

where i is the velocity of the cable, v_{PG} is the gear ratio of the gearbox, and ω_M is
the angular velocity of the motor. Contrary, the torque T_M generated by the motor is
transformed to the cable force through

$$f = \frac{2 v_{PG}}{\pi d_D} \eta_G \eta_W T_M \quad , \tag{8.28}$$

where η_G and η_W are the efficiency factors of the gearbox and the winch, respectively. Industrial gearboxes have efficiency factors of around $\eta_G = 0.95\ldots0.97$ at
nominal load. Care must be taken when the winch is operated below the nominal

load of the gearbox since the efficiency factor η_G decreases notably in this case. Such usage occurs frequently for cable robots especially when cables are uncoiled with small tension. Due to the small forces in the cables during uncoiling, the friction in the gearbox can prevent recuperation of energy due to insufficient efficiency in the gearbox.

This becomes evident if one considers the following example: The IPAnema 1 winch has a nominal cable force of $f_{max} = 720\,\mathrm{N}$, a gear ratio $\nu_{PG} = 12$, and a nominal motor torque $T_M = 3\,\mathrm{Nm}$. The two-stage planetary gearbox has an efficiency of $\eta_G = 0.95$ at nominal load, i.e. the friction force is approximately $F_F = (1-\eta_G)f_{max} = 36\,\mathrm{N}$. Typical values for the mechanical efficiency of such winches are determined to be between $0.8\ldots0.9$ depending on the load-state. Thus, one has to generate some torque with the motor if the cable is operated at low tension even if the cable is uncoiled.

The second important factor for the winch diameter results from the fact that the diameter of the winch largely effects the rotational moment of inertia. If the drum is modeled as a hollow cylinder, one finds the moment of inertia I_D to be

$$I_D = \frac{l_D \pi \varrho_D}{32} \left(d_D^4 - (d_D - t_D)^4 \right) \quad , \tag{8.29}$$

where ϱ_D is the density, t_D is the thickness, and d_D the diameter of the drum. One can conclude that the moment of inertia of the drum largely depends on its diameter. Especially for dynamic applications of cable robots, it is advantageous to keep the inertia of the winch low to exploit the full dynamic bandwidth of the motor.

Thirdly, the ratio from the cable diameter and the drum diameter d_D/d_C is a lower limit for the diameter of the drum. For synthetic fiber cable, a ratio $d_D/d_C = 8\ldots12$ is applied for short-term usage[6] where for steel cables a ratio of 25 and more is typical. As one can see from Fig. 8.22, also for synthetic fiber cables larger ratios are required for long-term usage. In most cases, a minimum diameter of the drum is governed by the requirements of the cable.

A short example is given to illustrate the cable criterion. A winch shall be designed for the usage of Dyneema cables with a diameter of $d_C = 2.5\,\mathrm{mm}$. Based on a ratio of 25, the drum (and the pulleys) should have a minimum diameter of $d_D = 62.6\,\mathrm{mm}$. In order to increase resistance to bending fatigue of the cable for long-term usage, a factor of 40 is advised leading to a minimum diameter of $d_D = 100\,\mathrm{mm}$.

8.6.2.2 Linear Systems

Wide-spread ways to mechanically realize the linear drives are direct driven linear motors as well as spindle nut drives used in machine tools. The former actuation requires specific customization and allows for higher dynamics but is most costly

---

[6]There are no strictly reliable measurements for the lower limit of ratio d_D/d_C available for synthetic fibers, also because there are huge amounts of different new materials on the market.

Fig. 8.31 Schematic sketch of the linear actuators for cable robot IPAnema 2 planar (left)   T: sliding carriage with pulleys,   U: rotor of the linear direct drive,   V: linear guideway,   W: housing for the drag chain,   X: drag chain with power and encoder cables,   Y: carrier for supply cable and drag chain,   Z: stator of linear direct drive.   The entirely assembled planar cable robot IPAnema 2 with the four linear motors (right)

and difficult to operate. Spindle nut drives are actuated using the electric servo motors as used in winches.

One can directly connect the linear drive (Fig. 8.31) to the cable as exemplified in the StringMan robot [448]. Such design requires only one panning pulley and thus allows for little friction losses as well as damage in the cable. However, the application is limited to smaller robots since the length of the actuation system is directly coupled to the size of the workspace. Using linear drives, one can determine precisely the effective cable length through the integrated sensors of the actuator. In contrast to winches, less disturbance is caused by ovalization and undefined cable coiling since the whole cable remains tensed during normal operation. Using pulley tackles [325, 327], one can realize a gear like behavior where gear ratios are achieved that correspond to gearbox ratios $v_{PG} < 1$. Thus, the cable velocities are generally larger than the actuator velocities and the cable forces are lower than the actuator forces. Taking into account that linear drives already provide good dynamic performance, one can realize ultra-high speed robot. Merlet conjectures that reaching the speed of sound is achievable [327], however, no experimental results are presented yet. By using pulley tackles in the transmission, the cable stroke of the actuator is multiplied by the number of pulleys thus reducing the installation size of the actuation system. Such designs compromise the lifetime of the cables since the cables are subject to excessive wear caused by the bending over many pulleys. Additionally, significant losses in the cable force are caused by friction in the pulleys. Roughly speaking, the overall losses are proportional $\eta_L = \eta_R^{n_R}$ where n_R is the number of pulleys and η_R is the efficient of each pulley.

8.6.2.3 Twisting of Cables

An exotic way of actuating cable-driven parallel robot is twisting the cables. If an axial rotation is applied to a cable, the cable begins to shorten and exerts also a pulling force on the platform. This actuation scheme for cable robots is proposed by Shoham [439]. In serial robots, twisting of cable is a known actuation principle [426, 497]. A large benefit of twisting the cable is that a very high transmission ratio around 400 times higher than a typical winch can be achieved without using gearboxes [497]. Contrary, only relatively small changes in the length can be achieved, the transmission ratio is highly nonlinear, and twisting the cable causes large wear on the cable [426]. There are little figures available for the lifetime of cables under twisting loads. Some preliminary values for synthetic fiber cables can be found from [426] where experimental tests are presented using cables as artificial muscles. Therefore, excessive testing is required before usage in production installations.

8.6.3 Selection of the Actuators

Cable robots are usually operated with electric servo motors due to good controllability. Many cable robots use synchronous direct current servo motors (see e.g. [210, 404]). Servo motors and suitable gearboxes are available in high quality by many vendors and can be selected in many variants ranging from rated powers of some watts to several hundreds kilowatts. Using gearboxes, the ratio of motor torque and speed can be precisely customized to applications' needs. Servo motors are both suitable for slow speed cable robots with high payload as well as for high dynamics applications. The technical limitations of servo motors are a constant maximum torque which is mostly independent from its speed, a maximum velocity being almost independent from the torque, and limitation in the total rated power. Servo motors allow for both position and force control. One of the main drawbacks of servo motors is that they are expensive compared to other types of electric drives. In turn, stepper motors can be employed for cost efficient actuation especially for small cable robots. Stepper motors have a complex characteristic curve relating the available torque to the motor speed. Due to the simple actuation scheme, stepper drives are preferably used with open-loop position control.

A conservative way to determine the required motors and gearboxes is to compute the worst case motor torque required to fulfill the application as well as to use the highest required velocity v_{max} defined for the application. Taking into account the characteristics of the winch, one can directly compute the required maximum speed of the motor from

$$n_{M,max} \geq \frac{v_{max} v_{PG}}{2\pi r_D} \quad , \tag{8.30}$$

where v_{PG} is the reduction ratio of the gearbox and r_D is the radius of the drum in the winch. In contrast to the velocity, one has to take into account losses in the

Table 8.19 Parameters for the drive section with typical values as well as lower and upper bounds for the parameter for the IPAnema 3 winch. Winch efficiency is determined at maximum rated power of the drive

Symbol	Description	Typical values	IPAnema 3 (lower)	(upper)
η_{P}	Efficiency factor of pulley	0.97	0.97	
η_{PG}	Efficiency factor of gearbox	0.95–0.97	0.97	
η_{W}	Efficiency factor of winch	0.80–0.90	>0.80	
n_{P}	Number of guiding pulleys	1–5	2	4
η	Overall mechanical efficiency	0.87–0.63	0.73	0.69

mechanical transmission to compensate for friction in the drive-train for selecting the motor torque. Similarly, one receives the required nominal torque of the motor from

$$ T_{M,max} \geq \frac{F_C}{\eta_P^{n_P} \eta_W \eta_{PG} v_{PG} r_D} \quad , \tag{8.31} $$

where η_P, η_W, η_{PG} are the efficiency factors of the pulley, the winch, and the gearbox, respectively, and n_P is the number of guiding pulleys. Thus, the overall mechanical transmission efficiency is

$$ \eta = \eta_P^{n_P} \eta_W \eta_{PG} \quad . \tag{8.32} $$

Typical values for these factors are given in Table 8.19. The numerical examples of the IPAnema winches show that one has to consider a buffer of around 35–47% of torque on the motor side to account for losses in the drive-train. However, using the dimensioning of the motor and gearbox as described above can lead to notably over-sized drive-trains when the robot is operated dynamically. Then, the scenario-based approach from Kraus [261, 263] can be applied to tailor the drive-train accurately to the duty cycle of the robot.

The basic idea of the scenario-based dimensioning of the drive-trains is based on a repetitive duty cycle of the robot. In practice, such cycles are defined by the robot program that defines the motion of the mobile platform and hence also velocities and accelerations. Therefore, one can use rigid body kinematics to determine the platform's inertia wrench $\mathbf{w}_I$ from

$$ \mathbf{w}_I = \begin{bmatrix} \mathbf{f}_I \\ \boldsymbol{\tau}_I \end{bmatrix} = \begin{bmatrix} m_P \mathbf{I}_3 & \mathbf{0} \\ \mathbf{0} & \mathbf{I}_P \end{bmatrix} \begin{bmatrix} \mathbf{a}_P \\ \boldsymbol{\alpha}_P \end{bmatrix} + \begin{bmatrix} \mathbf{0} \\ \boldsymbol{\omega}_P \times \mathbf{I}_P \boldsymbol{\omega}_P \end{bmatrix} \quad , \tag{8.33} $$

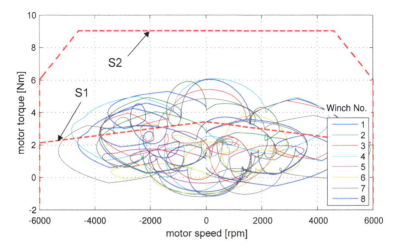

Fig. 8.32 Examples for the scenario-based dimensioning of the drive-train with the reference curves of the motor for its rated torque (S1) and the peak torque (S2)

where $\mathbf{a}_P$, $\boldsymbol{\alpha}_P$ are the linear and angular accelerations of the platform, respectively, and $\boldsymbol{\omega}_P$ is the angular velocity of the mobile platform. Furthermore, m_P and $\mathbf{I}_P$ are the platform mass and the platform's moment of inertia. The inertia wrench $\mathbf{w}_1$ and the applied wrench $\mathbf{w}_P$ is used in the structure equations to compute cable forces (see Sect. 3.6) and cable velocities (see Eq. (4.4) in Sect. 4.2.3). Computing these motor torques and the motor velocities along a trajectory provides a sequence of states in a speed-torque diagram required for the scenario (Fig. 8.32).

Having determined application scenario curves as given in the diagram, one may test for combinations of motors and gearboxes according to data sheets from vendors of drive-trains. Using peak values for the rated power that is computed from the respective product of speed and torque, one can pick candidates for the motors. Then, selecting appropriate gearboxes allows for shaping the motor's curve to enclose the characteristic trajectory of the scenario. Using a model of energy efficiency as discussed by Kraus [261, 266], one can additionally fine tune the drive selection for optimal exploitation of the thermal behavior of the motor.

8.6.4 Sensor Integration

Common control strategies require to measure the system states during operation in order to improve performance of cable robots. Furthermore, sensors are used to assure reliable performance. This requires the integration of sensors into the cable robot. The sensors are integrated into different hardware elements of the cable robot including the winches, the guiding pulleys, and the mobile platform. The most frequent use of sensors in the winch is measuring the kinetostatic state of the cable. This includes the determination of the current cable length or velocity as well as the current cable force.

8.6.4.1 Cable Position and Velocity Sensors

The current cable length l_i must be determined for accurate control of the robot's kinematic transformation. There are several possibilities to measure the length, which are elaborated in the following. When using servo motors, it is convenient to include an encoder or resolver into the motor which feeds back the current position or velocity to the motor inverter. This kind of position sensors are available from many manufacturers for drive-trains in good quality. If the winch has a well-defined cable guidance system, the effective length of the cable is computed from an offset and the overall transmission factor of the winch. Encoders are available as so-called multi-turn absolute encoders that allow the controller to directly recover the absolute length of the cable in each cycle also after loss of power or during startup of the control system. Relative encoders need to cross a reference marker that can be located on the cable, on the drum, or in the encoder to reconstruct the absolute position. For larger cable robots, it is not practical to do such referencing because it could be dangerous to move the winches before the position of the platform is known. Such initialization procedures require a separate control strategy to operate the robot before valid absolute length values are available, by e.g. uncoiling single cables and later recovering a tensed state.

Motor-integrated sensing is also applicable for linear actuation systems where the linear displacement of the actuator is measured with an angular encoder or resolver if a servo motor is used to move the pulley tackle.

Beside motor-integrated sensors, a direct length measurement of the cable is desirable. An incremental but relative length measurement system is achieved by applying an additional friction pulley on the cable preferably close the winch's exit point. Such measurement systems are subject to drift errors when the cables are moved with very high velocities and accelerations. Another approach to determine the cable length is optical tracking or magnetic markers [325, 338] on the surface or inside the cable. An optical or magnetic sensor is then installed at the outlet of the winch to record the markers. Using markers allows for significant improvements of the repeatability at reasonable costs as little efforts must be spent on the winches to reproduce a given position. Additionally, referencing with markers overcomes position errors caused by inaccurate coiling of the cable. Clearly, this requires purpose-made cables to perform such measurements. To the best of the author's knowledge, such cables are not available as ready to use products but have to be customized. Also the impact of markers on cable fatigue is unknown.

An important practical problem with all position and velocity measurements is that one receives information about perfectly rigid cables where changes in the cable length caused by elastic, plastic, or hysteretic effects cannot be measured. A second practical problem is to determine the initial cable length after installation. Even if the sensor provides absolute values, one has to perform referencing before initial operation and accurate referencing is difficult to achieve. If the absolute position of the platform pose is determined with an external measurement device, one executes the referencing for this known reference pose to match it with the sensor data.

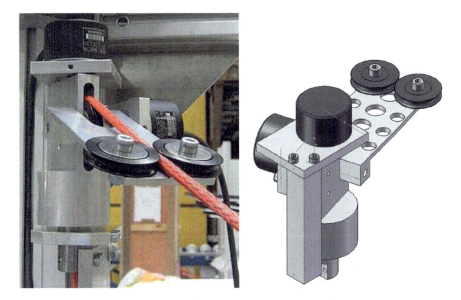

Fig. 8.33 Prototypic measurement device to sense the direction of the cable of the IPAnema 1 robot using guiding pulleys on the lever and two encoders for measuring the direction in spherical coordinates

8.6.4.2 Sensing the Cable Direction

Determining the current cable direction vector $\mathbf{u}_i$ allows to set up the structure matrix without estimating the pose through the kinematic transformation. Also, forward kinematics can be simplified, either by directly using the cable direction to compute the position of the distal anchor points or by selecting amongst ambiguous solutions in forward kinematics.

The direction of the cable is determined with encoders (Fig. 8.33) that are built into the cable guidance system [206, 306, 496], with cameras [108] and image processing, or with one-dimensional arrays of photo sensors. The accuracy of the measurement is typically only accurate up to some degree. Therefore, the measurement is only used as a rough estimate instead of an exact value. A prototypic realization of an encoder-based direction measurement is shown in Fig. 8.33. The drawback of the mechanism measurement is that the additional forces to move the measurement unit disturb the motion of the robot.

8.6.4.3 Measuring the Pose of the Platform

The pose $(\mathbf{r}, \mathbf{R})$ is usually only implicitly measured through the cable length $\mathbf{l}$ and by computing the forward kinematics code. If such transformation is not available or not accurate enough, the measurement of the platform position and orientation

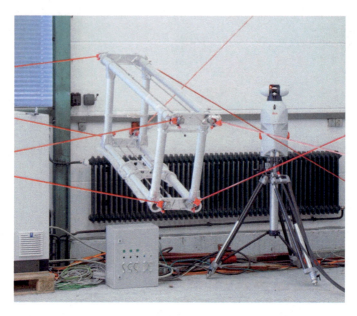

Fig. 8.34 Pose determination of the platform of the IPAnema 3 robot using a laser tracker

is required. To measure the platform pose, some approaches are proposed in the literature. Camera-based vision sensors are used to track markers on the mobile platform [96, 108] and their measurement signal is employed in the position control loop. Other approaches are inertial measurement units (IMU) [449], (differential) GPS, or laser tracking (Fig. 8.34) of the platform [264]. The pose measurement is required for external calibration (see Sect. 9.2) and motion control where the latter requires real-time sensor data processing. The latter is more involved since many three- and six-dimensional measurement systems are not designed to supply their data with real-time speed and efficiency. Even fast computer systems may fail to provide the position estimates with little latency time to allow for control cycles in the magnitude of milliseconds. To the best of the author's knowledge, no results are reported about feeding back such platform measurements into the control system at full cycle time making measured pose data available for control. Beside this, a couple of demonstrators used camera-based vision for validation purpose, see e.g. [25, 527].

8.6.4.4   Cable Force Measurement

Determination or estimation of the current cable force is a prerequisite for all kinds of cable force control. In the simplest case, one monitors the cable forces to keep them between the minimum and maximum limit or to detect failures such as slackness or even breaking of the cables. In contrast for the use in closed-loop control, five main properties for the force measurement are important:

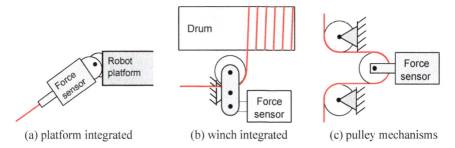

(a) platform integrated   (b) winch integrated   (c) pulley mechanisms

Fig. 8.35 Concepts for integrating a one-axis force sensor into a cable robot

- Accuracy of the measurement: Precise measurement simplifies the use of the force signal in control.
- Dynamic resolution: The maximum sampling frequency of the force signal is limited by the sensors, the A/D converter, and the cycle time of the control system. As expressed in [334], the sampling bandwidth should be higher than 1 kHz.
- Latency: Measurement, signal conversion, filtering, and transmission through the fieldbus lead to a dead time.
- Noise: Force sensors are subject to significant noise disturbing the quality of the measurement.
- Systematic errors: Parasitic inertia, friction, etc.

The effects listed above vary in quality and quantity depending on the sensor used in the robot as well as on the concept used to integrate the sensor.

A straightforward method for integrating a force sensor is to place the sensor between the distal end of the cable and the mobile platform (Fig. 8.35a). From mechanical engineering point of view, the installation is rather simple and the measurement signal is practically free of parasitic inertia and friction. Thus, based on the quality of the sensor, one measures directly the force applied to the mobile platform as assumed by the standard static model. When using a one-axial force sensor, an additional spherical or universal joint must be used to guarantee that the sensor is only loaded in axial direction to prevent damage to the sensor. A disadvantage of distal force sensing is that one has to transmit the signal from the platform to the controller making additional signal and energy connection to the mobile platform necessary. An example of such measurement is shown in Fig. 8.36.

A straightforward and frequently used technique to measure the cable force is to add a linear force sensor to one of the guiding pulleys inside the actuation system (Figs. 8.35c and 8.37). Application of this concept can be found in [261, 376, 413]. Assuming that the pulley has no friction, one can directly determine a linear estimate of the cable force. When using a linear motor with a pulley mechanism, there are a couple of possible locations for the force sensor. Since the full cable is under tension, the force sensor can be located at the proximal end of the cable where it does not move. This eases installation and electric wiring of the sensor. Measuring the cable force at the proximal end of the cable suffers from some disturbance caused by

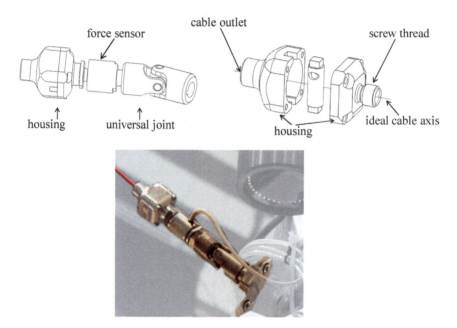

Fig. 8.36 A force sensor at the distal end of the cable. Using a universal joint between the platform and the force sensor allows to load the sensor only in axial direction. The lower photo shows the connector assembled with the robot according to patent DE 102012024451 (A1)

Fig. 8.37 A force sensor integrated into the guiding pulleys inside a winch. Care must be taken to protect the force sensor from shear forces

friction in the guiding pulleys. As a rule of thumb, 3% of the cable force is dissipated in each pulley and consequently cannot be measured by the sensors on the proximal end.

When installing the sensor in a winch, it is favorable that the cable's attack angle on the pulley do not change (see Fig. 8.35b). Then, one receives a constant multitude between a factor of 1 and 2 of the cable force depending on the wrapping angle around the cable. The advantages of this arrangement are a compact hardware design and ease of installation since the sensor does not move relative to the actuation system. Furthermore, there is little parasitic inertia in the mechanical system

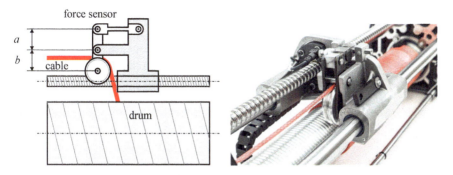

Fig. 8.38 Measurement principle for determining the cable force with a lever mechanism (left) and a typical implementation on the IPAnema 3 winch (right)

allowing for determination of the cable force at high bandwidths. However, some drawbacks remain. Especially for large-scale cable robots, the force measurement on the proximal side disregards the effects of pulley friction, the effect of sagging and elasticity, as well as for huge robots some delay in the measurement caused by limited wave velocity. A practical implementation is exemplified for the IPAnema 3 winch in Fig. 8.38.

If the drum is mounted on a lever, one can also add a force sensor to the lever to determine the cable force. This approach is applicable if the cable is directly guided from the drum to the workspace or to a distal pulley. The mechanical design is simple and by choosing an appropriate geometry for the lever, one can easily guarantee a single-axis load for the force sensor. In this design, sensors measuring pushing forces are applicable. A disadvantage is the high parasitic inertia of the lever and the drum making the measurement insensitive to higher frequencies. The disadvantages named for measuring at the guiding pulley also apply for measurements at the drum.

If the winch has a spooling mechanism, a lever can be used for the pulley (Fig. 8.35b). Again, this design is a simple but efficient way to receive only one-dimensional tension in force sensors. From the scheme in Fig. 8.38, it becomes apparent that balancing the distances a and b allows for fine tuning of the measurement range of the force sensor: Available force sensors have fixed maximum loads that may mismatch with the desired maximum load of the winch. Small changes in the length a and b allow to accurately exploit the measurement range of the force sensor, by e.g. changing the pivot point of the lever by less than 10%, one can vary the effective scaling factor for the force sensor by more than factor 2.

Another approach is using a torque sensor in the drive-train to measure the cable force. The torque sensor is placed between the drum and the clutch preferably after a possibly applied gearbox. Clutches with integrated torque sensors are also available. The mechanical integration is elegant and compact where some effort needs to be undertaken in transmitting the signal from the rotating sensors in the drive-train. Integrating the sensor in the winch mechanics is robust but comes at the cost of even higher parasitic inertia and disturbing effects such as friction in the pulleys, elasticity in the cables, and sagging as well as friction in the bearing of the drum.

Finally, motor-integrated force sensing (e.g. by evaluating the servo motor's current) can be applied but is subject to additional measurement errors induced by backlash and friction in the gearbox. Experimental evaluation with a planetary gearbox showed that friction force in the gearbox is higher than the cable force [259]. Therefore, force sensing on the motor-shaft is rather limited when using a gearbox. For high payload and gearless drive-trains, acceptable force measurements at the motor are obtained at the Cable Simulator.

For the sensor itself, one-axial load-cells can be applied for distal and proximal force measurement. Most of the sensors used in practice employ a metallic or silicon stain gauge where sensors with integrated amplifiers are available. Piezo sensors are less versatile for the use with cable robots since these sensors only detect changes in the force rather than absolute forces. Therefore, Piezo sensors are only applicable if the cable forces change at higher frequencies.

8.7 Conclusions

The design of cable robots is an engineering process with a number of phases that require iterative approaches. Firstly, the application requirements have to be compiled as a solid basis for the design decisions to be made. Based on this requirement specification, the system design is carried out. Considering the desired properties of the robot, one has to define the motion pattern of the robot, the kinematic classification, and decide if the robot is fully-constrained or suspended. In order to facilitate the geometric dimensioning, one defines a parametric model as template. A collection of such models for different archetypical applications is presented. For new or special applications, one needs to define a specific parameterization that fits this application. The parameterization defines the design space for the geometric parameter synthesis of the robot. Two automatic approaches are presented based on optimal design and constraint programming. For plain design problems, one might succeed by applying rule of thumb guidelines and manually tune the geometry of the robot. In this procedure, the engineer employs ad hoc the tools for analysis described in earlier chapters in order to investigate the feasibility of the robot to be designed. It must be concluded, that different techniques for tuning the geometric parameters are possible but all of them require expertise from the side of the engineer to be used efficiently.

After fixing the geometry of the robot, the mechanical design and the controller design of the robot has to be done. From the technical requirements, one has to first select the cable since the cable defines technical specifications for the winches, guidance elements, and end-connectors. Then, one has to perform the mechanical design of the robot including the selection or construction of the winches, the dimensioning of the machine frame, and the construction of the mobile platform. Furthermore, the cable guidance systems need to be selected and appropriate distal cable end connectors have to be chosen.

The mechanical design shall be accompanied by setting up a dynamic simulation model that allows to quickly validate the mechanical design as well as to prepare the controller design and parameterization. Furthermore, the simulation model can be used to tool up for the initial operation of the robot, since most control parameters can be prepared based on the simulation in order to allow for an accelerated initial operation of the robot.

The design of cable robots remains a rather challenging topic requiring considerable knowledge in different disciplines and is heavily influenced by practical experiences with earlier cable robot designs. Today, the theoretical foundation of this engineering process for cable robots is rather preliminary since little components have gained maturity from multiple iterations.

Chapter 9
Practice

Abstract In this section, operating experience and experimental results are presented related to building and running cable robots. Firstly, the basic procedure for calibration of cable robots is outlined. Then, the IPAnema robot family is introduced where the different demonstrator systems exemplify possible design decisions. Finally, some other cable robots are presented which are designed and built using the methodology described in this book.

9.1 Introduction

As robotics is an applied engineering science, one has eventually to proof the effectiveness of the theoretical findings in an experiment. The aim of using robots is to provide automation solutions to real world problems. One can approach this in two different ways. The *deductive* approach is based on deriving the complicated structure of an actual system from basic axioms and fundamental theorems. Mathematics is the very archetype of this approach and well-structured physical disciplines such as mechanics follow this paradigm. Here, it is mostly followed for deriving the kinematic and static foundation of cable robots. The second approach is based on *induction* and requires unconditional observation of systems to identify patterns and to conjecture general rules. If a scientist is not able to construct his experiments as desired, he must resign to observe the arrangements that are available. Typical examples of this approach are astronomy and also many aspects of life science. Although one can build in principle any variants of a cable robot, this is practically impossible due to limitation of resources. Therefore, observation of the physical robot is required to extract the essentials.

Theoretical approaches are required to solve some kinds of problems that cannot be attacked by just building and observing the robot. However, the deduction of a theorem or a model for cable robots from the natural laws requires decisions what effects must be addressed and what effects are neglected. For example, the validity and applicability of Newtonian mechanics are unconditionally accepted for modeling of cable robots where relativistic effects are neglected since such effects are anticipated to be very small. However, other effects such as friction in the cables

© Springer International Publishing AG, part of Springer Nature 2018
A. Pott, *Cable-Driven Parallel Robots*, Springer Tracts in Advanced
Robotics 120, https://doi.org/10.1007/978-3-319-76138-1_9

and pulleys, aerodynamic effects of the moving platform, and thermal effects in the materials are to be discussed and depend on the use-case. Furthermore, it can be hardly tested if one has overlooked something in the model or if one simply failed in its implementation.

A purely inductive point of view on cable robots is seldom followed. One can take the perspective of an observer who qualitatively and quantitatively determines the behavior of the cable robots. Is a certain pose of the robot stiff? Does the robot vibrate or sway, and with which frequencies and amplitudes? The observations can lead to a discovery of phenomena that need explanation and guide the direction for further development of the model. A veritable model can reliably predict and possibly explain the observable phenomena with acceptable errors and minimal effort. In contrast, adding complexity to the model without receiving a more accurate prediction is pointless. When one proposes an extension to a theoretical model without comparison with the experiment, one cannot assess correctness and validity unless the model is experimentally validated or already known observations can be explained with the extended model. However, one can do theoretical studies on a validated model to analyze its properties.

The research on cable robots is mostly driven through research in the field of kinematics, dynamics, and control. As robotics is an interdisciplinary science, the connection of the different domains is involved and complex. Providing and operating the test-beds required for the experimental validation is predominantly not the main objective of the research projects undertaken. Lately, the research on cable robots is more driven by applications but still lacks an established practice in building cable robots. Thus, there are no standard robots available making comparison amongst experimental results involved and sometimes also fuzzy.

The objective of this chapter is to provide accurate information on the test-beds used for the experimental validation of many of the results discussed so far. Furthermore, we present a number of solutions for practical challenges that occur if one meets the challenge of testing a hypothesis experimentally. The first step in that practical direction is the calibration where one tries to identify the actual geometry of the robot. Then, an overview on some demonstrators is given where some implementation details are elaborated and compared.

9.2   Calibration

Accuracy is a key performance indicator for all kinds of robotic devices. A detailed technical definition of accuracy of robots and how to measure it is subject in ISO 9283 [220]. The *position accuracy* of a robot describes its ability to move its reference points to the desired absolute position in space. In contrast, *repeatability* is the deviation between the actually reached position of the end-effector when approaching to the same position in configuration space in a number of times. Repeatability is influenced by the reproducibility of the motion in the actuators and in the mechanics including effects such as elastic reactions, control errors, sensor errors, hysteresis,

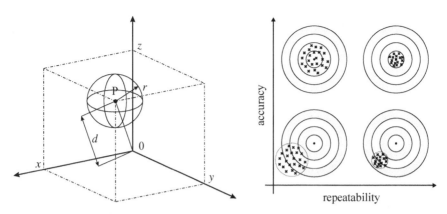

Fig. 9.1 Definition of position accuracy d and position repeatability r according to ISO 9283 [220]

cable creeping, thermal effects, and clearance. The effect of good or poor accuracy and repeatability is depicted in Fig. 9.1. Repeatability is in general equal or better than accuracy because accuracy suffers from bad repeatability. In contrast, a robot can have a good repeatability without having good accuracy, e.g. if the referencing of the actuators is poor or if the parameters in the kinematic transformation are imprecise. Especially, over-constrained cable robots suffer additionally from shortcomings in repeatability since violations in the constraints cause additional effects on the pose. By different means of correction, one can improve the accuracy towards the level of the robot's repeatability.

Although significant efforts are undertaken in the design, manufacturing, and assembly of robots to build the robot accurately according to the design specifications, one cannot prevent errors and uncertainties in the geometry of the robot. By using costly procedures, one can only reduce but not eliminate the errors. While machine parts such as drums, pulleys, and guidance systems can be manufactured with narrow tolerances, the setup of large machine frames suffers from significant errors. Thus, improving the accuracy by means of calibration during the initial operation of the robot is an important topic. *Calibration* is the procedure to estimate the *actual* numerical values of the geometrical design parameters of the robot. Usually, one considers the *nominal* parameters to be given, where the nominal parameters are the ideal values fixed as the results of the design procedure. Thus, calibration is a special kind of model parameter identification where the model to be identified is a kinematic model and the sought parameters relate to the geometry of the robot. However, in the context of cable robots, it makes sense to extend the calibration procedure to consider also the cable forces and optionally the elastic effects in the cables to efficiently increase the robot's accuracy. Merlet [322] argues that optimal design aims at making the robot insensitive to geometric changes where the calibration procedure requires such sensitivity to be effective.

Evaluating the effectiveness of the calibration procedure is a time-consuming procedure. Eventually only the results allow to benchmark the quality of the calibration after finishing the whole calibration procedure and experimentally measuring the accuracy. If the result of the calibration is not satisfactory, it is difficult to assess how to improve the quality of the procedure.

9.2.1 Review of Literature

The calibration of cable robots is highly related to the calibration of conventional parallel robots as both types of robots have closed-kinematic loops causing a highly coupled motion between the actuators and the mobile platform. Thus, a number of conventional strategies developed for machine tools and industrial robots [133] are not applicable due to the different structure of parallel robots.

Due to the coupled motion of the closed-kinematic loops, the calibration of parallel robots is based on measurement and matching of externally determined poses of the end-effector. This includes the problem of selecting the poses to be measured. The problem of optimal selection of measurement poses for a serial robot is addressed by Borm [44], who employs an observability index to rate the measurement pose set. Also, Khalil [242] discusses the impact of different measurement positions for the calibration of robots. Then, Wampler [484] generalizes a method to apply calibration procedures to parallel robots, and Hollerbach [212] introduces the well-known calibration index in order to assess the quality of the set of measurement poses. Nahvi [357, 358] proposes a method for calibration of a redundant parallel robot addressing also pose selection and the numerical conditioning of the least-square problem. Daney provides a series of reference papers on the calibration of parallel robots [109, 110] where also the notion of certified calibration is addressed. Boye and Verl propose the calibration of conventional parallel robots [63, 478] taking into account the problem of optimal measurement pose selection and estimation of the reachable accuracy of the calibration procedure.

A couple of authors address calibration of cable-driven parallel robots. Tadokoro [452] mentions calibration for a portable rescue crane where for a mobile cable robot the calibration is a part of the regular use. Joshi [236] presents a calibration procedure with inclinometer as proposed by Khalil for a 6-6 cable robot and presents simulation results from the calibration. Borgstrom [42] proposes a self-calibration method based on position and forces differences for a planar 2T robot with four cables. The error of the calibrated robot is determined to 19.8 mm for tension-based self calibration and 6.3 mm for the position based calibration. Miermeister [345, 347] develops a thorough differential kinematic model of cable robots and applies both conventional as well as self-calibration methods to spatial cable robots with eight cables. Since the behavior of cable robots does not only depend on the geometric parameters but notably on material parameters such as cable stiffness and masses, Miermeister proposes to use a multi-stage parameter identification [342]. Sandretto [8] proposes the calibration of the ReelAx8 robot using weighted least-square and self-calibration where the model of the robot consists of the proximal and distal anchor points as well as offsets of the cable length. Additionally, interval analysis is used for the calibration of a robot [9] where different strategies are proposed to gain robustness against measurement errors. Duan [132] presents the calibration of the 50 m test-bed of the FAST cable robot and the improvement of the accuracy of a three-level positioning device is tackled.

Calibration of cable robots is related to the usage of cable sensors for pose measurement and calibration of other manipulators. Williams [491] presents a pose measurement device based on cables and determines a translational error of 0.61 mm. The calibration of the device is based on fixing the end-effector at known reference positions. Legnani [285] proposes a cable-based measurement device with six cables for the calibration of industrial robots.

Varziri [470] addresses the calibration of a cable-driven hybrid robot arm with a Gauss-Newton and a Levenberg-Marquardt method to solve the optimization problem. A singular value decomposition is used on the identification matrix in order to identify the ill-conditioned parameters in the model.

Lately, some procedures for calibration related to conventional parallel robots are proposed. Gayral [159] proposes an index to take into account the accuracy of the sensors and the noise of the measurement devices to benchmark calibration procedures. Recently, Gottlieb [177] presents a non-parametric calibration approach that is a generic correction function. It is argued to be simple but highly efficient for improving the accuracy of conventional parallel robots. However, this approach is not yet applied to cable robots and it remains open if it is efficient for redundantly constrained robots.

Self-calibration or *auto-calibration* is studied for conventional parallel robots with and without actuation redundancy. Khalil [241] fixes some parts of a conventional parallel robot without actuator redundancy in order to obtain redundant measurements that can be used in a self-calibration procedure. Although the approach is elegant, it cannot be directly applied to cable robots since the cables do not constrain the robot in the same way as conventional joints do. However, a variant is possible if the motion of the mobile platform is constrained by an additional mechanical joint. Yiu [510] proposes a self-calibration procedure of a conventional parallel robot with three planar over-constrained RRR chains. A kinematic model identification is computed using the extra sensor information from the redundant leg of the robot. Müller [353, 354] proposes to calibrate redundantly actuated parallel robots without additional sensors by using so-called motion reversal points as landmarks. Patel [379] studies the calibration of conventional parallel robots with a redundant leg and Takeda [455] tackles self-calibration using a double-ball bar.

9.2.2 Principles and Aspects of Calibration

The main aspect in calibration is to fine-tune the kinematic transformation represented through the kinematic code such that the measured values on the available inputs (usually the installed sensors) relate perfectly to the measurable outputs of the robot embodied by the motion of the end-effector. In order to identify these relations, one distinguishes four main principles for calibration [322, 353]:

- *External calibration* is based on full or partial measurement of the mobile platform through an independent device that is usually attached to the robot only for the sake of the calibration routine. This is the conventional method used for parallel robots and it is well applicable for cable robots.
- In *constrained calibration*, one uses geometrically well-defined mechanical fixtures to generate a certain and exactly defined pattern of measurement poses. Also, mechanical constraints built into the cable robots are possible. Constrained calibration is of interest for small cable robots. Adequate fixtures with sufficient precision for the calibration can be cost-efficiently manufactured. For such a calibration procedure, the robot is manually displaced to the relevant poses on that fixture.
- In *self-calibration*, one exploits an integrated sensor redundancy to gain over-constrained measurements. This approach is highly interesting for over-constrained cable robots since such sensor redundancy is available in these robots anyway.
- Finally, one can use natural *landmarks* in the kinematic mapping in order to identify specific points within the workspace as reference points in order to tune the parameters. This concept is proposed for conventional robots [353, 354].

A calibration procedure is mainly defined by four aspects (Fig. 9.2): The choice of the measurement device, the number and distribution of poses to be measured, the kinematic model, and the numerical procedure to estimate the parameters:

Fig. 9.2 The main steps in the calibration procedure

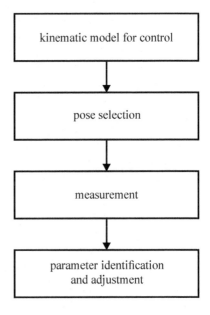

- *Kinematic model*: From a calibration point of view, the kinematic model mainly defines the geometric parameters to be determined in the calibration procedure. Although an advanced model exists as shown in Chap. 7, the relevant factor for calibration is which model is used e.g. as kinematic transformation within the controller.
- *Pose selection* defines the number of poses to be measured as well as the distribution of the nominal measurement poses within the workspace. The number of poses to be measured clearly scales the efforts to execute the measurement. Appropriate distributions heavily influence the quality and numerical conditioning of the calibration problem to be solved.
- *Measurement device*: The main properties of the measurement device are its accuracy and its fraction in determining a pose. Measurement devices capable of determining the full pose information consisting of position and orientation are rarely available and expensive. In contrast, partial pose determination is wide-spread and affordable. Laser interferometer as well as time-of-flight measurements provide only the distance between the measurement device and a single point on the robot. Some devices additionally include information on the direction of the measurement resulting in polar coordinates of the reference point. Also, full pose reconstruction is possible e.g. using three to six distinct reference points, sequential measurements, or by using stereo-vision. However, determination of full pose information with high accuracy remains time-consuming and expensive.

  Secondary properties of the measurement device are its measurement volume, the ease of setup and usage, its price, and the measurement speed.
- *Calibration algorithm*: The actual parameter identification algorithm takes the nominal measurement poses with the measured data, the data received from the internal sensors of the robot, and the respective kinematic model as input data and estimates the actual geometric parameters of the robot that fit best to the measured data. The calibration algorithm transforms this matching problem into a minimization problem that is solved with available optimization techniques.

9.2.3 Calibration Kinematic Model

Calibration of a robot only makes sense in view of a certain kinematic model. The kinematic transformation used in the robot controller is the only relevant model. The kinematic transformation is a model and thus a simplification of the real physical behavior. It is the core of a *model* to be a simplification of reality in order to achieve a trade-off between the computability and accuracy. In Sect. 3.2, the assumptions made for the standard model are detailed and more involved effects are tackled in Chap. 7. No matter which model is used, some simplifications must be accepted and thus, some errors remain. Eventually one chooses a model that compromises the available efforts (and applicable resources) with the required accuracy for an application. Since the kinematic model is an approximation of the real robot, one accepts by using the model that a certain choice in the geometric parameters of the model does not necessarily reflect a physical system property. Instead, the observed effect may be an artifact generated by the simplifications that are inherent in the model. For example, consider a real cable robot with guiding pulleys with a robot controller that employs a hard-coded standard kinematic model. This situation is typical for a couple of demonstrators in operation. If the cables are relatively short and sufficiently tensed, the cables form straight lines. However, the end of the line where the cable attacks the guiding pulley actually displaces in space on the surface of a torus as carried out in Sect. 7.2.1. The kinematic code in the controller uses a fixed point $\mathbf{a}_i$ in replacement for this toroidal constraint surface. Optimizing the choice for the point $\mathbf{a}_i$ to provide the best accuracy for the robot leads to a numerical value that has no physical relevance for the real robot. Although it is likely that this vector $\mathbf{a}_i$ is near the pulley, there is no guarantee that it actually is.

For the application of the kinematic model in the controller, it does not matter if the parameters have a physical relevance if the desired accuracy is reached. More generally speaking, the parameters of every simplified model are artificial in that they have no actual representations in the robot. The conclusion is that every applicable model includes some simplifications and therefore the sought optimal parameters are not necessarily quantities that can be found physically in the robot. The optimal parameters are just those settings that provide the optimal performance when applied to the used kinematic transformation.

In order to perform a meaningful parameter identification, one has to define the actual kinematic model $\mathbf{\Phi}^{\text{IK}}(\mathbf{y}, \mathbf{g})$ based on the pose $\mathbf{y}$ and a vector of geometric parameters $\mathbf{g}$. Therefore, the standard model presented in Sect. 4.2.1 is considered or one selects an advanced model as introduced in Chap. 7. The selected model $\mathbf{\Phi}^{\text{IK}}$ defines two primary aspects. Firstly, the type of mathematical equations to deal with including the algorithm to compute the solution of the equations, and secondly, the set of parameters $\mathbf{g}$ being specific for the actual robot to be calibrated. The model equations are prescribed by the kinematic model which is used in the controller. Moreover, one has to define which parameters can actually be changed in the controller. Sometimes not every parameter that influences the performance can actually be configured in the controller. For example, in a kinematic code that takes into account pulleys (Sect. 7.2.1), each pulley could have a different radius, however, it is quite common to set all radii to equal values in the implementation. Thus, only one parameter can be configured. For the sake of simplicity, the standard model is employed in the following which is derived in detail in Sect. 3.2. For the utilization in the kinematic code, the closure constraints v_i according to Eq. (3.1) are understood to be functions of the pose $\mathbf{y}$ and the cable lengths $\mathbf{l}$. In view of the calibration, one has to consider the geometric parameters $\mathbf{g}$ of the robot to be initially unknown parameters. Thus, the closure constraints are functions

$$v_i(\mathbf{a}_i, \mathbf{b}_i, \mathbf{y}, l_i) : \mathbf{a}_i - \mathbf{r} - \mathbf{R}\mathbf{b}_i - \mathbf{l}_i = \mathbf{0} \quad \text{for} \quad i = 1, \ldots, m \quad . \tag{9.1}$$

In the same way, the inverse kinematics function has to be understood as a function of the geometric parameters $\mathbf{g} = [\mathbf{a}_i^{\text{T}}, \mathbf{b}_1^{\text{T}}, \ldots, \mathbf{a}_m^{\text{T}}, \mathbf{b}_m^{\text{T}}]^{\text{T}} \in \mathbb{R}^{6m}$. Thus, the function can be written as

$$\mathbf{l} = \boldsymbol{\Phi}^{\text{IK}}(\mathbf{y}, \mathbf{g}) \quad . \tag{9.2}$$

The same procedure is applicable when using more involved kinematic models such as the pulley kinematic model given by Eq. (7.2.1). In this case, the geometric parameters $\mathbf{g}$ has to be extended by the radius r_{R} and the orientation $\mathbf{R}_{\text{A},i}$ of the proximal pulleys. Elastic cable models and hefty cable models add material parameters of the cable, center of gravity of the platform, platform mass etc. which are also subject to the calibration procedure. In practice, additional parameters of the robot may be unknown for the standard model. In a winch, one may have a highly precise encoder (see Sect. 8.6.4.1), however after the assembly of the robot, there is an uncertainty in the mapping between the measured data that is received from the controller and the actual absolute cable length. One can model this uncertainty be introducing initially unknown offset $l_{0,i}$ for each cable with

$$l_i = l_{0,i} + \Theta_{\text{M},i} \, \nu_{\text{w},i} \quad , \tag{9.3}$$

where Θ_i is the rotation angle of the actuator (assuming that the encoder is mounted on the motor shaft) and $\nu_{\text{w},i}$ is the overall transmission ratio of the winch including motor, gearbox, drum radius, and pitch of the drum for the i-th winch. Also, the exact value of this transmission factor $\nu_{\text{w},i}$ is unknown and one has to estimate it from the mechanical design of the winch according to Eq. (6.26). The transmission factor varies amongst the winches due to manufacturing and assembly errors. Thus, even for the standard model, one ends up with a larger number of parameters effecting the mapping between the measurable parameters and the platform pose $\mathbf{y}$ although one neglects most of the nontrivial effects such as elastics, sagging, thermal effects, and pulleys.

9.2.4 Pose Measurement

For external calibration, a full or partial determination of the platform pose is required. The type of measurement device used for the procedure defines if the full pose can be determined or only a partial pose measurement is made. For example, partial pose measurements are only the determination of the position of one point on the mobile platform (e.g. using a laser tracker), only orientation through inclinometer, or even only the distance between a platform pose and a reference point (e.g. through a double-ball bar or laser distance sensors). Furthermore, the accuracy of the measurement is important since one cannot improve the accuracy of the robot beyond the systematical error of the measurement device.

9.2.5   Parameter Fitting

In the following, it is assumed that full pose measurement is applied and a set of nominal measurement poses $\mathcal{P} = \{\mathbf{y}_1, \ldots, \mathbf{y}_{n_M}\}$ for the calibration has been chosen. Then, one receives n_M tuples of cable lengths $\mathbf{l}^{(i)} = \mathbf{\Phi}^{IK}(\mathbf{y}_i)$ and respective external measurements for the pose $\mathbf{y}^{(i)}$. If the model $\mathbf{\Phi}^{IK}$ is perfectly valid, then the measured poses exactly match the nominal poses in the set $\mathcal{P}$. In practice, the measured poses differ from the nominal poses and one computes the difference between the cable length predicted by the kinematic model for the nominal poses $\mathbf{y}_i$ and the cable length determined for this actual poses $\mathbf{y}^{(i)}$ from

$$\xi(\mathbf{l}^{(i)}, \mathbf{y}^{(i)}, \mathbf{g}) = \mathbf{\Phi}^{IK}(\mathbf{y}^{(i)}, \mathbf{g}) - \mathbf{l}^{(i)} \quad , \tag{9.4}$$

where the function ξ is called the residual function. Collecting the values of the residual function for each measured pose $\mathbf{y}^{(i)}$ yields the vector function

$$\boldsymbol{\xi}(\mathbf{g}) = [\xi(\mathbf{l}^{(1)}, \mathbf{y}^{(1)}, \mathbf{g})^T, \ldots, \xi(\mathbf{l}^{(n_M)}, \mathbf{y}^{(n_M)}, \mathbf{g})^T]^T \quad , \tag{9.5}$$

that only depends on the geometry $\mathbf{g}$ of the robot as the numerical values for the pose set from the $\mathcal{P}$ and the measured pose $\mathbf{y}^{(i)}$ are substituted into the equation. By searching for the geometric parameters $\mathbf{g}$ that minimizes the least-square problem

$$\mathbf{g}_{opt} = \min_{\mathbf{g}} \left(\frac{1}{2} \boldsymbol{\xi}(\mathbf{g})^T \boldsymbol{\xi}(\mathbf{g}) \right) \quad , \tag{9.6}$$

one finds the geometry $\mathbf{g}_{opt}$ of the robot that fits best with the observed measurements. Minimization of $\boldsymbol{\xi}^T\boldsymbol{\xi}$ is equivalent to the least-square method and this optimal configuration corresponds to the most likely robot geometry if the errors in the pose measurements are assumed to be Gaussian random variables. Formally speaking, one has to solve the minimization problem to receive the optimal geometric parameters $\mathbf{g}_{opt}$.

The measurements and the kinematic model are transformed into an optimization problem which can be solved by algorithms such as Levenberg-Marquardt method or interval analysis [9]. It must be noted that there are no fundamental results that such minima are unique. Furthermore, over-fitting may occur, i.e. making the robot very accurate in the measurement poses but inaccurate elsewhere. This inaccuracy might even occur at poses between the nominal measurement poses from the set $\mathcal{P}$.

For numerical solving of the minimization problem as well as for the analysis of the numerical stability, the Jacobian matrix of the objective function $\boldsymbol{\xi}$ has to be considered. Therefore, one computes the pose-dependent derivatives of the function $\mathbf{\Phi}^{IK}$ with respect to the geometric parameters $\mathbf{g}$ from

$$\mathbf{J}_G = \frac{\partial \mathbf{\Phi}^{IK}}{\partial \mathbf{g}} \quad , \tag{9.7}$$

where $\mathbf{g}$ collects all geometric parameters subject to the calibration procedure. Although $\mathbf{J}_G$ is related to the kinematic Jacobian matrix that occurs in the differential kinematic (Sect. 4.2.3), it has one column for each geometric parameter in the model, where the positions of the proximal and distal anchor points already contribute $6m$ parameters. The *identification matrix* $\mathbf{H}$ [212] is composed from the Jacobian matrices $\mathbf{J}_G(\mathbf{y}_i)$ taken at different measurement poses $\mathbf{y}_i \in \mathcal{P}$ and it reads

$$\mathbf{H} = \begin{bmatrix} \mathbf{J}_G^{(1)}(\mathbf{y}_1) \\ \vdots \\ \mathbf{J}_G^{(n_M)}(\mathbf{y}_{n_M}) \end{bmatrix} . \tag{9.8}$$

Using the identification matrix $\mathbf{H}$, one computes a first order estimate $\Delta\mathbf{g}$ for the actual geometric parameters by solving the linear equation

$$(\mathbf{H}\mathbf{H}^T)\Delta\mathbf{g} = \mathbf{H}^T \boldsymbol{\xi} \ , \tag{9.9}$$

where $\boldsymbol{\xi}$ is the collected difference between the measured and the estimated cable length.

9.2.6 Measurement Pose Selection

The result of the calibration notably depends on the selection of the poses in the set $\mathcal{P}$ as shown e.g. by Boye and Verl [63, 478]. To decide which poses are suited for calibration, a performance measure for the pose set is required. The *observability index* k_O is defined as the condition number of the identification matrix $\mathbf{H}$ and is computed from the ratio between the smallest and the largest singular value of $\mathbf{H}$ as follows

$$k_O = \frac{\sigma_{max}}{\sigma_{min}} \ , \tag{9.10}$$

where σ_{min} and σ_{max} are the smallest and largest singular value of $\mathbf{H}$, respectively. The observability index k_O has an optimal value of 1 when the maximum and minimum singular values are equal. This means that the transformation identified with the identification matrix induced no distortion in the parameter space. However, this optimal situation is unrealistic in practice. In turn, high values for the observability index are related to large numerical errors due to bad numerical conditioning of the related numerical problem. Since one can influence the observability index by choosing appropriate poses for the calibration procedure, one can assess the measurement pose set $\mathcal{P}$ using this index.

One could expect that the final error of the calibration procedure generally
decreases when using a larger number of poses since the influence of statistical
measurement errors and noise is canceled with higher numbers of measurements.
This intuition from conventional physical measurement is wrong since a large number of pose measurements used in the parameter fitting has a negative influence on
the numerical conditioning of the fitting algorithm. The numerical errors caused by
larger problems can exceed the benefits from the statistical effects. Therefore, one
searches for a trade-off between a high number of poses for averaging out measurement errors and a low number of poses to keep numerical errors small. Thus, a
well-distributed and minimal set of measurements reduces the efforts for executing
the measurement and increases the expectable accuracy of the calibration.

The effectiveness of the whole calibration procedure can only be assessed by
experimental verification. In contrast, one can improve the numerical conditioning
of the problem by proper selection of the poses for the measurement pose set. This
selection can be supported by numerical simulation. Furthermore, one has to take into
account that the improvements made through calibration are achieved in a section of
the workspace around the measurement pose set. As experiments show, the accuracy
may be improved in a small area around the selected pose set $\mathcal{P}$ whereas in other
regions of the reachable workspace, the accuracy gets poorer. Therefore, also the
desired workspace is a side condition for designing the measurement pose set $\mathcal{P}$.

An example for the determination of the condition number $k_0(\mathbf{H})$ depending on
the number of the measurement poses in the set $\mathcal{P}$ is shown in Fig. 9.3. The nominal
design of the IPAnema 1 is used in this case study. The basic pose set $\mathcal{P}$ is generated
from 250 randomly selected poses in the range $x, y \in [-1; 1]$ and $z \in [0.5; 1.5]$
and for randomly set orientations through Bryant angles in the range $[-\frac{\pi}{6}; \frac{\pi}{6}]$. As
observed in other contributions (see e.g. [345]), there is a minimum in the condition
number around 25 poses.

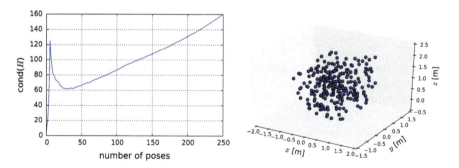

Fig. 9.3 Left: Condition number of the identification matrix $k_0(\mathbf{H})$ for different numbers of poses
in the measurement pose set $\mathcal{P}$ for the IPAnema 1 robot. Right: Distribution of random poses $\mathcal{P}$ in
the workspace

9.3 IPAnema Robot Family

Research on cable robot is not an end in itself but targets on enabling the construction and also the application in real world scenarios. Many of the theoretical considerations and the direction of research pursued in this work originate from the question what it takes to put such cable robots to practical work. In the following sections, the system architecture of the cable robot family IPAnema is presented which consists of multiple generations of demonstrator systems. As outline, the nominal parameters of these reference designs are given. These values are used throughout this work to provide the basis and rationale for the numerical examples. Since these robots are used for experimental evaluations also measurements are presented. All parameters given in the tables are nominal values, i.e. the ideal design parameters. These numbers reflect the physical dimensions and used components of the prototypes.

The robots of the IPAnema family as well as related other cable robots introduced below guided the line of research that led to this work. It is important to understand that the IPAnema robots are not the result of the design procedure presented in Chap. 8 but also the design procedure matured while building and configuring a dozen robots. Hence, the derivation of the design procedure and the development of this family of robots evolved simultaneously. The robots are therefore understood as solid experience with an elaborated design but we do not call these robots best practice as the robots are not optimal. The final results of the design procedure are given below in terms of the description of the demonstrator. All the erroneous trials undertaken are concealed for the sake of brevity. Thus, the robots are understood as some feasible practice and the short stories told about these robots should provide the reader with some hints for future projects (Fig. 9.4).

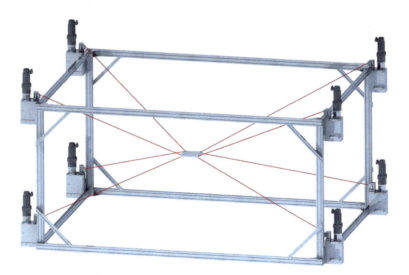

Fig. 9.4 CAD model of the cable robot IPAnema 1 in 8-4 configuration

The first reference design is the cable robot IPAnema 1 (and its slightly modified version IPAnema 1.5) that was used from 2007 to 2010 at Fraunhofer IPA (Germany) for experimental purpose. The parameters of the very first version of this robot are given in Table 9.1. It uses a flat platform with an 8-4 configuration where two cables share one distal anchor point. Later, the winches were reconfigured (IPAnema 1.5 in Table 9.1) such that the upper winches are connected to the lower side of the mobile platform and the lower winches are connected to the upper side of the mobile platform leading to a crossed cable configuration. At the same time, the winches are displaced in the xy-plane along the edges of the machine frame to avoid interference amongst the cables. In 2010, the robot geometry was completely redesigned for a handling application. The resulting robot is named IPAnema 2 and its geometric parameters are given in Table 9.1. The size of the frame is largely increased and a platform of reasonable size for the targeted application is added. By elaborating and refining the technical approaches from the IPAnema 1 and 2 generation, a third robot is developed which geometry is given at the bottom of Table 9.1.

9.3.1 IPAnema 1

The purpose of the IPAnema robot family is to bridge the gap between the laboratory test-beds to an industrial pilot installation by scaling up what is basically possible to what is feasible in an industrial application. The IPAnema 1 is the original demonstrator for the experimental evaluation in this work. It is an all new robot design based on newly constructed winches, an industrial real-time controller platform, and state of the art industrial drive-trains. The main objective in its design is the industrialization of the cable robot technology restricting the employed components to those that are available in industrial quality. Such design restrictions are considered as basis for the technology transfer targeting at commercial applications and fulfilling regulations such as safety norms and implications of the machine directive.[1] The IPAnema 1 robot provides a six degrees-of-freedom mobile platform constrained by seven or eight cables. The application focus is on industrial applications in the field of material handling as well as fast pick-and-place applications (Fig. 9.5 and Table 9.2).

9.3.1.1 Control System Architecture

The cable robot IPAnema 1 presents some technological innovations in 2007 which were needed to execute robotic motion programs. The controller is developed based on the open controller architecture for machine tools between 2007 and 2009 and seems to be the first computerized numerical control (CNC) based controller used for a cable robot. The basic of the controller is the motion kernel by Industrielle Steuerungstechnik GmbH (ISW, Germany). The controller architecture of

---

[1] European Machine Directive 2006/42/EC.

Table 9.1 Nominal geometric parameters: base vectors $\mathbf{a}_i$ and platform vectors $\mathbf{b}_i$ of four robots used for the reference examples: IPAnema 1, 1.5, 2, and 3

Robot	Cable i	Base vector $\mathbf{a}_i$	Platform vector $\mathbf{b}_i$
IPAnema 1	1	$[-2.0, 1.5, 2.0]^{\mathrm{T}}$	$[-0.06, 0.06, 0.0]^{\mathrm{T}}$
	2	$[2.0, 1.5, 2.0]^{\mathrm{T}}$	$[0.06, 0.06, 0.0]^{\mathrm{T}}$
	3	$[2.0, -1.5, 2.0]^{\mathrm{T}}$	$[0.06, -0.06, 0.0]^{\mathrm{T}}$
	4	$[-2.0, -1.5, 2.0]^{\mathrm{T}}$	$[-0.06, -0.06, 0.0]^{\mathrm{T}}$
	5	$[-2.0, 1.5, 0.0]^{\mathrm{T}}$	$[-0.06, 0.06, 0.0]^{\mathrm{T}}$
	6	$[2.0, 1.5, 0.0]^{\mathrm{T}}$	$[0.06, 0.06, 0.0]^{\mathrm{T}}$
	7	$[2.0, -1.5, 0.0]^{\mathrm{T}}$	$[0.06, -0.06, 0.0]^{\mathrm{T}}$
	8	$[-2.0, -1.5, 0.0]^{\mathrm{T}}$	$[-0.06, -0.06, 0.0]^{\mathrm{T}}$
IPAnema 1.5	1	$[-1.8, 1.5, 2.0]^{\mathrm{T}}$	$[-0.1, 0.1, 0.0]^{\mathrm{T}}$
	2	$[1.8, 1.5, 2.0]^{\mathrm{T}}$	$[0.1, 0.1, 0.0]^{\mathrm{T}}$
	3	$[1.8, -1.5, 2.0]^{\mathrm{T}}$	$[0.1, -0.1, 0.0]^{\mathrm{T}}$
	4	$[-1.8, -1.5, 2.0]^{\mathrm{T}}$	$[-0.1, -0.1, 0.0]^{\mathrm{T}}$
	5	$[-2.0, 1.3, 0.0]^{\mathrm{T}}$	$[-0.1, 0.1, 0.2]^{\mathrm{T}}$
	6	$[2.0, 1.3, 0.0]^{\mathrm{T}}$	$[0.1, 0.1, 0.2]^{\mathrm{T}}$
	7	$[2.0, -1.3, 0.0]^{\mathrm{T}}$	$[0.1, -0.1, 0.2]^{\mathrm{T}}$
	8	$[-2.0, -1.3, 0.0]^{\mathrm{T}}$	$[-0.1, -0.1, 0.2]^{\mathrm{T}}$
IPAnema 2	1	$[-4.0, 3.0, 5.0]^{\mathrm{T}}$	$[-0.65, 0.125, 0.25]^{\mathrm{T}}$
	2	$[4.0, 3.0, 5.0]^{\mathrm{T}}$	$[0.65, 0.125, 0.25]^{\mathrm{T}}$
	3	$[4.0, -3.0, 5.0]^{\mathrm{T}}$	$[0.65, -0.125, 0.25]^{\mathrm{T}}$
	4	$[-4.0, -3.0, 5.0]^{\mathrm{T}}$	$[-0.65, -0.125, 0.25]^{\mathrm{T}}$
	5	$[-4.0, 3.0, 1.0]^{\mathrm{T}}$	$[-0.75, 0.100, 0.75]^{\mathrm{T}}$
	6	$[4.0, 3.0, 1.0]^{\mathrm{T}}$	$[0.75, 0.100, 0.75]^{\mathrm{T}}$
	7	$[4.0, -3.0, 1.0]^{\mathrm{T}}$	$[0.75, -0.100, 0.75]^{\mathrm{T}}$
	8	$[-4.0, -3.0, 1.0]^{\mathrm{T}}$	$[-0.75, -0.100, 0.75]^{\mathrm{T}}$
IPAnema 3	1	$[8.185, 5.693, 3.203]^{\mathrm{T}}$	$[0.061, 0.649, -0.262]^{\mathrm{T}}$
	2	$[8.224, -5.492, 3.236]^{\mathrm{T}}$	$[0.061, -0.651, -0.262]^{\mathrm{T}}$
	3	$[-8.491, -5.322, 3.250]^{\mathrm{T}}$	$[-0.070, -0.652, -0.261]^{\mathrm{T}}$
	4	$[-8.545, 5.464, 3.221]^{\mathrm{T}}$	$[-0.070, 0.648, -0.261]^{\mathrm{T}}$
	5	$[7.208, 6.464, -0.590]^{\mathrm{T}}$	$[0.095, 0.749, 0.261]^{\mathrm{T}}$
	6	$[7.869, -5.558, -0.549]^{\mathrm{T}}$	$[0.095, -0.746, 0.261]^{\mathrm{T}}$
	7	$[-8.271, -5.546, -0.528]^{\mathrm{T}}$	$[-0.086, -0.746, 0.262]^{\mathrm{T}}$
	8	$[-8.192, 5.648, -0.583]^{\mathrm{T}}$	$[-0.086, 0.749, 0.262]^{\mathrm{T}}$

Fig. 9.5 Cable-driven parallel robot IPAnema setup with seven cables

the IPAnema 1 robot is depicted in Fig. 9.6. The motors in the winches are connected through an optical SERCOS II interface and set-point values are transferred with a cycle time of 2 ms to the servo amplifiers which run a decentralized position-velocity-torque cascaded control. The control system is implemented into a PC-based real-time operating system RTX by Interval Zero. A software programmable logic controller (PLC) is coupled to the NC-kernel at full cycle time. One of the major challenges in the development of the controller system is to integrate the custom kinematic transformation (Chap. 4) into the controller architecture. For path generation, it is necessary to calculate the set-point values for the cable length from given Cartesian coordinates in real-time and to allow supervision and correction of the generated set-values. This is essentially done through an inverse kinematics code based on the standard model. The algorithms described in Sect. 4.2.1 are implemented as real-time C-code. The Cartesian set-point coordinates of the mobile platform are generated from an NC-program written in G-Code (DIN 66025 [123]). Contrary, the actual values measured by multi-turn absolute encoders inside the winches are transformed into the current pose of the end-effector. The latter transformation is by far the more complicated one: it involves the forward kinematics and its real-time implementation for the standard model is described in Sect. 4.3. The computation time for the forward kinematics limits the cycle time to 2 ms.

Table 9.2 Overview of the IPAnema 1 specification

Parameter and symbol	Value/component	Unit
Number of cables m	8	–
Degrees-of-freedom n	6, 3R3T, fully-constrained	–
Parameterization Φ^G	IPAnema 1	–
Size of the robot frame	$4.0 \times 3.0 \times 2.0$	m
Size of the mobile platform	0.12×0.12	m
Rated cable force f_{min}, f_{max}	10–180	N
Max. cable velocity $\dot{l}_{max}$	10	m/s
Drive	MSK050B-0600-NN-M2-UG1-RNNN	–
Drive power	1.8	kW
Gear box	GTE120-NN1-003A-NN20	–
Gear ratio ν_{PG}	3:1	–
Drum diameter d_D	0.1	m
Cable type	Dyneema, LIROS D-Pro 01505-0150	–
Cable diameter d_C	1.5	mm
Specific cable stiffness k'_C	28500	N
Cable force sensor	Disynet XFTC-300-A1-2.000	–
Cable force measurement	±2000	N
Control system	RTX, ISG Motion Kernel	–

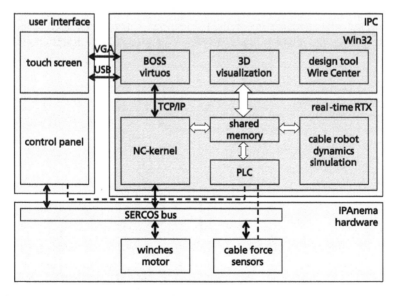

Fig. 9.6 System structure for the cable robot IPAnema 1

To ensure the reliable operation of the cable robot system, the tension in the cables is monitored by the PLC. This is done in two different ways: Firstly, the tension is estimated from the measured current in the motors. This kind of measurement is integrated into the motors but is rather inaccurate. A better measurement is received from force sensors which are integrated into the distal anchor points of the cables. For supervision, the comparison with the force limits f_{min} and f_{max} is done in the PLC.

9.3.1.2   Design of Winches and Drive-Trains

The mechanical design of the winches is derived from crane winches where some additional requirements are taken into account to control and operate the cable robot. A first requirement for permanent operation of the cable robot without excessive wear of the cables is bending the cable at most with a radius that is significantly larger than the diameter of the cables. Secondly, the direction of the cable changes continuously during operation of the cable robot. Therefore, it is necessary to include an omnidirectional guidance mechanism into the winch. The concept of the IPAnema 1 winch is shown in Fig. 9.7.

A synchronous servo motor IndraDyn S by Bosch-Rexroth with integrated multi-turn absolute encoder is coupled to a planetary gearbox with transmission ratio of $\nu_{PG} = 3$. This drive-train is connected to a drum with a diameter $d_D = 100$ mm. The winch can store a cable length of up to $\Delta l_{max} = 6$ m. The winches are equipped with multi-turn absolute encoders allowing to obtain the absolute cable length at the fully cycle time with a resolution of 50 μm. The drum is connected to an additional gearing that moves a cable guidance in parallel to the drum. Due to equal pitch of the drum and the spindle, the relative direction of the coiled cable is constant allowing for

Fig. 9.7 Concept for servo-controlled winch for cable robot IPAnema 1

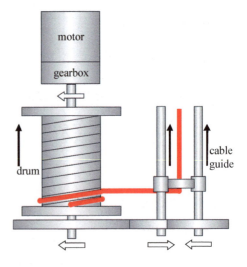

reliable coiling and uncoiling of the cable. This is especially important since the velocities and accelerations of the cables are high for the IPAnema 1 robot.

The cable guidance includes the spooling unit that redirects the cable to run parallel to the axis of the drum. A second panning pulley is mounted to the housing of the winch that allows for an omnidirectional redirection of the cable into the inner workspace of the cable robot.

9.3.1.3 Control Software and Kinematic Transformation

The control system of the cable robot consists of closed-loop position control algorithms which are integrated into the amplifier modules for each winches. On the position level, servo-controller from Bosch-Rexroth (power supply unit HMV01.1E-W0030 with axis inverter HMS01.1N-W0020) are used to execute the motion commands which are sent by the CNC-kernel through the optical SERCOS II bus. The decoupled cascaded closed-loop position and current control for each servo motor is tuned for each drive within the framework of the servo-controller.

The CNC-kernel runs within the real-time extension RTX of the Windows XP operating system. The CNC-kernel interprets robot programs written in G-Code and generates smooth trajectories according to the G-Code program in operational space coordinates. To translate the desired motion of the mobile platform to cable lengths and then to set values of the servo motors, a kinematic transformation based on the standard model is used (Sect. 4.2.1). Contrary, the measured cable length is read from the absolute encoders of the winches in order to estimate the current position of the mobile platform. The latter is done by a real-time capable forward transformation (Sect. 4.3).

For interfacing with standard industrial equipment, a programmable logic controller (CoDeSys PLC by 3S-Smart Software Solutions) is coupled with the NC-kernel through a shared memory interface. The data exchange is performed at every interpolation cycle time of 2 ms. The integration of additional sensor data, i.e. the cable forces read from the force sensors is processed in the PLC. Furthermore, the PLC can interface with additional IT-infrastructure such as a manufacturing execution system.

The user interface of the robot consists of two main components. Firstly, a touch-screen with a graphical user interface (GUI) is implemented in the Windows XP operating system and allows for visualization of complex data structures in textural and graphical form. Secondly, a control panel is integrated to provide hardware keys to switch operation modes, unlock the drives, and to provide an emergency stop button. A non real-time data exchange is possible between the GUI and the CNC-kernel where the latency of the Windows system is hardly recognized by the user when reading the display. The data exchange between the user interface and the CNC-kernel is realized through a TCP/IP stack. The control panel is coupled with the SERCOS II bus allowing to directly interact with the CNC-kernel and the PLC.

The multi-physics dynamic simulation of the cable robot dynamics presented in Chap. 6 is coupled to this control system. This hardware-in-the-loop simulation is used to verify and evaluate this robot as well as to design parts of the control system. Especially, the design of the force control of the IPAnema 1 is largely simplified as it is optimized with a simulation model.

9.3.1.4 Experimental Results

This hardware of the robot is used for experimental evaluation in a couple of studies. In the following, we focus on accuracy assessment according the norm ISO 9283 [220], which describes performance criteria for robots and which are important to evaluate if the robot's technique is adequate for practical applications. All accuracy measurements are performed using a Leica Absolute Tracker AT901-MR with a certified absolute accuracy of less than $25\,\mu m$.

As an example, the definition of the position repeatability is shown in Fig. 9.1. Similar definitions are detailed in ISO 9283 for the path repeatability and for the distance repeatability. Following the evaluation procedure given in the norm, 30 single measurements are taken at different velocities of the IPAnema 1 robot ($v_1 = 0.5\,m/s$, $v_2 = 2.5\,m/s$, $v_3 = 5\,m/s$). The pose repeatability is found to be smaller than $r_{Pose} = 0.75\,mm$ for all velocities where it largely depends on the nominal velocity. Interestingly, it is found that the best values could be achieved with highest velocity of the cable robot where a repeatability better than $r_{Pose} = 0.35\,mm$ is determined. For the distance repeatability, a value better than $r_{Dis} = 0.2\,mm$ is determined. For distance repeatability, we measured better values at lower velocities where the best values of less than $r_{Dis} = 0.06\,mm$ are measured for v_1. Finally, the path repeatability is investigated and the experiment yields a value of $r_{Path} = 0.5\,mm$. Again, the best values are measured for v_1 where the repeatability is $r_{Path} = 0.17\,mm$.

A summary of performance indices of different robots from the IPAnema family are given in Table 9.3. The frame size is the diagonal length of the respective robot frame giving a reference value for the robot's overall size as well as an indication of the average cable length. All experiments except for the IPAnema 2 planar are undertaken with synthetic fiber cables. The experimental evaluation of the cable robot

Table 9.3 Summary of the position accuracy and repeatability measurements of the IPAnema 1, IPAnema 2 planar, IPAnema 3, and IPAnema 3 Mini prototype

Robot	Frame size [m]	Repeatability [mm]	Accuracy [mm]
IPAnema 1	5.39	0.75 [401]	5.63 [229]
IPAnema 2	12.66	0.59 [434]	22.32 [434]
IPAnema 2 (pulley)	12.66	0.51 [434]	17.50 [434]
IPAnema 2 planar	4.47	0.1–0.2	8.5
IPAnema 3	21.2	5.237 [431]	40.423 [431]
IPAnema 3 Mini	1.69	0.081 [431]	1.384 [431]

accuracy and repeatability yields encouraging results to underline that the cable robot technology is capable of fulfilling industrial requirements.

9.3.2 IPAnema 2

The IPAnema 2 system was set up in 2010 and is a reconfiguration of some IPAnema 1 components, where a couple of improvements and extensions are integrated. The basic mechanical design of the winch is kept. The reconfiguration includes significant changes in the drive-trains including the exchange of the planetary gearboxes allowing to increase the maximum cable force by four. With a newly designed mobile platform, the rated payload of the robot is increased by a factor of ten while reducing the maximum velocities and accelerations by four. The increase in the payload beyond the change in the maximum cable velocity is achieved by reducing the nominal acceleration. At the same time, the volume of the workspace grows by ten thanks to the larger machine frame. The experimental retrofitting practically shows that reconfiguration of cable robots is feasible and allows to modify the technical parameters in a range that is not possible with conventional robots. The technical key data of the IPAnema 2 are given in Table 9.4.

9.3.2.1 Design for Assembly Task

The robot is designed to avoid collisions of the cables with the mobile platform and with the handled objects. Therefore, the robot geometry is checked for interference between cables following the geometric approach (Sect. 5.2.6). The region of cable interference for the IPAnema 2 robot is depicted in Fig. 9.8. It can be seen that no intersection between the region of interference and the workspace occurs. Note, that the upper and lower cables are crossed in the workspace, i.e. the upper winches are connected to the lower platform level and the lower winches are connected to a higher platform level. This design allows for improved stiffness and reduction of cable-environment collisions. Still, no cable-cable collisions occur inside of the machine frame and the workspace of this robot is thereby free of interference.

9.3.2.2 Mechanical Design

The machine frame of the IPAnema 2 robots is constructed from steel beams to realize the increased size to $9 \times 7 \times 5.5$ m. Using a steel frame shows better static and dynamic stiffness behavior compared to the aluminum frame used with the IPAnema 1 robot. In contrast to the much smaller IPAnema 1 frame, installation and also maintenance of the large frame requires a working platform to access the upper winches. In order to resign to such equipment for operating the robot and also to

Table 9.4 Overview of the IPAnema 2 specification

Parameter and symbol	Value/component	Unit
Number of cables m	8	–
Degrees-of-freedom n	6, 3R3T, fully-constrained	–
Parameterization $\mathbf{\Phi}^G$	IPAnema 2	–
Size of the robot frame	$9.0 \times 7.0 \times 5.5$	m
Size of the mobile platform	$1.5 \times 0.225 \times 0.5$	m
Rated cable force f_{min}, f_{max}	100–720	N
Max. cable velocity l_{max}	2.5	m/s
Drive	MSK050B-0600-NN-M2-UG1-RNNN	–
Drive power	1.8	kW
Gear box	Bosch Rexroth GTE160-NN2-012B-NN16	–
Gear ratio v_{PG}	12:1	–
Drum diameter d_D	0.1	m
Cable type	Dyneema, LIROS D-Pro 01505-0250	–
Cable diameter d_C	2.5	mm
Specific cable stiffness k_C'	55000	N
Cable force sensor	Disynet XFTC-300-A1-2.000	–
Cable force measurement	±2000	N
Control system	RTX, ISG Motion Kernel, v.262	–

simplify wiring of the winches, the later version of the IPAnema robots placed all winches on the floor and used fixed pulleys on the upper parts of the machine frame.

For the new robot, the end-effector is designed to match the footprint of the collector modules of the exemplary handling application. The mobile platform has a size of $1.5 \times 0.225 \times 0.5$ m with integrated grippers at the lower side for the handling application. Additionally, a light-weight robot is installed on the mobile platform to execute simple assembly tasks at different positions in the large workspace of the IPAnema 2 robot. A new version of the control software makes use of multiple CNC-channels to simultaneous control the cable robot and a light-weight robotic arm. To supply the serial robot with the CAN fieldbus as well as with electric and pneumatic energy, an energy chain is installed leading fieldbus, energy, and pressured air to the mobile platform. This gives also the opportunity to connect force sensors on the distal cable ends (Fig. 8.36). Thanks to the largely increased payload and the existing energy chain, the force sensors with their A/D converter are installed on the mobile platform and the measured forces are mapped into the PLC. However, the cable forces are only used for supervision and for statically controlling the pretension in the cables during referencing.

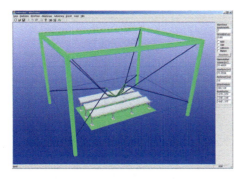 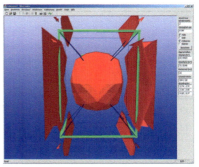

Fig. 9.8 Design tool. Left: Motion study for the application and geometry design of cable robots. Right: Top view of the IPAnema 2 robot: Interference between cables (red region around the machine frame) only occurs outside the workspace (round region in the center)

As the IPAnema winches are operated with the maximum stroke to exploit the full workspace, experimental inductive end-of-travel sensors are installed for sensory detection of the minimum and maximum cable length in the winches.

9.3.3 IPAnema 2 Planar

In a couple of applications such as warehousing, a robot system is needed that performs a planar motion with two translational degrees-of-freedom, where rotation may not be required. For that purpose, the planar IPAnema 2 system is designed using four linear direct drives to actuate the platform. The open space in the robot frame has a width of 4 m and a height of 2 m. All four linear actuators are placed in the lower part of the frame and the cables are guided by pulleys through the frame structure. The center of gravity of the overall robot system is close to the floor giving the robot a solid stability. Two arrangements for the cables are possible. In the conventional design, the cables are guided around four pulleys that are placed in the corners of the machine frame. This kind of design is largely studied as planar robot. In the second design, the pulleys are located on trolleys that move along a linear guiding on the inner side of the robot frame. In this setting, two pairs of the motors operate antagonistic by generating one translational degree-of-freedom at the end-effector. The trolleys are connected to the motors by additional cables such that the vertical trolley movement is mechanically coupled to the end-effector motion. Thus, one receives a cable-driven parallel robot with decoupled Cartesian motion (Fig. 9.9).

For this orthogonal arrangement of the cables, the kinematic transformation becomes trivial and the kinematic Jacobian matrix of the robot is isotropic in every tensed configuration. The motion of the actuator is mapped one-to-one to the horizontal or vertical motion of the platform, respectively. To improve the motion of the platform, a force control is applied to the pairs of actuators which is also very

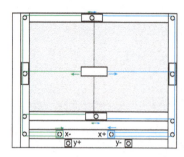

Fig. 9.9 Planar IPAnema 2. left: kinematic concept for the decoupled motion of the mobile platform.
Right: prototype with orthogonal cable arrangement

simple since there is only a one-dimensional coupling in the motor pairs. The orthogonal robot design has a rectangular workspace allowing the cable robot to reach the
very corners of the robot frame. Therefore, the robot has an excellent ratio between
workspace and installation space. The rectangular shape of the workspace is highly
favorable for warehousing applications.

All conventional planar cable robots suffer from low stiffness perpendicular to
the plane of the cables if no additional mechanical elements are used to constrain
the robot to the plane. However, such guiding largely compromise the minimalist
mechanical design of a cable robot. Also, experimental tests with pairs of cables
show little improvements in this behavior. One method to overcome this problem
is analyzed using the experimental setup of the IPAnema 2 planar robot. Instead
of cables, steel bands are used to actuate the robot. Using sufficiently thin strips,
one can coil and guide the steel band on drums and over rollers. The mechanical
element equivalent to a winch consists of a simple drum where the steel band is
reeled like a film spool making the mechanical design simple and straightforward.
In the experimental setup, the drum is aligned between two bearings and a simple
clutch is used to connect the drum to a servo motor without gearbox. As a result,
one receives a planar cable robot which is kinematically equivalent and that employs
bands as transmission elements. Using commercially available spring bands with a
thickness of 0.2 mm for the bands, one can use off-the-shelf components for the robot.
In contrast to steel or synthetic fiber cables, the steel bands yield a finite stiffness
against bending. Even more, if the robot is pretensed, acceptable stiffness values are
obtained for the 4×2 m prototype. Beside the gains in stiffness, steel bands offer
some advantages such as accurately coiling and a largely improved repeatability of
0.1 mm. The comparison of the accuracy of the IPAnema 2 planar with the other
robots is shown in Table 9.3.

Table 9.5 Overview of the IPAnema 3 specification

Parameter and symbol	Value/component	Unit
Number of cables m	8	–
Degrees-of-freedom n	6, 3R3T, fully-constrained	–
Parameterization $\boldsymbol{\Phi}^G$	IPAnema 3	–
Size of the robot frame	$17.0 \times 12.0 \times 4.5$	m
Size of the mobile platform	$1.5 \times 0.225 \times 0.5$	m
Rated cable force f_{min}, f_{max}	100–3000	N
Max. cable velocity $\dot{l}_{max}$	1.7	m/s
Drive	Bosch Rexroth MSK071E-0300-NN-M1-UG1-NNNN	–
Drive power	5.0	kW
Gear box	Bosch Rexroth GTE160-NN1-005B-NN16	–
Gear ratio ν_{PG}	12:1	–
Drum diameter d_D	0.1	m
Cable type	Dyneema, LIROS D-Pro 01505-0600	–
Cable diameter d_C	6.0	mm
Specific cable stiffness k_C'	900000	N
Cable force sensor	Tecsis F2301	–
Cable force measurement	Range 0–4000	N
Control system	TwinCAT 3.1, Beckhoff	–

9.3.4   IPAnema 3

The IPAnema 3 demonstrator is largely composed of new components. A new generation of winches is designed for higher payloads and larger cable stroke. The controller is ported to TwinCAT 3.1 (Beckhoff, Germany), which essentially contains a newer version of the ISG motion kernel. The required extension of the kinematic transformation for cable robots is implemented without proprietary extensions. The fieldbus is changed from SERCOS II to EtherCAT (Table 9.5).

9.3.4.1   Mechanical Design

To simplify mechanical design and installation, the eight winches are grouped in so-called winch batteries to pairs or even quadruples. These batteries are either located at the lower corners of the machine frame (four pairs) or in the middles of the head sides of the frame (quadruples). In order to realize the predominantly box shaped geometry of the points A_i, cable guidance units (see Sect. 8.6.1.2) are applied to the machine frame. The rationale behind is the simplification in mechanical design, transport, and setup of the system since the electro-mechanical components are concentrated in the compact winch batteries. A drawback of this mechanical packaging is the additional elasticity in the longer cables as well as uneven distribution of the elasticity amongst the winches due to the different overall lengths of the cables.

Compared to the first two generations of the IPAnema robot, significantly stronger motors and gearboxes are used to exploit the specifications of the winches. The combination of motor and gearbox allows for nominal cable force $f_{max} = 3\,\text{kN}$ and a nominal cable velocity of $\dot{l}_{max} = 1.7\,\text{m/s}$. Since the motors in the winches generate considerably higher forces in overload mode, the force sensors with thin film implant are chosen to measure cable forces of up to 4 kN with an accuracy of 0.2% according to the manufacturer. In addition, the winches are equipped with binary end-of-travel sensors which are simply connected to the servo amplifier in order to prevent mechanical damage when reaching the maximum and minimum position of the winch's spooling system.

9.3.4.2 Controller Architecture

The IPAnema 3 controller architecture is depicted in Fig. 9.10 and represents an evolution of the controller architecture used for the IPAnema 1 robot. Compared to the previous version, the control system is implemented on the TwinCAT 3.1 platform by Beckhoff, which allows to customize the kinematic transformation for inverse and forward kinematics within the real-time framework. Thus, the set-point pose $\mathbf{y}_s$ determined from the interpolation of the CNC program is fed to the custom inverse kinematics block and simultaneously mapped through a shared memory to the PLC. Using the inverse kinematics code (with or without pulley), the set-point cable lengths are computed. The drive interface maps geometric cable length to the respective angle of the motor taking into account gear ratio and overall transmission ratio of the winch and encoder resolution. Using the position control interface of the decentralized motor controllers, the CNC control sends the set-point values $\mathbf{\Theta}_s$ over the EtherCAT fieldbus to the drives with a 1 ms cycle time. The servo amplifier implements a vendor specific cascaded position-velocity-current control scheme where a finally pulse width modulation (PWM) is used to generate the current for the synchronous servo motors. Closed-loop control is performed only within the servo amplifier. The motor shaft angle is determined through motor integrated encoders which are both used for the closed-loop control and are sent over the fieldbus to the CNC controller. The angles $\mathbf{\Theta}$ are transformed by the drive interface to cable lengths which are in turn mapped through the forward kinematics to the estimation $\mathbf{y}_1$ of the current pose. The CNC uses this value for showing the actual position of the platform and to determine position lag. However, the difference between $\mathbf{y}_s$ and $\mathbf{y}_1$ is only supervised to remain under a threshold.

The controller architecture does not support force control. However, the force controller is implemented in the PLC which is synchronized at full cycle time with the motion controller of the CNC stack. The controller outputs $\Delta\mathbf{l}_F$ of the force controller are then super-positioned with the set-point values $\mathbf{l}_s$ computed from the

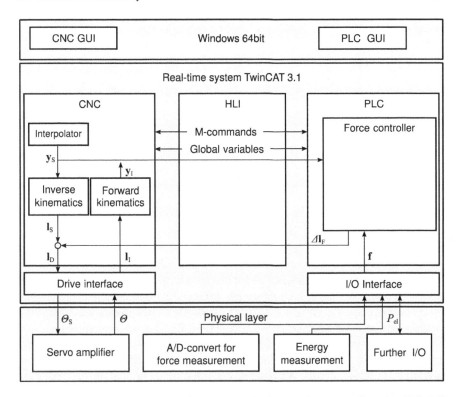

Fig. 9.10 The controller architecture of the IPAnema 3 robot with force control based on TwinCAT 3.1

inverse kinematics to yield the input $\mathbf{l}_D$ for the drive interface. In addition to the force measurement $\mathbf{f}$ that is fed into the PLC through analogue-to-digital (AD) converters and EtherCAT fieldbus, also the energy consumption of the robot is monitored and mapped to the PLC at full cycle time. When required, also further analogous and digital inputs and outputs (I/O) can be controlled by the PLC e.g. in order to trigger a gripper.

9.3.4.3 Reconfiguration

The IPAnema 3 robot is subject of a couple of reconfiguration experiments. The initial operation of the winches is executed in the steel frame already used for the IPAnema 2 robot. Different geometries both for the robot frame as well as for the mobile platform are tested. This includes a planar platform and a strut shaped platform. Also, an experimental evaluation of transportation and setup at a remote location are executed (Fig. 9.11). It is determined that transport, setup, initial operation, demonstration, and dismounting of the robot is possible within 10 days. Eventually, the robot is set up

Fig. 9.11 Examples for the reconfiguration of the cable robots. Usage of a strut platform (left) and installation of the cable robot in a large handling and logistics application (right)

with the parameters given in Table 9.1 using the 6 mm Dyneema cable where 2.5 mm cables are used in the earlier experiments.

For enabling efficient reconfiguration, it is essential to package the mechanical integration of winches in batteries and thereby to ease the electric wiring as well as rapid setup of the cables. Using double pulley units (see Sect. 8.6.1.2) for the upper anchor points A_i contributes to quick execution of reconfiguration. Firstly, the winches are kept accessible when installed on the floor. Secondly, the cable guidance units are constructed so that the cables can be led through with the bulky cable-end connector already installed, and finally using cable guidance units for the upper anchor points minimizes the installation work that needs to be done high over the ground. Especially, for large-scale robots, the additional time consumed for working safely at height must not be under estimated. Accurate referencing for both ends of the cable is also time-consuming to achieve in practice. Therefore, well-defined machine parts such as cable-end connector proof useful in reconfiguration. In terms of sensor support in the initial operation, winch integrated force sensors and end-of-travel sensors facilitate defined tension in the cables, support cable length referencing, and finally contribute to the machine safety.

9.3.5 IPAnema 3 Mini

The IPAnema 3 Mini is designed as a test-bed for developing kinematic codes, force control algorithms, and to test calibration methods (Fig. 9.12). The hardware design (Table 9.6) of the robot is reduced to the essential parts and built into a light-weight aluminum frame. Thus, also the winches and the control cabinet are built into the frame of the robot. The winches are simplified by directly connecting the drum to the motor shaft. Compared to the industrial IPAnema 3 winch, the gearboxes and the spooling unit are discarded. The cables are guided over a series of pulleys both to emulate the behavior of the IPAnema 3 robot and to minimize wear. Motor integrated multi-turn encoders allow for cable length estimation and force sensors on the platform as well as within the machine frame provide the cable forces at full

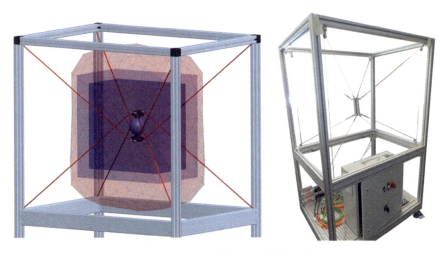

Fig. 9.12 Laboratory small-scale cable robot IPAnema 3 Mini. The robot uses industrial-grade drives with low-cost winches, and the IPAnema 3 controller architecture for development of kinematic codes, calibration, and force control

Table 9.6 Overview of the IPAnema 3 Mini specification

Parameter and symbol	Value/component	Unit
Number of cables m	8	–
Degrees-of-freedom n	6, 3R3T, fully-constrained	–
Parameterization Φ^G	IPAnema 3	-
Size of the robot frame	$1.1 \times 0.8 \times 1.0$	m
Size of the mobile platform	$0.044 \times 0.08 \times 0.166$	m
Rated cable force f_{min}, f_{max}	0–40	N
Max. cable velocity $\dot{l}_{max}$	2.0	m/s
Drive	Beckhoff AM3121-0201-0001	–
Drive power	0.2	kW
Gear box	gearless	–
Gear ratio ν_{PG}	1:1	–
Drum diameter d_D	0.02	m
Cable type	Dyneema, LIROS D-Pro 01505-0150	–
Cable diameter d_C	1.5	mm
Specific cable stiffness k_C'	28500	N
Cable force sensor	Futek LRM200	–
Cable force measurement	Range 0–111	N
Control system	TwinCAT 3.1, Beckhoff	–

cycle time through A/D converter. For the proximal force measurement, a pulley arrangement as shown in Fig. 8.35c is used. For the cables, 1.5 mm Dyneema cables are used which soft behavior is convenient for manual handling.

The control system is based on TwinCAT 3.1 by Beckhoff running on 64-bit Windows as host system. This control system is taken from the IPAnema 3 robot making the exchange of algorithms e.g. for kinematic transformation and force control straightforward. In fact, the testing of new algorithms is frequently done on the smaller robot for safety reasons before transferring the results to the larger, stronger, and more costly IPAnema 3. All sensor and actuator data including measured cable length from the encoder, cable tension from analogous force sensors, and motor set-point values for the amplifiers are transmitted through an EtherCAT fieldbus into the PLC. In contrast to all other IPAnema robots, cap rail sized amplifiers EL7201-0010 by Beckhoff are used for the decentralized cascaded velocity-current control where the position control is done by the NC-kernel [259]. The cycle time for open-loop motion generation in the CNC is 1 ms and the PLC runs force control algorithms with the same cycle time.

Beside testing of control algorithms, the IPAnema 3 Mini serves as test-bed for haptic human-robot interaction where different control algorithms for haptic displays are analyzed [259]. Therefore, the platform is designed to form a handle that can be grasped with a human hand (Fig. 9.12). The cable arrangement is chosen in the crossed configuration to allow for high torques in the available wrench set as well as for a larger orientation workspace. A number of performance results related to cable force and operational space force control have been published [259, 261].

9.4   Other Cable Robots

9.4.1   Copacabana

The Copacabana robots consist of 16 identical winches and drive-trains that are mounted within a common steel frame. Since all drive-trains are equal, the demonstrator offers rich possibilities of reconfiguration, both in terms of changing the geometry of the robot as well as changing the number of cables. The winch construction is based on the IPAnema 3 winch where the installed drive-trains allow for an actuation with maximum cable velocity of $\dot{l}_{max} = 1.5$ m/s and maximum cable forces of $f_{max} = 1200$ N. To the best of the author's knowledge, it is the only installation with two cooperating cable robots in one machine frame. Due to the setup, the robot allows to simultaneously operate two cable robots with different platform geometries in order to compare designs. The technical parameters of the two robots are identical and given in Table 9.7.

The robots are equipped with Dyneema cables with a diameter $d_c = 6$ mm. Thus, the cable stiffness is high compared to the size of the robot and the nominal cable force. The two robots are equipped with a standard industrial safety system. Thus,

Table 9.7 Overview of the Copacabana robot specification

Parameter and symbol	Value/component	Unit
Number of cables m	8	–
Degrees-of-freedom n	6, 3R3T, fully-constrained	–
Parameterization Φ^G	IPAnema 3	–
Size of the robot frame	$4.36 \times 2.59 \times 3.37$	m
Size of the mobile platform	$0.44 \times 0.43 \times 1.00$	m
Rated cable force f_{min}, f_{max}	100–1200	N
Max. cable velocity $\dot{l}_{max}$	1.5	m/s
Drive	Bosch Rexroth MSK061C-0600-NN-M1-UG1-NNNN	–
Drive power	4.7	kW
Gear box	Bosch Rexroth GTE120-NN1-005A-NN05	–
Gear ratio ν_{PG}	5:1	–
Drum diameter d_D	0.1	m
Cable type	Dyneema, LIROS D-Pro 01505-0600	–
Cable diameter d_C	6.0	mm
Specific cable stiffness k_C'	900000	N
Cable force sensor	Tecsis F2306	–
Cable force measurement	Range 0–2000	N
Control system	TwinCAT 3.1, Beckhoff	–

long term tests can be executed without operator supervision. The robots serve amongst others as test-bed for the validation of the dynamic simulation presented in Chap. 6.

9.4.2   Expo 2015

Beside applications in production engineering, cable robots are applied to entertainment installations such as acting in a media show. An example of this utilization was shown during the Universal Exposition EXPO 2015, which was hosted in Milan, Italy in the German Pavilion [463]. Here, two cable robots were installed with huge displays mounted to the mobile platform. Compared to industrial applications, high accuracy and rapid movements are second to the esthetic impression of the motion. The robots with their screens were actuated over the visitors, making safety of the robotic system a severe challenge. In order to address the safety issues, winches and controller systems from the field of stage equipment are used which implements redundant control system fulfilling safety integrity level 3 (SIL 3) according to the

Fig. 9.13 Cable robots in transporting film during commissioning at the German Pavilion at EXPO 2015

applicable norm IEC/EN 62061.[2] Engineering the system according to the strict regulations allows to playback pre-computed trajectories on drive level. However, it is not possible to certify the used kinematic codes for forward kinematics, inverse kinematics, and force control according to SIL 3. Therefore, a simple but robust safety PLC is used to playback pre-computed sequences of cable length and supervise the cable forces. These sequences are generated in a specific 3D programming environment that allows to arrange the desired motions of the platform under esthetic aspects and synchronized to the video, audio, and light show. Furthermore, the off-line programming system performs trajectory verification in the full six-dimensional operational space allowing the artists to exploit the general workspace which is a superset of the translational workspace and the orientation workspace. The pre-computed translational workspace only served as visual indicator for the artists to design trajectories in the six-dimensional Euclidian motion group SE_3 (Fig. 9.13).

9.4.2.1 Programming and Trajectory Verification

However, since the artists directing the show are not trained in robotic concepts, the design and verification of the trajectory is modeled using cubic Bezier-splines to parameterize both the translational motion as well as the orientation by Bryant angles. Thus, a sequence of poses $(\mathbf{r}, \mathbf{R})_i$ is generated that is tested for criteria relevant for executing the motion including wrench-feasibility (Sect. 3.4.2), interference, and reachability.

---

[2]IEC/EN 62061: Safety of machinery: Functional safety of electrical, electronic and programmable electronic control systems.

A suspended configuration is employed to mitigate the risk of excessive forces in the cables during emergency stop. Although each drive-train is carried out with a high safety performance level, it cannot be guaranteed that the drives move synchronously during emergency stop. Thus, a fully-constrained configuration could generated unpredictable internal forces. In contrast, the risk evaluation for the suspended configuration showed that emergency stop may lead to uncontrolled local motion of the platform. Since the motion is executed out of the reach of the audience, no risk was identified for this parasitic motion. This parasitic motion prevents overload in the cables and in the platform during emergency stop and the platform cannot fall down since it remains constrained by at least some of the eight cables. Clearly, the active cable configuration [330] (the subset of cables still under tension), is undefined in this case and a dedicated manual recovery procedure is required after emergency stop. To achieve the high safety standard for the winch hardware, each winch possesses two independent braking systems and two separated encoders. Furthermore, the overload safety factors for brakes and cables are chosen to withstand ten times of the nominal load.

9.4.2.2 Parameter Design

The geometrical design of the Expo cable robots is governed by the given layout of the building as well as the desired size of the projection screens on the mobile platform. In order to generate a large translational workspace in the lateral plane, the proximal anchor points are placed close to the surrounding walls. The footprint of the building is mostly square with 14×14 m and rounded corners. Two cable robots are installed mirror-inverted side by side. Thus, the layout of each robot has a trapezoidal shape (see Fig. 9.14). The anchor points are placed based on a mixture of the idea behind the CoGiRo parameterization for the machine frame and the IPAnema 3 parameterization for the mobile platform. The trapezoidal shape of the robot frame leads to distortion of the workspace as it can be seen from the workspace

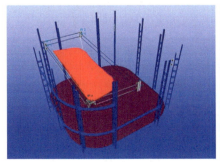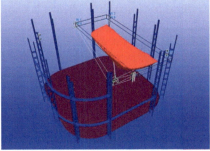

Fig. 9.14 Layout of the left (Lucia) and right (Roberto) cable robots installed at the Expo with the calculated wrench-feasible workspace integrated in the installation environment

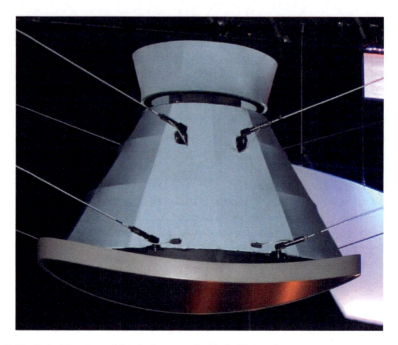

Fig. 9.15 One of the two mobile platforms used with the Expo robot

plots in Fig. 9.14. The numerical values for the frame are simply chosen to fill the available installation space of the building. The geometry of the platform is defined so that the distal anchor points are placed on the surface of the back-projection screen of the platforms. Then, some manual tuning is executed to compromise the size of the orientation workspace and the avoidance of cable-cable interference as well as cable-platform collisions. A remarkable feature in the mechanical design of the robot are the steel cables with embedded electrical fibers that provided electric power and HD-video signals to the moving screens. In Fig. 9.15, one can see the electric contacting as additional fibers between the cable ends and the mobile platform at the distal anchor points. The cable ends are fixed in jackets and connected to the mobile platform through swivel bolts. The proximal cable guidance is realized using panning pulleys where the normal axis of the panning is pointing downwards since the winches are installed in supporting framework under the roof of the building. An overview of the technical specifications of the pair of Expo robots is given in Table 9.8.

9.4.2.3   Conclusions

The installation and operation of cable robots in the German Pavilion shows that cable robots can be designed and safely operated in public environments. Although

Table 9.8 Overview of the specifications of the robots Expo Lucia/Roberto at the world exhibition 2015 in Milan

Parameter and symbol	Value/component	Unit
Number of cables m	8	–
Degrees-of-freedom n	6, 3R3T, suspended	–
Parameterization $\mathbf{\Phi}^G$	Modified CoGiRo frame, IPAnema 3 platform	–
Size of the robot frame	$5.39 \times 13.78 \times 7.08$	m
Size of the mobile platform	$1.36 \times 0.88 \times 0.88$	m
Rated cable force f_{min}, f_{max}	200–2400	N
Max. cable velocity $\dot{l}_{max}$	1.5 (2.0 max.)	m/s
Drive	Asynchronous motor	–
Drive power	5.5	kW
Gear box	Special worm gear TDA10	–
Gear ratio ν_{PG}	10.25:1	–
Drum diameter d_D	0.225	m
Cable type	Steel, Gustav Wolf PAWO F 4e	–
Cable diameter d_C	7.0	mm
Specific cable stiffness k_C'	~425000 (estimated)	N
Cable force sensor	N/A	–
Cable force measurement	N/A	N
Control system	F&P control, based on safety PLC (SIL 3) from Mitsubishi Electric	–

strict safety regulations needs to be met, the required techniques are available to put the robot show into operation. Around forty shows per day have been executed for six months showing robust performance of all involved components. The overall availability of the show was beyond 98.5%.

9.4.3 MPI CableSimulator

The motion simulator called *CableSimulator* is installed at the Max-Planck Institute for Biological Cybernetics and is developed in cooperation with Fraunhofer IPA [343]. Cable robots are a good choice for car driving and flight simulators, since cable robots are capable to generate smooth and high dynamic motion within a large workspace. The accurateness of the motion has to be achieved on the acceleration level where the positioning accuracy is second for the passenger. This is due to the fact the human passenger is sensitive to acceleration through his vestibular system where position and velocity can only be indirectly determined e.g. through vision.

Furthermore, the mobile platform can freely move through a large robot frame. A large cable span is no drawback for the motion simulator application since the environment around the passenger's cabin needs to be empty for safety reasons. Beside its application for car driving and flight simulator, the MPI motion simulator is designed for research on human senses as well as medical therapy.

9.4.3.1 Geometry Design

The geometrical design of the robot frame is governed by the available installation space in the building. Thus, the proximal anchor points are installed in the corners of the room where consoles are employed to induce the significant forces into the building structure.

The design of the mobile platform is driven by the requirements of having a large installation space for the passenger and other equipment for the targeted applications in virtual reality as well as car driving and flight simulators. Therefore, an icosahedron structure made from carbon fiber struts was chosen to achieve high rigidity along with a large installation space inside the icosahedron. The carbon structure offers a lightweight framework with 60 kg weight that allows to withstand the significant peak loads of the drive-trains. The outer radius of the platform is chosen large enough to allow for two passenger seats inside or to install an additional gimbal system to allow for full rotation as it is required for car driving and flight simulation. Originally, the endless Z12 design (Sect. 8.4.13) was foreseen for the robot design but is discarded because of technical complexity in a safety critical application.

9.4.3.2 Drive-Train Selection and Safety

The selection of the drive-trains is constrained from the safety requirements and compatibility with the controller. The application in motion simulation requires very smooth and also silent motion generation which is realized by a gearless drive-train. Synchronous servo drives with a rated power of 46 kW are chosen to achieve high accelerations for the significant platform loads. The payload of the robot is high enough to allow for two probands, virtual reality equipment, and instrumentation on the mobile platform.

The CableSimulator addresses even higher safety requirements compared to the Expo installation since the robot is designed to carry one or two passengers on its mobile platform. Furthermore, the control system is designed to react online to user-input prohibiting to evaluate the signal off-line before execution.

Again, stage technology is used as basis for the design of the safety critical drive-trains and low-level control system including dual-channel axis controller, redundant brakes, and cables with a very high safety factor against failure.

9.4.3.3 Results

The constant orientation workspace contains a $4 \times 5 \times 5$ m box with a maximum orientation of $\pm 40°$ for roll and pitch while the yaw axis is rather limited with around $\pm 5°$ due to singularities. The CableSimulator is amongst the largest available cable robots in a fully-constrained configuration using steel cables. The observed stiffness properties are outstanding. Due to the excellent stiffness of the used cables, the CableSimulator achieves very high Cartesian stiffness values. The bandwidth for motion simulation is determined to be in the range of 5–14 Hz depending on pretension and the platform's position in the workspace. Measurements with an inertial motion unit (IMU) on the platform verify that the large-scale simulation can actually yield ± 10 m/s^2 of acceleration to the passenger (Tables 9.9 and 9.10).

9.4.4 Segesta

The Segesta prototype was built at the University of Duisburg-Essen, Germany [68, 139, 210, 473] around 2003. It employs a rectangular machine frame and a flat

Table 9.9 Overview of the specifications of the MPI CableSimulator in Tubingen, Germany

Parameter and symbol	Value/component	Unit
Number of cables m	8	–
Degrees-of-freedom n	6, 3R3T, fully-constrained	–
Parameterization Φ^G	CableSimulator (Icosahedron)	–
Size of the robot frame	$15 \times 12 \times 8$	m
Size of the mobile platform	Diameter: 2.6	m
Rated cable force f_{min}, f_{max}	1000–14000	N
Max. cable velocity l_{max}	5.0	m/s
Drive	Synchronous servo	–
Drive power	46	kW
Gear box	Gearless	–
Gear ratio ν_{PG}	1:1	–
Drum diameter d_D	0.32	m
Cable type	Steel	–
Cable diameter d_C	14.0	mm
Specific cable stiffness k_C'	1690000 (estimated)	N
Cable force sensor	N/A	–
Cable force measurement	N/A	N
Control system	TwinCAT 3.1 with Waagner-Biro axis controller	–

Table 9.10 Geometry data of the MPI CableSimulator

Cable i	Base vector $\mathbf{a}_i$ [m]			Platform vector $\mathbf{b}_i$ [m]		
	x	y	z	x	y	z
1	0.8090	14.8620	7.9060	−1.3830	0.5280	−0.7470
2	0.8100	14.8500	0.6090	−0.7990	1.2448	0.7457
3	11.3190	14.5780	7.9530	1.4535	0.3697	−0.7482
4	11.3230	14.5650	1.1580	0.9539	1.1471	0.7450
5	11.1960	0.5820	7.9820	0.8187	−1.2676	−0.7465
6	11.2040	0.5670	1.1860	1.4027	−0.5507	0.7464
7	0.4180	0.4940	7.9700	−0.9346	−1.1698	−0.7458
8	0.4200	0.4800	0.6750	−1.432	−0.3925	0.7475

triangular platform with 7-3 configuration where groups of two or three cables share a common distal anchor point. This is realized by clamping thin cables in a common hole on the planar platform. Using brushless DC motors from Maxon to directly drive the winches allows for very high accelerations of the light-weight mobile platform. Each drive-train has a rated power of some $P_w = 430\,\text{W}$ and allows for maximum cable velocity $\dot{l} = 6.8\,\text{m/s}$ and cable forces of $f_{max} = 63\,\text{N}$. Position measurements are done through optical incremental encoders in the motors with a resolution of 2000 ticks per rotation. In recent years, the robot is redesigned to an 8-3 configuration and newer publications show a revised geometry partly also with more than eight cables. The main specifications are summarized in Table 9.11.

The original Segesta control system was based on a real-time Linux kernel with a complete motion controller stack. Later, the controller system was fully rewritten based on a dSpace DS1005 real-time controller system where the implementation of the controller is performed using a MATLAB/Simulink tool chain with code generation for the controller.

In Table 9.12, the coordinates of the anchor points of the cable robot Segesta (Fig. 9.16) are given.

9.4.5   Storage Retrieval Machine CABLAR

The storage retrieval machine CABLAR (Fig. 2.12) is designed and implemented by the University of Duisburg-Essen, Germany [70, 71, 273]. The goal of the design is to proof the feasibility of a cable robot for operation in the field of warehousing to load and unload high-bay shelf. Thus, the geometry of this robot is flat with the main motion in one horizontal and one vertical degree-of-freedom. The robot uses drive-trains by SEW Eurodrive and a TwinCAT 3 controller from Beckhoff. The robot employs a laser scanner on the mobile platform to determine its platform position for referencing and control. A remarkable technical feature of the robot is the use

Table 9.11 Overview of the specifications of the Segesta cable robot, University of Duisburg-Essen, Germany

Parameter and symbol	Value/component	Unit
Number of cables m	7/8	–
Degrees-of-freedom n	6, 3R3T, fully-constrained	–
Parameterization Φ^G	Segesta 7	–
Size of the robot frame	$0.83 \times 0.63 \times 1$	m
Size of the mobile platform	$0.105 \times 0.2 \times 0$	m
Rated cable force f_{min}, f_{max}	0–63	N
Max. cable velocity $\dot{l}_{max}$	6.8	m/s
Drive	Servo drives Maxon EC60	–
Drive power	0.43	kW
Gear box	Gearless	–
Gear ratio ν_{PG}	1:1	–
Drum diameter d_D	0.0242	m
Cable type	Dyneema	–
Cable diameter d_C	N/A	mm
Specific cable stiffness k_C'	N/A	N
Cable force sensor	Custom strain gage	–
Cable force measurement	N/A	N
Control system	Custom design based on MATLAB/Simulink	–

Table 9.12 Nominal geometric parameters: base vectors $\mathbf{a}_i$ and platform vectors $\mathbf{b}_i$ of the Segesta 8 robot

Cable i	Base vector $\mathbf{a}_i$ [m]	Platform vector $\mathbf{b}_i$ [m]
1	$[0.0, 0.0, 0.0]^T$	$[-0.0525, -0.0760, 0.0]^T$
2	$[0.0, 0.0, 1.0]^T$	$[-0.0525, -0.0760, 0.0]^T$
3	$[0.83, 0.0, 1.0]^T$	$[0.0525, -0.0760, 0.0]^T$
4	$[0.83, 0.0, 0.0]^T$	$[0.0525, -0.0760, 0.0]^T$
5	$[0.83, 0.63, 0.0]^T$	$[0.0, 0.1240, 0.0]^T$
6	$[0.0, 0.63, 1.0]^T$	$[0.0, 0.1240, 0.0]^T$
7	$[0.0, 0.63, 0.0]^T$	$[0.0, 0.1240, 0.0]^T$
8	$[0.83, 0.63, 1.0]^T$	$[0.0, 0.1240, 0.0]^T$

Fig. 9.16 Schematic setup
of the Segesta prototype with
eight cables in 8-3
configuration

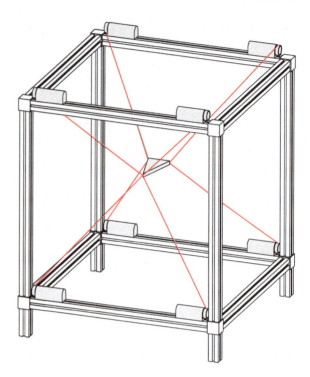

of a long energy chain to provide electrical signals and power supply to the mobile
platform through a passively balanced connection.

A second robot is designed for the operation in wind tunnels e.g. for testing
mockups of ship hulls (Fig. 2.17). This design uses cables with constant effective
length while moving the proximal anchor points in horizontal guideways [77]. A
main benefit from using cables with constant length is that cable tension can be
controlled with a very high energy efficiency. This can be understood when one
considers the mapping between the cable forces and actuator forces. Due to the
attack angle between the cable and the guideway, a portion of the cable force is
distributed between the horizontal guideways below and above the platform leading
to higher tension in the cables than in the actuators. Therefore, this tension is passively
maintained without energy consumption in the drives.

9.4.6 CoGiRo

The cable robot CoGiRo is developed by Laboratoire d'Informatique, de Robo-
tique et de Microélectronique de Montpellier (LIRMM) and Tecnalia, Montpellier, France and is one of the largest suspended cable robots when it was set up
(Fig. 9.17). Although the robot uses eight cables and is of the redundantly restrained

Fig. 9.17 CoGiRo: a large-dimension suspended cable-driven parallel robot with six degrees-of-freedom by LIRMM and Tecnalia. (Courtesy of LIRMM, Montpellier, France)

type (RRPM), the platform of the CoGiRo is always in a suspended configuration making use of gravity to keep the cables under tension. An overview of the technical specifications is given in Table 9.13. The special arrangement of the cables on the platform (see Sect. 8.4.9) allows for a large orientation workspace where especially the yaw rotation about the z-axis is maximized. Furthermore, the geometry of the platform is optimized in order to increase stiffness of the robot since little stiffness is gained from controlling the cable force distribution. Since the robot operates in a suspended configuration, the load is efficiently shared amongst the cables and makes high payloads of more than 300 kg kilogram possible throughout the workspace. A maximum payload of some 500 kg is possible in the center of the workspace. The mass of the unloaded platform is around $m_P = 100$ kg which is important for the cable pretension in unloaded operation.

The machine frame of the CoGiRo prototype is constructed from modular framework structures as used for stage equipment. This kind of structure allows for reconfiguration of the machine frame in different geometrical shapes. The size of the frame is $15.24 \times 11.24 \times 5.93$ m. The winches are grouped in pairs at each corner of the robot's footprint, where the cables are routed over panning pulleys with a diameter of $d_R = 100$ mm at the top side of the robot frame. In contrast to the IPAnema robots, steel cables according to DIN 3069 with a diameter of $d_C = 4$ mm are employed. Although this robot has a similar size as the IPAnema 3 robot, sagging of the cables is more important due to the notably higher cable weight. However, the steel cables show a reduced creeping in turn reducing efforts for referencing of the cable length.

The winches of the cable robot use a single layer drum without spooling. The motors from B&R (8LSA85) have a rated power of around 12 kW each allowing for

Table 9.13 Overview of the CoGiRo specification according to different sources from LIRMM, Montpellier, France

Parameter and symbol	Value/component	Unit
Number of cables m	8	–
Degrees-of-freedom n	6, 3R3T, suspended fully-constrained	–
Parameterization Φ^G	CoGiRo	–
Size of the robot frame	$15.24 \times 11.24 \times 5.93$	m
Size of the mobile platform	$1.00 \times 1.00 \times 1.00$	m
Rated cable force f_{min}, f_{max}	5000	N
Max. cable velocity $\dot{l}_{max}$	3.5	m/s
Drive	B&R 8LSA85	–
Drive power	12.0	kW
Gear box	Geared belt drive	–
Gear ratio ν_{PG}	3:1	–
Drum diameter d_D	0.135	m
Cable type	Steel, DIN 3069	–
Cable diameter d_C	4.0	mm
Specific cable stiffness k'_C	N/A	N
Cable force sensor	N/A	–
Cable force measurement	N/A	N
Control system	B&R Automation Studio	–

a nominal cable force of some 4200 N. Since the winches use a belt transmission instead of a gearbox, the inertia in the drive-train is largely reduced compared to other winches.

The control system is based on Automation Studio by B&R. The controller run at a cycle time of 1.2 ms. The controller architecture is similar but not identical with the one used for the IPAnema robots. The robot is subject to a number of calibration experiments where geometry data is also available from direct measurement with a laser tracker.

9.5 Conclusions

The practice in the field of cable robots is currently in a period of intensive consolidation and maturing. While the first twenty years after Landsberger's original study on the subject [280], the practice was governed by proof-of-concept works. Different laboratory robots show that motion generation with cables is feasible, finite stiffness can be generated through unilateral constraints, and cable robots can indeed generate a huge workspace and perform ultra fast motion. However, a lot of practically relevant issues were postponed or left unsolved. Therefore, a couple of problems caused by the prototypic components of the laboratory test-beds needs to be tackled in order to apply cable robots is serious applications. Excessive wear of the cables is addressed using sophisticated cable guidance systems with pulleys and sufficiently large drum. The reliability of controller systems for flexible and long-term application is solved by adopting conventional controller architectures from machine tools and adapting the kinematic codes for cable robots to run on these controllers. The simple winches are exchanged by new developments or adoptions from related fields like stage technology in order to provide safe and reliable operation. Finally, systems such as the EXPO robots and the CableSimulator proofed that high standards on safety can be met using cable robot technology.

All these developments require to combine the involved mathematical description of kinematics, statics, workspace, and control to work along with conventional controllers and drive-trains. What seems quite simple from a theoretical point of view, e.g. using a singular value decomposition, turns out to be a project of several person-months if the implementation shall be run under real-time constraints in a PLC.

The main objective in the further development of cable robots from a practical point of view is the improvement of accuracy, a standardization of force control in the controller architecture, rapidly deployable solutions for parameter identification. Calibration and customization of components must be fostered towards the domain specific needs of a variety of applications.

Chapter 10
Summary

Since the first research in 1985 on cable-driven parallel robots, this field has been notably structured and shaped. Now, a number of fundamental problems are intensively studied and both a theoretical foundation and applicable practice are developed. The scope of this book is to structure the field, to discuss the foundation, and to present a solid theory of cable robots.

A consistent terminology is a prerequisite for the scientific development within a field of research. Based on the introduced technical terms, different classification schemes are recalled and introduced that allow for accurate assessment of the algorithms sketched and developed throughout this book. The distinction between under-constrained cable robots (IRPM) and fully- or redundantly-constrained cable robots (CRPM, RRPM) is fundamental for the development of efficient kinematics codes. The presented fields of application as well as the robot archetypes serve as a knowledge base from examples in a field that still lacks stringent design rules for choosing the robot architecture.

The kinematics and statics of fully- and redundantly-constrained cable robots based on the standard cable model is a understood and thorough theory that is mature for application in different fields. The basic equations, relations, and concepts are introduced and discussed in detail in this work. Based on this solid background, a number of algorithms and methods are available to efficiently compute the kinematics transformations, to evaluate statics including cable force distributions, and to characterize the stiffness of the robot. In contrast, the equations of under-constrained robots can be set up but it lacks efficient methods for the computation of kinematics and the development of such tools remains a challenging problem, especially under real-time constraints.

The determination of the wrench-feasible workspace of cable robots is solved. For a cable robot, the most important questions are in which poses the platform can be operated with positive tension and to compute the feasible tension in the cables. Using the presented methods, such computations are done efficiently within some seconds. Additionally, the problem of cable-cable interference is related to the

© Springer International Publishing AG, part of Springer Nature 2018
A. Pott, *Cable-Driven Parallel Robots*, Springer Tracts in Advanced
Robotics 120, https://doi.org/10.1007/978-3-319-76138-1_10

workspace and is efficiently computed. Therefore, workspace determination is an applicable tool for both robot analysis as well as robot design.

The structure of the dynamic model of a cable robot is well understood and in line with other simulation tools. The proposed approach and other results in the literature show that the effects of the robot mechanics, elastic cables, winch dynamics, friction in the pulleys, and the cascaded control can be taken into account. Based on the standard model, the simulation is even possible under real-time constraints. The main problem for simulation of cable robots is to identify the robot's physical parameters. Especially the material parameters of the cables are difficult to determine and vary amongst cables made from the same material.

Many advances have been made in the field of cable models in the recent years. Three main topics are addressed: The effect of guiding pulleys on kinematics and friction, the disturbance from elastic cables, and the influence of the cable mass leading to sagging of the cable. All three effects significantly change the modeling equations used for kinematics, statics, and dynamics. Since more sophisticated cable models change the mathematical classification of problems, such as inverse kinematics, statics, and dynamics, many known results for the standard model cannot be extended to the respective cable models. A large number of problems remain open on both the theoretical as well as the practical side. How many solution has the forward kinematics of a cable robot with sagging cables? How can one compute such solutions efficiently under real-time constraints?

In spite of some dozens of laboratory prototypes and demonstrator systems for special purposes, the design of cable robots remains a tricky procedure. A mechatronical approach to system design is sketched from requirement specification through selection of an architecture, parameter synthesis, and mechanical design. However, the procedure is rather lengthy and an engineer still has to rely on his expert knowledge to cope with a couple of problems. Some design decisions become necessary that are not backed by cable robot knowledge. This can lead to a couple of iterations in the procedure. However, a number of robots are successfully implemented showing that it is possible to build productive and safe cable robots.

Finally, the overview of the work on applied cable robots and practice reflects the state of the art of what one can actually do with cable robots today. Many questions around the operation of the cable robots are at the frontier between research and application. However, also for experimental scientific investigations, one has eventually to make decisions on the hardware to be used in the test-bed and, due to lack of information in this field, some choices must be made without prior knowledge and accepted guidelines. In spite of a complicated theory and a number of numerically ill-conditioned problems, the prototypes have a tendency to work smoothly although the theoretical evaluation predicts some instabilities.

10.1 Open Issues

The list of open problems is long and not everything is named or even discussed here. The *dynamic simulation* of cable robots is tackled by a large number of authors. However, the state of the art in this field is hardly consolidated and bringing together the description of the different subsystems remains an open topic. Since cables are special machine elements and their use as freely moving transmission elements is seldom used in other applications, the commercially established engineering tools have significant blind spots when used to simulate cable robots. This notably slows down research efforts on cable robots since a scientist has to introduce new elements in an existing software framework which is complicated and often frustrating. In contrast, one can rely on custom software tools or very special purpose tools with support for cable models where one has to resign from a number of wide-spread functions available in the general purpose software. Therefore, the main problem is the consolidation and reuse of simulation tools for cable robots, that allow to integrate the efforts undertaken in the past on a common platform or through improved interpretability amongst different platforms.

Under-constrained cable robots lately attracted more interest as the reduced number of actuators make them more profitable. However, the additional theoretical challenges in kinematics, workspace, and control are usually under-estimated when starting projects on such prototypes. The current state of the art lacks efficient real-time capable kinematic codes to control such robots. Also, there is a need for applicable controller and design methods of under-constrained cable robots.

The development and application of better *cable models* attracted a lot of attention over the past years and bears a number of further problems for the coming years. Beyond the standard model, the kinematics of the anchor points, including pulleys and guiding surfaces, are tackled. Also, linear and nonlinear elastic reactions in the cables and sagging of the cables are described. However, only some handpicked algorithms are extended to include these effect, whereas the majority of the applied methods is still based on the standard model. Extending the cable robot models with advanced cable models raises the question of the number of solutions (if any) for the inverse and forward kinematics. Furthermore, it raises the question of efficient kinematic codes to compute the kinematic transformations, on their respective numerical stability, on existence and unique for cable force distributions, on evaluation of the stiffness, and location as well as nature of singularities. Even new kinds of singularities are conjectured that may result from the cable model.

An important open field in applied cable robots is the improvement of both *repeatability and accuracy*. Although some improvements have been made in recent studies, many applications demand for better performance. Efforts in different fields of cable robots are undertaken for improving the accuracy: Advanced kinematic codes and cable models allow to take more sophisticated effects into account. Practically, all of these approaches increase the number of parameters which need to be identified to make the models effective. Therefore, advances in parameter identification and calibration are required to put the improved cable models to productive work.

Calibration of cable robots is still at a very early stage and applications of efficient techniques remain tasks for a small group of experts. Thus, the possible benefits in accuracy are rarely exploited for application. It remains an open issue to develop methods that are simple to apply. Many cable robots are of the over-constrained type and cable forces sensors are also frequently installed. Both properties allow basically to apply auto-calibration procedures which can at time or even continuously estimate the robot's parameter for better performance. Beside improvements in accuracy, one might even employ such methods to changing geometry or aging components.

The *design procedure* for cable robots is in its infancy. In the understanding of the author, design is mastering the art of the cable robots as it includes aspects from all other problems and fields discussed in this book. A number of important steps in the design procedure of cable robots are understood but there is hardly an established procedure or widely agreed results for the steps. The proposed design procedure is far from being holistic, however, it seems to be the first outline covering all steps from specification to initial operations. Therefore, future works on design shall elaborate on the connection between the clear mathematical solution of single steps, such as optimal design of the geometry, and the structural decisions made in the design procedure. The questions remain which geometry template shall be used, which cables are applicable, and how to choose from the known mechanical and electrical components. Such optimization problems are known to be difficult and their solution is anticipated to remain open for quite some time.

A problem related to design is *reconfiguration*. Compared to other robots, the physical and module structure of cable robots is highly suitable to adopt the robot to changed application requirements. However, except for some use case studies, no methodology for this problem was proposed yet.

10.2   Outlook

Today, cable-driven parallel robots are at the crossroads to applications. Cable robots possess a number of properties that are superior to any other robotic device. At its core, the advantages of cable robots rely on the ultra light-weight nature of the used cables and the mechanically simple winches to operate the cables. Thus, a threefold scalability is possible. Firstly, one can efficiently scale the payload in a realistic range of grams to hundreds of tons, i.e. in eight orders of magnitude. Secondly, one can scale the size of the robot from millimeters to some hundreds of meters, i.e. in five orders of magnitude, and, finally, the manipulation through cables is efficiently done from quasi-static structures with an ability to reconfigure, as known from civil engineering, up to ultra-fast and dynamic motion with accelerations of up to 420 m/s and velocities of some 20 m/s.

Some pilot applications impressively show the potential of the technology where the cable cameras are yet the most successful one. The world largest telescope FAST is initially operated in the near future. The installations on the Expo 2015 in Milan as entertainment system proofed that a safe and reliable operation with the audience is feasible. Most recently, the flight simulation *CableRobotSimulator* impressively highlights dynamical capabilities along with the possibility to transport people. However, other usage in entertainment, production engineering, measurement devices, rehabilitation, elderly care, and energy generation have gained good positions to enter the market.

The possibility to highly customize the properties of the cable robot is perhaps its biggest strength as well as weakness at the same time. Where the apparently simple principle of cable robots allows for configuring the robot towards many applications' needs, the complexity of the underlying theory requires considerable expert's knowledge to exploit the potential. In this sense, this work is contributed in the hope that its comprehensive overview of the theory helps to leverage the application potentials of cable robots.

Appendix A
Notation and Definitions

This works is meant to present an unified theory of cable-driven parallel robots. As part of this attempt both the terminology and the mathematical description shall be harmonized. However, due to the number of fields touched, it is a challenge to keep symbols through all chapters and to maintain common notion where possible. In the following, the systematics for notation are described.

Scalar real values and natural numbers are noted in italic letters s. Vectors are noted in bold as $\mathbf{x}$ and their symbols are usually lower case letters. Where necessary, information on the dimension of the vector is given when introducing the vector. The components of vectors and matrices are noted in square brackets as $\mathbf{r} = [x, y, z]^T$. Position vectors, velocities, and accelerations as well as forces and torques are understood to be elements of $\mathbb{R}^3$ for spatial robots and $\mathbb{R}^2$ for planar robots. If not stated otherwise, vectors are understood to be columns. The zero vector with all elements vanishing is denoted by $\mathbf{0} \in \mathbb{R}^n$ and its dimension n shall be selected from the context. When comparing two vectors by using the operators $<, >, \leq, \geq$ the comparison has to be done component-wise. Let $\mathbf{a} = [a_1, \ldots, a_n]^T \in \mathbb{R}^n$ and $\mathbf{b} = [b_1, \ldots, b_n]^T \in \mathbb{R}^n$, then

$$\mathbf{a} > \mathbf{b} \quad \text{holds true if and only if} \quad a_i > b_i \ \forall \ i = 1, \ldots, n \ . \tag{A.1}$$

Sometimes vectors are also compared with scalar values. This comparison is also understood to be executed component-wise. Let $\mathbf{a} = [a_1, \ldots, a_n]^T$ and $s \in \mathbb{R}$, then

$$\mathbf{a} > s \quad \text{holds true if, and only if,} \quad a_i > s \ \forall \ i = 1, \ldots, n \ . \tag{A.2}$$

The scalar product of two vectors $\mathbf{a} \cdot \mathbf{b}$ of the same dimension is the sum of the product of its respective components. The scalar product is equivalent to

$$\mathbf{a} \cdot \mathbf{b} = \mathbf{a}^T \mathbf{b} = \sum_i a_i b_i \ . \tag{A.3}$$

© Springer International Publishing AG, part of Springer Nature 2018
A. Pott, *Cable-Driven Parallel Robots*, Springer Tracts in Advanced
Robotics 120, https://doi.org/10.1007/978-3-319-76138-1

The notion $\mathbf{a}^2$ is a shorthand for

$$\mathbf{a}^2 = \mathbf{a} \cdot \mathbf{a} = \mathbf{a}^\mathsf{T} \mathbf{a} = \sum_i a_i a_i = ||\mathbf{a}||_2^2 \qquad (A.4)$$

and the result is a scalar which is the squared length of the vector or the square of the Euclidean norm of vector $\mathbf{a}$. Matrices are noted with bold letters as $\mathbf{M}$ and their symbols are usually capital letters. If not said otherwise, the matrices here are all real-valued. The square identify matrix $\mathbf{I} = \mathrm{diag}(1, \ldots, 1) \in \mathbb{R}^{n \times n}$ as well as the zero matrix $\mathbf{0} \in \mathbb{R}^{m \times n}$ with all elements being zero have dimensions m, n fitting to the context of the equations, respectively. A set is denoted with calligraphic letters such as $\mathcal{S}$ which also applies to the notation of the workspace $\mathcal{W}$ that is basically also a set of poses. Sets may have a finite or infinite number of elements, such as $\mathbb{R}, \mathbb{R}^n$, and SO_3, and we use curly brackets to enumerate elements $\mathcal{S} = \{1, 2, 5\}$. Lists $\mathcal{L}$ are special finite sets and bridge the way from mathematics to a computer implementation. In addition to mathematical sets, lists are assumed to have a well-defined sequence allowing for indexing where sets are unordered.

Interval variables are written with a hat like $\widehat{a}$. Consequently, vectors of intervals are denoted as bold letters with a hat $\widehat{\mathbf{b}}$ and interval matrices $\widehat{\mathbf{M}}$. When written in form of the lower interval bounds a and the upper bounds b, square brackets and a semicolon are used for the interval $[a; b]$. Note that this notation is used both for the application with interval analysis as well as for ordinary notation of parameter ranges.

Coordinate frames are abbreviates with a calligraphic $\mathcal{K}$, however, coordinate systems are not understood to be sets. A spatial coordinate frame is equivalent to a pose and one possible parameterization is composed from the position $\mathbf{r} \in \mathbb{R}^3$ and the orientation matrix $\mathbf{R} \in SO_3 \subset \mathbb{R}^{3 \times 3}$ where the special orthogonal group SO_3 is defined as follows:

$$SO_3 = \left\{ \mathbf{R} \in \mathbb{R}^{3 \times 3} \mid \mathbf{R}\mathbf{R}^\mathsf{T} = \mathbf{I}, \quad \det(\mathbf{R}) = 1 \right\} . \qquad (A.5)$$

Subscripts in italic letters are understood to symbolize indices taking natural numbers, e.g. to select components from vectors and sets. Sequences of subscripts represent multiple indexing, e.g. to name the components of a matrix like $\mathbf{A}_{ij}$. Subscripts in normal letters are names, multiple normal letters without comma separation also form a name, for example $\mathcal{K}_{\mathrm{TCP}}$ for the coordinate frame of the TCP. When names and indices are combined as subscripts, the index is separated with an additional comma, e.g. the frame $\mathcal{K}_{\mathrm{A},i}$ denotes the ith proximal anchor point frame.

Derivatives with respect to time t are noted with dots over the letter. The rule applies both for scalar and vectors. Let s be a length, then $\dot{s} = \frac{ds}{dt}$ is the linear velocity and $\ddot{s} = \frac{d^2 s}{dt^2}$ is the linear acceleration. For the position vector $\mathbf{r}$, one gets the velocity vector $\mathbf{v} = \dot{\mathbf{r}} = \frac{d\mathbf{v}}{dt}$ and the acceleration vector $\mathbf{a} = \dot{\mathbf{v}} = \ddot{\mathbf{r}} = \frac{d^2 \mathbf{r}}{dt^2}$.

The usage of poses consisting of a position $\mathbf{r} \in \mathbb{R}^3$ and an orientation $\mathbf{R} \in SO_3$ needs special treatment. Such a pose represents an unique state of a coordinate frame $\mathcal{K}$ in the Euclidian motion group SE_3 which is a six-dimensional manifold composed from the product $\mathbb{R}^3 \times SO_3$. Therefore, one can represent a pose by pair of the positions $\mathbf{r}$ and $\mathbf{R}$ that we denote with $\mathbf{y} = (\mathbf{r}, \mathbf{R})$. For many computer codes it is required to choose a parameterization for $\mathbf{R}$ such as Euler angles, Bryant angles, roll-pitch-yaw, Rodriguez parameters, Quaternion, or simply the nine coefficients $[r_{11}, \ldots, r_{33}]^T$ of the rotation matrix. The notation $(\mathbf{r}, \mathbf{R})$ is used whenever the method is independent from the parameterization used. Parameterizations are avoided where possible for the sake of generality. However, some operations cannot be carried out without choosing a certain parameterization of rotation. The pose vector $\mathbf{y}$ is written as a tuple of parameters. Note that such parameter vectors are restricted in their mathematical operations since common operations as plus and minus, have no physical meaning if the components of the vector $\mathbf{y}$ are e.g. $\mathbf{y} = [x, y, z, a, b, c]^T$ where x, y, z are the Cartesian coordinates and a, b, c are the Euler angles then adding or subtracting two such vectors has no physical meaning.

Appendix B
Introduction to Interval Analysis

Interval Arithmetic was firstly introduced by Ramon E. Moore [410] and was originally used to propagate computation and round-off errors in numerical computations. This is achieved by determining guaranteed bounds on computations in a robust way. Beside the handling of round-off errors, interval analysis have been proven a valuable tool in many other numerical problems such as linear algebra, solving of nonlinear equations, constraint programming, and optimization. A major property of interval analysis is its ability to derive guaranteed bounds for the values of an analytic function in a given interval. This can be done even if the coefficients of the equation are subject to uncertainties as long as one can give ranges (intervals) for these coefficients. Interval algorithms were developed for a couple of numerical problems such as solving nonlinear systems of equations, enclosing the roots of polynomials, and finding all solutions of systems of inequalities. During the last decades, interval algorithms were developed for constrained global optimization [40, 196, 359]. These methods were successfully applied to problems where conventional methods were hardly able to deal with. Especially the inherent property to deal with round-off errors in a robust way and to compute strict bounds for the numerical error of the algorithms are superior to conventional computations with real values. However, there are some additional numerical costs for the interval evaluation and for certain problems interval algorithms are rigorous but rather inefficient.

An *interval* $\widehat{x}$ is an ordered pair $[a; b]$ of two real numbers

$$\widehat{x} = [a; b] = \{x \in \mathbb{R} \mid a \leq x \leq b\} \quad , \tag{B.1}$$

where a is called *infimum* and b is called *supremum* of $\widehat{x}$. The difference between infimum and supremum is called *width* (diameter) of the interval and the mean value is called center (middle). Thus, the following functions are defined

$$\inf \widehat{x} = a \quad , \tag{B.2}$$

$$\sup \widehat{x} = b \quad , \tag{B.3}$$

© Springer International Publishing AG, part of Springer Nature 2018
A. Pott, *Cable-Driven Parallel Robots*, Springer Tracts in Advanced
Robotics 120, https://doi.org/10.1007/978-3-319-76138-1

$$\operatorname{diam} \widehat{x} = b - a \ ,$$ (B.4)

$$\operatorname{mid} \widehat{x} = \frac{1}{2}(a + b) \ .$$ (B.5)

The set of all real valued intervals is denoted with $\mathbb{I}$. A vector of interval is called a *box*. Analogously to the arithmetics of real numbers, the elementary operations $+, -, *, /$ are declared for the set of intervals $\mathbb{I}$ as follows:

$$\widehat{x} \circ \widehat{y} = [a; b] \circ [c; d] = \{x \circ y \mid a \leq x \leq b, c \leq y \leq d\} \ ,$$ (B.6)

where $\circ$ is any of the elementary operations $+, -, *, /$. The following rules apply for the elementary operations

$$[a; b] + [c; d] = [a + c; b + d] \ ,$$ (B.7)

$$[a; b] - [c; d] = [a - d; b - c] \ ,$$ (B.8)

$$[a; b] * [c; d] = [\min(ac, ad, bc, bd); \max(ac, ad, bc, bd)] \ ,$$ (B.9)

$$[a; b] / [c; d] = [a; b] * [1/d; 1/c] \quad \text{if} \ \ 0 \notin [c, d] \ .$$ (B.10)

The result of any such operation is an interval, i.e. the set of intervals is closed with respect to the arithmetic operations $+, -, *$. Only for the division, the expression $\widehat{x}/\widehat{y}$ is undefined if $0 \in \widehat{y}$.[1] The degenerated intervals of the form $[a; a]$ are associated with the real numbers and the interval operations yield the same results. Furthermore, the interval operations converge towards the results for real values arithmetics, if the width of all intervals converges towards zero. Therefore, interval analysis can be understood as a generalization of the arithmetics of real numbers [410].

An interval is called *positive* (*negative*) if $\inf \widehat{x} \geq 0$ ($\sup \widehat{x} \leq 0$) and *strictly positive* (*strictly negative*) if $\inf \widehat{x} > 0$ ($\sup \widehat{x} < 0$). Two intervals $\widehat{x}, \widehat{y}$ are *equal* if $\inf \widehat{x} = \inf \widehat{y}$ and $\sup \widehat{x} = \sup \widehat{y}$. Intervals are partially sorted and $[a; b] < [c; d]$ holds true only if $b < c$.

B.1   Interval Evaluation of a Function

Interval analysis can be applied to ordinary continuous[2] functions that are composed of the elementary operations as introduced in the previous section. This is achieved by exchanging the real-values variables $(x_1, \ldots, x_n)$ of the function $f : \mathbb{R}^n \to \mathbb{R}$

---

[1]It is possible to extended interval arithmetics with the values $\pm\infty$ as limits of an interval where the division by 0 is allowed. Such an extended interval arithmetics is also closed with respect to division, see [196].

[2]We restrict ourselves to continuous functions because it serves well for the purpose of this work. Anyway, there are extended techniques that allow to deal with non-continuous functions as well, see e.g. [196].

by real-values intervals $(\widehat{x}_1, \ldots, \widehat{x}_n)$. This results in a function $f^1 : \mathbb{I}^n \to \mathbb{I}$ that maps the interval vector $\widehat{x}$ onto an interval $\widehat{y}$. Since the interval function f^1 is equal to the real-valued function f except for the type of its arguments, we omit a special notation for such interval functions. We assume that interval operations have to be applied for evaluation if at least one operand is in an interval. Calculating the interval value of a function is called *interval evaluation* of the function. From the definition of the interval operations, it follows

$$\widehat{z} = f(\widehat{x}) \quad \Leftrightarrow \quad \inf \widehat{z} \leq f(\mathbf{x}) \leq \sup \widehat{z} \quad \forall \, \mathbf{x} \in \widehat{\mathbf{x}} \ , \tag{B.11}$$

i.e. the interval evaluation of a function yields guaranteed bounds $\widehat{\mathbf{z}}$ of the image space of the function f over the interval box $\widehat{\mathbf{x}}$.

B.2 Over-Estimation

Beside many similarities between interval arithmetics and arithmetics of real numbers, there are some important differences that needs to be taken into account. Both commutative and associative property holds true for addition and multiplication of intervals. Contrary, the distributive property cannot be used with intervals in its common form

$$\widehat{a}(\widehat{b} + \widehat{c}) \neq \widehat{a}\widehat{b} + \widehat{a}\widehat{c}, \quad \widehat{a}, \widehat{b}, \widehat{c} \in \mathbb{I} \ . \tag{B.12}$$

Sub-distributivity is a weak form of the distributive property and it holds true for every interval $\widehat{a}, \widehat{b}, \widehat{c} \in \mathbb{I}$

$$\widehat{a}(\widehat{b} + \widehat{c}) \subset \widehat{a}\widehat{b} + \widehat{a}\widehat{c} \ . \tag{B.13}$$

To receive an interval evaluation with as strict as possible bounds, it is favorable to evaluate the left hand side of Eq. (B.13) since it yields stricter bounds. In general, it can be stated that an interval evaluation yields closer bounds if every variable occurs only once in the function. If the same variable occurs multiple times in the same function, the so-called *interval identity* is lost, i.e. it cannot be taken into account that each instance of the variable x has the same value $x \in \widehat{x}$. Therefore, we may receive an over-estimation for the function's image if interval identity is not fulfilled. Even for very simple expressions such as $\widehat{a}^2 \subset \widehat{a} * \widehat{a}$ and $0 \subset \widehat{a} - \widehat{a} \neq 0$, we find a significant over-estimation. For example, evaluating the former expressions for $\widehat{x} = [-1; 1]$ yields $\widehat{x} * \widehat{x} = [-1; 1]$ where the strict result is $\widehat{x}^2 = [0; 1]$. The source of the overestimation comes from dealing with $\widehat{x} * \widehat{x}$ as with $\widehat{x} * \widehat{y}$, where the ranges for $\widehat{y} = \widehat{x}$ are identical just by chance. Functions with complicated expressions cannot be factored or rearranged so that every variable occurs only once. Therefore, over-estimation cannot be avoided in general. But still one can often find equivalence transformations leading to a smaller over-estimation. On the other hand, if interval identity holds true, one can conclude from the evaluation of the function

$$\widehat{y} = f(\widehat{\mathbf{x}}) \tag{B.14}$$

that f is surjective in the interval $\widehat{y}$, i.e. for every value $y \in \widehat{y}$ in the image space it exists at least one $\mathbf{x} \in \widehat{\mathbf{x}}$ in the domain.

B.3   Software and Implementation

The inclusion of a value in an interval is mathematically justified under the assumption that any number can be exactly represented with round-off errors. In practice, this is hardly possible due to finite accuracy of real computer hardware. Since the number of digits in limited at least by the amount of memory, we have to deal with some kind of round-off errors in any computation. Most microcomputers allow to control the direction of the round-off effect so that one receives a range of values where the exact value is enclosed. A systematic control called outward round-off is supported by many computers and allows to enclose the real value in an interval. An important application to interval analysis is therefore to keep track of all round-off errors during a computation. These errors can be a consequence of uncertainties in the initial data as well as method errors caused by the algorithm. These round-off errors cannot be avoided by interval analysis but unlike standard real-valued computations we get a rigorous estimation of these errors. Therefore, the result of an interval computation might be an interval with an inadequate large width and thus little practical use. In any case, standard algorithms would have reported one single but completely wrong value without any indication of a catastrophic round-off effect. Due to their special ability to deal with round-off and method errors interval analysis is called *robust* or *reliable* computation and a whole branch of numerical mathematics was developed around this property.

There are a number of computer libraries and development environments for interval analysis. Results presented in this work mostly used BIAS/Profil by Knüppel [249, 250], since this library is platform-independent and work both with Windows and Linux. Other implementations such as PASCAL-XSC [194], C-XSC [248] and Sun Forte [447] offer similar functions. An extension for MATLAB (MathWorks Inc.) for interval analysis was developed by Rump [427, 428]. A notable collection of algorithms for interval linear algebra called VERSOFT is available from Rohn.

Based on basic implementation for interval arithmetics, different tools were developed to do practical interval computations with advanced algorithms. An example with many applications in mechanism science is ALIAS [101] and its extension ALIAS/Maple [102] with an interface for the computer algebra system Maple (Waterloo Maple Inc.). The basics of the interval algorithms implemented there can be found in Moore [410], Neumaier [359] and Hansen [195, 196].

A typical library for a high level computer language such as C, C++, or Python as well for the scripting languages of numerical packages provides amongst the basic arithmetics operations $+, -, *, /$ a selection of elementary functions such as $\sin(\cdot)$, $\cos(\cdot)$, $\sqrt{\cdot}$, etc. These functions are efficiently implemented from their real-valued counterparts by making use of the individual properties such as monotony in order to compute largely improved bounds. Some packages additionally include hardware-based control of round-off errors through directed rounding. Briefly speaking, this instructs the computer to conservatively select the bounds for the result of an arithmetic operation.

References

1. Abbasnejad, G., & Carricato, M. (2012). Real solutions of the direct geometrico-static problem of under-constrained cable-driven parallel robots with 3 cables: A numerical investigation. *Meccanica, 47*(7), 1761–1773.
2. Abbasnejad, G., & Carricato, M. (2014). Direct geometrico-static problem of undercon-strained cable-driven parallel robots with five cables. In *Computational Kinematics* (vol. 15, pp. 59–66). Berlin: Springer.
3. Afshari, A., & Meghdari, A. (2007). New jacobian matrix and equations of motion for a 6 d.o.f. cable-driven robot. *International Journal of Advanced Robotic Systems, 4*(1), 63–68.
4. Agrawal, S. K., & Alp, A. B. (2002). Cable suspended robots: design, planning and control. In *IEEE International Conference on Robotics and Automation* (vol. 4, pp. 4275–4280).
5. Albus, J. S. (1989). Cable arrangement and lifting platform for stabilized load lifting. *U. S. Patent No. 4, 883, 184*, Nov 28, 1989.
6. Albus, J. S., Bostelman, R. V., & Dagalakis, N. G. (1992). The NIST ROBOCRANE. *Journal of Research at the National Institute of Standards and Technology, 97*(3), 373–385.
7. Albus, J. S., Bostelman, R. V., & Dagalakis, N. G. (1993). The NIST ROBOCRANE. *Journal of Robotic Systems, 10*(5), 709–724.
8. Alexandre dit Sandretto, J., Daney, D., & Gouttefarde, M. (2013). Calibration of a fully-constrained parallel cable-driven robot. *Romansy 19 - Robot Design, Dynamics and Control, CISM international centre for mechanical Sciences* (pp. 77–84). Berlin: Springer.
9. Alexandre dit Sandretto, J., Trombettoni, G., Daney, D., & Chabert, G. (2014). Certified calibration of a cable-driven robot using interval contractor programming. In *Computational Kinematics* (vol. 15, pp. 209–217). Berlin: Springer.
10. Alikhani, A., Behzadipour, S., Alasty, A., & Sadough Vanini, S. A. (2011). Design of a large-scale cable-driven robot with translational motion. *Robotics and Computer-Integrated Manufacturing, 27*(2), 357–366.
11. Alikhani, A., Behzadipour, S., Vanini, S. A. S., & Alasty, A. (2009). Workspace analysis of a three dof cable-driven mechanism. *Journal of Mechanisms and Robotics, 1*(4), 1–7.
12. Aref, M. M., Gholami, P., & Taghirad, H. D. (2008). Dynamic and sensitivity analysis of KNTU CDRPM: A cable driven redundant parallel manipulator. In *IEEE/ASME International Conference on Mechatronic and Embedded Systems and Applications (MESA)* (pp. 528–533).
13. Aref, M. M., Oftadeh, R., & Taghirad, H. D. (2009). Kinematics and jacobian analysis of the KNTU CDRPM: A cable driven redundant parallel manipulator. *Iranian Conference on Electrical Engineering, 7*, 319–324.
14. Aref, M. M., & Taghirad, H. D. (2008). Geometrical workspace analysis of a cable-driven redundant parallel manipulator: KNTU CDRPM. *IEEE/RSJ International Conference on Intelligent Robots and Systems*, 1958–1963.

© Springer International Publishing AG, part of Springer Nature 2018
A. Pott, *Cable-Driven Parallel Robots*, Springer Tracts in Advanced
Robotics 120, https://doi.org/10.1007/978-3-319-76138-1

15. Aref, M. M., Taghirad, H. D., & Barissi, S. (2009). Optimal design of dexterous cable driven parallel manipulators. *International Journal of Robotics and Automation, Vol. 14, No. 4, 2009*(1), 29–47.
16. Arsenault, M. (2006). *Développement et analyse de mécanismes de tenségrité.* PhD thesis, Québec: Université Laval.
17. Arsenault, M. (2010). Optimization of the prestress stable wrench closure workspace of planar parallel three-degree-of-freedom cable-driven mechanisms with four cables. *IEEE International Conference on Robotics and Automation (ICRA), 2010*, 1182–1187.
18. Arsenault, M. (2012). Stiffness analysis of a planar 2-DoF cable-suspended mechanism while considering cable mass. In Bruckmann, T., & Pott, A. (Eds.), *Cable-Driven Parallel Robots. Mechanisms and Machine Science* (vol. 12, pp. 405–421). Berlin: Springer.
19. Azizian, K. (2012). *Optimum-synthesis methods for cable-driven parallel mechanisms.* PhD thesis, Québec: Université Laval.
20. Azizian, K., & Cardou, P. (2012). The constant-orientation dimensional synthesis of planar cable-driven parallel mechanisms through convex relaxations. In Bruckmann, T., & Pott, A. (Eds.), *Cable-Driven Parallel Robots. Mechanisms and Machine Science* (vol. 12, pp. 215–230). Berlin: Springer.
21. Azizian, K., & Cardou, P. (2012). The dimensional synthesis of planar parallel cable-driven mechanisms through convex relaxations. *Journal of Mechanisms and Robotics, 4*(3), 031011.
22. Azizian, K., & Cardou, P. (2013). The dimensional synthesis of spatial cable-driven parallel mechanisms. *Journal of Mechanisms and Robotics, 5*(4), 044502.
23. Azizian, K., Cardou, P., & Moore, B. (2010). On the boundaries of the wrench-closure workspace of planar parallel cable-driven mechanisms. *Proceedings of the ASME Design Engineering Technical Conference, 2*, 203–212.
24. Azizian, K., Cardou, P., & Moore, B. (2012). Classifying the boundaries of the wrench-closure workspace of planar parallel cable-driven mechanisms by visual inspection. *Journal of Mechanical Design, 4*(2)
25. Babaghasabha, R., Khosravi, M. A., & Taghirad, H. D. (2015). Adaptive control of KNTU planar cable-driven parallel robot with uncertainties in dynamic and kinematic parameters. *Mechanisms and Machine Science, 32*, 145–159.
26. Bahrami, A., & Bahrami, M. N. (2011). Optimal design of a spatial four cable driven parallel manipulator. *IEEE International Conference on Robotics and Biomimetics (ROBIO), 2011*, 2143–2149.
27. Baoyan, D., Qiu, Y.-Y., Fushun, Z., & Zi, B. (2008). Analysis and experiment of the feed cable-suspended structure for super antenna. *IEEE/ASME International Conference on Advanced Intelligent Mechatronics, 2008*, 329–334.
28. Barrette, G., & Gosselin, C. (2000). Kinematic analysis and design of planar parallel mechanism actuated with cables. *ASME 26th Biennial Mechanisms and Robotics Conference.*
29. Barrette, G., & Gosselin, C. (2005). Determination of the dynamic workspace of cable-driven planar parallel mechanisms. *ASME Journal of Mechanical Design, 127*, 242–248.
30. Bauer, C. (2011). Device and method for detecting the inventory of a selling and/or storing device, and a storage-managing system equipped with said device. *Patent, WO 2012 101248*, Jan. 28, 2011.
31. Bedoustani, Y. B., Bigras, P., Taghirad, H. D., & Bonev, I. A. (2011). Lagrangian dynamics of cable-driven parallel manipulators: A variable mass formulation. *Transactions of the Canadian Society for Mechanical Engineering, 35*(4), 529–542.
32. Bedoustani, Y. B., Taghirad, H. D., & Aref, M. M. (2008). Dynamics analysis of a redundant parallel manipulator driven by elastic cables. In *10th International Conference on Control, Automation, Robotics and Vision* (pp. 536–542).
33. Behzadipour, S., & Khajepour, A. (2006). Stiffness of cable-based parallel manipulators with application to stability analysis. *Journal of Mechanical Design, 128*(1), 303–310.
34. Berti, A. (2015). *Kinematics and statics of cable-driven parallel robots by interval-analysis-based methods.* PhD thesis, Italy: University of Bologna.

35. Berti, A., Merlet, J.-P., & Carricato, M. (2012). Solving the direct geometrico-static problem of 3–3 cable-driven parallel robots by interval analysis: preliminary results. In Bruckmann, T., & Pott, A. (Eds.), *Cable-Driven Parallel Robots. Mechanisms and Machine Science* (vol. 12, pp. 251–268). Berlin: Springer.
36. Berti, A., Merlet, J.-P., & Carricato, M. (2014). Workspace analysis of redundant cable-suspended parallel robots. In Bruckmann, T., & Pott, A. (Eds.), *Cable-Driven Parallel Robots. Mechanisms and Machine Science* (vol. 32, pp. 41–53). Berlin: Springer.
37. Berti, A., Merlet, J.-P., & Carricato, M. (2015). Solving the direct geometrico-static problem of underconstrained cable-driven parallel robots by interval analysis. *International Journal of Robotics Research, 35*(6), 723–739.
38. Billette, G., & Gosselin, C. (2009). Producing rigid contacts in cable-driven haptic interfaces using impact generating reels. In *IEEE International Conference on Robotics and Automation (ICRA)* (pp. 307–312).
39. Blanchet, L., & Merlet, J.-P. (2014). Interference detection for cable-driven parallel robots (CDPRs). *IEEE/ASME International Conference on Advanced Intelligent Mechatronics (AIM)* (pp. 1413–1418).
40. Bliek, C., Spellucci, P., Vicente, L. N., Neumaier, A., Granvilliers, L., Monfroy, E., Benhamou, F., Huens, E., van Hentenryck, P., Sam-Haroud, D., Faltings, B. (2001). COCONUT Deliverable D1: Algorithms for solving nonlinear constrained and optimization problems: The state of the art.
41. Borgstrom, P. H. (2009). Novel cable-driven robotic platforms and algorithms for environmental sensing applications. *ProQuest Dissertations and Theses, 3405592*, 201.
42. Borgstrom, P. H., Jordan, B. L., Borgstrom, B. J., Stealey, M. J., Sukhatme, G. S., Batalin, M. A., et al. (2009). NIMS-PL: A cable-driven robot with self-calibration capabilities. *IEEE Transactions on Robotics, 25*(5), 1005–1015.
43. Borgstrom, P. H., Jordan, B. L., Sukhatme, G. S., Batalin, M. A., & Kaiser, W. J. (2009). Rapid computation of optimally safe tension distributions for parallel cable-driven robots. *IEEE Transactions on Robotics, 25*(6), 1271–1281.
44. Borm, J.-H., & Menq, C.-H. (1991). Determination of optimal measurement configurations for robot calibration based on observability measure. *International Journal of Robotics Research, 10*(1), 51–63.
45. Bosscher, P. (2004). *Disturbance robustness measures and wrench-feasible workspace generation techniques for cable-driven robots.* PhD thesis, Georgia: Georgia Institute of Technology.
46. Bosscher, P., & Ebert-Uphoff, I. (2004). A stability measure for underconstrained cable-driven robots. In *IEEE International Conference on Robotics and Automation* (vol. 5, pp. 4943–4949).
47. Bosscher, P., & Ebert-Uphoff, I. (2004). Wrench-based analysis of cable-driven robots. In *IEEE International Conference on Robotics and Automation* (pp. 4950–4955). New Orleans.
48. Bosscher, P., Riechel, A. T., & Ebert-Uphoff, I. (2006). Wrench-feasible workspace generation for cable-driven robots. *IEEE Transactions on Robotics, 22*(5), 890–902.
49. Bosscher, P., Williams II, R. L., Bryson, L. S., & Castro-Lacouture, D., (2007). Cable-suspended robotic contour crafting system. *Automation in Construction, 17*(1), 45–55.
50. Bosscher, P., Williams II, R. L., & Tummino, M. (2005). A concept for rapidly-deployable cable robot search and rescue systems. In *Proceedings of the ASME International Design Engineering Technical Conferences & Computers and Information in Engineering Conference, DETC2005* (vol. 7, pp. 589–598).
51. Bosscher, P. M., & Williams II, R. L. (2010). Apparatus and method associated with cable robot system. *U.S. Patent No. 7, 753, 642 B2*, July 13, 2010.
52. Bostelman, R. V., Dagalakis, N. G., & Albus, J. S. (1992). A robotic crane system utilizing the Stewart platform configuration. In *Proceedings of 4th International Symposium on Robotics and Manufacturing (ASME)*.
53. Bostelman, R. V., & Albus, J. S. (1993). Stability of an underwater work platform suspended from an unstable reference. *Proceedings of Engineering in Harmony with Ocean (OCEAN), 2*, 321–325.

54. Bostelman, R. V., Albus, J. S., Dagalakis, N. G., Jacoff, A. S., & Gross, J. (1994). Applications of the NIST ROBOCRANE. In *Fifth International Symposium on Robotics and Manufacturing: Research, Education, and Applications (ASME)* (pp. 403–410).

55. Bostelman, R. V., Albus, J. S., Murphy, K., Tsai, T., & Amatucci, E. (1994). A Stewart platform lunar rover. In *Engineering Construction and Operations in Space IV*.

56. Bostelman, R. V., Albus, J. S., Dagalakis, N. G., & Jacoff, A. S. (1996). Robocrane project: An advanced concept for large scale manufacturing. In *Proceedings of Association for Unmanned Vehicles Systems Int*. FL: Orlando.

57. Bostelman, R. V., Jacoff, A. S., Dagalakis, N. G., & Albus, J. S. (1996). RCS-based robocrane integration. In *International Conference on Intelligent Systems*: A Semiotic Perspective.

58. Bottema, O., & Roth, B. (1979). *Theoretical kinematics*. New York: Dover Publications.

59. Bouchard, S. (2008). *Géométrie des robots parallèles entraînés par des câbles*. PhD thesis, Québec: Université Laval.

60. Bouchard, S., & Gosselin, C. (2006). Kinematic sensitivity of a very large cable-driven parallel mechanism. In *ASME International Design Engineering Technical Conferences & Computers and Information in Engineering Conference, DETC2006* (pp. 851–858).

61. Bouchard, S., & Gosselin, C. (2008). Workspace optimization of a very large cable-driven parallel mechanism for a radiotelescope application. *ASME International Design Engineering Technical Conferences*, 8, 963–970.

62. Bouchard, S., Moore, B., & Gosselin, C. (2010). On the ability of a cable-driven robot to generate a prescribed set of wrenches. *Journal of Mechanisms and Robotics*, 2(1), 1–10.

63. Boye, T., Verl, A., & Pott, A. (2006). Optimal tolerance, model and pose selection for calibration of parallel manipulators. In *37th International Symposium on Robotics*. Germany.

64. Brackbill, E. A., Mao, Y., Agrawal, S. K., Annapragada, M., & Dubey, V. N. (2009). Dynamics and control of a 4-dof wearable cable-driven upper arm exoskeleton. In *IEEE International Conference on Robotics and Automation (ICRA)* (pp. 2300–2305).

65. Brau, E., Gosselin, F., & Lallemand, J. P. (2005). Design of a singularity free architecture for cable driven haptic interfaces. *Symposium on Haptic Interfaces for Virtual Environment and Teleoperator Systems, 2005*, 208–213.

66. Brau, E., Lallemand, J. P., & Gosselin, F. (2005). Analytic determination of the tension capable workspace of cable actuated haptic interfaces. *ACM International Conference Proceeding Series, 157*, 195–200.

67. Brown, G. W. (1987). Suspension system for supporting and conveying equipment, such as a camera. *U.S. Patent 4, 710, 819*, Dec 1, 1987.

68. Bruckmann, T. (2010). *Auslegung und Betrieb redundanter paralleler Seilroboter*. PhD thesis, Germany: University of Duisburg-Essen.

69. Bruckmann, T., Hiller, M., & Schramm, D. (2010). An active suspension system for simulation of ship maneuvers in wind tunnels. In *New Trends in Mechanism Science. Mechanisms and Machine Science* (vol. 5, pp. 537–544). Berlin: Springer.

70. Bruckmann, T., Lalo, W., Nguyen, K., & Salah, B. (2012). Development of a storage retrieval machine for high racks using a wire robot. In *ASME International Design Engineering Technical Conferences & Computers and Information in Engineering Conference* (p. 771).

71. Bruckmann, T., Lalo, W., Schramm, D., & Hiller, M. (2013). Design and realization of a high rack storage and retrieval machine based on wire robot technology. In *Proceedings of the XV International Symposium on Dynamic Problems of Mechanics* (pp. 771–780).

72. Bruckmann, T., Mikelsons, L., Brandt, T., Hiller, M., & Schramm, D. (2008). Wire robots Part I – Kinematics, analysis and design. In *Parallel Manipulators*. Vienna: I-Tech Education and Publishing.

73. Bruckmann, T., Mikelsons, L., Brandt, T., Hiller, M., & Schramm, D. (2008). Wire robots Part II – Kinematics, analysis and design. In *Parallel Manipulators*. Vienna: I-Tech Education and Publishing.

74. Bruckmann, T., Mikelsons, L., Brandt, T., Hiller, M., & Schramm, D. (2009). Design approaches for wire robots. *ASME Conference Proceedings, 2009*(49040), 25–34.

75. Bruckmann, T., Mikelsons, L., & Hiller, M. (2011). A design-to-task approach for wire robots. In *Interdisciplinary Applications of Kinematics* (pp. 83–97). Dordrecht: Springer.

76. Bruckmann, T., Mikelsons, L., Hiller, M., & Schramm, D. (2007). A new force calculation algorithm for tendon-based parallel manipulators. In *IEEE/ASME International Conference on Advanced Intelligent Mechatronics* (pp. 1–6).

77. Bruckmann, T., Mikelsons, L., Pott, A., Abdel-Maksoud, M., Brandt, T., & Schramm, D. (2009). A novel tensed mechanism for simulation of maneuvers in wind tunnels. In *33rd ASME Mechanics and Robotics Conference (MECH 2009)* (pp. 17–24).

78. Bruckmann, T., Mikelsons, L., Schramm, D., & Hiller, M. (2007). Continuous workspace analysis for parallel cable-driven Stewart-Gough platforms. *PAMM, 7*(1), 4010025–4010026.

79. Bruckmann, T., Pott, A., & Hiller, M. (2006). Calculating force distributions for redundantly actuated tendon-based stewart platforms. *Advances in Robot Kinematics (ARK)* (pp. 403–412). Berlin: Springer.

80. Bruckmann, T., Sturm, C., Fehlberg, L., & Reichert, C. (2013). An energy-efficient wire-based storage and retrieval system. In *IEEE/ASME International Conference on Advanced Intelligent Mechatronics* (pp. 631–636).

81. Buterbaugh, A., Kent, B. M., Mentzer, C., Scott, M., & Forster, W. (2007). Demonstration of an inverted stewart platform target-suspension system using lightweight, high-tensile strings. *IEEE Antennas and Propagation Magazine, 49*(5), 185–190.

82. Campbell, P. D., Swaim, P. L., & Thompson, C. J. (1995). *In Charlotte™ Robot Technology for Space and Terrestrial Applications*. Warrendale, PA: SAE International.

83. Capua, A., Shapiro, A., & Shoval, S. (2009). Motion analysis of an underconstrained cable suspended mobile robot. *IEEE International Conference on Robotics and Biomimetics (ROBIO)* (pp. 788–793).

84. Capua, A., Shapiro, A., & Shoval, S. (2010). Motion planning algorithm for a mobile robot suspended by seven cables. *IEEE Conference on Robotics Automation and Mechatronics (RAM)* (pp. 504–509).

85. Caro, S., Wenger, P., & Chablat, D. (2012). Non-singular assembly mode changing trajectories of a 6-DOF parallel robot. In *ASME International Design Engineering Technical Conferences & Computers and Information in Engineering Conference* (p. 1245).

86. Carricato, M., & Abbasnejad, G. (2012). Direct geometrico-static analysis of under-constrained cable-driven parallel robots with 4 cables. In Bruckmann, T., & Pott, A. (Eds.), *Cable-Driven Parallel Robots. Mechanisms and Machine Science* (vol. 12, pp. 269–285). Berlin: Springer.

87. Carricato, M., Abbasnejad, G., & Walter, D. (2012). Inverse geometrico-static analysis of under-constrained cable-driven parallel robots with four cables. In *Advances in Robot Kinematics (ARK)* (pp. 365–372).

88. Carricato. M., & Merlet, J.-P. (2010). Geometrico-static analysis of under-constrained cable-driven parallel robots. In *Advances in Robot Kinematics (ARK)* (pp. 309–319).

89. Carricato, M., & Merlet, J.-P. (2011). Direct geometrico-static problem of under-constrained cable-driven parallel robots with three cables. In *IEEE International Conference on Robotics and Automation (ICRA)* (pp. 3011–3017).

90. Carricato, M., & Merlet, J.-P. (2011). Inverse geometrico-static problem of under-constrained cable-driven parallel robots with three cables. In *13th World Congress in Mechanism and Machine Science* (pp. 1–10).

91. Carricato, M., & Merlet, J.-P. (2013). Stability analysis of underconstrained cable-driven parallel robots. *IEEE Transactions on Robotics, 29*(1), 289–296.

92. Castelli, G., & Ottaviano, E. (2009). Modelling and simulation of a cable-based parallel manipulator as an assisting device. In *Computational Kinematics* (pp. 17–24). Berlin: Springer.

93. Castelli, G., & Ottaviano, E. (2014). A cartesian cable-suspended robot for aiding mobility. In *Computational Kinematics* (vol. 15, pp. 369–376). Berlin: Springer.

94. Castelli, G., Ottaviano, E., & Rea, P. (2014). A cartesian cable-suspended robot for improving end-users' mobility in an urban environment. *Robotics and Computer-Integrated Manufacturing, 30*(3), 335–343.

95. Chablat, D., & Wenger, P. (1998). Working modes and aspects in fully parallel manipulators. *Proceedings - IEEE International Conference on Robotics and Automation, 3*, 1964–1969.

96. Chellal, R., Cuvillon, L., & Laroche, E. (2014). A kinematic vision-based position control of a 6-DoF cable-driven parallel robot. In Bruckmann, T., & Pott, A. (Eds.), *Cable-Driven Parallel Robots. Mechanisms and Machine Science* (vol. 32, pp. 213–225). Berlin: Springer.

97. Choe, W., Kino, H., Katsuta, K., & Kawamura, S. (1996). Design of parallel wire driven robots for ultrahigh speed motion based on stiffness analysis. *Proceedings of the Japan/USA Symposium on Flexible Automation, 1*, 159–166.

98. Clavel, R. (1988). DELTA, a fast robot with parallel geometry. In *18th International Symposium on Industrial Robots* (pp. 91–100).

99. Collard, J.-F., & Cardou, P. (2013). Computing the lowest equilibrium pose of a cable-suspended rigid body. *Optimization and Engineering, 14*(3), 457–476.

100. Cone, Lawrence L. (1985). Skycam, an aerial robotic camera system. *Byte Magazine, 10*, 122–132.

101. COPRIN. (2003). A C++ Algorithm library of interval analysis for equation systems. *The COPRIN project.*

102. COPRIN. (2003). The maple interface for ALIAS. *The COPRIN project (Manual).*

103. Corbel, D., Gouttefarde, M., Company, O., & Pierrot, F. (2010). Towards 100G with PKM. Is actuation redundancy a good solution for pick-and-place? In *IEEE International Conference on Robotics and Automation (ICRA)* (pp. 4675–4682).

104. Crawford, D. W., & Nemeth, E. A. (2012). Amusement park ride with cable-suspended vehicles. *U.S. Patent No. 8, 147, 344 B2*, April 3, 2012.

105. Cunningham, D., & Asada, H. H. (2009). The Winch-Bot: A cable-suspended, under-actuated robot utilizing parametric self-excitation. In *IEEE International Conference on Robotics and Automation* (pp. 1844–1850).

106. Dagalakis, N. G., Albus, J. S., Wang, B.-L., Unger, J., & Lee, James D. (1989). Stiffness study of a parallel link robot crane for shipbuilding applications. *ASME Journal of Mechanical Design, 111*(3), 183–193.

107. Dallej, T., Gouttefarde, M., Andreff, N., Dahmouche, R., & Martinet, P. (2012). Vision-based modeling and control of large-dimension cable-driven parallel robots. In *IEEE/RSJ International Conference on Intelligent Robots and Systems (IROS)* (pp. 1581–1586).

108. Dallej, T., Gouttefarde, M., Andreff, N., Michelin, M., & Martinet, P. (2011). Towards vision-based control of cable-driven parallel robots. In *IEEE/RSJ International Conference on Intelligent Robots and Systems (IROS)* (pp. 2855–2860).

109. Daney, D. (2002). Optimal measurement configurations for gough platform calibration. In *IEEE International Conference on Robotics and Automation* (vol. 1, pp. 147–152).

110. Daney, D., Papegay, Y., & Neumaier, A. (2004). Interval methods for certification of the kinematic calibration of parallel robots. In *Proceedings of the IEEE International Conference on Robotics and Automation.*

111. Darwin, L., Jonathan, E., Ying, T., & Denny, O. (2016). CASPR: A comprehensive cable-robot analysis and simulation platform for the research of cable-driven parallel robots. In *IEEE International Conference on Robotics and Biomimetics (ROBIO)* (pp. 3004–3011).

112. Dasgupta, B., & Mruthyunjaya, T. S. (1998). Force redundancy in parallel manipulators: Theoretical and practical issues. *Mechanism and Machine Theory, 33*(6), 727–742.

113. Dekker, R., Khajepour, A., & Behzadipour, S. (2006). Design and testing of an ultra-high-speed cable robot. *International Journal of Robotics and Automation, 21*(1), 25–34.

114. Denavit, J., & Hartenberg, R. S. (1955). A kinematic notation for lower pair mechanisms based on matrices. *Journal of Applied Mechanics, 22*, 215–221.

115. Diao, X. (2007). *Study of Cable Robots for Hardware-in-the-Loop Contact Dynamics Simulation.* New Mexico State University.

116. Diao, X., & Ma, O. (2005). Dynamics analysis of a cable-driven parallel manipulator for hardware-in-the-loop dynamic simulation. In *IEEE/ASME International Conference on Advanced Intelligent Mechatronics* (pp. 837–842).

117. Diao, X., & Ma, O. (2006). Workspace analysis of a 6-DOF cable robot for hardware-in-the-loop dynamic simulation. In *IEEE International Conference on Intelligent Robots and Systems* (pp. 4103–4108).

118. Diao, X., & Ma, O. (2006). Workspace analysis of a 6-DOF cable robot for hardware-in-the-loop dynamic simulation. In *Proceedings of the IEEE International Conference on Intelligent Robots and Systems (IROS)* (pp. 4103–4108). China.

119. Diao, X., & Ma, O. (2007). Force-closure analysis of general 6-dof cable manipulators. In *Proceedings of the IEEE International Conference on Intelligent Robots and Systems (IROS)* (pp. 3931–3936).

120. Diao,X., & Ma, O. (2007). Vibration analysis of cable-driven parallel manipulators for hardware-in-the-loop contact-dynamics simulation. In *Proceedings of the ASME International Design Engineering Technical Conferences & Computer and Information in Engineering Conference (IDET/CIE 2007)*.

121. Diao, X., & Ma, O. (2009). Vibration analysis of cable-driven parallel manipulators. *Multibody System Dynamics, 21*(4), 347–360.

122. Dietmaier, P. (1998). The Stewart-Gough platform of general geometry can have 40 real postures. *Advances in Robot Kinematics (ARK)* (pp. 7–16). Austria: Kluwer Academic Publishers.

123. DIN:66025:1983. Programmaufbau für numerisch gesteuerte Arbeitsmaschinen; Allgemeines. Berlin: Beuth.

124. DIN 3068:1972-03. Drahtseile aus Stahldrähten. Berlin: Beuth.

125. DIN:EN:ISO 8373:2010–2011. Manipulating industrial robots – Vocabulary. Berlin: Beuth.

126. Du, J., Ding, W., & Bao, H. (2012). Cable vibration analysis for large workspace cable-driven parallel manipulators. In Bruckmann, T., & Pott, A., (Eds.), *Cable-Driven Parallel Robots. Mechanisms and Machine Science* (vol. 12, pp. 437–449). Berlin: Springer.

127. Duan, B., Qiu, Y.-Y., Zhang, F., & Zi, B. (2009). On design and experiment of the feed cable-suspended structure for super antenna. *Mechatronics, 19*(4), 503–509.

128. Duan, Q. J., Du, J., Duan, B., Li, T. J., & Tang, A. (2010). Modeling of variable length cable driven parallel robot. In *IEEE/ASME International Conference on Mechatronics and Embedded Systems and Applications (MESA)* (pp. 545–548).

129. Duan, Q. J., Du, J., Duan, B., & Tang, A. (2010). Deployment/retrieval modeling of cable-driven parallel robot. *Mathematical Problems in Engineering, 2010*, 1–10.

130. Duan, Q. J., & Duan, X. C. (2011). Analysis of cable-actuated parallel robot with variable length and velocity cable. *Procedia Engineering, 15*, 2732–2737.

131. Duan, Q., Vashista, V., & Agrawal, S. K. (2015). Effect on wrench-feasible workspace of cable-driven parallel robots by adding springs. *Mechanism and Machine Theory, 86*, 201–210.

132. Duan, X., Qiu, Y., Duan, Q., & Jingli, D. (2014). Calibration and motion control of a cable-driven parallel manipulator based triple-level spatial positioner. *Advances in Mechanical Engineering, 1–10*, 2014.

133. Duelen, G., & Schröer, K. (1991). Robot calibration - Method and results. *Robotics and Computer-Integrated Manufacturing, 8*(4), 223–231.

134. Ebert-Uphoff, I., & Voglewede, P. A. (2004). On the connections between cable-driven parallel manipulators and grasping. In *IEEE International Conference on Robotics and Automation* (pp. 4521–4526). New Orleans.

135. El-Ghazaly, G., Gouttefarde, M., & Creuze, V. (2015). Hybrid cable-thruster actuated underwater vehicle-manipulator systems: A study on force capabilities. In *IEEE/RSJ International Conference on Intelligent Robots and Systems* (pp. 1672–1678).

136. Emmens, A. R., Spanjer, S. A. J., & Herder, J. L. (2014). Modeling and control of a large-span redundant surface constrained cable robot with a vision sensor on the platform. In Bruckmann, T., & Pott, A., (Eds.), *Cable-Driven Parallel Robots. Mechanisms and Machine Science* (vol. 32, pp. 249–260). Berlin: Springer.

137. Fahham, H. R., & Farid, M. (2010). Optimum design of planar redundant cable-suspended robots for minimum time trajectory tracking. In *International Conference on Control Automation and Systems (ICCAS)* (pp. 2156–2163).

138. Fahham, H. R., & Farid, M. (2010). Minimum-time trajectory planning of spatial cable-suspended robots along a specified path considering both tension and velocity constraints. *Engineering Optimization, 42*(4), 387–402.

139. Fang, S. (2005). *Design, modeling and motion control of tendon-based parallel manipulators.* PhD thesis, Germany: University of Duisburg-Essen.

140. Fassi, I., Legnani, G., & Magnani, P. L. (2010). Design of a tendon driven parallel manipulator for micro-factory applications. In *International Symposium on Robotics (ISR) and German Conference on Robotics (ROBOTIK)* (pp. 1–6).

141. Fattah, A., & Agrawal, S. K. (2002). Design of cable-suspended planar parallel robots for an optimal workspace. In *Proceedings of the Workshop on Fundamental Issues and Future Research Directions for Parallel Mechanisms and Manipulators* (pp. 195–202).

142. Fattah, A., & Agrawal, S. K. (2002). Workspace and design analysis of cable-suspended planar parallel robots. *Proceedings of the ASME Design Engineering Technical Conference, 5,* 1095–1103.

143. Fattah, A., & Agrawal, S. K. (2005). On the design of cable-suspended planar parallel robots. *Journal of Mechanical Design, 127*(5), 1021–1028.

144. Fehlberg, L., Reichert, C., Zitzewitz, J. V., & Bruckmann, T. (2013). Ausnutzung energiespeichernder Elemente zur Effizienzsteigerung seilbasierter Regalbediengeräte. In *Fachtagung Mechatronik* (pp. 171–176).

145. FEM 1.001, (1998). *Rules for the design of hoisting appliances.* Frankfurt: VDMA Verlag

146. Ferraresi, C., Paoloni, M., Pastorelli, S., & Pescarmona, F. (2004). A new 6-DOF parallel robotic structure actuated by wires: The WiRo-6.3. *Journal of Robotic Systems, 21*(11), 581–595.

147. Ferraresi, C., Paoloni, M., & Pescarmona, F. (2007). A new methodology for the determination of the workspace of six-DOF redundant parallel structures actuated by nine wires. *Robotica, 25*(01), 113.

148. Ferraresi, C., & Pescarmo, F. (2010). Cable driven devices for telemanipulation. In *Remote and Telerobotics* (pp. 171–190).

149. Feyrer, K. (2015). *Wire ropes. Tension, Endurance, Reliability* (2nd ed.). Berlin, Heidelberg: Springer.

150. Fiedler, M., Nedoma, J., Ramik, J., Rohn, J., & Zimmermann, K. (2006). *Linear optimization problems with inexact data.* Boston: Kluwer Academic Publishers.

151. Filipovic, M., Djuric, A., & Kevac, L. (2014). The rigid S-type cable-suspended parallel robot design, modelling and analysis. *Robotica,* 1–13.

152. Fischer, R. (2003). *Elektrische Maschinen.* Berlin: Springer.

153. Gagliardini, L., Caro, S., & Gouttefarde, M. (2015). Dimensioning of cable-driven parallel robot actuators, gearboxes and winches according to the twist feasible workspace. In *IEEE International Conference on Automation Science and Engineering (CASE)* (pp. 99–105).

154. Gagliardini, L., Caro, S., Gouttefarde, M., & Girin, A. (2016). Discrete reconfiguration planning for Cable-Driven Parallel Robots. *Mechanism and Machine Theory, 100,* 313–337.

155. Gagliardini, L., Caro, S., Gouttefarde, M., & Girin, A. (2015). A reconfiguration strategy for Reconfigurable cable-driven parallel robots. In *IEEE International Conference on Robotics and Automation (ICRA)* (pp. 1613–1620).

156. Gagliardini, L., Caro, S., Gouttefarde, M., Wenger, P., & Girin, A. (2014). Optimal design of cable-driven parallel robots for large industrial structures. In *IEEE International Conference on Robotics and Automation (ICRA)* (pp. 5744–5749).

157. Gallina, P., Rossi, A., & Williams II, R. L. (2001). Planar cable-direct-driven robots, Part II: Dynamics and control. In *ASME Design Technical Conference.*

158. Gao, B., et al. (2012). Combined inverse kinematic and static analysis and optimal design of a cable-driven mechanism with a spring spine. *Advanced Robotics, 26,* 923–946.

159. Gayral, T., & Daney, D. (2014). A sufficient condition for parameter identifiability in robotic calibration. In *Computational Kinematics* (vol. 15, pp. 131–138). Berlin: Springer.

160. German, J. J., Jablokow, K. W., & Cannon, D. J. (2001). The cable array robot: theory and experiment. In *IEEE International Conference on Robotics and Automation (ICRA)* (vol. 3, pp. 2804–2810).

161. Ghasemi, A., Eghtesad, M., & Farid, M. (2008). Workspace analysis of planar and spatial redundant cable robots. In *American Control Conference* (pp. 2389–2394).
162. Ghasemi, A., Eghtesad, M., & Farid, M. (2008). Workspace analysis of redundant cable robots. In *World Automation Congress* (pp. 1–6).
163. Ghasemi, A., Eghtesad, M., & Farid, M. (2009). Workspace analysis for planar and spatial redundant cable robots. *Journal of Mechanisms and Robotics, 1*(4), 044502.
164. Ghasemi, A., Eghtesad, M., & Farid, M. (2010). Neural network solution for forward kinematics problem of cable robots. *Journal of Intelligent and Robotic Systems, 60*(2), 201–215.
165. Ghasemi, A., Farid, M., & Eghtesad, M. (2008). Interference free workspace analysis of redundant 3D cable robots. In *World Automation Congress* (pp. 1–6).
166. Gogu, G. (2008). *Structural synthesis of parallel robots*. Dordrecht: Springer.
167. Gonzalez-Rodríguez, A., Ottaviano, E., Castillo-García, F. J., & Rea, P. (2015). A novel design to improve pose accuracy for cable robots. In *Proceedings of the 14th IFToMM World Congress* (pp. 264–269). Taipei.
168. Gosselin, C. (2008). On the determination of the force distribution in overconstrained cable-driven parallel mechanisms. In *Proceedings of the Second International Workshop on Fundamental Issues and Future Research Directions for Parallel Mechanisms and Manipulators* (pp. 9–17). France.
169. Gosselin, C. (2012). Global planning of dynamically feasible trajectories for three-DOF spatial cable-suspended parallel robots. In Bruckmann, T., & Pott, A. (Eds.), *Cable-Driven Parallel Robots. Mechanisms and Machine Science* (vol. 12, pp. 3–22). Berlin: Springer.
170. Gosselin, C. (2014). Cable-driven parallel mechanisms: state of the art and perspectives. *Mechanical Engineering Reviews, 1*(1), 1–17.
171. Gosselin, C., & Angeles, J. (1990). Singularity analysis of closed-loop kinematic chains. *IEEE Transactions on Robotics and Automation, 4*, 281–290.
172. Gosselin, C., & Angeles, J. (1991). A global performance index for the kinematic optimization of robotic manipulators. *Journal of Mechanical Design, 113*(3), 220.
173. Gosselin, C., & Bouchard, S. (2010). A gravity-powered mechanism for extending the workspace of a cable-driven parallel mechanism: Application to the appearance modelling of objects. *International Journal of Automation Technology, 4*(4), 372–379.
174. Gosselin, C., & Grenier, M. (2011). On the determination of the force distribution in over-constrained cable-driven parallel mechanisms. *Meccanica, 46*(1), 3–15.
175. Gosselin, C., Lefrancois, S., Zoso, N., Angeles, J., Kovecses, J., Boulet, B., et al. (2010). Underactuated cable-driven robots: Machine, control and suspended bodies. *Advances in Intelligent and Soft Computing, 83*, 311–323.
176. Gosselin, C., Ren, P., & Foucault, S. (2012). Dynamic trajectory planning of a two-DOF cable-suspended parallel robot. In *IEEE International Conference on Robotics and Automation (ICRA)*.
177. Gottlieb, J. (2014). Non-parametric calibration of a stewart platform. In *Proceedings of the Workshop on Fundamental Issues and Future Research Directions for Parallel Mechanisms and Manipulators*.
178. Gouttefarde, M. (2005). *Analyse de l'espace des poses polyvalentes des mécanismes parallèles entraînés par câbles*. PhD thesis, Québec: Université Laval.
179. Gouttefarde, M., Collard, J.-F., Riehl, N., & Baradat, C. (2012). Simplified static analysis of large-dimension parallel cable-driven robots. In *IEEE International Conference on Robotics and Automation* (pp. 2299–2305).
180. Gouttefarde, M., Collard, J.-F., Riehl, N., & Baradat, C. (2015). Geometry selection of a redundantly actuated cable-suspended parallel robot. *IEEE Transactions on Robotics, 31*(2), 501–510.
181. Gouttefarde, M., Company, O., & Pierrot, F. (2013). On the simplifications of cable model in static analysis of large-dimension cable-driven parallel robots. In *IEEE/RSJ International Conference on Intelligent Robots and Systems* (pp. 928–934).
182. Gouttefarde, M., Daney, D., & Merlet, J.-P. (2011). Interval-analysis-based determination of the wrench-feasible workspace of parallel cable-driven robots. *IEEE Transactions on Robotics, 27*(1), 1–13.

183. Gouttefarde, M., & Gosselin, C. (2004). On the properties and the determination of the wrench-closure workspace of planar parallel cable-driven mechanisms. *Proceedings of the ASME Design Engineering Technical Conference, 2*, 337–346.
184. Gouttefarde, M., & Gosselin, C. (2005). Wrench-closure workspace of six-dof parallel mechanisms driven by 7 cables. *Transactions of the Canadian Society for Mechanical Engineering, 29*(4), 541–552.
185. Gouttefarde, M., & Gosselin, C. (2006). Analysis of the wrench-closure workspace of planar parallel cable-driven mechanisms. *IEEE Transactions on Robotics, 22*(3), 434–445.
186. Gouttefarde, M., Krut, S., Company, O., Pierrot, F., & Ramdani, N. (2008). On the design of fully constrained parallel cable-driven robots. *Advances in Robot Kinematics (ARK)* (pp. 71–78). Berlin: Springer.
187. Gouttefarde, M., Merlet, J.-P., & Daney, D. (2006). Determination of the Wrench-Closure workspace of 6-dof parallel cable-driven mechanisms. In *Advances in Robot Kinematics (ARK)* (pp. 315–322). Springer.
188. Gouttefarde, M., Merlet, J.-P., & Daney, D. (2007). Wrench-Feasible workspace of parallel cable-driven mechanisms. In *IEEE International Conference on Robotics and Automation* (pp. 1492–1497).
189. Gouttefarde, M., Nguyen, D. Q., & Baradat, C. (2014). Kinetostatic analysis of cable-driven parallel robots with consideration of sagging and pulleys. In *Advances in Robot Kinematics (ARK)* (pp. 213–221).
190. Guilin, Y., Pham, C. B., & Yeo, S. H. (2006). Workspace performance optimization of fully restrained cable-driven parallel manipulators. In *IEEE/RSJ International Conference on Intelligent Robots and Systems* (pp. 85–90).
191. Guilin, Y., Yeo, S. H., & Pham, C. B. (2004). Kinematics and singularity analysis of a planar cable-driven parallel manipulator. In *IEEE/RSJ International Conference on Intelligent Robots and Systems* (vol. 4, pp. 3835–3840).
192. Hadian, H., & Fattah, A. (2008). Best kinematic performance analysis of a 6–6 cable-suspended parallel robot. In *IEEE/ASME International Conference on Mechatronic and Embedded Systems and Applications* (pp. 510–515).
193. Hamedi, J., & Zohoor, H. (2008). Kinematic modeling and workspace analysis of a spatial cable suspended robot as incompletely restrained positioning mechanism. *International Journal of Aerospace and Mechanical Engineering, 2*(2), 109–118.
194. Hammer, R., Hocks, M., Kulisch, U., & Ratz, D. (1993). *Numerical toolbox for verified computing. Basic numerical problems*. New York: Springer.
195. Hansen, E. (1992). *Global optimization using interval analysis*. New York: Marcel Dekker.
196. Hansen, E., & Walster, G. W. (2004). *Global optimization using interval analysis*. New York: Marcel Dekker.
197. Hao, F., & Merlet, J.-P. (2005). Multi-criteria optimal design of parallel manipulators based on interval analysis. *Mechanism and Machine Theory, 40*(2), 157–171.
198. Hassan, M., & Khajepour, A. (2006). *Optimum connection positions for the redundant limb in cable-based parallel manipulators*. In *Proceedings of the ASME International Engineering Congress and Exposition (IMECE)*.
199. Hassan, M., & Khajepour, A. (2007). Minimum-norm solution for the actuator forces in cable-based parallel manipulators based on convex optimization. In *IEEE International Conference on Robotics and Automation* (pp. 1498–1503).
200. Hassan, M., & Khajepour, A. (2008). Minimization of bounded cable tensions in cable-based parallel manipulators. *2007 Proceedings of the ASME International Design Engineering Technical Conferences and Computers and Information in Engineering Conference, DETC2007, 8*, 991–999.
201. Hassan, M., & Khajepour, A. (2008). Optimization of actuator forces in cable-based parallel manipulators using convex analysis. *IEEE Transactions on Robotics, 24*(3), 736–740.
202. Hassan, M., & Khajepour, A. (2010). Analysis of a large-workspace cable-actuated manipulator for warehousing applications. *Proceedings of the ASME International Design Engineering Technical Conferences and Computers and Information in Engineering Conference DETC2009, 7(PART A)*, 45–53.

203. Hassan, M., & Khajepour, A. (2011). Analysis of bounded cable tensions in cable-actuated parallel manipulators. *IEEE Transactions on Robotics, 27*(5), 891–900.
204. Hearle, J. W. S. (2001). *High-performance fibres*. Boca Raton: CRC Press.
205. Hebsacker, M. (2000). *Entwurf und Bewertung Paralleler Werkzeugmaschinen - das Hexaglide*. PhD thesis, Zurich: ETH Zurich.
206. Heyden, T. (2006). *Bahnregelung eines seilgeführten Handhabungssystems mit kinematisch unbestimmter Lastführung*. Fortschritt-Berichte VDI, Reihe 8, Nr. 1100. Düsseldorf: VDI Verlag.
207. Heyden, T., & Woernle, C. (2004). Flatness-based trajectory tracking control of an under-constrained cable suspension manipulator. *Applied Mathematics and Mechanics (PAMM), 4*, 129–130.
208. Heyden, T., & Woernle, C. (2006). Dynamics and flatness-based control of a kinematically undetermined cable suspension manipulator. *Multibody System Dynamics, 16*(2), 155–177.
209. Higuchi, T., Ming, A., & Jiang-yu, J. (1988). Application of multi-dimensional wire cranes in construction. In *5th International Symposium on Robotics in Construction* (pp. 661–668).
210. Hiller, M., Fang, S., Mielczarek, S., Verhoeven, R., & Franitza, D. (2005). Design, analysis and realization of tendon-based parallel manipulators. *Mechanism and Machine Theory, 40*(4), 429–445.
211. Hiller, M., & Kecskeméthy, A. (1988). Equations of motion of complex multibody systems using kinematical differentials. In *Proceedings of 9th Symposium on Engineering Applications of Mechanics* (pp. 425–430). Canada.
212. Hollerbach, J. M., & Wampler, C. W. (1996). The calibration index and taxonomy for robot kinematic calibration methods. *International Journal of Robotics Research, 15*(6), 573–591.
213. Hui, L. I. (2015). A giant sagging-cable-driven parallel robot of fast telescope: Its tension-feasible workspace of orientation and orientation planning. In *Proceedings of the 14th IFToMM World Congress* (pp. 373–381). Taipei.
214. Husty, M. L. (1996). An algorithm for solving the direct kinematic of stewart-gough-type platforms. *Mechanism and Machine Theory, 31*(4), 365–380.
215. Husty, M. L. (2009). Non-singular assembly mode change in 3-RPR-parallel manipulators. In *Computational Kinematics* (pp. 51–60). Heidelberg: Springer.
216. Husty, M. L., Mielczarek, S., & Hiller, M. (2001). Constructing an overconstrained planar 4RPR manipulator with maximal forward kinematics solution set. In *Proceedings of the RAAD'01 Robotics in the Alpe-Adria-Danube Region*.
217. Husty, M. L., Mielczarek, S., & Hiller, M. (2002). A redundant spatial Stewart-Gough platform with maximal forward kinematics solution set. *Advances in Robot Kinematics (ARK)* (pp. 147–154). Spain: Kluwer Academic Publishers.
218. Irvine, H. (1981). *Cable structures*. Cambridge MA: MIT Press.
219. Ishii, M., & Sato, M. (1994). A 3D Spatial interface device using tensed strings. *Presence: Teleoperators and Virtual Environments, 3*(1), 81–86.
220. ISO:9283:1998. Manipulating industrial robots – Performance criteria and related test methods. Berlin: Beuth.
221. ISO/IEC 14772-1:1997. Information technology – Computer graphics and image processing – The virtual reality modeling language – Part 1: Functional specification and UTF-8 encoding. Berlin: Beuth.
222. ISO 4308-1:3003. Cranes and lifting appliances – Selection of wire ropes – Part 1: General. Berlin: Beuth.
223. Izard, J.-B., Gouttefarde, M., Baradat, C., Culla, D., & Sallé, D. (2012). Integration of a parallel cable-driven robot on an existing building façade. In Bruckmann, T., & Pott, A. (Eds.), *Cable-Driven Parallel Robots. Mechanisms and Machine Science* (vol. 12, pp. 149–164). Berlin: Springer.
224. Izard, J.-B., Gouttefarde, M., Michelin, M., Tempier, O., & Baradat, C. (2012). A reconfigurable robot for cable-driven parallel robotic research and industrial scenario proofing. In Bruckmann, T., & Pott, A. (Eds.), *Cable-Driven Parallel Robots. Mechanisms and Machine Science* (vol. 12, pp. 135–148). Berlin: Springer.

225. Jeong, J. W., Kim, S. H., & Kwak, Y. K. (1998). Design and kinematic analysis of the wire parallel mechanism for a robot pose measurement. *IEEE International Conference on Robotics and Automation, 4*, 2941–2946.

226. Jeong, J. W., Kim, S. H., & Kwak, Y. K. (1998). Development of a parallel wire mechanism for measuring position and orientation of a robot end-effector. *Mechatronics, 8*(8), 845–861.

227. Jeong, J. W., Kim, S. H., & Kwak, Y. K. (1999). Kinematics and workspace analysis of a parallel wire mechanism for measuring a robot pose. *Mechanism and Machine Theory, 34*(6), 825–841.

228. Jiang, Q., & Kumar, V. (2010). The direct kinematics of objects suspended from cables. In *34th Annual Mechanisms and Robotics Conference* (vol. 2, pp. 193–202).

229. Jing, J. (2010). *Inbetriebnahme eines Seilroboters und Vermessung seiner Betriebseigenschaften*. Studienarbeit: Universität Stuttgart, Germany.

230. Jingli, D., Bao, H., & Chen, G. (2010). Nonlinear PD control for a curved cable driven parallel robot. *Journal of Vibration and Shock, 29*(2), 141–144.

231. Jingli, D., Duan, X., & Bao, H. (2010). Static stiffness of a cable-supporting system with the cable sag effects considered. *Jixie Gongcheng Xuebao/Journal of Mechanical Engineering, 46*(17), 29–34.

232. Jingli, D., Bao, H., & Duan, B. (2010). Tracking control of cable-driven parallel robots considering cable sag effects. *Jixie Gongcheng Xuebao/Journal of Mechanical Engineering, 46*(3), 17–21.

233. Jingli, D., Bao, H., Duan, X., & Cui, C. (2010). Jacobian analysis of a long-span cable-driven manipulator and its application to forward solution. *Mechanism and Machine Theory, 45*(9), 1227–1238.

234. Jingli, D., Bao, H., Zong, Y., & Cui, C. (2011). Dynamic analysis of cables with varying-length in cable-driven parallel robots. *Journal of Vibration and Shock, 30*(8), 19–23.

235. Jingli, D., Bao, H., Cui, C., & Yang, D. (2012). Dynamic analysis of cable-driven parallel manipulators with time-varying cable lengths. *Finite Elements in Analysis and Design, 48*(1), 1392–1399.

236. Joshi, S. A., & Surianarayan, A. (2003). Calibration of a 6-dof cable robot using two inclinometers. *Performance Metrics for Intelligent Systems* (pp. 3660–3665).

237. Kawamura, S., Choe, W., Tanaka, S., & Pandian, S. R. (1995). Development of an ultrahigh speed robot FALCON using wire drive system. In *IEEE International Conference on Robotics and Automation* (pp. 1764–1850).

238. Kawamura, S., & Ito, K. (1993). New type of master robot for teleoperation using a radial wire drive system. In *IEEE International Conference on Intelligent Robots and Systems (IROS)* (pp. 55–60).

239. Kawamura, S., Kino, H., & Won, C. (2000). High-speed manipulation by using parallel wire-driven robots. *Robotica, 18*(1), 13–21.

240. Khakpour, H., Birglen, L., & Tahan, S. A. (2014). Synthesis of differentially driven planar cable parallel manipulators. *IEEE Transactions on Robotics, 30*(3), 619–630.

241. Khalil, W., & Besnard, S. (1999). Self calibration of Stewart-Gough parallel robots without extra sensors. *IEEE Transactions on Robotics and Automation, 15*(6), 1116–1121.

242. Khalil, W., Gautier, M., & Enguehard, Ch. (1991). Identifiable parameters and optimum configurations for robots calibration. *Robotica, 9*(1), 63–70.

243. Khosravi, M. A., & Taghirad, H. D. (2011). Dynamic analysis and control of cable driven robots with elastic cables. *Transactions of the Canadian Society of Mechanical Engineering, 35*(4), 543–557.

244. Khosravi, M. A., & Taghirad, H. D. (2011). On the modelling and control of fully constrained cable driven robots with flexible cables. In *IEEE 2nd International Conference on Control, Instrumentation and Automation* (pp. 1030–1035).

245. Khosravi, M. A., & Taghirad, H. D. (2014). Dynamic modeling and control of parallel robots with elastic cables: singular perturbation approach. *IEEE Transactions on Robotics, 30*(3), 694–704.

246. Kim, J., Park, F. C., Ryu, S. J., Kim, J., Hwang, J. C., Park, C., et al. (2001). Design and analysis of a redundantly actuated parallel mechanism for rapid machining. *IEEE Transactions on Robotics and Automation, 17*(4), 423–434.

247. Kino, H., & Kawamura, S. (2015). Mechanism and control of parallel-wire driven system. *Journal of Robotics and Mechatronics, 27*(6), 599–607.

248. Klatte, R., Kulisch, U., Wiethoff, A., Lawo, C., & Rauch, M. (1993). *C-XSC - A C++ Class Library for Extended Scientific Computing.* New York: Springer.

249. Knüppel, O. (1994). PROFIL / BIAS - A fast interval library. *Computing, 53,* 277–287.

250. Knüppel, O. (1999). Profil/Bias v2.0, Manual.

251. Korayem, M. H., Bamdad, M., & Bayat, S. (2009). Optimal trajectory planning with maximum load carrying capacity for cable suspended robots. In *6th International Symposium on Mechatronics and its Applications* (pp. 1–6).

252. Korayem, M. H., Bamdad, M., & Saadat, M. (2007). Workspace analysis of cable-suspended robots with elastic cable. In *IEEE International Conference on Robotics and Biomimetics* (pp. 1942–1947).

253. Korayem, M. H., Bamdad, M., Tourajizadeh, H., Shafiee, H., Zehtab, R. M., & Iranpour, A. (2013). Development of ICASBOT: A Cable-suspended robot's with six DOF. *Arabian Journal for Science and Engineering, 38*(5), 1131–1149.

254. Korayem, M. H., Maddah, S. M., Taherifar, M., & Tourajizadeh, H. (2014). Design and programming a 3d simulator and controlling graphical user interface of ICaSbot, a cable suspended robot. *Scientia Iranica B, 21*(3), 663–681.

255. Kossowski, C. D. (2001). *A novel wire-driven parallel robot, design, analysis and simulation of the CAT4 (Cable Actuated Truss, 4 degrees of freedom).* Master Thesis, Kingston, Ontario: Queen's University.

256. Kozak, K., Zhou, Q., & Wang, J. (2004). Static analysis of cable-driven manipulators with non-negligible cable mass. In *IEEE Conference on Robotics, Automation and Mechatronics* (vol. 2, pp. 886–891).

257. Kozak, K., Zhou, Q., & Wang, J. (2006). Static analysis of cable-driven manipulators with non-negligible cable mass. *IEEE Transactions on Robotics, 22*(3), 425–433.

258. Kraft, M., & Schäper, E. (2005). Simulation and optimisation of a tendon-based Stewart platform. In *Intelligent Production Machines and Systems* (pp. 405–410).

259. Kraus, W. (2015). *Force control of cable-driven parallel robots.* PhD thesis, Germany: Universität Stuttgart.

260. Kraus, W., Kessler, M., & Pott, A. (2015). Pulley friction compensation for winch-integrated cable force measurement and verification on a cable-driven parallel robot. In *IEEE International Conference on Robotics and Automation (ICRA)* (pp. 1627–1632).

261. Kraus, W., Mangold, A., Ho, W. Y., & Pott, A. (2014). Haptic interaction with a cable-driven parallel robot using admittance control. In Bruckmann, T., & Pott, A. (Eds.), *Cable-Driven Parallel Robots. Mechanisms and Machine Science* (vol. 2, pp. 201–212). Berlin: Springer.

262. Kraus, W., Miermeister, P., & Pott, A. (2012). Investigation of the influence of elastic cables on the force distribution of a parallel cable-driven robot. In Bruckmann, T., & Pott, A. (Eds.), *Cable-Driven Parallel Robots. Mechanisms and Machine Science* (vol. 12, pp. 103–115). Berlin: Springer.

263. Kraus, W., & Pott, A. (2013). Scenario-based dimensioning of the actuator of parallel cable-driven robots. *New Trends in Mechanisms and Machine Science* (pp. 131–139). Berlin: Springer.

264. Kraus, W., Schmidt, V., Pott, A., & Verl, A. (2012). Investigation on a planar cable-driven parallel robot. In *German Conference on Robotics (ROBOTIK)* (pp. 193–198).

265. Kraus, W., Schmidt, V., Rajendra, P., & Pott, A. (2014). System identification and cable force control for a cable-driven parallel robot with industrial servo drives. In *IEEE International Conference on Robotics and Automation (ICRA)* (pp. 5921–5926).

266. Kraus, W., Spiller, A., & Pott, A. (2013). Energieeffizienz von parallelen Seilrobotern. In *SPS IPC DRIVES* (pp. 297–306).

267. Kurtz, R., & Hayward, V. (1991). Dexterity measure for tendon actuated parallel mechanisms. *Proceedings of International Conference on Advanced Robot* (pp. 1141–1146).
268. Kurtz, R., & Hayward, V. (1994). Dexterity measures with unilateral actuation constraints: the n+1 case. *Advanced Robotics, 9*(5), 561–577.
269. Lafourcade, P., & Llibre, M. (2003). First steps toward a sketch-based design methodology for wire-driven manipulators. In *IEEE/ASME International Conference on Advanced Intelligent Mechatronics* (vol. 1, pp. 143–148).
270. Lafourcade, P., Llibre, M., & Reboulet, C. (2002). Design of a parallel wire-driven manipulator for wind tunnels. In *Workshop on Fundamental Issues and Future Research Directions for Parallel Mechanisms and Manipulators*. Canada.
271. Lafourcade, P., Zheng, Y. -Q., & Liu, X. (2003). Stiffness analysis of wire-driven parallel kinematic manipulators. In *Proceedings of 11th World Congress on Theory of Machines and Mechanisms (IFToMM)*.
272. Lahouar, S., Ottaviano, E., Zeghoul, S., Romdhane, L., & Ceccarelli, Marco. (2009). Collision free path-planning for cable-driven parallel robots. *Robotics and Autonomous Systems, 57*(11), 1083–1093.
273. Lalo, W., Bruckmann, T., & Schramm, D. (2013). Optimal control for a wire-based storage retrieval machine. *Mechanisms and Machine Science* (pp. 631–639). Berlin: Springer.
274. Lamaury, J. (2013). *Contribution à la commande des robots parallèles à cables à redondance d'actionnement*. PhD thesis, Montpellier, France: Université Montpellier II.
275. Lamaury, J., & Gouttefarde, M. (2012). A tension distribution method with improved computational efficiency. In Bruckmann, T., & Pott, A. (Eds.), *Cable-Driven Parallel Robots. Mechanisms and Machine Science* (vol. 12, pp. 71–85). Berlin: Springer.
276. Lamaury, J., & Gouttefarde, M. (2013). Control of a large redundantly actuated cable-suspended parallel robot. In *IEEE International Conference on Robotics and Automation* (pp. 4659–4664).
277. Lamaury, J., Gouttefarde, M., Michelin, M., & Tempier, O. (2012). Design and control of a redundant suspended cable-driven parallel robot. In: Lenarčič, J., & Husty, M. L. (Eds.), *Latest Advances in Robot Kinematics* (pp. 237–244). Dordrecht: Springer.
278. Lambert, C., Nahon, M., & Chalmers, D. (2007). Implementation of an aerostat positioning system with cable control. *IEEE/ASME Transactions on Mechatronics, 12*(1), 32–40.
279. Lamine, H., Bennour, S., & Romdhane, L. (2016). Design of cable-driven parallel manipulators for a specific workspace using interval analysis. In *Advanced robotics* (pp. 1–10).
280. Landsberger, S. E. (1984). *Design and construction of a cable-controlled, parallel link manipulator*. Master thesis, Cambridge MA: Massachusetts Institute of Technology.
281. Landsberger, S. E. (1985). *A new design for parallel link manipulators*. Cambridge MA: Sea Grant College Program, Massachusetts Institute of Technology.
282. Landsberger, S. E., & Sheridan, T. B. (1987). Parallel link manipulators. *U.S. Patent No. 4, 666, 362*, May 19, 1987.
283. Lazard, D. (1993). On the representation of rigid-body motions and its application to generalized platform manipulators. In *Computational Kinematics* (pp. 175–182). Kluwer Academic Publishers.
284. Lefrancois, S., & Gosselin, C. (2010). Point-to-point motion control of a pendulum-like 3-dof underactuated cable-driven robot. In *IEEE International Conference on Robotics and Automation (ICRA)* (pp. 5187–5193).
285. Legnani, G., & Tiboni, M. (2014). Optimal design and application of a low-cost wire-sensor system for the kinematic calibration of industrial manipulators. *Mechanism and Machine Theory, 73*, 25–48.
286. Li, H., Zhang, X., Yao, R., Sun, J., Pan, G., & Zhu, W. (2012). Optimal force distribution based on slack rope model in the incompletely constrained cable-driven parallel mechanism of FAST telescope. In Bruckmann, T., & Pott, A. (Eds.), *Cable-Driven Parallel Robots. Mechanisms and Machine Science* (vol. 12, pp. 87–102). Berlin: Springer.
287. Lim, W. B., Yeo, S. H., Guilin, Y., & Mustafa, S. K. (2009). Kinematic analysis and design optimization of a cable-driven universal joint module. In *IEEE/ASME International Conference on Advanced Intelligent Mechatronics* (pp. 1933–1938).

288. Lim, W. B., Yeo, S. H., Guilin, Y., Mustafa, S. K., & Zhang, Z. (2011). Tension optimization for cable-driven parallel manipulators using gradient projection. In *IEEE/ASME International Conference on Advanced Intelligent Mechatronics (AIM)* (pp. 73–78).

289. Lin, X., Yi, C., Jin, C., & Shuai, J. (2010). Design and workspace optimization of a 6/6 cable-suspended parallel robot. In *International Conference on Computer Application and System Modeling (ICCASM)* (vol. 10, pp. V10–610–V10–614).

290. Lindemann, R., & Tesar, D. (1989). Construction and demonstration of a 9-string 6 DOF force reflecting joystick for telerobotics. In *NASA Techdocs* (pp. 55–63).

291. Liu, H., & Gosselin, C. (2009). A planar closed-loop cable-driven parallel mechanism. *Transactions of the Canadian Society for Mechanical Engineering, 33*(4), 587.

292. Liu, H. (2012). *Conceptual design, static and dynamic analysis of novel cable-loop-driven parallel mechanisms.* PhD thesis, Québec: Université Laval.

293. Liu, H., Gosselin, C., & Laliberté, T. (2012). Conceptual design and static analysis of novel planar spring-loaded cable-loop-driven parallel mechanisms. *Journal of Mechanisms and Robotics, 4*(2), 021001.

294. Liu, P., Zhang, L., Wang, K. Y., & Zhang, J. (2009). Dynamics analysis and control of wire-driven rehabilitative robot. *Zhongguo Jixie Gongcheng/China Mechanical Engineering, 20*(11), 1335–1339.

295. Liu, P., Zhang, L. X., Wang, K. Y., & Zhang, J. (2009). Dynamic modeling and control of a wire-driven rehabilitation robot. *Journal of Harbin Engineering University, 30*(7), 811–815.

296. Liu, X., Qiu, Y.-Y., & Sheng, Y. (2010). Proofs of existence conditions for workspaces of wire-driven parallel robots and a uniform solution strategy for the workspaces. *Jixie Gongcheng Xuebao/Journal of Mechanical Engineering, 46*(7), 27–34.

297. Liwen, G., Huayang, X., & Zhihua, L. (2015). Kinematic analysis of cable-driven parallel mechanisms based on minimum potential energy principle. *Advances in Mechanical Engineering, 7*(12),

298. Loloei, A. Z., Aref, M. M., & Taghirad, H. D. (2009). Wrench feasible workspace analysis of cable-driven parallel manipulators using LMI approach. In *IEEE/ASME International Conference on Advanced Intelligent Mechatronics* (pp. 1034–1039).

299. Loloei, A. Z., & Taghirad, H. D. (2011). Controllable workspace of general cable driven redundant parallel manipulator based on fundamental wrench. In *CCToMM M3 Symposium*.

300. Loloei, A. Z., & Taghirad, H. D. (2012). Controllable workspace of cable-driven redundant parallel manipulators by fundamental wrench analysis. *Transactions of the Canadian Society for Mechanical Engineering, 36*(3), 297–314.

301. Lourakis. M. I. A. (2004). levmar: Levenberg-Marquardt nonlinear least square algorithms in C/C++.

302. Lutz Medical Engineering. (2015). The float rehabilitation training. www.thefloat.ch.

303. Ma, O., & Angeles, J. (1991). Architecture singularities of platform manipulators. In *IEEE International Conference on Robotics and Automation* (pp. 1542–1547).

304. Madsen, K., Nielsen, H. R., & Tingleff, O. (2004). *Methods for non-linear least square problems* (2nd ed.). Lyngby: Technical University of Denmark.

305. Maeda, K., Tadokoro, S., Takamori, T., Hiller, M., & Verhoeven, R. (1999). On design of a redundant wire-driven parallel robot WARP manipulator. In *Proceedings of IEEE International Conference on Robotics and Automation* (pp. 895–900).

306. Maier, T. (2004). *Bahnsteuerung eines seilgeführten Handhabungssystems.* Fortschritt-Berichte VDI, Reihe 8, Nr. 1047. Düsseldorf: VDI Verlag.

307. Maier, T., & Woernle, C. (1997). Kinematic control of cable suspension robots. In *Proceedings of the NATO-ASI Computational Methods in Mechanisms* (vol. 2, pp. 421–430). Varna.

308. Maier, T., & Woernle, C. (1998). Kinematische Steuerung von seilgeführten Handhabungssystemen. *Zeitschrift für angewandte Mathematik und Mechanik, 78*(Suppl. 2), 603–604.

309. Maier, T., & Woernle, C. (1999). Flatness-based control of underconstrained cable suspension manipulators. In *Advances in Multibody Systems and Mechatronics* (pp. 277–290). Technische Universitaet Graz.

310. Maier, T., & Woernle, C. (2001). Dynamics and control of a cable suspension manipulator. In *Proceedings of the 9th German-Japanese Seminar on Nonlinear Problems in Dynamical Systems*.

311. Mammitzsch, J., Kunz, M., Michael, M., & Nendel, K. (2009). Anwendungsspezifische Beschichtungen für Faserseile im Maschinenbau. In *12. Chemnitzer Textiltechnik-Tagung* (pp. 199–205). Chemnitz.

312. Mao, Y., & Agrawal, S. K. (2010). Wearable cable-driven upper arm exoskeleton - motion with transmitted joint force and moment minimization. In *IEEE International Conference on Robotics and Automation (ICRA)* (pp. 4334–4339).

313. Mao, Y., & Agrawal, S. K. (2011). A cable driven upper arm exoskeleton for upper extremity rehabilitation. In *IEEE International Conference on Robotics and Automation (ICRA)* (pp. 4163–4168).

314. Masouleh, M. T., Husty, M. L., & Gosselin, C. (2002). Forward kinematic problem of 5-PRUR parallel mechanisms using study parameters. *Advances in Robot Kinematics: Analysis and Control* (pp. 211–221). Dordrecht: Kluwer Academic Publishers.

315. Maza, I., Kondak, K., Bernard, M., & Ollero, A. (2010). Multi-UAV cooperation and control for load transportation and deployment. *Journal of Intelligent and Robotic Systems, 57*(1–4), 417–449.

316. McColl, D., & Notash, L. (2009). Extension of antipodal theorem to workspace analysis of planar wire-actuated manipulators. In *Computational Kinematics* (pp. 9–16). Heidelberg: Springer.

317. Merlet, J.-P. (1989). Singular configurations of parallel manipulators and grassmann geometry. *International Journal of Robotics Research, 8*(5), 45–56.

318. Merlet, J.-P. (1999). The importance of optimal design for parallel structures. In *Parallel Kinematic Machines* (pp. 99–110). London: Springer.

319. Merlet, J.-P. (2012). Unsolved Issues in Kinematics and Redundancy of Wire-driven Parallel Robots. Keynote Presentation in *1st International Conference on cable-driven parallel robots*. Stuttgart, Germany.

320. Merlet, J.-P. (2004). Analysis of the influence of wires interference on the workspace of wire robots. *Advances in Robot Kinematics (ARK)* (pp. 211–218). Italy: Kluwer Academic Publishers.

321. Merlet, J.-P. (2004). Solving the forward kinematics of a gough-type parallel manipulator with interval analysis. *International Journal of Robotics Research, 23*(3), 221–235.

322. Merlet, J.-P. (2006). *Parallel robots* (2nd ed.). Berlin: Springer.

323. Merlet, J.-P. (2008). Analysis of wire elasticity for wire-driven parallel robots. In *2nd European Conference on Mechanism Science (EuCoMeS)* (pp. 471–478).

324. Merlet, J.-P. (2008). Kinematics of the wire-driven parallel robot MARIONET using linear actuators. In *Proceedings of the IEEE International Conference on Robotics and Automation*.

325. Merlet, J.-P. (2010). MARIONET, A family of modular wire-driven parallel robots. *Advances in Robot Kinematics (ARK)* (pp. 53–61). Berlin: Springer.

326. Merlet, J.-P. (2012). Managing the redundancy of $N-1$ wire-driven parallel robots. In Lenarčič, J., & Husty, M. L. (Eds.), *Latest Advances in Robot Kinematics* (pp. 405–412). Dordrecht: Springer.

327. Merlet, J.-P. (2013). Comparison of actuation schemes for wire-driven parallel robots. In *Mechanisms and Machine Science* (vol. 7, pp. 245–254). Berlin: Springer.

328. Merlet, J.-P. (2013). Kinematic analysis of the 4-3-1 and 3-2-1 wire-driven parallel crane. In *IEEE International Conference on Robotics and Automation (ICRA)* (pp. 4635–4640).

329. Merlet, J.-P. (2013). Wire-driven Parallel Robot: Open Issues. In *Romansy 19 – Robot Design, Dynamics and Control of CISM International Centre for Mechanical Sciences* (vol. 544, pp. 3–10). Berlin: Springer.

330. Merlet, J.-P. (2014). Checking the cable configuration of cable-driven parallel robots on a trajectory. In *IEEE International Conference on Robotics and Automation (ICRA)* (pp. 1586–1591).

331. Merlet, J.-P. (2014). The influence of discrete-time control on the kinematico-static behavior of cable-driven parallel robot with elastic cables. In *Advances in Robot Kinematics (ARK)* (pp. 113–121).
332. Merlet, J.-P. (2015). On the inverse kinematics of cable-driven parallel robots with up to 6 sagging cables. In *IEEE/RSJ International Conference on Intelligent Robots and Systems* (pp. 4356–4361).
333. Merlet. J.-P. (2015). The kinematics of cable-driven parallel robots with sagging cables: preliminary results. In *IEEE International Conference on Robotics and Automation (ICRA)* (pp. 1593–1598).
334. Merlet, J.-P. (2017). Simulation of discrete-time controlled cable-driven parallel robots on a trajectory. In *IEEE Transactions on Robotics, 33*(3), 675–688.
335. Merlet, J.-P., & Alexandre dit Sandretto, J. (2014). The forward kinematics of cable-driven parallel robots with sagging cables. In Bruckmann, T., & Pott, A. (Eds.), *Cable-Driven Parallel Robots. Mechanisms and Machine Science* (vol. 32, pp. 3–15). Berlin: Springer.
336. Merlet, J.-P., & Daney, D. (2007). A new design for wire-driven parallel robot. In *2nd International Congress, Design and Modelling of Mechanical systems*. Monastir.
337. Merlet, J.-P., & Daney, D. (2009). Kinematic analysis of a spatial four-wire driven parallel crane without constraining mechanism. In *Computational Kinematics*. Heidelberg: Springer.
338. Merlet, J.-P., & Daney, D. (2010). A portable, modular parallel wire crane for rescue operations. In *Proceedings of the IEEE International Conference on Robotics and Automation* (pp. 2834–2839).
339. Merlet, J.-P., Gosselin, C., & Huang, T. (2016). Parallel mechanisms. In Siciliano, B., & Khatib, O. (Eds.), *Springer Handbook of Robotics* (2nd ed, pp. 443–461). Berlin, Heidelberg: Springer.
340. Michelin, M., Baradat, C., Nguyen, D. Q., & Gouttefarde, M. (2014). Simulation and control with XDE and matlab/simulink of a cable-driven parallel robot (CoGiRo). In Bruckmann, T., & Pott, A. (Eds.), *Cable-Driven Parallel Robots. Mechanisms and Machine Science* (vol. 32, pp. 71–83). Berlin: Springer.
341. Michael, M. (2010). *Beitrag zur Treibfähigkeit von hochfesten synthetischen Faserseilen.* PhD thesis, Chemnitz, Germany: Techn. Univ. Chemnitz.
342. Miermeister, P., Kraus, W., & Pott, A. (2012). Differential kinematics for calibration, system investigation, and force based forward kinematics of cable-driven parallel robots. In Bruckmann, T., & Pott, A. (Eds.), *Cable-Driven Parallel Robots. Mechanisms and Machine Science* (vol. 12, pp. 319–333). Berlin: Springer.
343. Miermeister, P., Lächele, M., Boss, R., Masone, C., Schenk, C., Tesch, J., Kerger, M., Teufel, H., Pott, A., & Bulthoff, H. H. (2016). The CableRobot simulator: large scale motion platform based on cable robot technology. In *IEEE/RSJ International Conference on Intelligent Robots and Systems (IROS)* (pp. 3024–3029).
344. Miermeister, P., & Pott, A. (2010). Modelling and real-time dynamic simulation of the cable-driven parallel robot IPAnema. In *European Conference on Mechanism Science (EuCoMeS 2010)* (pp. 353–360). Romania.
345. Miermeister, P., & Pott, A. (2012). Auto calibration method for cable-driven parallel robots using force sensors. In *Advances in Robot Kinematics (ARK)* (pp. 269–276).
346. Miermeister, P., & Pott, A. (2013). Design of cable-driven parallel robots with multiple platforms and endless rotating axes. In *Proceedings of the Second Conference on Interdisciplinary Applications in Kinematics* (pp. 21–29). Berlin: Springer.
347. Miermeister, P., Pott, A., & Verl, A. (2012). Auto-calibration method for overconstrained cable-driven parallel robots. In *German Conference on Robotics (ROBOTIK)* (pp. 301–306).
348. Mikelsons, L., Bruckmann, T., Schramm, D., & Hiller, M. (2008). A real-time capable force calculation algorithm for redundant tendon-based parallel manipulators. In *IEEE International Conference on Robotics and Automation (ICRA)* (pp. 3869–3874).
349. Ming, A., & Higuchi, T. (1994). Study on multiple degree-of-freedom positioning mechanism using wires (Part 1). *International Journal of the Japan Society for Precision Engineering, 28*(2), 131–138.

350. Ming, A., & Higuchi, T. (1994). Study on multiple degree-of-freedom positioning mechanism using wires (Part 2) - development of a planar completely restrained positioning mechanism. *International Journal of the Japan Society for Precision Engineering, 28*(3), 235–242.

351. Morris, M. M. (2007). *A planar cable-driven robotic device for physical therapy assistance.* Master Thesis, Florida: Florida Atlantic University.

352. Mroz, G., & Notash, L. (2004). Design and prototype of parallel, wire-actuated robots with a constraining linkage. *Journal of Robotic Systems, 21*(12), 677–687.

353. Müller, A., & Ruggiu, M. (2012). Self-calibration of redundantly actuated PKM based on motion reversal points. In *Advances in Robot Kinematics (ARK)* (pp. 75–82).

354. Müller, A., & Ruggiu, M. (2014). Self-calibration of redundantly actuated PKM exploiting kinematic landmarks. In *Computational Kinematics* (vol. 15, pp. 93–102). Berlin: Springer.

355. Müller, K., Reichert, C., & Bruckmann, T. (2014). Analysis of a real-time capable cable force computation method. In Bruckmann, T., & Pott, A. (Eds.), *Cable-Driven Parallel Robots. Mechanisms and Machine Science* (vol. 32, pp. 227–238). Berlin: Springer.

356. Müller, K., Reichert, C., & Bruckmann, T. (2014). Analysis of geometrical force calculation algorithms for cable-driven parallel robots with a threefold redundancy. In *Advances in Robot Kinematics (ARK)* (pp. 203–211).

357. Nahvi, A., & Hollerbach, J. M. (1996). The noise amplification index for optimal pose selection in robot calibration. In *IEEE International Conference on Robotics and Automation* (pp. 647–654).

358. Nahvi, A., Hollerbach, J. M., & Hayward, V. (1994). Calibration of a parallel robot using multiple kinematic closed loops. In *IEEE International Conference on Robotics and Automation (ICRA)* (vol. 1, pp. 407–412).

359. Neumaier, A. (1990). *Interval methods for systems of equations.* Encyclopedia of mathematics and its application. New York: Cambridge University Press.

360. Nguyen, D. Q., Gouttefarde, M., Company, O., & Pierrot, F. (2013). On the simplifications of cable model in static analysis of large-dimension cable-driven parallel robots. In *IEEE International Conference on Intelligent Robots and Systems (IROS)* (pp. 928–934).

361. Nguyen, D. Q. (2014). *On the Study of Large-Dimension Reconfigurable Cable-Driven Parallel Robots.* PhD thesis, France: Université Montpellier.

362. Nguyen, D. Q., & Gouttefarde, M. (2014). Stiffness matrix of 6-DOF cable-driven parallel robots and its homogenization. In *Advances in Robot Kinematics (ARK)* (pp. 181–191).

363. Nguyen, D. Q., Gouttefarde, M., Company, O., & Pierrot, F. (2014). On the analysis of large-dimension reconfigurable suspended cable-driven parallel robots. In *IEEE International Conference on Robotics and Automation (ICRA)* (pp. 5728–5735).

364. Notash, L., & Kamalzadeh, A. (2007). Inverse dynamics of wire-actuated parallel manipulators with a constraining linkage. *Mechanism and Machine Theory, 42*(9), 1103–1118.

365. Oh, S.-R., & Agrawal, S. K. (2004). Nonlinear sliding mode control and feasible workspace analysis for a cable suspended robot with input constraints and disturbances. *Proceedings of the American Control Conference, 5*, 4631–4636.

366. Oh, S.-R., & Agrawal, S. K. (2005). Cable suspended planar robots with redundant cables: Controllers with positive tensions. In *IEEE Transactions on Robotics, 21*(3), 457–465.

367. Oh, S.-R., & Agrawal, S. K. (2006). The feasible workspace analysis of a set point control for a cable-suspended robot with input constraints and disturbances. *IEEE Transactions on Control Systems Technology, 14*(4), 735–742.

368. Oh, S.-R., Mankala, K., Agrawal, S. K., & Albus, J. S. (2004). Dynamic modeling and robust controller design of a two-stage parallel cable robot. In *IEEE International Conference on Robotics and Automation, 2004* (vol. 4, pp. 3678–3683).

369. Oh, S.-R., Mankala, K., & Albus, J. S. (2004). Robust control of dual-stage cable suspended robots with input constraints for cargo handling. In *ASME International Mechanical Engineering Congress Exposition* (pp. 1221–1230).

370. Oh, S.-R., Ryu, J.-C., & Agrawal, S. K. (2006). Dynamics and control of a helicopter carrying a payload using a cable-suspended robot. *Journal of Mechanical Design, 128*(5), 1113–1121.

371. Otis, M. J.-D. (2009). *Analyse, commande et intégration d'un mécanisme parallèle entraîné par câbles pour la réalisaton d'une interface haptique comme métaphore de navigation dans un environnement virtuel*. PhD thesis, Québec: Université Laval.

372. Otis, M. J.-D., Comtois, S., Laurendeau, D., & Gosselin, C. (2010). Human safety algorithms for a parallel cable-driven haptic interface. *Advances in Intelligent and Soft Computing, 83*, 187–200.

373. Otis, M. J.-D., Mokhtari, M., Tremblay, C. D., Laurendeau, D., Michelle de Rainville, F., & Gosselin, C. (2008). Hybrid control with multi-contact interactions for 6DOF haptic foot platform on a cable-driven locomotion interface. In *Symposium on haptic interfaces for virtual environment and teleoperator systems (HAPTICS)* (pp. 161–168).

374. Otis, M. J.-D., Perreault, S., Nguyen Dang, T.-L., Lambert, P., Gouttefarde, M., Laurendeau, D., et al. (2009). Determination and management of cable interferences between two 6-DOF foot platforms in a cable-driven locomotion interface. *Man and Cybernetics Systems, 39*(3), 528–544.

375. Ottaviano, E., & Ceccarelli, M. (2002). Optimal design of CaPaMan (Cassino parallel manipulator) with a specific orientation workspace. *Robotica, 20*(2), 159–166.

376. Ottaviano, E., Ceccarelli, M., & de Ciantis, M. (2007). A 4–4 cable-based parallel manipulator for an application in hospital environment. In *Mediterranean Conference on Control & Automation* (pp. 1–6).

377. Ottaviano, E., Gattulli, V., Potenza, F., & Rea, P. (2015). Modeling a planar point mass sagged cable-suspended manipulator. In *Proceedings of the 14th IFToMM World Congress* (pp. 491–496). Taipei.

378. Ouyang, B., & Shang, W. (2014). Wrench-feasible workspace based optimization of the fixed and moving platforms for cable-driven parallel manipulators. *Robotics and Computer-Integrated Manufacturing, 30*(6), 629–635.

379. Patel, A. J., & Ehmann, K. F. (2000). Calibration of a hexapod machine tool using a redundant leg. *International Journal of Machine Tools and Manufacture, 40*(4), 489–512.

380. Park, J., Chung, W.-K., & Moon, W. (2005). Wire-suspended dynamical system: stability analysis by tension-closure. *IEEE Transactions on Robotics, 21*(3), 298–308.

381. Perreault, S., Cardou, P., Gosselin, C., & Otis, M. J.-D. (2010). Geometric determination of the interference-free constant-orientation workspace of parallel cable-driven mechanisms. *ASME Journal of Mechanisms and Robotics, 2*(3), 031016

382. Perreault, S., & Gosselin, C. (2008). Cable-driven parallel mechanisms: Application to a locomotion interface. *Journal of Mechanical Design, 130*(10), 102301.

383. Pham, C. B., Guilin, Y., & Yeo, S. H. (2005). Dynamic analysis of cable-driven parallel mechanisms. In *IEEE/ASME International Conference on Advanced Intelligent Mechatronics* (pp. 612–617).

384. Pham, C. B., Yeo, S. H., & Guilin, Y. (2004). Workspace analysis and optimal design of cable-driven planar parallel manipulators. In *IEEE Conference on Robotics, Automation and Mechatronics* (vol. 1, pp. 219–224).

385. Pham, C. B., Yeo, S. H., Guilin, Y., Kurbanhusen, M. S., & Chen, I.-M. (2006). Force-closure workspace analysis of cable-driven parallel mechanisms. *Mechanism and Machine Theory, 41*(1), 53–69.

386. Pham, C. B., Yeo, S. H., & Yang, G. (2005). Tension analysis of cable-driven parallel mechanisms. In *IEEE/RSJ International Conference on Intelligent Robots and Systems* (pp. 2601–2606).

387. Pham, H. H., & Chen, I.-M. (2003). Optimal synthesis for workspace and manipulability of parallel flexure mechanism. In *Proceedings of the 11th World Congress in Mechanism and Machine Science*. Tianjin, China.

388. Pott, A. (2007). *Analyse und Synthese von Werkzeugmaschinen mit paralleler Kinematik*. Fortschritt-Berichte VDI, Reihe 20, Nr. 409. Düsseldorf: VDI Verlag.

389. Pott, A. (2008). Forward kinematics and workspace determination of a wire robot for industrial applications. In Lenarčič, J., & Wenger P. (Eds.), *Advances in Robot Kinematics* (pp. 451–458). Berlin: Springer.

390. Pott, A. (2010). An algorithm for real-time forward kinematics of cable-driven parallel robots. In Lenarčič, J., & Stanisic, M. M. (Eds.), *Advances in Robot Kinematics* (pp. 529–538). Dordrecht: Springer.

391. Pott, A. (2012). Influence of pulley kinematics on cable-driven parallel robots. In Lenarčič, J., & Husty, M. L. (Eds.), *Latest Advances in Robot Kinematics* (pp. 197–204). Dordrecht: Springer.

392. Pott, A. (2013). An improved force distribution algorithm for over-constrained cable-driven parallel robots. In *Computational Kinematics* (pp. 139–146). Dordrecht: Springer.

393. Pott, A. (2014). On the limitations on the lower and upper tensions for cable-driven parallel robots. In *Advances in Robot Kinematics (ARK)* (pp. 243–251). Cham: Springer.

394. Pott, A., Boye, T., & Hiller, M. (2005). Parameter synthesis for parallel kinematic machines from given process requirements. In *IEEE/ASME International Conference on Advanced Intelligent Mechatronics* (pp. 753–758).

395. Pott, A., Boye, T., & Hiller, M. (2006). Design and optimization of parallel kinematic machines under process requirements. In *Proceedings of the 5th Chemnitz Parallel Kinematics Seminar* (pp. 193–212).

396. Pott, A., Bruckmann, T., & Mikelsons, L. (2009). Closed-form force distribution for parallel wire robots. In *Computational Kinematics* (pp. 25–34). Berlin: Springer.

397. Pott, A., Franitza, D., & Hiller, M. (2004). Orientation workspace verification for parallel kinematic machines with constant leg length. In *Proceedings of Mechatronics and Robotics* (pp. 984–989).

398. Pott, A., & Hiller, M. (2006). A framework for analysis, synthesis and optimization of parallel kinematic machines. *Advances in Robot Kinematics (ARK)* (pp. 103–112). Slovenia: Springer.

399. Pott, A., & Hiller, M. (2006). A parallel implementation for the optimization of parallel kinematic machines under process requirements. In *Proceedings of EuCoMeS, 1st European Conference on Mechanism Science*.

400. Pott, A., & Kraus, W. (2016). Determination of the wrench-closure translational workspace in closed-form for cable-driven parallel robots. In *IEEE International Conference on Robotics and Automation (ICRA)* (pp. 882–888).

401. Pott, A., Meyer, C., & Verl, A. (2010). Large-scale assembly of solar power plants with parallel cable robots. In *41st International Symposium on Robotics (ISR) and 6th German Conference on Robotics (ROBOTIK)* (pp. 999–1004).

402. Pott, A., & Miermeister, P. (2016). Workspace and interference analysis of cable-driven parallel robots with an unlimited rotation axis. In Lenarčič, J., & Merlet, J.-P. (Eds.), *Advances in Robot Kinematics* (pp. 346–355). Warschau: Springer.

403. Pott, A., Miermeister, P., & Kraus, W. (2014). Roboteranordnung nach Art eines parallelen Seilroboters sowie Verfahren zum Antreiben der Roboteranordnung. *Patent, DE 102012025432 B3*, Dec. 21, 2012.

404. Pott, A., Mütherich, H., Kraus, W., Schmidt, V., Miermeister, P., & Verl, A. (2012). IPAnema: A family of cable-driven parallel robots for industrial applications. In Bruckmann, T., & Pott, A. (Eds.), *Cable-Driven Parallel Robots. Mechanisms and Machine Science* (vol. 12, pp. 119–134). Berlin: Springer.

405. Pusey, J. L. (2006). *Cable suspended parallel robots: design, workspace, and control*. PhD thesis, Newark, USA: University of Delaware.

406. Pusey, J. L., Fattah, A., Agrawal, S., Messina, E., & Jacoff, A. (2004). Design and workspace analysis of a 6–6 cable-suspended parallel robot. *Mechanism and Machine Theory, 39*(7), 761–778.

407. Qin, H., Bin, L., Zheng, Y.-Q., & Lin, Q. (2010). Kinematical and dynamical simulation for the aircraft model at 6-DOF in WDPSS. *International Conference on Mechanic Automation and Control Engineering (MACE)* (pp. 291–294).

408. Rahimi, M. A., Hemami, H., & Zheng, Y. F. (1999). Experimental study of a cable-driven suspended platform. In *IEEE International Conference on Robotics and Automation* (vol. 3, pp. 2342–2347).

409. Ramadour, R., & Merlet, J.-P. (2014). Computing safe trajectories for an assistive cable-driven parallel robot by selecting the cables under tension and using interval analysis. In *IEEE/ASME International Conference on Advanced Intelligent Mechatronics (AIM)* (pp. 1349–1354).

410. Ramon, R. E. (1966). *Interval analysis*. New Jersey: Prentice-Hall.

411. Rauter, G., Zitzewitz, J. V., Duschau-Wicke, A., Vallery, H., & Riener, R. (2010). A tendon-based parallel robot applied to motor learning in sports. In *Proceedings of the 3rd IEEE RAS & EMBS* (pp. 82–87).

412. Rezazadeh, S., & Behzadipour, S. (2011). Workspace analysis of multibody cable-driven mechanisms. *Journal of Mechanisms and Robotics*, *3*(2), 021005.

413. Richter, T., Lorenc, S. J., & Bernold, L. E. (1998). Cable based robotic work platform for construction. In *Proceedings of the 15th ISARC* (pp. 137–144).

414. Riechel, A. T., & Ebert-Uphoff, I. (2004). Force-feasible workspace analysis for underconstrained, point-mass cable robots. In *IEEE International Conference on Robotics and Automation* (vol. 5, pp. 4956–4962).

415. Riehl, N. (2011). *Modélisation et design de robots parallèles à câbles de grande dimension*. PhD thesis, Montpellier, France: Montpellier II.

416. Riehl, N., Gouttefarde, M., Pierrot, F., & Baradat, C. (2010). On the static workspace of large dimension cable-suspended robots with non negligible cable mass. In *Proceedings of the ASME Design Engineering Technical Conference* (vol. 2, pp. 261–270).

417. Roberts, R. G., Graham, T., & Lippitt, T. (1998). On the inverse kinematics, statics, and fault tolerance of cable-suspended robots. *Journal of Robotic Systems*, *15*(10), 581–597.

418. Roberts, R. G., Graham, T., & Trumpower, J. M. (1997). On the inverse kinematics and statics of cable-suspended robots. In *IEEE International Conference on Systems, Man, and Cybernetics* (vol. 5, pp. 4291–4296).

419. Rodnunsky, J., & Bayliss, T. (1993). Aerial cableway and method for filming subjects in motion. *U. S. Patent No. 5, 224, 426*, Nov. 13, 1991.

420. Ronga, F., & Vust, T. (1995). Stewart Platforms without Computer? In *Real Analytic and Algebraic Geometry* (pp. 196–212).

421. Rosati, G., Andreolli, M., Biondi, A., & Gallina, P. (2007). Performance of cable suspended robots for upper limb rehabilitation. In *2007 IEEE 10th International Conference on Rehabilitation Robotics, ICORR'07* (pp. 385–392).

422. Rosati, G., Gallina, P., & Masiero, S. (2007). Design, implementation and clinical tests of a wire-based robot for neurorehabilitation. *IEEE Transactions on Neural Systems and Rehabilitation Engineering*, *15*(4), 560–569.

423. Rosati, G., Gallina, P., Masiero, S., & Rossi, A. (2005). Design of a new 5 d.o.f. wire-based robot for rehabilitation. In *Proceedings of IEEE 9th International Conference on Rehabilitation Robotics* (pp. 430–433).

424. Rosati, G., & Zanotto, D. (2008). A novel perspective in the design of cable-driven systems. In *Proceedings of the ASME International Mechanical Engineering Congress & Exposition* (pp. 617–662).

425. Rosati, G., Zanotto, D., Secoli, R., & Rossi, A. (2009). Design and control of two planar cable-driven robots for upper-limb neurorehabilitation. In *IEEE International Conference on Rehabilitation Robotics* (pp. 560–565).

426. Rost, A. (2013). *Untersuchung von Antrieben mit Kunststoff-Faserseilen für den Einsatz in Leichtbau-Gelenkarmrobotern*. PhD thesis, Germany: Universität Stuttgart.

427. Rump, S. M. (1999). Fast and parallel interval arithmetic. *BIT Numerical Mathematics*, *39*(3), 539–560.

428. Rump, S. M. (1999). INTLAB - INTerval LABoratory. In Csendes, T. (Ed.), *Developments in reliable computing* (pp. 77–104). Dordrecht: Springer.

429. Ryu, J.-H. (Ed.). (2008). *Parallel manipulators*. Vienna: I-Tech Education and Publishing.

430. Schielen, W. (1986). *Technische Dynamik*. Stuttgart: B. G. Teubner.

431. Schmidt, V. (2016). *Modeling techniques and reliable real-time implementation for cable-driven parallel robots using plastic fibre ropes*. PhD thesis, Germany: Universität Stuttgart.

432. Schmidt, V., Kraus, W., & Pott, A. (2014). Presentation of experimental results on stability of a 3 DOF 4-cable-driven parallel robot without constraints. In Bruckmann, T., & Pott, A. (Eds.), *Cable-Driven Parallel Robots. Mechanisms and Machine Science* (vol. 32, pp. 87–99). Berlin: Springer.

433. Schmidt, V., Müller, B., & Pott, A. (2014). Solving the forward kinematics of cable-driven parallel robots with neural networks and interval arithmetic. In *Computational Kinematics* (vol. 15, pp. 103–110). Berlin: Springer.

434. Schmidt, V., & Pott, A. (2012). Implementing extended kinematics of a cable-driven parallel robot in real-time. In Bruckmann, T., & Pott, A. (Eds.), *Cable-Driven Parallel Robots. Mechanisms and Machine Science* (vol. 12, pp. 287–298). Berlin: Springer.

435. Schmidt, V., & Pott, A. (2015). Investigating the effect of cable force on winch winding accuracy for cable-driven parallel robots. *Proceedings of the Institution of Mechanical Engineers, Part K: Journal of Multi-body Dynamics* (pp. 237–241).

436. Sheng, Z., Park, J. -H., Stegall, P., & Agrawal, S. K. (2015). Analytic determination of wrench closure workspace of spatial cable driven parallel mechanisms. In *ASME 2015 International Design Engineering Technical Conferences & Computers and Information in Engineering Conference (DECT 2015)* (vol. 5C, pp. V05CT08A048).

437. Shiang, W. -J., Cannon, D., & Gorman, J. (1999). Dynamic analysis of the cable array robotic crane. In *IEEE International Conference on Robotics and Automation* (vol. 4, pp. 2495–2500).

438. Shiang, W. -J., Cannon, D., & Gorman, J. (2000). Optimal force distribution applied to a robotic crane with flexible cables. In *IEEE International Conference on Robotics and Automation* (vol. 2, pp. 1948–1954).

439. Shoham, M. (2005). Twisting wire actuator. *ASME Journal of Mechanical Design, 127*(3), 441–445.

440. Shoham, M. (2006). Twisting wire actuator. *Patent, EP 1685452 A2*, Oct. 27, 2003.

441. Snyman, J. A., & Hay, A. M. (2004). Analysis and optimization of a planar tendon-driven parallel manipulator. *Advances in Robot Kinematics (ARK)* (pp. 303–312). Italy: Kluwer Academic Publishers.

442. Staffetti, E., Bruyninckx, H., & De Schutter, J. (2002). On the invariance of manipulability indices. *Advances in Robot Kinematics (ARK)* (pp. 57–66). Spain: Kluwer Academic Publishers.

443. Stump, E., & Kumar, V. (2004). Workspace delineation of cable-actuated parallel manipulators. In *Proceedings of DETC 2004* (pp. 1303–1310).

444. Stump, E., & Kumar, V. (2006). Workspaces of cable-actuated parallel manipulators. *Journal of Mechanical Design, 128*(1), 159–167.

445. Sturm, C., Bruckmann, T., Schramm, D., & Hiller, M. (2011). Optimization of the wire length for a skid actuated wire based parallel robot. In *Proceedings of 13th World Congress on Mechanism and Machine Science (IFToMM2011)* (pp. 19–25).

446. Su, Y. X., Duan, B., Nan, R., & Peng, B. (2001). Development of a large parallel-cable manipulator for the feed-supporting system of a next-generation large radio telescope. *Journal of Robotic Systems, 18*(11), 633–643.

447. Sun, M. I. (2002). *Forte Developer 7*. C++ Interval Arithmetic Programming Reference. Sun Microsystems Inc.

448. Surdilovic, D., & Bernhardt, R. (2004). STRING-MAN: A new wire robot for gait rehabilitation. In *IEEE International Conference on Robotics and Automation* (pp. 2031–2036).

449. Surdilovic, D., Zhang, J., & Bernhardt, R. (2007). STRING-MAN: Wire-robot technology for safe, flexible and human-friendly gait rehabilitation. In *International Conference on Rehabilitation Robotics (ICORR)* (pp. 446–453).

450. Surdilovic, D., Radojicic, J., & Krüger, J. (2012). Geometric stiffness analysis of wire robots: A mechanical approach. In Bruckmann, T., & Pott, A. (Eds.), *Cable-Driven Parallel Robots. Mechanisms and Machine Science* (vol. 12, pp. 389–404). Berlin: Springer.

451. Tadokoro, S., Nishioka, S., Kimura, T., Hattori, M., Takamori, T., & Maeda, K. (1996). On fundamental design of wire configurations of wire-driven parallel manipulators with redundancy. *Proceedings of the Japan/USA Symposium on Flexible Automation, 1*, 151–158.

452. Tadokoro, S., Verhoeven, R., Hiller, M., & Takamori, T. (1999). A portable parallel manipulator for search and rescue at large-scale urban earthquakes and an identification algorithm for the installation in unstructured environments. In *Proceedings of IEEE International Conference on Intelligent Robots and Systems IROS 1999*.

453. Taghirad, H. D., & Nahon, M. (2008). Dynamic analysis of a macro-micro redundantly actuated parallel manipulator. *Advanced Robotics, 22*(9), 949–981.

454. Taghirad, H. D., & Nahon, M. (2008). Kinematic analysis of a macro-micro redundantly actuated parallel manipulator. *Advanced Robotics, 22*(6–7), 657–687.

455. Takeda, Y., Shen, G., & Funabashi, H. (2004). A DBB-Based kinematic calibration method for in-parallel actuated mechanisms using a fourier series. *Journal of Mechanical Design, 126*(5), 856–865.

456. Tanaka, M., Seguchi, Y., & Shimada, S. (1988). Kineto-statics of skycam-type wire transport system. In *Proceedings USA-Japan Symposium on Flexible Automation* (pp. 689–694).

457. Tang, A., Li, Y., Qu, H., & Xiao, J. (2011). Dynamics modeling and simulating analysis of a wire driven parallel mechanism. *Procedia Engineering* (pp. 788–794).

458. Tang, X., Tang, L., Wang, J., & Sun, D. (2012). Configuration synthesis for fully restrained 7-cable-driven manipulators. *International Journal of Advanced Robotic Systems, 142*(9), 1–10.

459. Tang, X., & Yao, R. (2010). Dimensional optimization of completely restrained positioning cable driven parallel manipulator with large span. In *Robot Manipulators New Achievements*.

460. Tang, X., & Yao, R. (2011). Dimensional design on the six-cable driven parallel manipulator of FAST. *Journal of Mechanical Design, 133*(11), 111012.

461. Tavolieri, C. (208). *Design of a cable-based parallel manipulator for rehabilitation applications*. PhD thesis, Italy: University of Cassino.

462. Tempel, P., Miermeister, P., Lechler, A., & Pott, A. (2015). Modelling of kinematics and dynamics of the IPAnema 3 cable robot for simulative analysis. *Applied Mechanics and Materials, 794*, 419–426.

463. Tempel, P., Schnelle, F., Pott, A., & Eberhard, P. (2015). Design and programming for cable-driven parallel robots in the German pavilion at the EXPO 2015. *Machines, 3*(3), 223–241.

464. Thomas, F., Ottaviano, E., Ros, L., & Ceccarelli, M. (2003). Coordinate-free formulation of a 3-2-1 wire-based tracking device using Cayley-Menger determinants. In *IEEE International Conference on Robotics and Automation* (vol. 1, pp. 355–361).

465. Thomas, F., Ottaviano, E., Ros, L., & Ceccarelli, M. (2002). Uncertainty model and singularities of 3-2-1 wire-based tracking systems. In *Advances in Robot Kinematics (ARK)*. Spain: Kluwer Academic Publishers.

466. Thompson, C. J., & Campbell, P. D. (1996). Tendon suspended platform robot. *U.S. Patent No. 5, 585, 707*, Dec. 17, 1996.

467. Uhlmann, E., Kraft, M., & Tonn, N. (2008). Entwicklung von Werkzeugmaschinen mit Parallelkinematik unter Verwendung von Seilantrieben. *Fertigungsmaschinen mit Parallelkinematik* (pp. 37–62). Aachen: Shaker.

468. Usher, K., Winstanley, G., & Carnie, R. (2005). Air vehicle simulator: An application for a cable array robot. In *IEEE International Conference on Robotics and Automation* (pp. 2241–2246).

469. Vallery, H., Lutz, P., Zitzewitz, J. V., Rauter, G., Fritschi, M., Everarts, C., Ronsse, R., Curt, A., & Bolliger, M. (2013). Multidirectional transparent support for overground gait training. In *IEEE International Conference on Rehabilitation Robotics (ICORR)* (pp. 1–7).

470. Varziri, M. S., & Notash, L. (2007). Kinematic calibration of a wire-actuated parallel robot. *Mechanism and Machine Theory, 42*(8), 960–976.

471. VDI. (1993). *Methodik zum Entwickeln und Konstruieren technischer Systeme und Produkte (VDI Richtlinie 2221)*. Berlin: Beuth.

472. VDI. (2004). *Design methodology for mechatronic systems (VDI Richtlinie 2206)*. Berlin: Beuth.

473. Verhoeven, R. (2004). *Analysis of the workspace of tendon-based stewart platforms*. PhD thesis, Germany: University of Duisburg-Essen.

474. Verhoeven, R., & Hiller, M. (2000). Estimating the controllable workspace of tendon-based stewart platforms. In *Advances in Robot Kinematics (ARK)* (pp. 277–284).

475. Verhoeven, R., & Hiller, M. (2002). Tension distribution in tendon-based Stewart platforms. *Advances in Robot Kinematics (ARK)* (pp. 117–124). Spain: Academic Publishers.

476. Verhoeven, R., Hiller, M., & Tadokoro, S. (1998). Workspace of tendon-driven stewart platforms: basics, classification, details on the planar-2-dof class. In *Proceedings of the 4th International Conference on Motion and Vibration Control (MOVIC98)* (vol. 3, pp. 871–876).

477. Verhoeven, R., Hiller, M., & Tadokoro, S. (1999). Workspace, stiffness, singularities and classification of tendon driven stewart platforms. *Advances in Robot Kinematics (ARK)* (pp. 105–114). Austria: Kluwer Academic Publishers.

478. Verl, A., Boye, T., & Pott, A. (2008). Measurement pose selection and calibration forecast. *CIRP Annals - Manufacturing Technology, 57*, 425–428.

479. Verl, A., De-Go, G., Dietz, T., & Pott, A. (2014). Amusement Ride. *U.S. Patent US 8, 920, 251 B2*, Dec 30, 2014.

480. Voglewede, P. A., & Ebert-Uphoff, I. (2005). Application of the antipodal grasp theorem to cable-driven robots. *IEEE Transactions on Robotics, 21*(4), 713–718.

481. Voss, K. H. J., Wijk, V., & Herder, J. L. (2012). Investigation of a cable-driven parallel mechanism for interaction with a variety of surfaces, applied to the cleaning of free-form buildings. *Advances in Robot Kinematics (ARK)* (pp. 261–268). Dordrecht: Springer.

482. Voss, K. H. J., Wijk, V., & Herder, J. L. (2013). A cable-driven parallel mechanism for the interaction with hemispherical surfaces. *Mechanisms and Machine Science* (pp. 409–417). Berlin: Springer.

483. Walker, M., & Orin, D. (1982). Efficient dynamic computer simulation of robotic mechanisms. *Journal of Dynamical Systems, Measurement and Control, 104*(3), 205–211.

484. Wampler, C. W., Hollerbach, J. M., & Arai, T. (1995). An implicit loop method for kinematic calibration and its application to closed-chain mechanisms. *IEEE Transactions on Robotics and Automation, 11*(5), 710–724.

485. Wang, X. (2005). Volumes of generalized unit balls. *Mathematics Magazine, 78*(5), 390–395.

486. Wang, K. Y., Zhang, L., Zhang, J., & Liu, F. (2010). Dynamics on pelvis for a 1R2T wire-driven parallel robot. *Jiangsu Daxue Xuebao (Ziran Kexue Ban) / Journal of Jiangsu University (Natural Science Edition), 31*(2), 131–135.

487. Wang, K. Y., Zhang, L. X., & Meng, H. (2010). Elasticity of 1R2T wire-driven parallel rehabilitation robots. *Nanjing Li Gong Daxue Xuebao/Journal of Nanjing University of Science and Technology, 34*(5), 602–607.

488. Wehr, M. (2017). *Beitrag zur Untersuchung von hochfesten synthetischen Faserseilen unter hochdynamischer Beanspruchung.* PhD thesis, Germany: Universität Stuttgart.

489. Weis, J. C., Ernst, B., & Wehking, K.-H. (2015). Use of high strength fibre ropes in multi-rope kinematic robot systems. In Bruckmann, T., & Pott, A. (Eds.), *Cable-Driven Parallel Robots. Mechanisms and Machine Science* (vol. 12, pp. 185–199). Berlin: Springer.

490. Williams II, R. L. (1999). Planar cable-suspended haptic interface: Design for wrench exertion. In *ASME Design Technical Conference 1999.*

491. Williams II, R. L., Albus, J. S., & Bostelman, R. V., (2004). 3D Cable-based cartesian metrology system. *Journal of Robotic Systems, 21*(5), 237–257.

492. Williams II, R. L., & Gallina, P., (2001). Planar cable-direct-driven robots, part I. *Proceedings of the ASME Design Engineering Technical Conference, 2*, 1233–1240.

493. Williams II, R. L., & Gallina, P., (2002). Planar cable-direct-driven robots: Design for wrench exertion. *Journal of Intelligent and Robotic Systems, 35*(2), 203–219.

494. Williams II, R. L., Xin, M., & Bosscher, P. (2008). Contour-crafting-cartesian-cable robot system concepts: workspace and stiffness comparisons. *Proceedings of the ASME International Design Technical Conferences.*

495. Wischnitzer, Y., Shvalb, N., & Shoham, M. (2008). Wire-driven parallel robot: Permitting collisions between wires. *International Journal of Robotics Research, 27*(9), 1007–1026.

496. Woernle, C. (2012). Trajectory tracking for a three-cable suspension manipulator by nonlinear feedforward and linear feedback control. In Bruckmann, T., & Pott, A. (Eds.), *Cable-Driven Parallel Robots. Mechanisms and Machine Science* (vol. 12, pp. 371–386). Berlin: Springer.

497. Wurtz, T., May, C., Holz, H., Natale, C., Palli, G., & Melchiorri, C. (2010). The twisted string actuation system. In *IEEE/ASME International Conference on Advanced Intelligent Mechatronics (AIM)* (pp. 1215–1220).

498. Xiaoling, J., Yijian, H., & Zheng, Y.-Q. (2011). Dimension optimization design of an under-restrained 6-DOF four-cable-driven parallel manipulator based on least square-support vector regression. In *International Conference on Mechatronics and Automation (ICMA)* (pp. 458–463).

499. Ya-fei, L., Fan, D.-P., Liu, H., & Hei, M. (2015). Transmission capability of precise cable drive including bending rigidity. *Mechanism and Machine Theory, 94*, 132–140.

500. Yamamoto, M., Yanai, N., & Mohri, A. (1999). Inverse dynamics and control of crane-type manipulator. *IEEE International Conference on Intelligent Robots and Systems, 2*, 1228–1233.

501. Yanai, N., Yamamoto, M., & Mohri, A. (2001). Inverse dynamics analysis and trajectory generation of incompletely restrained wire-suspended mechanisms. In *IEEE International Conference on Robotics and Automation, 2001* (vol. 4, pp. 3489–3494).

502. Yanai, N., Yamamoto, M., & Mohri, A. (2002). Anti-sway control for wire-suspended mechanism based on dynamics compensation. In *IEEE International Conference on Robotics and Automation, 2002* (vol. 4, pp. 4287–4292).

503. Yang, G., Pham, C. B., & Yeo, S. H. (2006). Workspace performance optimization of fully restrained cable-driven parallel manipulators. In *IEEE International Conference on Intelligent Robots and Systems* (pp. 85–90).

504. Yang, Y., & Zhang, Y. (2009). A new cable-driven haptic device for integrating kinesthetic and cutaneous display. In *ASME/IFToMM International Conference on Reconfigurable Mechanisms and Robots* (pp. 386–391).

505. Yangmin, L., & Qingsong, X. (2006). GA-based multi-objective optimal design of a planar 3-DOF cable-driven parallel manipulator. In *IEEE International Conference on Robotics and Biomimetics* (pp. 1360–1365).

506. Yao, R., Li, H., & Zhang, X. (2012). A modeling method of the cable driven parallel manipulator for FAST. In Bruckmann, T., & Pott, A. (Eds.), *Cable-Driven Parallel Robots. Mechanisms and Machine Science* (vol. 12, pp. 423–436). Berlin: Springer.

507. Yao, R., Tang, X., Li, T., & Ren, G. (2007). Analysis and design of 3T cable-driven parallel manipulator for the feedbacks orientation of the large radio telescope. *Chinese Journal of Mechanical Engineering, 43*(11), 105–109.

508. Yao, R., Tang, X., Wang, J., & Huang, P. (2010). Dimensional optimization design of the four-cable-driven parallel manipulator in FAST. *IEEE/ASME Transactions on Mechatronics, 15*(6), 932–941.

509. Yeo, S. H., Yang, G., & Lim, W. B. (2013). Design and analysis of cable-driven manipulators with variable stiffness. *Mechanism and Machine Theory, 69*, 230–244.

510. Yiu, Y. K., Meng, J., & Li, Z. X. (2003). Auto-calibration for a parallel manipulator with sensor redundancy. In *IEEE International Conference on Robotics and Automation* (pp. 3660–3665). Taipei.

511. You, X., Chen, W., Yu, S., & Wu, X. (2011). Dynamic control of a 3-DOF cable-driven robot based on backstepping technique. In *6th IEEE Conference on Industrial Electronics and Applications (ICIEA)* (pp. 1302–1307).

512. YU, K., Lee, L., Krovi, V. N., & Pei, T. C. (2010). Enhanced trajectory tracking control with active lower bounded stiffness control for cable robot. *IEEE International Conference on Robotics and Automation*, 669–674.

513. Yuan, H. (2015). *Static and dynamic stiffness analysis of cable-driven parallel robots.* PhD thesis, Rennes: INSA.

514. Yuan, H., Courteille, E., & Deblaise, D. (2014). Elastodynamic analysis of cable-driven parallel manipulators considering dynamic stiffness of sagging cables. In *IEEE International Conference on Robotics and Automation (ICRA)* (pp. 4055–4060).

515. Yuan, H., Courteille, E., & Deblaise, D. (2015). Static and dynamic stiffness analyses of cable-driven parallel robots with non-negligible cable mass and elasticity. *Mechanism and Machine Theory, 85*, 64–81.

516. Zanotto, D., Rosati, G., & Agrawal, S. K. (2011). Modeling and control of a 3-DOF pendulum-like manipulator. In *IEEE International Conference on Robotics and Automation (ICRA)* (pp. 3964–3969).

517. Zanotto, D., Rosati, G., Minto, S., & Rossi, A. (2014). Sophia-3: A semiadaptive cable-driven rehabilitation device with a tilting working plane. *IEEE Transactions on Robotics, 30*(4), 974–979.

518. Zarebidoki, M., Lotfavar, A., & Fahham, H. R. (2011). Dynamic modeling and adaptive control of a cable-suspended robot. *Proceedings of the World Congress on Engineering 2011, WCE 2011, 3*, 2469–2473.

519. Zein, M., Wenger, P., & Chablat, D. (2008). Non-singular assembly-mode changing motions for 3-RPR parallel manipulators. *Mechanism and Machine Theory, 43*(4), 480–490.

520. Zhang, Y., Agrawal, S. K., & Piovoso, M. J. (2006). Coupled dynamics of flexible cables and rigid end-effector for a cable suspended robot. In *American Control Conference* (p. 6).

521. Zhang, L., Liu, P., Wang, K. Y., & Zhang, J. (2009). Workspace analysis and kinematics study of a wire-driven rehabilitative robot. *Gaojishu Tongxin/Chinese High Technology Letters, 19*(2), 157–161.

522. Zhang, L. X., Liu, P., Wang, K. Y., & Zhang, J.-Y. (2009). Modeling and control of a wire-driven parallel rehabilitation robot. *Harbin Gongcheng Daxue Xuebao/Journal of Harbin Engineering University, 30*(1), 81–85.

523. Zhang, Y., Zhang, Y., Dai, X., & Yang, Y. (2009). Workspace analysis of a novel 6-dof cable-driven parallel robot. In *IEEE International Conference on Robotics and Biomimetics (ROBIO)* (pp. 2403–2408).

524. Zheng, Y.-Q., Qi, L., Jian, P. W., & Mitrouchev, P. (2009). Analysis of inverse kinematics and dynamics of a novel 6-degree-of-freedom wire-driven parallel gantry crane robot. In *IEEE/ASME International Conference on Advanced Intelligent Mechatronics, 2009* (pp. 1786–1791).

525. Zheng, Y., Lin, Q., & Liu, X. (2007). Initial test of a wire-driven parallel suspension system for low speed wind. In *12th IFToMM World Congress in Mechanism and Machine Science*.

526. Zhou, X., Tang, C. P., & Krovi, V. (2012). Analysis framework for cooperating mobile cable robots. *IEEE International Conference on Robotics and Automation*, 3128–3133.

527. Zhou, X., Jun, S.-K., & Krovi, V. (2014). Tension distribution shaping via reconfigurable attachment in planar mobile cable robots. *Robotica, 32*(02), 245–256.

528. Zi, B., Duan, B., Jingli, D., & Bao, H. (2008). Dynamic modeling and active control of a cable-suspended parallel robot. *Mechatronics, 18*(1), 1–12.

529. Zi, B., Qian, S., & Kecskeméthy, A. (2015). Design and development of a reconfigurable cable parallel robot. In *Proceedings of the 14th IFToMM World Congress*. Taipei.

530. Zi, B., Zhu, Z., & Jingli, D. (2011). Analysis and control of the cable-supporting system including actuator dynamics. *Control Engineering Practice, 19*(5), 491–501.

531. Zitzewitz, J. V., Rauter, G., Steiner, R., Brunschweiler, A., & Riener, R. (2009). A versatile wire robot concept as a haptic interface for sport simulation. In *Proceedings of the 2009 IEEE International Conference on Robotics and Automation (ICRA)* (pp. 313–318).

532. Zitzewitz, J. V., Rauter, G., Vallery, H., Morger, A., & Riener, R. (2010). Forward kinematics of redundantly actuated, tendon-based robots. In *IEEE/RSJ International Conference on Intelligent Robots and Systems (IROS)* (pp. 2289–2294).

533. Zitzewitz, J. V., Fehlberg, L., Bruckmann, T., & Vallery, H. (2012). Use of passively guided deflection units and energy-storing elements to increase the application range of wire robots. In Bruckmann, T., & Pott, A. (Eds.), *Cable-Driven Parallel Robots. Mechanisms and Machine Science* (vol. 12, pp. 167–184). Berlin: Springer.

534. Zoso, N., & Gosselin, C. (2012). Point-to-point motion planning of a parallel 3-DOF underactuated cable-suspended robot. In *IEEE International Conference on Robotics and Automation* (pp. 2325–2330).

Index

© Springer International Publishing AG, part of Springer Nature 2018
A. Pott, *Cable-Driven Parallel Robots*, Springer Tracts in Advanced
Robotics 120, https://doi.org/10.1007/978-3-319-76138-1

Printed by Printforce, the Netherlands